Current Topics in Developmental Biology
Volume 27

Series Editor

Roger A. Pedersen
Laboratory of Radiobiology
and Environmental Health
University of California
San Francisco, CA 94143

Editorial Board

John C. Gerhart
University of California, Berkeley

Peter Gruss
Max-Planck-Institute
of Biophysical Chemistry, Göttingen-Nikolausberg

Philip Ingham
Imperial Cancer Research Fund, Oxford

Story C. Landis
Case Western Reserve University

David R. McClay
Duke University

Gerald Schatten
University of Wisconsin, Madison

Virginia Walbot
Stanford University

Mitsuki Yoneda
Kyoto University

Founding Editors

A. A. Moscona
Alberto Monroy

Current Topics in Developmental Biology

Volume 27

Edited by

Roger A. Pedersen
Laboratory for Radiobiology
and Environmental Health
University of California, San Francisco
San Francisco, California

Academic Press, Inc.
Harcourt Brace Jovanovich, Publishers
San Diego   New York   Boston   London   Sydney   Tokyo   Toronto

Front cover photograph: An abnormal inflorescence produced by a rooted axillary bud of *Nicotiana tabacum* cv. Maryland Mammoth. After laying down 10 nodes the meristem divided. A portion of the cells in each half of the meristem formed a terminal flower while other cells in the meristem altered their developmental fate each forming a vegetative shoot instead of floral branches. Courtesy of Carl N. McDaniel, author of Chapter 1.

This book is printed on acid-free paper.

Copyright © 1992 by ACADEMIC PRESS, INC.

All Rights Reserved.

No part of this publication may be reproduced or transmitted in any form or by any means, electronic or mechanical, including photocopy, recording, or any information storage and retrieval system, without permission in writing from the publisher.

Academic Press, Inc.
1250 Sixth Avenue, San Diego, California 92101-4311

United Kingdom Edition published by
Academic Press Limited
24–28 Oval Road, London NW1 7DX

Library of Congress Catalog Number: 66-28604

International Standard Book Number: 0-12-153127-9

PRINTED IN THE UNITED STATES OF AMERICA
92 93 94 95 96 97   EB   9 8 7 6 5 4 3 2 1

Contents

Contributors ix
Preface xi

1
Determination to Flower in *Nicotiana*
Carl N. McDaniel
  I. Introduction 1
  II. Brief Historical Introduction to the Regulation of Flowering 2
  III. Floral Determination in *Nicotiana* 4
  IV. Floral Determination in Other Species 24
  V. Flowering as a Developmental Process 30
     References 33

2
Cellular Basis of Amphibian Gastrulation
Ray Keller and Rudolf Winklbauer
  I. Introduction 40
  II. Function of Bottle Cells in Gastrulation 40
  III. Mesodermal Cell Migration 51
  IV. Convergence and Extension Movements 66
  V. The Ignored Movement: Epiboly 76
  VI. Involution 77
  VII. Some Other Issues to Be Explored 78
     References 79

3
Role of the Extracellular Matrix in Amphibian Gastrulation
Kurt E. Johnson, Jean-Claude Boucaut, and Douglas W. DeSimone
  I. Introduction 92
  II. Fibrillar Extracellular Matrix 98

- III. Extracellular Matrix Gene Expression during Development   101
- IV. Synthesis and Distribution of Other Extracellular Matrix and Cell Surface Glycoconjugates   105
- V. Experimental Evidence for the Role of the Fibrillar Extracellular Matrix   106
- VI. Probes to Disrupt Mesodermal Cell–Fibrillar Matrix Interaction   113
- VII. Future Directions   119
  - References   121

4
Role of Cell Rearrangement in Axial Morphogenesis
Gary C. Schoenwolf and Ignacio S. Alvarez
- I. Introduction   129
- II. Cell Rearrangement during Neurulation   133
- III. Cell Rearrangement during Gastrulation   151
- IV. Summary and Model   164
  - References   165

5
Mechanisms Underlying the Development of Pattern in Marsupial Embryos
Lynne Selwood
- I. Introduction   175
- II. Initial Polarity   177
- III. Definitive Polarity   185
- IV. Disappearance and Renewal of Polarity   205
- V. Discussion and Conclusions   221
  - References   228

6
Experimental Chimeras: Current Concepts and Controversies in Normal Development and Pathogenesis
Y. K. Ng and P. M. Iannaccone
- I. Introduction   235
- II. Normal Development   239
- III. Pathogenesis   255
- IV. Conclusion   263
  - References   264

7
Genetic Analysis of Cell Division in *Drosophila*
Pedro Ripoll, Mar Carmena, and Isabel Molina
- I. Introduction 275
- II. Effect of Mitotic Mutations on Embryonic Development 279
- III. Mitosis during Postembryonic Development 291
- IV. Tubulin and Kinesin Gene Families 300
- V. General Consideration 302
  References 303

8
Retinoic Acid Receptors: Transcription Factors Modulating Gene Regulation, Development, and Differentiation
Elwood Linney
- I. Introduction 309
- II. Modular Structure of the Steroid Receptor-like Retinoic Acid and Retinoid X Receptors 310
- III. Molecular Specificity of Retinoic Acid Receptors and Retinoid X Receptors 314
- IV. Biological Roles of Retinoic Acid 324
- V. Relationships between Retinoic Acid Receptors and Other Transcription Factors and DNA Binding Proteins 330
- VI. Current and Future Approaches for Dissecting Developmental Specificity and Function of Retinoic Acid Receptors and Retinoid X Receptors 334
  References 339

9
Transcription Factors and Mammalian Development
Corrinne G. Lobe
- I. Introduction 351
- II. Mouse Embryogenesis Reviewed 352
- III. Regulatory Factors in Early Embryogenesis 354
- IV. Transcription Factors through Midembryogenesis 357
- V. Conclusions 373
  References 376

Index 385

Contributors

Numbers in parentheses indicate the pages on which the authors' contributions begin.

Ignacio S. Alvarez, Department of Anatomy, University of Utah School of Medicine, Salt Lake City, Utah 84132 (129)

Jean-Claude Boucaut, Laboratoire de Biologie Experimentale, Université Pierre et Marie Curie (Paris VI), 75005 Paris, France (91)

Mar Carmena, Centro de Biología Molecular (CSIC-UAM), Campus de Cantoblanco, 28049 Madrid, Spain (275)

Douglas W. DeSimone, Department of Anatomy and Cell Biology and the Molecular Biology Institute, University of Virginia Medical Center, Charlottesville, Virginia 22408 (91)

P. M. Iannaccone, Department of Pathology and Markey Program in Developmental Biology, Northwestern University, Chicago, Illinois 60611 (235)

Kurt E. Johnson, Department of Anatomy, The George Washington University Medical Center, Washington, D.C. 20037 (91)

Ray Keller, Department of Molecular and Cell Biology, University of California, Berkeley, Berkeley, California 94720 (39)

Elwood Linney, Department of Microbiology, Duke University Medical Center, Durham, North Carolina 27710 (309)

Corrinne G. Lobe, Department of Molecular Cell Biology, Max-Planck-Institute for Biophysical Chemistry, D-3400 Göttingen, Germany (351)

Carl N. McDaniel, Department of Biology, Plant Science Group, Rensselaer Polytechnic Institute, Troy, New York 12180 (1)

Isabel Molina, Centro de Biología Molecular (CSIC-UAM), Campus de Cantoblanco, 28049 Madrid, Spain (275)

Y. K. Ng, Department of Pathology and Markey Program in Developmental Biology, Northwestern University, Chicago, Illinois 60611 (235)

Pedro Ripoll, Centro de Biología Molecular (CSIC-UAM), Campus de Cantoblanco, 28049 Madrid, Spain (275)

Gary C. Schoenwolf, Department of Anatomy, University of Utah School of Medicine, Salt Lake City, Utah 84132 (129)

Lynne Selwood, Department of Zoology, La Trobe University, Bundoora, 3083 Victoria, Australia (175)

Rudolf Winklbauer, Max-Planck-Institute for Developmental Biology, Tübingen, Germany (39)

Preface

Recent advances in cell and molecular biology have opened new vistas in developmental biology. There has been rapid progress in defining the transcriptional regulation, structure, and role of many tissue- and stage-specific genes. This insight brings us to the threshhold of a truly profound understanding of the problem that has been recognized since the turn of the century as central to development: How is the genetic information that is inherited from the parents via the gametes expressed to achieve orderly differentiation and growth? Attainment of this goal will have an impact that transcends developmental biology, extending as well into the diagnosis and treatment of birth defects, genetic diseases, and cancers.

Equipped with the instruments of modern biology, developmental biologists are increasingly returning to investigations on germ cell and embryonic differentiation, gastrulation, pattern formation, neural development and other classical issues, with incisive results. In support of this renaissance, *Current Topics in Developmental Biology* continues to provide a forum for discussion and critical reviews of developmental issues ranging from molecules to mammals. These reviews should be valuable to researchers in the fields of plant and animal development, as well as students and other professionals desiring an introduction to current topics of cellular and molecular approaches to developmental biology.

In this volume, the topics range from pattern formation in plant and animal development to molecular aspects of gene regulation. The chapter by McDaniel summarizes recent experimental approaches to the factors, such as leaf–root interactions, that determine the transition from vegetative to floral growth in tobacco. The chapter by Keller and Winklbauer evaluates current knowledge about cellular behavior during gastrulation in amphibian embryos, emphasizing bottle cell formation, mesodermal cell migration, convergence and extension, and epiboly. The chapter by Johnson, Boucaut, and DeSimone examines the role of specific extracellular matrix components and their cell surface receptors in regulating mesodermal cell migration and directing cell movements during amphibian gastrulation. Schoenwolf and Alvarez discuss cell rearrangement during neurulation in avian embryos and during gastrulation in several experimental systems, focusing on the role of changes in cell shape, cell position, and cell number during these morphogenetic processes. Selwood reviews the unique features of early development in marsupials, from oogenesis and fertilization up to the trilaminar blastocyst stage, exploring the relationship between egg and embryo polarity. The chapter by Ng and Iannaccone examines the contributions of mammalian chimeras in understanding organ development as well as neoplasia and other pathologies, using computer simulation in the analysis of mosaic growth patterns. The chapter by Ripoll, Carmena, and Molina reviews the genetic control of the cell cycle in *Drosophila,* including both mutations affecting embryonic and postembryonic development. In his chapter, Linney reviews the retinoic acid receptors (RARs and RXRs) and their role as transcription factors that respond to developmentally active retinoids. The final chapter, by Lobe, reviews transcription factors with emphasis on those expressed during early mammalian development, including homeobox-containing and other *Drosophila*-like genes. Together, these chapters comprise a cross-section of current work at several levels of biological organization, from molecular to organismal.

As Series Editor for *Current Topics in Developmental Biology,* I have taken responsibility for maintaining a venerable institution that has served the field of developmental biology for the past three decades. I was willing to undertake this task only in partnership with a supportive group consisting of the authors, the Editorial Board, and other individuals. I thank the authors for their excellent chapters and their patience in accommodating my extensive revisions; I thank each of the members of my Editorial Board, John Gerhart, Peter Gruss, Philip Ingham, Story Landis, David McClay, Gerry Schatten, Ginny Walbot, and Mitsuki Yoneda, and also Yolanda Cruz, Carol Erickson, Richard Mullen, and Keith Yamamoto for their suggestions and reviews; I thank the previous Series Editor, Hans Bode, for his guidance; I thank Phyllis Moses, Acquisitions Editor for Academic Press, for sharing her experience and providing valuable direction; I thank Liana Hartanto in particular, and the Laboratory of Radiobiology and Environmental Health in general, for outstanding administrative and logistical support; I thank Raymond Pedersen for editorial assistance; and I thank my wife, Carmen Arbona, for her wise counsel throughout this enterprise.

Roger A. Pedersen

1
Determination to Flower in *Nicotiana*

Carl N. McDaniel
Department of Biology
Plant Science Group
Rensselaer Polytechnic Institute
Troy, New York 12180

I. Introduction
II. Brief Historical Introduction to the Regulation of Flowering
III. Floral Determination in *Nicotiana*
  A. Number of Nodes Produced Prior to Flower Formation
  B. Assays for Floral Determination
  C. Floral Determination in Organized Buds and Meristems
  D. Floral Determination in Tissues Other Than Shoot Apical Meristems
  E. Relationship between Flowering in Organized Meristems and in Meristems Regenerated from Stem Tissues
IV. Floral Determination in Other Species
  A. *Pharbitis nil*
  B. *Lolium temulentum*
  C. *Helianthus annuus*
  D. *Pisum sativum*
V. Flowering as a Developmental Process
References

I. Introduction

The transition from vegetative to floral growth requires a substantial alteration in the spatial patterns of cell expansion and cell division in the meristem, followed by the elaboration of unique lateral organs and cellular specialization. Some investigators have considered this transition to be such an enormous modification of structure and function as to be analogous to metamorphosis, exhibited by members of many animal phyla (Lang, 1987). The organization and functioning of a shoot apical meristem results from determination and differentiation events during embryogenesis (Christianson, 1985; McDaniel, 1984a,b) as does the formation of a larva from an egg (Slack, 1983). Competent shoot apical meristems initiate flower formation in response to a developmental signal(s) from the leaves as metamorphosis is elicited by a hormone. Flowers form from cells derived from an organized shoot apical meristem much as adult structures of an animal that appear after metamorphosis arise from cells in the larva. Although there are limitations to direct comparisons between the two kingdoms, the analogy may be valuable if only from a conceptual perspective.

The processes involved in making a plant are conceptually parallel to those associated with animal development (McDaniel, 1984a,b, 1989). Certainly the timing of the various developmental processes and events as well as the specific molecular mechanisms might be different. The stability of determined states and subsequently of differentiated states is clearly different because some plant cell types can be manipulated to form whole plants (see Steeves and Sussex, 1989). This capacity for totipotency has yet to be demonstrated in animals, for which only nuclear totipotency has been established (Gurdon, 1962). Although states of determination and differentiation appear to be very stable in animals, determination does not mean irreversible commitment to some specific developmental fate. Transdetermination in *Drosophila* imaginal disks (Gehring, 1972), iris to lens differentiation in amphibians (Reyer, 1962), and numerous other examples cited in Volume 20 of this series (Moscona and Monroy, 1986) establish that determination and differentiation in animals do not necessarily reflect irreversible states. In general developmental states are more plastic in plants than in animals; however, stable developmental states are critical in such plant developmental processes as flowering.

II. Brief Historical Introduction to the Regulation of Flowering

In the second decade of this century photoperiod was identified as a major factor in the control of flowering in some plants, and the leaves were identified as the receptor of the photoperiodic signal (see Evans, 1969a). The physiology of the photoperiodic response of the leaves was extensively researched in the subsequent decades, elucidating phytochrome as the receptor pigment, establishing the interplay between light and dark periods, and identifying endogenous rhythms that interacted with the photoperiodic processes. Grafting experiments established that an apparently stimulatory signal, coined "florigen" by Chailakhyan (1937; cf. Chailakhyan, 1988), was transmitted from photoperiodically induced leaves to the apex, where it acted to initiate floral development. Florigen was considered to be a common flowering hormone because of the following observations:

1. Leaves of long-day plants caused short-day recipient plants to flower in long days.
2. Leaves of short-day plants caused long-day recipient plants to flower in short days.
3. Leaves of day-neutral plants caused long-day and short-day plants to

flower when kept under conditions that were noninductive for the photoperiodic recipient (see Lang, 1965).

Attempts to isolate florigen have been unsuccessful. [*Note:* Because of this history and specific tenets associated with the florigen hypothesis, the term "floral stimulus" will be used to refer to the inductive signal(s) from the leaves.] More recent studies have identified floral inhibitors from leaves; however, they are also chemically uncharacterized (Lang, 1987; Murfet, 1989). These types of data have demonstrated that the processes that lead to the export of floral stimulus or inhibitor from the leaves can be regulated by the photoperiod, thereby permitting photoperiodic control of flowering time. In day-neutral plants the time of flowering is controlled in less obvious ways as discussed below. However, it should be realized that even though the points of regulatory control may vary in different plants the overall developmental process of flowering is most likely common among angiosperms (McDaniel, 1984a, 1989).

Although grafting experiments clearly indicate communication between the leaves and the apex, the lack of chemical identification of the compounds involved has led some to question the hormonal basis of this communication. It has been proposed that the apex is nutritionally deprived and that flowering results from a reallocation of nutrients to the apex (Sachs, 1977; Sachs and Hackett, 1977). Although this may prove to be the case, the control of nutrient allocation by the floral stimulus must then be explained.

During the transition from vegetative growth to floral development, the patterns of cell division and the organization of the meristem exhibit marked changes such that cells that had rarely divided begin to divide much more rapidly (see Bernier *et al.*, 1981). These observations led researchers in the 1950s to propose the existence of a group of meristem cells that is passively carried until the time of reproductive development and that floral initiation involves the activation of these cells, which then form the reproductive structures (see Steeves and Sussex, 1989). This hypothesis is not consistent with clonal analysis studies in corn and sunflower, which establish that descendents of the same meristem cell can participate in the formation of both vegetative and reproductive structures (McDaniel and Poethig, 1988; Jegla and Sussex, 1989). Thus, cells of the meristem function through the vegetative to floral transition. This direct evidence coupled with other data has led most researchers to abandon the *meristeme d'attente* hypothesis as a viable explanation of meristem behavior during flowering (see Steeves and Sussex, 1989).

We have an immense literature on flowering, and it has defied simple codification (Bernier *et al.*, 1981; Halevy, 1985). This wealth of data has made it difficult to find unifying patterns because exceptions abound. The conceptual framework that has unified animal development (Gurdon, 1985; Slack, 1983) may very well serve a similar role in the field of flowering (McDaniel, 1984a,b; 1989). In this chapter commitment in plants, buds, meristems, and cells to form flowers is discussed, focusing first on *Nicotiana tabacum*. These observations are then related to several other plant species for which relevant data are available.

III. Floral Determination in *Nicotiana*

The making of a plant involves numerous developmental processes that lead to different cell, tissue, and organ types (Fig. 1). During the differentiation of cells and structures one of the early events appears to be programming of a competent cell or group of cells for a future developmental fate, or determination. A cell, or group of cells, in which determination has taken place is said to be committed to whatever their particular developmental fate will be. The complete programming does not appear to take place in a single determination event; rather, sequences of determination events are responsible for terminal phenotypes. Determination is an operationally defined concept. In general terms, a cell, or a group of cells, is said to be determined when it exhibits the same developmental fate whether grown *in situ,* in isolation, or at a new place in the organism (Slack, 1983; McDaniel, 1984b).

Each determination event represents a change in the developmental state of the cells involved, and the change in state is often the end result of an inductive process. Induction involves the interaction between a developmental signal(s) and competent cells, resulting in a specific modification of developmental fate. Following induction, some determined states may be expressed in ways measur-

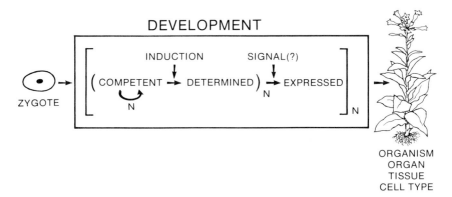

Fig. 1 This diagram illustrates some of the possible events that may occur during the ontogeny of an individual. Arrows indicate the normal direction of events but do not imply irreversibility. Competent cells/tissues are those that have the ability to respond to a developmental signal in a specific way. Cells and tissues can acquire different competences either via endogenous or exogenous means. Induction occurs when a developmental signal acts on competent cells/tissues to determine them for a specific developmental fate. The determined state is subsequently expressed. The "N" indicates that the bracketed sequence may be reiterated. [From McDaniel (1989) with permission.]

able by current techniques. For example, the cells may change morphology, synthesize a new protein, or gain the capacity to respond to a hormone. In other cases, the determined state may be cryptic, in that expression of a new phenotype does not directly follow induction but requires an additional developmental signal for expression.

Our experimental analysis of flowering in *Nicotiana* has identified one stable developmental state, as exemplified by buds or regenerated buds, whose meristems produce a terminal flower after initiating a limited number of nodes. This developmental pattern contrasts with that of other buds that initiate many nodes before making a terminal flower. We have called this state "floral determination," because our assay assesses developmental fate in several environments, the classical operation performed to establish the existence of a determined state (Slack, 1983; McDaniel, 1984b).

A. Number of Nodes Produced Prior to Flower Formation

1. Terminal Buds

The terminal meristems of the *Nicotiana* species we have studied first produce leaves, nodes, and internodes. A terminal meristem then ends its existence by development into a terminal flower (see Lang, 1989). The number of nodes produced by the terminal meristem of a day neutral tobacco plant before it forms a terminal flower is a function of the environmental conditions in which the plant is grown; under a given set of conditions the number of nodes produced by a population of plants is quite uniform (Seltman, 1974; Thomas *et al.,* 1975; McDaniel and Hsu, 1976; Table I). This same developmental pattern is expressed by photoperiodic cultivars (cv.) and species of *Nicotiana* when grown continuously under inductive conditions (Dennin and McDaniel, 1985; Gebhardt and McDaniel, 1987).

The uniform number of nodes produced by a meristem might result from an endogenous timing mechanism within the meristem that leads to the production of node-internode units at a certain rate for a specified period of time. Several observations appear to rule out endogenous integration of environmental cues by the meristem itself without inputs from other parts of the plant. When groups of *Nicotiana tabacum* cv. Wisconsin 38 plants are grown under the same conditions except for pot size, they make the same number of nodes but flower after different periods of time (Table II). The terminal meristem of *N. tabacum* cv. Wisconsin 38 can be forced to produce at least three times the normal number of nodes by rooting the terminal bud each time the plant produces from 6 to 10 leaves greater than 10 cm in length (McDaniel, 1980). Thus the rate of initiation can vary without altering the number of nodes produced, and the number of nodes a tobacco meristem can produce is not limited.

Table I Number of Nodes Produced by Various Species and Cultivars of *Nicotiana*

Species/cultivar	Number of nodes (± SD)	n	Range[a]	Growth conditions
Day-neutral plants				
N. tabacum cv. Wisconsin 38	35.5 ± 1.1	26	34–38	Field, summer
N. tabacum cv. Wisconsin 38	32.2 ± 1.6	17	29–35	Growth room, long day[b]
N. tabacum cv. Samsun	27.6 ± 1.6	32	25–32	Growth room, long day
N. tabacum cv. Hicks	25.6 ± 1.2	8	24–27	Growth room, long day
N. tabacum cv. Hicks	23.7 ± 1.5	22	22–28	Growth room, short day[b]
N. rustica	36.8 ± 1.8	11	34–39	Greenhouse, summer
Long-day plants				
N. silvestris	29.6 ± 1.7	14	27–33	Growth room, long day
N. silvestris	34.8 ± 2.2	8	32–38	Field, summer
Short-day plants				
N. tabacum cv. Hicks Maryland Mammoth	31.7 ± 2.3	21	27–35	Growth room, short day
N. tabacum cv. Maryland Mammoth	31.4 ± 1.6	12	29–34	Growth room, short day

[a]The first node counted was that with the first leaf greater than 10 cm in length. Usually there were four or five small leaves between the cotyledons and the first 10-cm leaf. All floral nodes were counted.

[b]Long days consisted of 16 hr of light and 8 hr of dark. Short days consisted of 8 hr of light and 16 hr of dark. Light source and intensity as well as day and night temperatures were not the same in all cases.

Leaf removal experiments have established that a normal number of nodes is produced so long as several mature leaves, located at any position on the main axis, are left on the plant (Hopkinson and Ison, 1982; McDaniel, 1980; Gebhardt and McDaniel, 1991). Thus, the basal leaves of a young plant have sufficient inductive capacity to cause the initiation of flowering. But young plants do not flower immediately; rather, they produce more leaves. This fact, coupled with the observation that roots induced close to the shoot tip increase vegetative growth, led us to propose that there is an interplay between leaves and roots such that a shoot meristem needs to be separated by some distance from the roots, as measured in node-internode units, before the meristem will initiate flowers (McDaniel, 1980).

Table II Relationship between Time to Flower and Number of Nodes Produced When *Nicotiana tabacum* cv. Wisconsin 38 Plants Are Grown in Different-Sized Pots

Soil volume (cm^3)	Nodes			Anthesis		
	Number of nodes ($\pm$ SD)	n	Range[a]	Days to anthesis ($\pm$ SD)	n	Range
90	31.4 $\pm$ 4.3	10	27–36	131.3 $\pm$ 21.7	10	106–177
1420	32.1 $\pm$ 0.7	10	31–33	86.2 $\pm$ 6.2	10	77–99

[a]Plants were transplanted from seedling pots directly into standard 5- or 15-cm pots after about 1 month. Plants were grown in a growth room under the following conditions: daily 16-hr light and 8-hr dark periods, a day temperature of 27.0 $\pm$ 1.0°C and a night temperature that decreased to a low of 18.0 $\pm$ 1.0°C, light provided by sodium-vapor and metal halide lamps with a photosynthetic photon flux density of 400–500 $\mu mol \cdot m^2 \cdot sec^{-1}$ at bench level, plants watered daily with tap water and fertilized weekly with Ra-Pid-Gro. Nodes counted as in Table I.

A more extensive characterization of this interplay between roots and leaves in two day-neutral and two short-day cultivars of *N. tabacum* has supported this original hypothesis, but also revealed genotypic influences on the root-shoot interplay (Gebhardt and McDaniel, 1991). The roots appear to play an inhibitory role in both day-neutral and short-day tobaccos, but in some genotypes the strength of the floral stimulus from the leaves varies as a function of when a leaf is initiated. In some cultivars under some environmental conditions the inductive capacity appears to be similar in all leaves that attained a length of at least 10 cm while in other cultivars the first several leaves initiated have little inductive capacity compared with the higher inductive capacity of leaves initiated later. It appears that this higher inductive capacity can override the root inhibitory influence, enabling these genotypes to flower in rerooting experiments even though the shoot meristem is separated from the roots by a relatively small number of node-internode units.

These results indicate that inductive capacity of leaves of different genotypes can vary. Could there also be differences among genotypes in the capacity of roots to prolong vegetative growth? In the species *N. tabacum* there is a wide range in the number of nodes produced by various cultivars or tobacco introductions (TIs) as listed in the U.S. Department of Agriculture (USDA) *Nicotiana* germplasm collection (Chaplin *et al.*, 1982). We hypothesized that if there is an interaction between roots and leaves that establishes the number of nodes produced by a cultivar or TI, then reciprocal grafts between many-noded and few-noded TIs should allow one to identify TIs that have roots or leaves with different signal strengths. For example, a plant could be few noded because it has a relatively weak inhibitory signal from the roots or because it has a relatively strong stimulatory signal from the leaves. We have selected 17 few-noded and 17 many-noded TIs to characterize via reciprocal grafts. In all pairs analyzed to date

Table III Number of Nodes Produced by Reciprocal Grafts between *Nicotiana tabacum* Tobacco Introductions That Make Many and Few Nodes

		Nodes		
Stock (root)	Scion (shoot)[a]	Number of nodes (± SD)	n	Range[b]
—	112[c] (not grafted)	55.8 ± 4.4	5	50–62
112	112	52.3 ± 4.4	4	48–57
—	1227 (not grafted)	19.0 ± 1.9	5	16–21
1227	1227	19.9 ± 0.8	8	19–21
112	1227	19.3 ± 0.5	10	19–20
1227	112	55.4 ± 4.4	12	51–66
—	1347 (not grafted)	16.6 ± 1.1	5	15–18
1347	1347	18.1 ± 1.3	9	17–20
112	1347	16.6 ± 1.0	11	16–19
1347	112	55.7 ± 4.2	7	51–63

[a]Shoot tips from seedlings that had expanded two or three true leaves were grafted onto seedling root stocks of the same age such that no stock leaves remained.
[b]Growth conditions as stated in Table II. All nodes counted on scion.
[c]This number refers to the tobacco introduction number in Chaplin et al. (1982).

the genotype of the shoot has determined the number of nodes produced by the shoot (Table III). Thus, mutations that affect the root signal or its activity appear to be rare compared to those that affect the leaf signal or its activity.

2. Axillary Buds

Nicotiana tabacum cv. Wisconsin 38 exhibits strong apical dominance such that all of the axillary buds below the inflorescence are inhibited after producing from seven to nine leaf primordia, with the longest primordium being less than 1.0 cm. When the shoot above a given axillary bud is removed, apical dominance is removed and the axillary bud will initiate growth, producing a shoot similar to the main shoot. The number of nodes produced by an axillary bud released from apical dominance is position dependent, with the bud at a given node position producing about one less node than the bud at the position directly below it (McDaniel et al., 1989b; Fig. 2). This position-dependent growth pattern is observed in buds on vegetative plants as well as on plants at various stages of flowering (McDaniel and Hsu, 1976). When an apical bud is grafted to the base of the main axis or a basal bud is grafted to an apical position, the buds develop according to their new positions (Chailakhyan and Khazhakyan, 1975; McDaniel and Hsu, 1976). When axillary buds are removed from the main axis and rooted, they produce a number of nodes similar to the number produced by plants grown from seed. These results establish that the number of nodes produced by an axillary bud released from apical dominance is not an endogenous property of the

1. Determination to Flower in *Nicotiana*

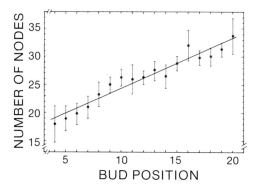

Fig. 2 Position-dependent growth of axillary buds on field-grown *N. tabacum* cv. Wisconsin 38 plants. At opening of the terminal flower, plants were decapitated in the internode above the indicated bud. The average number of nodes (± SD) produced by four buds at each position was plotted. Buds were numbered basipetally, with bud 1 being the first axillary bud immediately below the inflorescence. Linear regression of the means indicated a line with a slope of 0.89 ($R^2 = 0.94$). [From McDaniel *et al.* (1989b) with permission.]

bud but rather a function of the position of the bud on the main axis. In this regard and in the type of inflorescence produced, axillary buds released from apical dominance mimic the developmental pattern of the terminal bud (McDaniel *et al.*, 1989b). [*Note:* As considered below in Section III,C, the pattern of development exhibited by the uppermost axillary buds just below the inflorescence is different from the above-described pattern (McDaniel, 1978; McDaniel *et al.*, 1989b).]

Although isolated axillary buds produce a number of nodes similar to that produced by plants grown from seed, Chailakhyan and Khazhakyan (1974) reported that when stem pieces were rooted, the flowering time of the attached axillary bud was a function of the original position of the bud on the main axis. Thus isolated buds behaved differently than buds that remained attached to isolated stem pieces. We have confirmed this observation, establishing that the number of node-internode units on the stem explant influences the number of nodes produced by the attached axillary bud and that the response is position dependent. This interesting phenomenon requires more investigation before it can be related to what we know about node production and floral initiation.

The above results obtained with terminal and axillary buds establish that the number of nodes produced by an apical meristem of a day-neutral tobacco is not a property of the meristem alone. The roots have an inhibitory influence on flowering, and this may be important in maintaining the vegetative state in the meristems of young plants. The leaves provide a floral stimulus that, as the plant adds more node-internode units, leads to the initiation of an inflorescence by the apex. In a way still not understood the root signal(s) and the leaf signal(s) interact, apparently as a function of the number of node-internode units that separate the terminal meristem from the roots, to control when the inflorescence is initiated.

B. Assays for Floral Determination

As discussed above the number of nodes produced by a shoot apical meristem of the *Nicotiana* species we have studied is uniform under a given set of environmental conditions. Although it is unclear how the plant employs node-internode units as a means of establishing when the terminal meristem is to make a flower (Khait, 1986), this measuring process can be employed to assess the developmental fate of a meristem.

We have employed two different types of isolation assays to assess the state of floral determination in *Nicotiana*. In the first, the bud-rooting assay, an axillary or terminal bud is removed from the rest of the plant, rooted, and then transferred to a pot to grow and flower. A modification of this assay is to isolate a meristem with several leaf primordia and have it grow, root, and then mature to flowering. The number of nodes produced by the resulting plant is counted and compared to the number that would have been produced had the bud or meristem been permitted to flower *in situ*. If the number of nodes produced *in situ* and in isolation is the same, then the meristem in association with the bud or the leaf primordia is considered to be florally determined at the time of the manipulation. The second isolation assay assesses the developmental fate of cells and tissues that are not organized as a shoot apical meristem at the time of manipulation. Stem tissues taken from flowering, day-neutral tobacco plants form two types of *de novo* shoots when cultured on hormone-free medium: floral shoots, which have from zero to about eight nodes below a terminal flower, and vegetative shoots, which have produced only leaves in culture (Aghion-Prat, 1965a; Tran Thanh Van, 1973). We have assumed that *de novo* shoots forming a terminal flower after producing fewer than 10 nodes were derived from cells that were florally determined while those shoots that remain vegetative were derived from cells that were not florally determined (McDaniel *et al.*, 1989a).

In addition to the isolation assays, floral determination has also been assessed via the grafting of buds (McDaniel, 1978; McDaniel *et al.*, 1987). In this assay a bud is grafted to the base of a flowering plant and, after the graft has taken, the main axis is cut off just above the grafted bud. After the grafted bud has grown out and flowered, the number of nodes present on the graft is counted. If the number of nodes is similar to the number of nodes that would have been produced by the bud in its original location and if this number is substantially less than would be produced by a basal axillary bud released from apical dominance, then the bud is considered to have been florally determined at the time of grafting.

C. Floral Determination in Organized Buds and Meristems

Because the terminal meristem of the *Nicotiana* plant makes the terminal flower after producing a predictable number of node-internode units, it is reasonable to assume that at some time prior to actually making the flower, the meristem becomes programmed to make the flower or at least initiate floral development. Our ability to establish when this programming first occurs depends on at least two requirements. First, a new environment must be identified in which floral development can occur, or at least be initiated, so that floral development can be distinguished from vegetative development. Second, the programming must be sufficiently stable so that the floral programming is not disrupted by the manipulations associated with assessment of the developmental fate in the new environment. We have been fortunate in *Nicotiana* because these conditions have been met by three assays as described above.

It has been established by shifting intact, photoperiodic, tobacco plants from inductive to noninductive conditions that the plant, as a whole, becomes committed to flowering prior to floral development (Waterkeyn *et al.*, 1965; Hopkinson and Hannam, 1969). At what level of organization does this commitment to flower reside? Is it a property only of the whole plant? Or can this commitment be expressed at some lower level of organization? Employing the bud-rooting assay, it has been possible to establish that floral determination is a property of the terminal bud in the three photoperiodic types of *Nicotiana* we have examined (McDaniel *et al.*, 1985; Singer and McDaniel, 1986; Gebhardt and McDaniel, 1987; Fig. 3).

Terminal buds of *N. tabacum* cv. Maryland Mammoth, a short-day tobacco, exhibit floral determination and morphological patterns similar to those shown by terminal buds of long-day *Nicotiana silvestris* (McDaniel *et al.*, 1985) and day-neutral *N. tabacum* (Thomas *et al.*, 1975; Singer and McDaniel, 1986). Under conditions that permit flowering, these three photoperiodic response types of *Nicotiana* have a long period of vegetative growth. Over a period of no more than several days the terminal bud becomes florally determined, after which the terminal meristem produces from zero to four more nodes and the terminal flower. Although the processes in the leaf that ultimately control the export of floral stimulus in these three photoperiodic response types must exhibit differences because of the different photoperiodic requirements, these results indicate that the action of the floral stimulus from the leaves on the apex appears to be similar. These observations also make it clear that the bud-rooting assay we have employed to assess floral determination is, in fact, characterizing the change of state in the terminal bud that results from the inductive action of the floral stimulus.

The terminal bud employed in the rooting assay is a rather large structure containing the terminal meristem and about a dozen leaves and leaf primordia (the largest leaf being from 3 to 5 cm in length) as well as node and internode

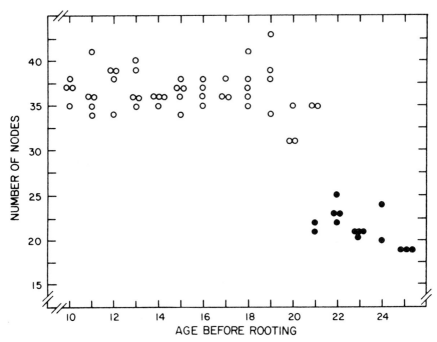

Fig. 3 Timing of floral determination in terminal buds of field-grown *N. tabacum* cv. Wisconsin 38 plants. Terminal buds were rooted at the indicated age. The age of the plant is defined as the number of leaves >3 cm in length. Number of nodes present on the plant after rooting: ○, undetermined bud; ●, determined bud. [From Singer and McDaniel (1986) with permission.]

tissues. It is possible to produce whole plants from explants composed of the much smaller terminal meristem (Smith and Murashige, 1970). Preliminary results from experiments employing day-neutral and short-day cultivars indicate that explants of meristems with only three or four leaf primordia exhibit floral determination at a time similar to that of isolated buds (Smith, 1990). The short-day cultivar may have a florally determined state that is less stable than that of the day-neutral cultivar, but further investigation is required to establish this firmly.

Axillary buds of day-neutral tobacco, when released from apical dominance, grow out and exhibit a pattern of development identical to that of the terminal bud. Employing axillary buds released from apical dominance, the bud-rooting assay established that they become florally determined in a pattern identical to that of main-axis terminal buds (Singer, 1985; McDaniel et al., 1987).

As noted earlier the developmental behavior of axillary buds just below the inflorescence is often different from that of the more basal axillary buds, although these upper buds are morphologically similar to more basal buds. When the uppermost axillary buds are removed from the main axis and rooted or grafted, they often produce a number of nodes similar to what they would have produced had they been permitted to grow out *in situ* (McDaniel, 1978). This pattern of growth is not a function of being near the apex of the plant because axillary buds isolated from the upper nodes of plants prior to flowering always produce a number of nodes similar to the number made by plants grown from seed (Singer and McDaniel, 1986). In *N. tabacum* cv. Wisconsin 38 only the upper three axillary buds exhibit this limited growth pattern, and they become determined for this pattern several days after the terminal bud becomes florally determined. In isolation, these florally determined axillary buds produce between 4 and 13 new nodes and a terminal flower. Thus an arrested axillary bud that is florally determined prior to being released from apical dominance can produce substantially more nodes before forming a terminal flower than the several produced by a growing, terminal, shoot apical meristem after floral determination. This difference is not understood, but it will be discussed below when we consider the growth pattern of regenerated shoots.

One of the more intriguing observations we have made has been the pattern of

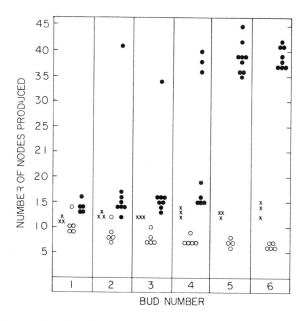

Fig. 4 Developmental fate of terminal meristems of axillary buds of *N. silvestris*. The developmental fate of rooted (●) and *in situ* (×) axillary buds was assessed in terms of the total number of nodes (vegetative plus floral) produced below the terminal flower. The number of nodes present on axillary buds at anthesis (the time of rooting) (○) was based on histological analysis of similarly positioned buds. Data for buds below the sixth bud are not presented because the pattern was the same as that reported for fifth and sixth buds. [From Dennin and McDaniel (1985) with permission.]

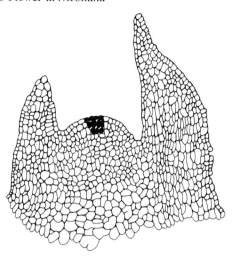

Fig. 6 Two populations of cells in a determined axillary bud of *N. silvestris* prior to rooting. This is a tracing of a median longitudinal section of a third axillary bud (× 480). Shaded cells are stably determined for floral development and will ultimately form the terminal flower. The number and positions of determined cells were arbitrarily selected. Other cells will form the vegetative structures surrounding the terminal flower. [From McDaniel *et al.* (1987) with permission.]

morphogenesis exhibited by the uppermost axillary buds of *N. silvestris* when rooted and grown to maturity (Dennin and McDaniel, 1985). Like axillary buds of day-neutral Wisconsin 38, each bud of this long-day plant exhibited one of two distinct developmental fates. A bud either produced below its terminal flower a number of nodes similar to the number made by an *in situ* bud released from apical dominance (i.e., a florally determined bud) or a number similar to that made by a plant grown from seed (i.e., a vegetative bud) (Fig. 4). However, the inflorescences of almost all of the florally determined axillary buds were abnormal. Normal inflorescences have six to eight floral branches located below the terminal flower. Most inflorescences of florally determined rooted buds did not contain floral branches but rather vegetative, axillary branches or main axes (Fig. 5). It thus appears that a subpopulation of cells in the meristems of these axillary buds became florally determined and expressed this state when forming the terminal flower while neighboring cells never acquired or reverted from a florally determined state to a more vegetative condition (Fig. 6). This pattern of

---

Fig. 5 Abnormal inflorescences produced by rooted axillary buds of *N. silvestris*. (A) After laying down 15 nodes the meristem produced a terminal flower (arrow). Cells directly below the terminal flower gave rise to two vegetative axillary buds, which produced 10 (left) and 12 (right) nodes before forming a terminal flower. (B) After laying down 14 nodes the meristem produced a terminal bud (arrow) that remained immature. Cells directly below the terminal flower bud formed a vegetative shoot instead of floral branches. [From Dennin and McDaniel (1985) with permission.]

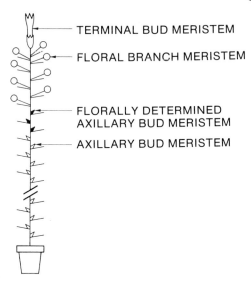

Fig. 7 The four developmentally distinct shoot apical meristems present on *N. tabacum* plants. Each meristem, except for the terminal bud meristem and sometimes the uppermost floral branch meristem, is located in the axil of a leaf or a bract. The developmental stage depicted in the diagram is the opening of the terminal flower. Terminal bud meristem has completed its growth as a meristem and has formed the terminal flower. Floral branch meristems are still growing and producing flowers. Florally determined axillary bud meristems and axillary bud meristems are developmentally arrested. [From Singer and McDaniel (1986) with permission.]

inflorescence morphogenesis has also been observed in rooted buds of day-neutral tobacco (Singer, 1985) and short-day tobacco (Dennin, 1985), although at lower frequencies.

The foregoing discussion makes it possible for us to identify, developmentally, four types of shoot apical meristems in *Nicotiana* plants (Fig. 7). The terminal bud meristem is formed during embryogenesis and, after germination, gives rise to all of the shoot system. It ends its existence by forming the terminal flower. Below the terminal flower are axillarily positioned floral branch meristems, most of which have an iterative sequence of individual flower, bract, and axillary meristem that can in turn form secondary floral branches. In function most floral branch meristems exhibit an unlimited pattern of growth because they continue to initiate flowers. Directly below the lowermost floral branch meristem are several florally determined axillary meristems whose pattern of development have been discussed above. Below the florally determined axillary meristems are the axillary bud meristems, whose pattern of development is similar to that of the terminal bud meristem when they are released from apical dominance.

The foregoing has discussed the commitment to flower that occurs in plants, buds, and meristems prior to actual flower initiation. Hicks and Sussex (1971) have cultured meristems of day-neutral tobacco at the time of the morphological initiation of the flower. Their results establish that once floral morphogenesis has begun there is a strong tendency for the isolated meristem to complete flower formation independently.

D. Floral Determination in Tissues Other Than Shoot Apical Meristems

Observations of abnormal inflorescences produced by root axillary buds of *N. silvestris* as well as similar, although less frequent, observations with buds of short-day and day-neutral tobaccos indicate that subpopulations of cells within a meristem can exist in different developmental states. Related to this is the observation that tissues taken from flowering day-neutral tobacco plants can form *de novo* flowers and floral shoots (Aghion-Prat, 1965a). When the explants are from the floral branches or the pedicel of a flower (the structure just below the flower proper), many of the meristems develop directly into a flower without any leaves, nodes, or internodes below the flower (Tran Thanh Van, 1973; Van den Ende *et al.*, 1984). Floral shoots can also form *de novo* on explants from leaves, although this process in leaves has been poorly characterized (Chailakhyan *et al.*, 1977).

On permissive medium explants from the main axis of the plant make two developmentally different types of meristems: (1) those that produce from 0 to about 10 nodes, usually less than 8, and then a terminal flower (i.e., floral shoots) and (2) those that produce few to many nodes but no flower (i.e., vegetative shoots) (McDaniel *et al.*, 1989a). The number of nodes produced by floral shoots is not tightly correlated with the source position along the main axis, although there is a tendency for explants from the upper positions to produce floral shoots with several fewer nodes (McDaniel *et al.*, 1987, 1989a; Table IV). Because the meristems that form the floral shoots exhibit a developmental pattern identical to that of florally determined terminal and axillary buds, it seems reasonable to assume that these meristems are florally determined.

Most cells in an explant do not participate in the formation of a shoot, and it is not known exactly why some cells form shoots while others do not (Meins *et al.*, 1982). A critical assumption we have made is that the probability that a cell will participate in the formation of a shoot meristem is not influenced by the cell being florally determined.

It has been known for a long time that the ratio of floral to vegetative shoots is a function of explant position, with more apical explants giving a higher ratio of floral to total shoots (Aghion-Prat, 1965a). Recently we reported that explants from vegetative plants as well as explants from the most basal portions of flowering plants can give rise to floral shoots (McDaniel *et al.*, 1989a; Table V) in contrast to what had been previously reported (Tran Thanh Van, 1973; Chailakhyan *et al.*, 1977; Lang, 1989). The frequency of floral shoots on explants from vegetative plants is approximately 0.1 to 0.3% (floral shoots/total

Table IV Number of Nodes Produced by *de Novo* Floral Shoots Regenerated on Explants from the Main Axis of *Nicotiana tabacum* cv. Wisconsin 38 as a Function of Explant Source Position

Explant source[a]	Plant 1			Plant 2			Plant 3			Plant 4		
	Number of nodes (± SD)	n	Range	Number of nodes (± SD)	n	Range	Number of nodes (± SD)	n	Range	Number of nodes (± SD)	n	Range
Inflorescence	1.4 ± 0.7	54	0–3	3.1 ± 1.2	22	1–5	2.0 ± 0.8	7	1–3	2.4 ± 1.1	8	1–4
Below inflorescence	3.7 ± 1.1	49	1–7	3.9 ± 0.6	15	3–5	5.0 ± 1.8	11	2–7	4.0 ± 1.7	22	2–7

[a] Stem segments about 2 mm thick were cultured on hormone-free medium for 6 weeks as described in McDaniel *et al.* (1989a). Data from McDaniel *et al.* (1987, unpublished) with permission.

Table V Percentage *de Novo* Floral Shoots Produced on Explants from the Main Axis of Vegetative and Floral *Nicotiana tabacum* cv. Wisconsin 38 Plants

Plant age[a]	Floral shoots[b] (%)
5	0.1
8	0.3
11	2.1
13	2.7
15	14.2
Inflorescence visible	14.2
Anthesis[c]	21.0

[a] Plant age is defined as the number of leaves greater than 10 cm in length. Data for ages 11, 13, 15, and inflorescence visible from Singer and McDaniel (1987) and for ages 5, 8, and anthesis from McDaniel *et al.* (1989a) with permission.

[b] The entire main axis was cut into 2-mm segments and cultured on hormone free medium as described in Singer and McDaniel (1987).

[c] Opening of terminal flower.

shoots) while the frequency on explants taken from the bottom quarter of flowering plants (i.e., the same nodes and internodes explanted from vegetative plants) is about 1%. We and other workers have shown that explants from the inflorescence can form 100% floral shoots (Tran Thanh Van *et al.*, 1974; Van den Ende *et al.*, 1984). Thus in vegetative plants and in the base of flowering plants the frequency of florally determined cells is low, while in the inflorescence almost all cells appear to be florally determined (Fig. 8).

Why do cells in a young vegetative plant become florally determined while the apical meristems do not? Although the answer eludes us, we do know from leaf-removal experiments that the basal leaves on a vegetative plant have inductive capacity and must therefore be a source of floral stimulus (Hopkinson and Ison, 1982; Gebhardt and McDaniel, 1991). In the young plant the level of the floral stimulus is insufficient to induce the terminal meristem into a florally determined state, perhaps because of the root influence. On the other hand, it would appear that a small fraction of the cells in the vegetative plant is sufficiently sensitive to the floral stimulus such that they can be induced into a florally determined state.

Explants from flowering day-neutral tobaccos regenerate floral shoots, but Rajeevan and Lang (1987) established only recently that explants of the short-day *N. tabacum* cv. Maryland Mammoth could regenerate floral shoots. In this cultivar, however, only explants from the pedicel formed floral shoots. Explants from floral branches or more basal tissues gave only vegetative shoots. This difference between day-neutral and photoperiodic tobaccos could reflect a dif-

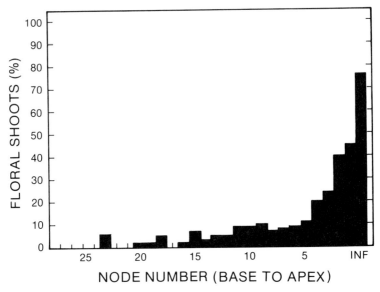

Fig. 8 Percentage of floral shoots formed by main-axis explants excised from *N. tabacum* cv. Wisconsin 38 plants at anthesis. Entire main axes of 29 plants were cut into 2-mm thick sections and cultured on hormone-free medium for 6 weeks. All regenerated shoots were dissected and classified as floral or vegetative. INF, inflorescence [Data from McDaniel *et al.* (1989a) with permission.]

ference in competence to respond to the floral stimulus or a difference in the stability of the florally determined state.

It is also possible to demonstrate the presence of florally determined cells in the plant by permitting internode cells to regenerate shoots on the plant (Aghion-Prat, 1965b; McDaniel *et al.*, 1989a). If a day-neutral plant is topped, and the exposed internode tissues are treated with an auxin and a cytokinin in a lanolin paste, treated cells in the cambial area will divide and subsequently form shoots. We know that axillary buds released from apical dominance will produce a number of nodes that is position dependent (Chailakhyan and Khazhakyan, 1974; McDaniel and Hsu, 1976; McDaniel *et al.*, 1989b). One would assume that *de novo* shoots would also respond to their position on the main axis, producing a number of nodes correlated to that of similarly positioned axillary buds. For most regenerated shoots this predicted pattern of development was observed (Table VI). For a substantial group of regenerated shoots, however, the number of nodes produced was at variance with the prediction. The question becomes, why do some *de novo* shoots fail to produce a number of nodes appropriate for their position on the main axis?

To address this question we need to consider the importance of place during the development of an organism. It is clear that development is position depen-

1. Determination to Flower in *Nicotiana*

Table VI Number of Nodes Produced on *Nicotiana tabacum* cv. Wisconsin 38 Plants by Axillary Buds *in Situ* and by Shoots Organized *de Novo* from Internode Tissues *in Situ*

	Axillary shoots[b]			*De novo* shoots[b]		
Position[a]	Number of nodes (± SD)	n	Range	Number of nodes (± SD)	n	Range
1	11.8 ± 1.7	17	8–14	5.2 ± 2.4	109	1–13
6	14.3 ± 1.8	15	12–17	9.2 ± 5.8	82	2–38
12	19.8 ± 0.9	16	18–23	16.7 ± 8.0	110	5–39
18	25.8 ± 1.9	18	23–29	26.2 ± 8.6	92	7–43

[a]Nodes-internodes were numbered basipetally, with 1 being the first node-internode below the lowest floral branch node.

[b]Plants were grown to anthesis of the terminal flower and then topped in the middle of the indicated internode. For axillary shoots only the axillary bud directly below the cut internode was permitted to grow. At the time of topping, axillary buds were arrested and had initiated seven to nine leaf primordia. For *de novo* shoots all axillary buds were removed and a thin layer of lanolin paste containing 1% indole-3-acetic acid and 2% N^6-benzyladenine was applied to the cut surface immediately after topping. Plants employed for axillary shoots had produced 33.6 ± 2.6 (66) nodes, while those employed for *de novo* shoots had produced 33.7 ± 2.4 (87) nodes. Table 3 from McDaniel *et al.* (1989a) is reproduced with permission.

dent: leaves form at specific places on the shoot apical meristem rather than on roots, fingers form on hands rather than on knees, and so on. At some time during the development of a structure, however, its developmental fate becomes independent of position. Leaf primordia gain independence early in their ontogeny (Steeves and Sussex, 1957; Smith, 1984). Amphibian limb fields likewise become position independent before limb formation (Harrison, 1918). Initially place is critical, but as ontogeny progresses history becomes more and more important. This process can be illustrated with *N. tabacum* cv. Wisconsin 38 axillary buds. The place where an axillary bud develops (i.e., in a position on the main axis or in isolation as an independent shoot after being rooted) can influence its developmental fate as measured by the number of nodes produced before forming a terminal flower (Chailakhyan and Khazhakyan, 1975; McDaniel and Hsu, 1976). However, if the positional history of a bud is such that it is located about four to six nodes below the terminal meristem during the time when the apex of the plant is being induced into a florally determined state by the floral stimulus, the history of the bud is now critical, not the place where it grows out and flowers (McDaniel, 1978; Singer and McDaniel, 1986). That is, as a result of previous inputs (i.e., influence of the floral stimulus during its formation), the fate of a bud reflects its original place and it produces atypically few nodes for the new environment where it matures (see Fig. 5).

The same logic can explain the atypical number of nodes produced by some of the regenerated shoots *in situ*. Axillary buds at positions 12 and 18 nodes below the inflorescence produced 19.8 and 25.8 nodes, respectively. *De novo* shoots at these internode locations produced 16.7 and 26.2 nodes, respectively. Although the averages are similar, the ranges are much larger for the *de novo* shoots. The regenerated shoots producing fewer than 10 nodes behaved like florally determined axillary and terminal buds. Thus, they are assumed to have been derived from florally determined cells. At internode position 12, about 25% of the shoots produced fewer than 10 nodes, while at position 18 about 3% did so (Table VII). When main-axis segments from these same locations were cultured, those from around position 18 produced 2 to 5% floral shoots, while those from around position 12 produced 3 to 9% floral shoots. Thus both the *in situ* and the isolation assays indicate that the developmental capacities of all main-axis cells are not the same. Some cells appear to have been induced into a florally determined state by the floral stimulus from the leaves.

We observed a similar percentage of florally determined cells at position 18 in both assays, while at position 12 the percentage was more than two times greater with the *in situ* assay. With the *in situ* assay we also observed at both positions between 20 and 30% of the shoots, which produced fewer nodes than predicted for that position but more than the number expected from a florally determined meristem. There are several explanations for these unpredicted numbers of nodes and the different frequencies of florally determined cells observed in the two assays. Two possible explanations will be considered.

As discussed earlier the developmental fate of a group of cells to form a specific structure or organ apparently is not fixed in a single determination event; rather, there appear to be many such events (Fig. 1). Developmental biologists have used such terms as *specification* and *canalization* to indicate different degrees or stabilities of determination for a specific developmental fate (Slack, 1983). Because we know so little about developmental states in plants, it is preferable not to use such terms to distinguish among types of determination events. It is also clear that determined states can change such that the developmental fate of a group of cells becomes more or becomes less restricted (i.e., a group of cells can move forward or backward in a developmental pathway). Because each assay employed to assess the developmental state of a group of cells presents the cells with a different set of environmental conditions, different results from different assays should not be unexpected. Thus, the *in situ* assay for floral determination may be less disruptive than the assay in culture and may thereby reveal states of floral determination that are less stable. For example, the *in situ* environment may permit all levels of floral determination to be expressed while isolation of the cell from the whole plant may permit only those cells that have attained a certain level of floral commitment to retain and express the florally determined state. Two observations support the interpretation that floral determination as measured by our assays is not an all-or-none state. First, floral shoots derived from inflorescence tissues tend to produce fewer nodes than floral shoots produced from more basal internodes (Tran Thanh Van, 1973;

Table VII Comparison as a Function of Main-Axis Position on *Nicotiana tabacum* cv. Wisconsin 38 Plants of the Percentage of Floral Shoots Produced in Culture with the Percentage of Shoots Regenerated *in Situ* That Produced Fewer Than 10 Nodes[a]

	In culture	In situ	
Explant source	Floral shoots (%)	Internode position	Shoots with fewer than 10 nodes (%)
Inflorescence to node 3	76 to 24	1	95
Node 4 to node 8	20 to 7	6	75
Node 10 to node 14	9 to 3	12	25
Node 16 to node 20	5 to 2	18	3

[a]Data from McDaniel *et al.* (1989a) with permission.

McDaniel *et al.*, 1987; Table IV). Second, tissues from the pedicel of a flower form meristems that produce a flower directly, that is, they produce no leaves (Van den Ende *et al.*, 1984). Thus, in an "early" florally determined state a meristem will produce a limited number of nodes and then a flower, while in a "late" florally determined state a meristem just makes a flower.

An alternative explanation would be based on the assumption that floral determination was a property of single cells and that a cell was either florally determined or not. In this case it would be the fraction of florally determined cells that were contained within a meristem that established the number of nodes formed. If a meristem were composed of exclusively florally determined cells, then a flower would form immediately. A smaller fraction of florally determined cells would lead to the formation of some leaves, nodes, and internodes before a flower was formed.

E. Relationship between Flowering in Organized Meristems and in Meristems Regenerated from Stem Tissues

Cells, organized as a shoot apical meristem, form lateral organs in specific spatial patterns and temporal sequences. A vegetative meristem expresses one pattern and sequence while a floral meristem expresses a different pattern and sequence. It is difficult to know exactly what type of information is responsible for such morphogenetic patterns and how the expression of such information is programmed. It is not possible for a single cell to form a flower, so how can it be possible for a cell to be florally determined, that is, programmed to form a flower? Clearly a cell can only be programmed to participate in the formation of a flower as a part of a group of cells that have been organized as a meristem. The fact that explants from several sources in the day-neutral tobacco plant have the capacity to organize developmentally different types of meristems, including those that form a flower directly, indicates that an early state of floral determination can be induced in cells, perhaps individual cells. This early state of floral determination appears to be related to the communication among cells within a meristem to allow for the expression of the spatial and temporal patterns of lateral organ initiation associated with flower formation. There are no compelling reasons to believe that the induction process leading to floral determination in cells outside of a shoot apical meristem would be different than that occurring in cells of the meristem.

IV. Floral Determination in Other Species

Students of flowering have long been interested in the unique events in the apex that commit an apex to a floral pattern of development (Lang, 1965; Bernier et al., 1981; Bernier, 1988). There have been numerous histological, cytological, and biochemical analyses attempting to establish the sequence of events during the transition from vegetative to floral development. Most of these studies have been carried out on plants whose flowering is controlled by an environmental cue, usually the photoperiod. Most of the work has been carried out from a physiological perspective with few researchers considering developmental states. The time of commitment to floral development was in most cases established by two types of assays. One was to grow plants under noninductive conditions and then establish the amount of exposure to inductive conditions required for the plant to flower when returned to noninductive conditions. In a small number of plants this period was less than 1 day (Evans, 1969b). The second assay was to expose the apex of the intact plant to various metabolic inhibitors to establish sensitive and insensitive periods during and after photoperiodic induction. In general, when the metabolic inhibitor specifically blocked flowering the sensitive period was of short duration, occurring on the day of and/or the day after photoperiodic induction (Bernier et al., 1981). Floral determination has been characterized by grafting and isolation assays in only a few of the many plants in which floral initiation has been investigated. Considered below are four species for which there are sufficient data for comparisons and discussion.

B. *Pharbitis nil*

The various aspects of the flowering physiology of *Pharbitis nil*, the Japanese morning glory, have been well documented (Takimoto, 1969; Vince-Prue and Gressel, 1985). One strain of this short-day plant, *P. nil* strain Violet, flowers in response to a single long night and has been extensively employed for floral initiation studies. Three studies have directly addressed the question of when the meristem becomes florally determined. Bhar (1970) cultured the terminal meristem, including two leaf primordia, at various times after the inductive night. Apices cultured 24 hr after the end of the night period did not form flowers, while those cultured after 36 hr produced terminal flowers. In a later study Matsushima and co-workers (1974) reported that apices cultured 5 hr after the end of the dark period did form terminal flowers. Unfortunately, the number of leaf primordia on the excised apices was not reported so it is difficult to compare and evaluate the two sets of results. If the Matsushima group used apices with just two primordia, as Bhar did, then it would appear that the meristem became florally determined several hours after dawn on the day following the inductive night. On the other hand, if larger explants were employed, then Bhar's results indicate that floral determination in the meristem of such small explants can only be established about 36 hr after the end of the inductive night.

Recently, Larkin and co-workers (1990) grafted terminal buds (i.e., all of the bud above the cotyledons, which consisted of six leaf primordia with the largest being 2–3 mm in length) from induced to noninduced plants at various times after the long night. The flowering patterns of grafted buds were compared to those of induced plants that had their cotyledons removed. (*Note:* The cotyledons are the source of the floral stimulus under these conditions.) (Fig. 9). Because the rate of signal movement from the cotyledons to the apex has been calculated to be greater than 20 cm/hr (Takeba and Takimoto, 1966; King *et al.,* 1968), transit time to the apex should be measured in minutes. Thus, cotyledon removal should establish not only when sufficient floral stimulus has exited the cotyledons to cause the plant to flower, but also approximately when sufficient floral stimulus has arrived at the apex to induce a florally determined state in an apical meristem. A fully induced *P. nil* plant usually forms a terminal flower and single flowers in the upper six axillary positions (i.e., at nodes three to eight). When the cotyledons were removed at the end of the dark period, no terminal flowers formed while over half of the plants produced axillary flowers at nodes three and four. When terminal buds were grafted 1 to 2 hr after the end of dark period, over 50% eventually formed axillary flowers but not terminal flowers. Cotyledon removal as well as bud grafting 4 to 5 hr after the end of the dark period permitted 100% of the plants and buds to form axillary flowers at least at nodes three and four. These results indicate that the floral stimulus acted within a period of about an hour or less to induce some axillary bud meristems into a florally determined state.

Four hours after dawn, over 70% of the plants from which cotyledons were removed ultimately formed terminal flowers; for grafted buds over 70% produced terminal flowers when grafted 8 to 9 hr after dawn. These results indicate that the cells that ultimately formed the terminal flower were induced into a florally determined state some hours after the meristems at some axillary bud positions had attained a florally determined state. As discussed by Larkin and co-

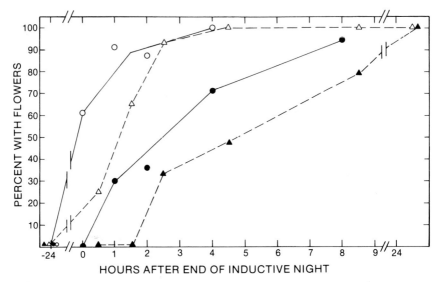

Fig. 9 Flowering response of *Pharbitis nil* Chois cv. Violet cotyledonless plants and grafted terminal buds. Percentage with axillary flowers (○, plants; △, buds) and terminal flower (●, plants; ▲, buds) plotted as a function of time of cotyledon removal or bud grafting. Cotyledons were removed and terminal buds grafted onto uninduced plants before and at various times after the inductive night. Plants were subsequently grown in noninductive conditions and scored for flowering. [Data from Larkin *et al.* (1990) with permission.]

workers (1990) marginal inductive treatments of whole plants employing various photoperiods and temperatures also indicate that the axillary bud meristems respond first to the floral stimulus. All buds that were grafted 1 day after the end of the inductive night formed inflorescences that were indistinguishable from those of plants that were induced with one long night. Thus, as a unit, all terminal buds had attained a state of floral determination comparable to that of the intact plant during the 24-hr period after the inductive night. These researchers did not graft any buds between 9 and 24 hr after dawn following the inductive night. Thus a normal pattern of floral determination may have been attained by the entire population of terminal buds at a time prior to the beginning of the second day after induction; as extrapolated in Fig. 9, the time would be about 12 hr after the end of the inductive night.

Increases in RNA synthesis are observed in apices around dawn after the inductive night while increases in the mitotic index and in the rate of primordium initiation are not observed until more than a day after the inductive night (see Vince-Prue and Gressel, 1985). Apex induction and the resulting floral determination clearly precede any gross cellular or morphological changes and are coincident with the earliest biochemical changes reported in the apex.

B. Lolium temulentum

Evans and colleagues have extensively characterized the flowering physiology of *Lolium temulentum*, strain Ceres, a long-day monocot that can be induced to flower with one long day (Evans, 1969c; Evans and King, 1985). Recently we have developed a culture procedure for apices of *L. temulentum* in which the time at which the apex becomes florally determined can be established (McDaniel *et al.*, 1991).

Shoot apices of induced and short-day grown plants are not visually distinguishable until about 2 or 3 days after the end of the inductive day. These apices are about 0.7 mm tall and have from 9 to 12 unexpanded leaf primordia below the meristem proper. When excised for culture, an apex had about 0.1 mm of stem tissue at its base; one expanding leaf of 1- to 4-mm length covered the meristem and the unexpanded leaf primordia. Explants from short-day grown plants did not initiate spikelet primordia (the structures that will form flowers) during the 3-week culture period but grew vegetatively. Essentially all apices excised on the day after the inductive long day initiated spikelet primordia during the culture period.

The time of floral determination in the apex of the *L. temulentum* plant has been established by culturing apices at various times during and after the inductive long day (Fig. 10). Leaf removal showed that sufficient floral stimulus to initiate spikelets is exported from the leaf blades about 4 hr prior to the time that some isolated apices gained the capacity to initiate spikelets in culture. At a rate of movement of 1–2.4 cm/hr (Evans and Wardlaw, 1966), 4 hr is thus about 1 hr longer than it should take the floral stimulus to reach the apex from the base of the leaf blades. About 12 hr after all of the plants had exported sufficient floral stimulus from their leaf blades to initiate spikelets, all excised apices gained the capacity to initiate spikelets in isolation. Thus, the maximum time required for the floral stimulus to induce some spikelet sites into a florally determined state in all apices appears to be about 9 hr. Because some apices isolated at about the estimated arrival time of the floral stimulus at the apex initiated spikelets, it is likely that the floral stimulus acts in less time than 9 hr to induce a state of floral determination in competent spikelet sites. However, floral determination at unique spikelet sites has been assessed in the context of an explanted apex that includes tissues and organs other than spikelet sites. To establish that a spikelet site does, in fact, become florally determined, the spikelet site itself must exhibit spikelet initiation in an environment other than that of an induced apex. This demonstration is not yet technically feasible.

As with *P. nil*, floral determination occurred prior to any cellular or morphological changes in the apex and coincided with the earliest reported biochemical changes (Rijven and Evans, 1967; Evans and Rijven, 1967).

Although apices excised 1 day after the standard long-day induction develop relatively normal inflorescences, apices given less than 24-hr inductive day or

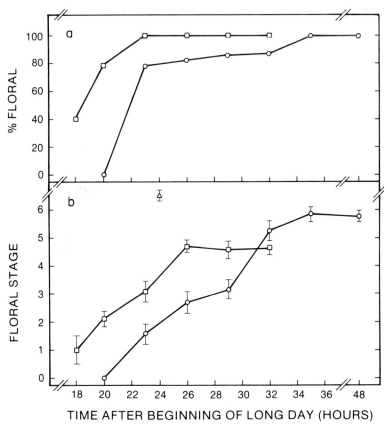

Fig. 10 Flowering response 3 weeks after the removal of leaf blades or excision and culture of shoot apices from plants of *Lolium temulentum* strain "Ceres" exposed to one long day when 48 days old. (A) Percentage exhibiting floral determination, i.e., percentage of plants (□) or of cultured apices (○) reaching at least the double ridges stage of floral development. (B) Average floral stage of plants (□) or culture apices (○) based on the following scale: 0, vegetative; 1, shoot apex elongated; 2, double ridges; 3, advanced double ridges; 4, glume primordia; 5, lemma primordia; 6, floret primordia. All apices excised from control plants grown in short days remained vegetative during the 3-week culture period (data not plotted). The flowering stage of intact plants given one long day is indicated for comparison (△). The vertical bars represent 2 × standard error, with 18 replicates per treatment. [Data from McDaniel et al. (1991) with permission.]

apices excised earlier than 1 day after induction often form inflorescences with fewer spikelets than normal. The extreme case is an apex with only a few spikelets. Similar patterns of inflorescence differentiation have been observed in whole plants after marginal inductive treatments (Evans and King, 1985). These observations indicate that as was found with *P. nil*, different potential sites of flower formation exhibit different sensitivities to the floral stimulus; however, unlike *P. nil* the most sensitive cells are those that will form the terminal spikelet rather than cells at axillary sites.

In both *P. nil* and *L. temulentum* the induction of the apex into a florally determined state occurs in a period of about 1 to 9 hr. These induction periods are similar to those observed for other processes. In animal systems, for example, neural induction in amphibians occurs over a 30-min to 2-hr period while kidney tubule induction occurs in a 12- to 24-hr period (see Saxen, 1985). Although there are few temporally characterized inductive processes in plants, the induction of epidermal cell dedifferentiation in carpel fusion of *Catharanthus roseus* takes only several hours, as the total process of induction and dedifferentiation is completed within just 9 hr (Verbeke and Walker, 1985).

C. *Helianthus annuus*

Sunflower is a day-neutral plant that, like *Nicotiana*, forms a uniform number of nodes in a given environment before initiating reproductive development. However, unlike *Nicotiana*, grafting and rooting of terminal buds has established that the number of leaves produced by a terminal meristem cannot be increased but only decreased (Habermann and Sekulow, 1972; Table VIII). Rooting shoot tips

Table VIII Floral Determination in the Terminal Meristem of *Helianthus annuus*[a]

Treatment	Number of leaf pairs produced		
	Before rooting	After rooting	Total
Rooting of shoot tips after production of different numbers of leaf pairs	0	17	17
	3	13	16
	9	7	16
	Before grafting	After grafting	Total
Grafting of shoot tips from plants of various ages onto seedling stocks	0	16	16
	4	13	17
	8	9	17
	On stock	Produced by scion	On stock and scion
Grafting of seedling shoot tips onto plants with various numbers of leaf pairs	3	15	18
	6	11	17
	9	8	17

[a]This sunflower cultivar produces about 16 leaf pairs before forming an inflorescence. Data from Habermann and Sekulow (1972) as presented in Table I of McDaniel (1984b) with permission.

of plants that had produced various numbers of leaf pairs did not increase or decrease the total number of leaf pairs produced by the meristem. Grafting of shoot tips from plants that had produced different numbers of leaf pairs onto seedling stocks did not increase the total number of nodes produced by the grafted meristem. Grafting seedling shoot tips onto the tops of stocks that had produced increasing numbers of leaf pairs caused a reduction in the number of nodes produced by the grafted seedling meristem. When seedling apices and half apices were cultured, they and the new meristems that formed initiated reproductive development after producing just a few leaves (Paterson, 1984).

The simplest interpretation of the data from these grafting and isolation experiments is that the sunflower terminal meristem as well as cells not located in an organized shoot apical meristem are florally determined soon after germination. For reasons not understood this commitment is not normally expressed until after the meristem has produced some specified number of leaves.

D. *Pisum sativum*

Pea is a plant that forms only lateral flowers as the terminal meristem produces leaf, node, and internode units and then ceases to grow (Murfet, 1985; Singer *et al.*, 1990). Each variety of pea forms flowers at a specific node position as a function of its genotype and the environment in which it is grown. Culturing of terminal buds has established that the cells that will form a flower at a particular node position become fated to form the flower before the node is initiated by the apical meristem (Ferguson *et al.*, 1991). The meristems of terminal buds isolated from plants that had three expanding leaves produced eight nodes after isolation and then ceased growth. In the axil of each of the upper three nodes a flower formed. Had the bud not been isolated, the meristem would have produced 11 nodes, with the upper 3 being floral. Thus, the terminal meristem within the context of the isolated bud was determined for a limited growth pattern including the differentiation of axillary flowers. Interestingly, the isolation accelerated the node of first flower by about three nodes.

V. Flowering as a Developmental Process

Making a coherent picture of the immense database now available on flowering will be impossible unless a general conceptual framework can be established (McDaniel, 1989). The above data on floral initiation in six different species has been presented in such a conceptual framework. The focus has been on an event that commits a meristem or a group of cells to a pattern of floral development. Determination is a concept that does not yet have an explicit mechanistic basis. It merely allows one to describe a change in the developmental behavior of a group of cells without knowing the basis for that change. Ultimately, when we know the molecular basis for a specific determined state, it will no longer be necessary to use the term "determined" because it will be just another step in the differentiation process, a step that can be identified, such as the presence of hormone receptors in a target cell or the expression of a specific transcriptional regulator. However, while we still have so many developmental phenomena that can be described but not mechanistically understood, a conceptual framework is necessary for codification, discussion, and analysis. Competence, induction, and determination have served the field of animal development extremely well (Gurdon, 1985; Slack, 1983), and there is no reason that this conceptual framework could not serve a similar function in plant development (McDaniel, 1984a,b) and flowering (McDaniel, 1989).

The patterns of flowering in the six species discussed here exhibit clear differences that indicate that critical developmental processes are occurring at different times in the ontogeny of the plants. From a physiological point of view the most obvious difference is that some have an explicit photoperiodic requirement. Clearly the regulation of the various biochemical processes in leaf tissues of each species that ultimately result in the transmission of the floral stimulus out of the leaf is different. Developmentally, however, even though *P. nil* is a short-day plant and *L. temulentum* is a long-day plant, it is the competence of the leaves to respond to the inductive photoperiod that is regulated. Once the floral stimulus has been exported from the leaves, its action at the apex appears to be identical. In each case the floral stimulus induces competent cells into a relatively stable state of floral determination over a time period measured in hours. Thus competence of leaves to respond to a photoperiodic signal and induction of competent apices into a florally determined state are examples of unifying concepts facilitated by a developmental framework.

It is not known whether the individual groups of cells in the meristem that will ultimately develop into flowers are themselves florally determined at the same time as the apex is florally determined. It is clear that within the context of the isolated apex their developmental fate is to form a flower. It is also clear that the floral stimulus acts to uniquely change the developmental fates of specific groups of cells, and that these cells are distributed within the apex in a precise spatial pattern. That is, presumptive leaf primordium cells do not form flowers, while presumptive axillary bud primordium cells do. In fact, in *L. temulentum* the action of the floral stimulus leads to changes in the developmental fate of presumptive and existent, but immature, leaf primordia: they no longer grow out to form leaves. Even groups of cells that have the capacity to become florally determined exhibit different degrees of competence to respond. In 6-day-old seedlings of *P. nil* the most sensitive cells are those about to initiate axillary buds three and four while in *L. temulentum* plants the first responses are in the cells that will form the terminal spikelet. Bud rooting has established that in the *N. silvestris* axillary bud meristem the cells that form the terminal flower can acquire and express a florally determined state while neighboring cells can express a vegetative pattern of development. These results indicate that the apex of a plant is composed of a mosaic of cells in terms of their developmental states. Unanswered is the question of the molecular basis for the different developmental states.

Sunflower and *N. tabacum* cv. Wisconsin 38, in contrast to *L. temulentum* and *P. nil*, are day neutral. Thus the time of flowering is not controlled in a precise way by photoperiodically regulated processes in the leaves. The regulatory control of flowering in day-neutral tobacco is different from that in sunflower and similar, at least in some respects, to that in *P. nil* and *L. temulentum*. In sunflower a florally determined state is acquired soon after germination but the state is not expressed. An analogy might be imaginal disk determination in *Drosophila melanogaster* where the basic determination as to disk type occurs during embryogenesis but is expressed days later when a third instar larva undergoes metamorphosis (Nothiger, 1972). The hormone ecdysone elicits in third instar larvae and pupae the expression of the previously determined state; however, in earlier molts this expression is inhibited. In sunflower it appears that expression of the florally determined state is also inhibited because regenerated and isolated apices initiate reproductive development without the normal prerequisite amount of vegetative growth. On the other hand, a sunflower terminal meristem cannot be made to make more than the normal complement of nodes. There appears to be some internal counting mechanism that can be short circuited but not easily reset. Day-neutral tobacco is quite different. The number of nodes made by a meristem before flower formation can be increased or decreased and when isolated a bud, unless florally determined, makes the full complement of nodes before flowering. Except when derived from inflorescence tissues, almost all *de novo* apices are vegetative and when rooted form a full complement of nodes before flowering (Singer, 1985). A shoot apical meristem of day-neutral tobacco responds in a manner similar to an *L. temulentum* or a *P. nil* meristem. It maintains itself in a vegetative state until it is induced into a florally determined state by the floral stimulus. Once florally determined it initiates floral development almost immediately, that is, in several days.

In pea the timing of floral determination has commonalities with that observed in tobacco and sunflower but the spatial pattern of cells in the meristem that will express flower formation is closest to that of *P. nil*. Like tobacco, the node of first flower can be increased or decreased by grafting, and this property appears to be a function, at least in part, of a graft-transmissible inhibitor primarily produced in the cotyledons (see Murfet, 1985). Like sunflower, floral determination occurs early in ontogeny and, in addition, the node to first flower is decreased when a bud is isolated. As in *P. nil* groups of cells are fated to form axillary flowers before the nodes have been separated from the meristem, and the terminal meristem has a different sensitivity to the floral stimulus than the axillary sites. In pea the terminal portion of a shoot apical meristem is incompetent to initiate a flower, while in *P. nil* a stronger or more prolonged inductive signal is required to elicit floral determination in the apical portion of the terminal meristem.

The major point at which regulatory control is exerted in the flowering process of the species discussed here therefore appears to differ among the species. The overall developmental process of flowering in all of the plants, however, is remarkably similar. After all, how many independent times did angiosperms evolve? If there was one common ancestor, then it is that archetype of the flowering process as modified in the various angiosperms that we are trying to understand. To emphasize the physiology of the flowering process is to highlight all of the modifications. Clearly one must characterize in detail each modification to understand the specific mechanisms employed for that modification. However, such a physiological focus coupled with the inclination to produce an integrated, mechanistic explanation of all flowering runs the risk of obscuring the flowering process itself. Competence, induction, and determination as depicted in Fig. 1 form the simplest conceptual basis for understanding developmental processes, including flowering.

Acknowledgments

I thank Susan Singer (Carleton College) and Susan Smith (Rensselaer) for their critical reviews of this article and Mary Ann Sessions for unfailing secretarial competence. I am most appreciative of the financial support provided for our research over the past 10 years by the NSF Developmental Biology and USDA Competitive Grants programs.

References

Aghion-Prat, D. (1965a). Neoformation de fleurs *in vitro* chez *Nicotiana tabacum* L. *Physiologie Vegetale* **3**, 229–303.
Aghion-Prat, D. (1965b). Floral meristem-organizing gradient in tobacco stems. *Nature (London)* **207**, 1211.
Bernier, G. (1988). The control of floral evocation and morphogenesis. *Annu. Rev. Plant Physiol. Plant Mol. Biol.* 39, 175–219.
Bernier, G., Kinet, J. M., and Sachs, R. M. (1981). "The Physiology of Flowering," Vols. 1 and 2. CRC Press, Boca Raton, Florida.
Bhar, D. S. (1970). *In vitro* studies of floral shoot apices of *Pharbitis nil*. *Can. J. Bot.* **48**, 1355–1358.
Chailakhyan, M. Kh. (1988). 50 years of the hormonal theory of plant development. *Fiziol. Rast. (Moscow)* **35**, 591–608.
Chailakhyan, M. Kh., and Khazhakyan, Kh. K. (1974). Interactions between leaves and shoots during efflorescence of plants of photoperiodically neutral species. *Dokl. Akad. Nauk SSSR* **217**, 1214–1217.
Chailakhyan, M. Kh., and Khazhakyan, Kh. K. (1975). Role of stem zones of different tiers in

the flowering of plants of photoperiodically neutral species. *Dokl. Akad. Nauk SSSR* **224**, 1445–1448.

Chailakhyan, M. Kh., Bavrina, T. V., Konstantinova, T. N., and Aksenova, N. P. (1977). Morphogenetic potentials of tobacco calluses of stem and leaf origin. *Dokl. Akad. Nauk SSSR* **232**, 500–503.

Chaplin, J. R., Stavely, J. R., Litton, C. C., Pittarelli, G. W., and West, W. H., Jr. (1982). Catalog of the tobacco introductions in the U.S. Department of Agriculture's tobacco germplasm collection (*Nicotiana tabacum*). *Agric. Rev. Man.* **S-27**, 1–48.

Christianson, M. L. (1985). An embryogenic culture of soybean: Towards a general theory of somatic embryogenesis. *In* "Tissue Culture in Forestry and Agriculture" (R. R. Henke, K. W. Hughes, M. P. Constantin, and A. Hollaender, eds.), pp. 83–103. Plenum, New York.

Dennin, K. A. (1985). "Determination for Floral Development in Photoperiodic Tobacco." M.Sc. Thesis, Rensselaer Polytechnic Institute, Troy, New York.

Dennin, K. A., and McDaniel, C. N. (1985). Floral determination in axillary buds of *Nicotiana silvestris*. *Dev. Biol.* **112**, 377–382.

Evans, L. T. (1969a). A short history of the physiology of flowering. *In* "The Induction of Flowering" (L. T. Evans, ed.), pp. 1–13. Macmillan, Melbourne.

Evans, L. T., (ed.) (1969b). "The Induction of Flowering." Macmillan, Melbourne.

Evans, L. T. (1969c). *Lolium temulentum* L. *In* "The Induction of Flowering" (L. T. Evans, ed.), pp. 328–349. Macmillan, Melbourne.

Evans, L. T., and King, R. W. (1985). *Lolium temulentum*. *In* "CRC Handbook of Flowering" (A. H. Halevy, ed.), Vol. 3, pp. 306–323. CRC Press, Boca Raton, Florida.

Evans, L. T., and Rijven, A. H. G. C. (1967). Inflorescence initiation in *Lolium temulentum* L. XI. Early increases in the incorporation of ^{32}P and ^{35}S by shoot apices during induction. *Aust. J. Biol. Sci.* **20**, 1033–1042.

Evans, L. T., and Wardlaw, I. F. (1966). Independent translocation of ^{14}C-labelled assimilates and of the floral stimulus. *Planta* **68**, 310–326.

Ferguson, C. J., Huber, S. C., and Singer, S. R. (1991). Determination for a pattern of limited growth occurs early in *Pisum sativum* buds. *Planta* **185**, 518–522.

Gebhardt, J. S., and McDaniel, C. N. (1987). Induction and floral determination in the terminal bud of *Nicotiana tabacum* L. cv. Maryland Mammoth, a short-day plant. *Planta* **172**, 526–530.

Gebhardt, J. S., and McDaniel, C. N. (1991). Flowering response of *Nicotiana tabacum* L. day-neutral and short-day cultivars: Interactions among roots, genotype, leaf ontogenic position and growth conditions. *Planta* **185**, 513–517.

Gehring, W. J. (1972). The stability of the determined state in cultures of imaginal disks in *Drosophila*. *In* "The Biology of Imaginal Disks. Results and Problems in Cell Differentiation" (H. Ursprung and R. Nothiger, eds.), Vol. 5, pp. 35–58. Springer-Verlag, Berlin.

Gurdon, J. B. (1962). The developmental capacity of nuclei taken from intestinal epithelium cells of feeding tadpoles. *J. Embryol. Exp. Morphol.* **10**, 622–640.

Gurdon, J. B. (1985). Introductory comments. *In* "Cold Spring Harbor Symposium on Quantitative Biology, Molecular Biology of Development" (J. Sambrook, ed.), Vol. L, pp. 1–10. Cold Spring Harbor Laboratory, Cold Spring Harbor, New York.

Habermann, H. M., and Sekulow, D. B. (1972). Development and aging in *Helianthus annuus* L. Effects of the biological *milieu* of the apical meristem on patterns of development. *Growth* **36**, 339–349.

Halevy, A. H. (ed.) (1985). "CRC Handbook of Flowering," Vols. 1–5. CRC Press, Boca Raton, Florida.

Harrison, R. (1918). Experiments on the development of the forelimb of *Amblystoma*, a self-differentiating equipotential system. *J. Exp. Zool.* **25**, 413–461.

Hicks, G. S., and Sussex, I. M. (1971). Organ regeneration in sterile culture after median bisection of the flower primordia of *Nicotiana tabacum*. *Bot. Gaz.* **132**, 350–363.

Hopkinson, J. M., and Hannam, R. V. (1969). Flowering in tobacco: The course of floral induction under controlled conditions and in the field. *Aust. J. Agric. Res.* **20,** 279–290.
Hopkinson, J. M., and Ison, R. L. (1982). Investigations of ripeness to flower in tobacco. *Field Crop Res.* **5,** 335–348.
Jegla, D. E., and Sussex, I. M. (1989). Cell lineage patterns in the shoot meristem of the sunflower in the dry seed. *Dev. Biol.* **131,** 215–225.
Khait, A. (1986). Hormonal mechanisms for size measurement in living organisms in the context of maturing juvenile plants. *J. Theor. Biol.* **118,** 471–484.
King, R. W., Evans, L. T., and Wardlaw, I. F. (1968). Translocation of floral stimulus in *Pharbitis nil* in relation to that of assimilates. *Z. Pflanzenphysiol.* **59,** 377–388.
Lang, A. (1965). Physiology of flower initiation. *In* "Encyclopedia of Plant Physiology" (W. Ruhland, ed.), Vol. 15, Pt. 1, pp. 1380–1536. Springer-Verlag, Berlin.
Lang, A. (1987). Perspectives in flowering research. *In* "Plant Gene Systems and Their Biology" (J. L. Key and L. McIntosh, eds.), pp. 3–24. Alan R. Liss, New York.
Lang, A. (1989). *Nicotiana*. *In* "CRC Handbook of Flowering" (A. H. Halevy, ed.), Vol. 6, pp. 427–483. CRC Press, Boca Raton, Florida.
Larkin, J. C., Felsheim, R., and Das, A. (1990). Floral determination in the terminal bud of the short-day plant *Pharbitis nil*. *Dev. Biol.* **137,** 434–443.
McDaniel, C. N. (1978). Determination for growth pattern in axillary buds of *Nicotiana tabacum* L. *Dev. Biol.* **66,** 250–255.
McDaniel, C. N. (1980). Influence of leaves and roots on meristem development in *Nicotiana tabacum* L. cv. Wisconsin 38. *Planta* **148,** 462–467.
McDaniel, C. N. (1984a). Shoot meristem development. *In* "Positional Controls in Plant Development" (P. Barlow and D. J. Carr, eds.), pp. 319–347. Cambridge Univ. Press, Cambridge.
McDaniel, C. N. (1984b). Competence, determination and induction in plant development. *In* "Pattern Formation: A Primer in Developmental Biology" (G. Malacinski, ed.), pp. 393–412. Macmillan, New York.
McDaniel, C. N. (1989). Floral initiation as a developmental process. *In* "Plant Reproduction: From Floral Induction to Pollination" (E. Lord and G. Bernier, eds.), pp. 51–57. American Society of Plant Physiologists, Rockville, Maryland.
McDaniel, C. N., and Hsu, F. C. (1976). Position-dependent development of tobacco meristems. *Nature (London)* **259,** 564–565.
McDaniel, C. N., and Poethig, R. S. (1988). Cell lineage patterns in the shoot apical meristem of the germinating maize embryo. *Planta* **175,** 13–22.
McDaniel, C. N., Singer, S. R., Dennin, K. A., and Gebhardt, J. S. (1985). Floral determination: Timing, stability, and root influence. *In* "Plant Genetics" (M. Freeling, ed.), pp. 73–87. Alan R. Liss, New York.
McDaniel, C. N., Singer, S. R., Gebhardt, J. S., and Dennin, K. A. (1987). Floral determination: A critical process in meristem ontogeny. *In* "The Manipulation of Flowering" (J. G. Atherton, ed.), pp. 109–120. Butterworths, London.
McDaniel, C. N., Sangrey, K. A., and Jegla, D. E. (1989a). Cryptic floral determination: Explants from vegetative tobacco plants have the capacity to form floral shoots *de novo*. *Dev. Biol.* **134,** 473–478.
McDaniel, C. N., Sangrey, K. A., and Singer, S. R. (1989b). Node counting in axillary buds of *Nicotiana tabacum* cv. Wisconsin 38, a day-neutral plant. *Am. J. Bot.* **6,** 403–408.
McDaniel, C. N., King, R. W., and Evans, L. T. (1991). Floral determination and *in vitro* floral differentiation in isolated shoot apices of *Lolium temulentum* L. *Planta* **185,** 9–16.
Matsushima, H., Itoyama, T., Mashiko, Y., and Mizukoshis, T. (1974). Critical time of floral differentiation in *Pharbitis nil* shoot apex. *In* "Plant Growth Substance," pp. 967–973. Hirokawa, Tokyo.

Meins, F., Jr., Foster, R., and Lutz, T. (1982). Quantitative studies of bud initiation in cultured tobacco tissues. *Planta* **155**, 473–477.

Moscona, A. A., and Monroy, A., (eds.). (1986). "Commitment and instability in cell differentiation." *Curr. Top. Dev. Biol.* **20**.

Murfet, I. C. (1985). *Pisum sativum*. In "CRC Handbook of Flowering" (A. H. Halevy, ed.), Vol. 4, pp. 97–126. CRC Press, Boca Raton, Florida.

Murfet, I. C. (1989). Flowering genes in *Pisum*. In "Plant Reproduction: From Floral Induction to Pollination" (E. Lord and G. Bernier, eds.), pp. 10–18. American Society of Plant Physiologists, Rockville, Maryland.

Nothiger, R. (1972). The larval development of imaginal disks. In "The Biology of Imaginal Disks. Results and Problems in Cell Differentiation" (H. Ursprung and R. Nothiger, eds.), pp. 1–34. Springer-Verlag, Berlin.

Paterson, K. E. (1984). Shoot tip culture of *Helianthus annuus*—Flowering and development of adventitious and multiple shoots. *Am. J. Bot.* **71**, 925–931.

Rajeevan, M. S., and Lang, A. (1987). Comparison of *de-novo* flower-bud formation in a photoperiodic and a dayneutral tobacco. *Planta* **171**, 560–564.

Reyer, R. W. (1962). Regeneration in the amphibian eye. In "Regeneration" (D. Rudnik, ed.), pp. 211–265. Ronald Press, New York.

Rijven, A. H. G. C., and Evans, L. T. (1967). Inflorescence initiation in *Lolium temulentum* L. X. Changes in ^{32}P incorporation into nucleic acids of the shoot apices at induction. *Aust. J. Biol. Sci.* **20**, 13–24.

Sachs, R. M. (1977). Nutrient diversion: An hypothesis to explain the chemical control of flowering. *HortScience* **12**, 220–222.

Sachs, R. M., and Hackett, W. P. (1977). Chemical control of flowering. *Acta Hortic.* **68**, 29–49.

Saxen, L. (1985). Alternative implementation mechanisms of embryonic induction. In "Molecular Determinants of Animal Form" (G. M. Edelman, ed.), pp. 1–13. Alan R. Liss, New York.

Seltman, H. (1974). Effect of light periods and temperatures on plant form of *Nicotiana tabacum* L. cv. Hicks. *Bot. Gaz.* **135**, 196–200.

Singer, S. R. (1985). "Spatial and Temporal Aspects of Floral Determination in *Nicotiana tabacum* L." Ph.D. Thesis, Rensselaer Polytechnic Institute, Troy, New York.

Singer, S. R., and McDaniel, C. N. (1986). Floral determination in the terminal and axillary buds of *Nicotiana tabacum* L. *Dev. Biol.* **118**, 587–592.

Singer, S. R., and McDaniel, C. N. (1987). Floral determination in internode tissues of dayneutral tobacco first occurs many nodes below the apex. *Proc. Natl. Acad. Sci. U.S.A.* **84**, 2790–2792.

Singer, S. R., Hsiung, L. P., and Huber, S. C. (1990). *Determinate (det)* mutant of *Pisum sativum* (leguminosae:papilionoideae) exhibits an indeterminate growth pattern. *Am. J. Bot.* **77**, 1330–1335.

Slack, F. M. V. (1983). "From Egg to Embryo: Determinative Events in Early Development." Cambridge Univ. Press, Cambridge.

Smith, R. H. (1984). Developmental potential of excised primordial and expanding leaves of *Coleus blumei* Benth. *Am. J. Bot.* **71**, 1114–1120.

Smith, R. H., and Murashige, T. (1970). *In vitro* development of isolated shoot apical meristems of angiosperms. *Am. J. Bot.* **57**, 562–568.

Smith, S. E. M. (1990). "Leaf Process and Meristem Determination in *Nicotiana*." M.S. Thesis, Rensselaer Polytechnic Institute, Troy, New York.

Steeves, T. A., and Sussex, I. M. (1957). Studies on the development of excised leaves in sterile culture. *Am. J. Bot.* **44**, 665–673.

Steeves, T. A., and Sussex, I. M. (1989). "Patterns in Plant Development, Second Edition." Cambridge Univ. Press, Cambridge.

Takeba, G., and Takimoto, A. (1966). Translocation of the floral stimulus in *Pharbitis nil*. *Bot. Mag.* **79**, 811–814.

Takimoto, A. (1969). *Pharbitis nil* Chois. *In* "The Induction of Flowering" (L. T. Evans, ed.), pp. 90–115. Macmillan, Melbourne.

Thomas, J. F., Anderson, C. E., Raper, C. D., and Downs, R. J., Jr. (1975). Time of floral initiation in tobacco as a function of temperature and photoperiod. *Can. J. Bot.* **53**, 1400–1410.

Tran Thanh Van, M. (1973). Direct flower neoformation from superficial tissue of small explants of *Nicotiana tabacum* L. *Planta* **115**, 87–92.

Tran Thanh Van, M., Dien, N. T., and Chlyah, A. (1974). Regulation of organogenesis in small explants of superficial tissue of *Nicotiana tabacum* L. *Planta* **119**, 149–159.

Van den Ende, G., Croes, A. F., Kemp, A., and Barendse, G. W. M. (1984). Development of flower buds in thin-layer cultures of flower stalk tissue from tobacco: Role of hormones in different stages. *Physiol. Plant.* **61**, 114–118.

Verbeke, J. A., and Walker, D. B. (1985). Rate of induced cellular dedifferentiation in *Catharanthus rosea*. *Am. J. Bot.* **72**, 1314–1317.

Vince-Prue, D., and Gressel, T. (1985). *Pharbitis nil*. *In* "Handbook of Flowering" (A. H. Halevy, ed.), Vol. 4, pp. 47–81. CRC Press, Boca Raton, Florida.

Waterkeyn, L., Martens, P., and Nitsch, J. P. (1965). The induction of flowering in *Nicotiana* I. Morphological development of the apex. *Am. J. Bot.* **52**, 264–270.

2
Cellular Basis of Amphibian Gastrulation

Ray Keller
Department of Molecular and Cell Biology
University of California, Berkeley
Berkeley, California 94720

Rudolf Winklbauer
Max-Planck-Institute for Developmental Biology
Tübingen, Germany

I. Introduction
II. Function of Bottle Cells in Gastrulation
    A. Bottle Cells and Invagination
    B. Bottle Cell Ingression and Mesoderm Formation
    C. Patterning of Bottle Cell Formation
    D. Experimental Tests of Function during Gastrulation
    E. Evolutionary Implications
III. Mesodermal Cell Migration
    A. Description of the Process
    B. Extracellular Matrix of the Blastocoel Roof and the Substrate of Mesodermal Cell Migration
    C. Motile Activity and Behavior of Migratory Mesoderm Cells
    D. Role of Cell Interactions in Mesoderm Cell Migration
    E. Function of Mesoderm Cell Migration in Gastrulation
IV. Convergence and Extension Movements
    A. Convergence and Extension in *Xenopus*: Demonstration of Two Regions of Convergence and Extension in Sandwich Explants
    B. Convergence and Extention in Other Amphibians
    C. Roles of the Epithelial Cells and the Deep Mesodermal Cells in Convergence and Extension
    D. Convergence and Extension by Radial and Mediolateral Intercalation
    E. Cell Behavior during Mediolateral Intercalation
    F. Function of Convergence and Extension in Gastrulation
V. The Ignored Movement: Epiboly
    A. Epiboly in *Xenopus*
    B. Epiboly in Urodeles
    C. Evidence for Autonomy of Epiboly
VI. Involution
VII. Some Other Issues to Be Explored
    A. Evolution and the Design of Gastrulation: Variation among Amphibians
    B. Significance of Cytomechanics and Cell Interactions in Populations
    References

I. Introduction

The objective of this article is to evaluate current knowledge about how cells produce gastrulation in amphibians. We will organize our discussion around the major regional processes and the cell behaviors underlying them. These regional processes include (1) bottle cell function in invagination and ingression, (2) mesodermal cell migration, (3) convergence and extension of the marginal zone, and (4) epiboly of the animal cap. We have reviewed this subject several times in the past 5 years, once focusing on the cellular basis of the regional morphogenetic movements and their function in gastrulation (Keller, 1986) and again focusing on the function of extracellular matrix in gastrulation (Keller and Winklbauer, 1990). We discussed convergence and extension exhaustively, with emphasis on cell motility and function in gastrulation (Keller *et al.*, 1991a) and with emphasis on cell interactions and patterning (Keller *et al.*, 1991b). Thus we will provide less detail here, particularly on those aspects not concerning gastrulation. More emphasis will be given to bottle cell function and mesodermal cell migration. We will build on previous discussions, providing essential background, and then focus on what is not known, on what is new, or on what is in conflict.

Integration of force-generating cell behaviors, such as change in cell shape or protrusive activity, into mechanisms of multicellular tissue morphogenesis is extraordinarily complex and requires understanding of the physiological, behavioral, and mechanical interactions of cells in populations. It is fundamental that we know where cells move and what cells generate the forces bringing about these movements. It is also important to know specifically what cell behavior generates these forces, not in terms of general catogories, such as cell crawling or cell shape change, but in terms of specific, geometrically defined changes in shape or detailed, spatiotemporal patterns of protrusive activity. We must also be able to demonstrate the mechanism by which these specific cell behaviors generate the appropriate pattern of forces. This task is particularly difficult because a given cell behavior can have a variety of results depending on the mechanical and geometric environment in which it acts. Finally we must attack the question of what role or contribution each of these regional movements makes to the whole process of gastrulation. This again is not a simple task, because there is much variation in balance between regional autonomy and regional interactions in various species.

II. Function of Bottle Cells in Gastrulation

"Bottle" cells (also known as "flask" cells), described in the classical literature and so named because of their similarity to the products of glass blowers in those days, represent two different, fundamental morphogenetic processes in gastrulation of amphibians—invagination of the archenteron and ingression of mesodermal cells. Their function as an invaginating cell population is best represented in *Xenopus* archenteron formation (Keller, 1975, 1981; Hardin and Keller, 1988), and their function as an ingressing cell population is best represented by mesoderm ingression in urodeles and anurans other than *Xenopus* (Vogt, 1929; Holtfreter, 1944; J. Smith and Malacinski, 1983; Lundmark, 1986; Purcell, 1990; 1992). We will discuss the *Xenopus* bottle cells first.

A. Bottle Cells and Invagination

Invagination of bottle cells and their subsequent respreading figures heavily in formation of the archenteron in *Xenopus*, whereas in other anurans and urodeles studied thus far only the formation of the pharyngeal endoderm involves these processes.

1. Behavior of Invaginating Bottle Cells

Invaginating bottle cells show at least two distinct phases of behavior: (1) formation of their unique shape and (2) respreading, in which they return to their original cuboidal shape. Prospective bottle cells encompass the first five or six rows of endodermal epithelial cells at the lower edge of the involuting marginal zone (IMZ) of the *Xenopus* late blastula stage embryo. They appear no different from the remaining 8 to 10 rows of prospective endodermal cells lying above them in the IMZ. The apices of these cells constrict and concentrate their sparsely distributed pigment into a small area next to the vegetal, subblastoporal endoderm, forming the thin, dark *blastoporal pigment line*. Concurrently, the cells become elongated in the apical–basal direction, and their basal ends dilate. This process begins on the dorsal side, marking the onset of gastrulation, passes to lateral regions by early midgastrula, and reaches the ventral midline at the midgastrula. Bottle cells are moved inside as the archenteron deepens, and there they respread to form a large area on the periphery of the archenteron, beginning middorsally in the late gastrula and progressing to the ventral midline during neurulation. Thus respreading of bottle cells recapitulates the order of their formation. This analysis of their behavior is based on time-lapse recordings, correlated scanning electron micrographic and histological studies, tracing cells with Nile blue marks and with fluorescein dextran amine (FDA), and cell extirpations (see Keller, 1975, 1981; Hardin and Keller, 1988).

2. Apical Constriction and Respreading: Intrinsic, Autonomous Properties

Are these behaviors due to forces generated from within the bottle cells, or are they passive responses to external forces? Apical contraction and subsequent relaxation of these apices and respreading of bottle cells are autonomous and intrinsic to the prospective bottle cells at the late blastula stage, because these events occur in explanted prospective bottle cells on roughly the same schedule as *in vivo* (Hardin and Keller, 1988). Thus signals from surrounding tissues are not necessary for bottle cell apical constriction or respreading, and the mechanical forces driving these processes must be generated by bottle cells themselves rather than resulting from mechanical interactions with adjacent tissues. However, not all aspects of the shape change or the function of bottle cells is displayed in isolation (see Section II,A,4 below).

3. Structural and Physiological Basis of Apical Constriction

We have no direct evidence on this subject from *Xenopus* and must rely on other amphibians and even on other cells and other systems for clues as to what might happen with bottle cells. The pioneering work of Baker (1965) and Perry and Waddington (1966) showed that the contracted apices contain electron-dense material that probably represents microfilaments (also see Perry, 1975). Numerous vesicles occupy the subapical region, the nucleus is in the dilated basal region, and microtubules are oriented lengthwise along the apical–basal axis. Better electron microscopic methods applied to bottle cells of the neural plate in succeeding years (Baker and Schroeder, 1967; Burnside, 1971, 1973; Karfunkel, 1971; Schroeder, 1973) showed circumapical microfilaments, presumably functioning in contraction, and microtubules oriented parallel to the long axis. But the improved methods available as electron microscopy came of age were never applied to gastrula bottle cells and thus they remain poorly characterized structurally, during both their formation and respreading.

On the basis of extensive work on the similar, but by no means identical, neural plate cells, Burnside (1971) proposed that the elongation or columnarization of the neural plate cells was due to microtubules and that the subsequent wedging was due to participation of the apical microfilament system in contraction. Agents disrupting microtubules cause rounding of elongated amphibian neural plate cells (Karfunkel, 1971; Burnside, 1973; Löfberg and Jacobson, 1974), but Schoenwolf and Powers (1987) found that in the chick embryo, complete rounding of cells under these conditions occurred only at the arrested metaphase, implying that normally microtubules make only a partial contribution to the elongate shape. Bottle cells of *Xenopus* do not maintain an elongate shape in isolation and although bottle cells of some species do so (see Section II,A,5 below), these have not been tested for dependence of this shape on microtubules.

Apical constriction of the type shown by bottle cells may be due to several mechanisms, one of which is the contraction of an apical belt of actin microfilaments, mentioned above, presumably by interaction with myosin. There is direct evidence for such a mechanism in other systems in which contractile proteins have been localized and characterized immunologically and biochemically, and where regulation of contraction has been studied in permeabilized model systems and isolated circumferential bundles (e.g., see Lee *et al.*, 1977, 1983; Nagele and Lee, 1980; Nagele *et al.*, 1981; Lee and Nagele, 1985; also see Owaribe and Masuda, 1982; Brady and Hilfer, 1982). However, not all systems showing cell wedging or "bottle" cell formation are dependent on microfilaments (see Schoenwolf *et al.*, 1988) and cell cycle-dependent shape changes may also be involved (Smith and Schoenwolf, 1988; Schoenwolf and Smith, 1990). We know comparatively little about the structural and physiological basis of bottle cell formation in amphibians, both of the *Xenopus* type of invaginating cell and the ingressing type, discussed below. Moreover, the diversity in other systems (see Ettensohn, 1985; Schoenwolf and Smith, 1990; Lane, 1990, 1992) suggests amphibian bottle cells of both types should be investigated directly.

Membrane organization is another poorly understood aspect of bottle cell behavior. The numerous subapical vesicles could reflect membrane uptake as the apices contract, and perhaps these are reinserted as bottle cells respread. Membrane uptake during apical constriction and reinsertion in the spreading phase could be studied by labeling the membrane with ferritin and following its distribution with electron microscopy, using the methods developed by Betchaku and Trinkaus (1986) for the teleost fish, *Fundulus*.

4. Bottle Cell Behavior and Function: Context Dependence

The formal model by which bottle cell shape produces an invagination, defined as bending of an epithelium, goes back to the work of Rhumbler (1902) and was "simulated" first by a rubber and brass model by Lewis (1947) and later with computers (Odell *et al.*, 1981; Hardin and Keller, 1988). The mechanics of how changes in cell shape are related to invagination are complex and it cannot be assumed that an observed change in shape is active or has the effect supposed unless experimentally demonstrated (see Ettensohn, 1985; Smith and Schoenwolf, 1989; Schoenwolf and Smith, 1990; Lane, 1990, 1992).

Microsurgical manipulations analyzed with high-resolution time-lapse recordings, correlated scanning electron microscopy, and mechanical simulations show that specific features of bottle cell behavior and function in *Xenopus* are dependent on the mechanical properties of surrounding tissues and on their position around the large subblastoporal endodermal mass (Hardin and Keller, 1988). Isolated populations of bottle cells contract equally in all directions to form small, round apices, and they do not elongate in the apical–basal direction or dilate at their basal ends as they do *in vivo*. *In vivo* nearly all contraction is directed vegetally, toward the subblastoporal endoderm, yielding apices elongated mediolaterally. The subblastoporal endoderm does not appear to stretch as easily as the IMZ, and thus the outer IMZ is stretched vegetally as the bottle cells contract and compress against the subblastoporal endoderm. Abutment against the stiff subblastoporal endoderm probably results in their wedge shape, which tends to rotate the lower edge of the IMZ. These two effects probably conspire to initiate involution of the lower IMZ in much more dramatic fashion than would occur in a uniform environment (Hardin and Keller, 1988). Direct measurements of mechanical properties of these tissues should be done to determine if these ideas are correct.

Bottle cell shape changes as tension builds in the IMZ. Bottle cells that form early contract uniformly, forming small, rounded apices and stretching adjacent prospective bottle cells, enlarging their apical areas. When the remaining ones contract later, circumferential tension has already built up around the subblastoporal endoderm, and so these late contractions occur primarily in the direction of least resistance—the animal–vegetal direction—accounting for the elongated shapes aligned around the circumference of the subblastoporal endodermal mass. This biasing of an intrinsically uniform contraction by surrounding tissues efficiently directs all movement vegetally (Hardin and Keller, 1988).

To summarize, the bottle cells of *Xenopus* are good examples of the emergence of a morphogenetic mechanism by integration of discrete processes. Apical constriction of individual cells, by whatever cytoskeletal mechanism, is not a morphogenetic mechanism in itself but a force-generating process that is transformed into morphogenesis by virtue of the mechanical and geometric environment in which it is exercised. Bottle cells contracting in isolation produce a structure very different from that produced *in situ*; nearly all the specific effects of bottle cell formation on the IMZ are not resident within the cells but arise from mechanical interactions within the cell population (see Hardin, 1990).

5. Invasiveness of Bottle Cells

Holtfreter favored the idea that bottle cells functioned primarily by being invasive rather than by change in shape (Holtfreter, 1944; also see Holtfreter, 1943a,b). The blastocoel is alkaline, and bottle cells cultured on glass in alkaline medium migrate aggressively with their deep (basal) ends functioning as the leading edge. Likewise bottle cells explanted onto endodermal cells invade the endodermal substrate leading with their basal ends (Holtfreter, 1944). But scanning electron microscopy of *Xenopus* bottle cells (Keller and Schoenwolf, 1977) did not reveal the filopodia or lamellipodia that would be expected on these ends of invasive cells, despite preservation of such protrusions on adjacent cells. On the other hand, bottle cells appear to be more tightly attached than other suprablastoporal endodermal cells to the underlying mesoderm, and when bottle cells are removed, the remaining endodermal epithelium of the IMZ often slips relative to the underlying mesoderm (Keller, 1981). Thus it appears that bottle cells serve as an anchorage for the entire IMZ epithelium to the moving mesodermal cells beneath. The invasiveness described by Holtfreter was from urodele and anuran species having ingressing mesodermal bottle cells, in addition to the endodermal, invaginating ones like those of *Xenopus*, leaving open the possibility that invasiveness is a property of the ingressing type in these other species.

2. Cellular Basis of Amphibian Gastrulation

Polarized protrusive activity of bottle cells on glass in Holtfreter's experiments not only indicates an apical–basal polarity but also suggests an intrinsic elongation in this axis. What Holtfreter called the "coated" outer surface of the bottle cells represents the nonadhesive apical surfaces of epithelial cells and would not attach and show protrusive activity in any case. But if this is the *only* constraint on protrusive activity, the entire basal surface should have spread in all directions on the substratum, the cell taking on a "fried egg" morphology as a result. To lay on their sides, long axes parallel to the glass as depicted in Holtfreter's drawings, the cells must have an intrinsic apical–basal elongation that defines a long axis parallel to the substratum. As pointed out above, *Xenopus* bottle cells do not maintain apical–basal elongation in culture; either this shape is not intrinsic in this type of bottle cell or we have not provided conditions for its expression.

6. Pattern of Bottle Cell Formation: Not a Reflection of a Propagated Signal

There is an overall progression of bottle cell formation from dorsal to ventral, but locally individual bottle cells initiate contraction not as a progressive wave but as individuals seemingly stochastically distributed within the population of prospective bottle cells (Keller, 1981; Hardin and Keller, 1988). Some bottle cells initiate contraction in complete isolation, four or five cell diameters from ones already formed, and others contract very late, and until they do, they form isolated islands of uncontracted apices among those already contracted (Keller, 1981; Hardin and Keller, 1988). Moreover, experiments show that bottle cell formation is not dependent on formation of previous ones; removal of all dorsal bottle cells does not prevent subsequent formation of lateral and ventral ones (Keller, 1981). Thus their pattern of formation is not consistent with the interesting idea of a stretch-mediated signal propagated, from contracted cells to adjacent cells, as the initiator of their contraction (Odell *et al.*, 1981).

7. Bottle Cell Respreading

We know little about bottle cell respreading. Disappearance of bottle cells was once thought due to their death, but tracing them with vital dye (Keller, 1975) and lineage tracer (Hardin and Keller, 1988) shows that in *Xenopus* they disappear by losing their distinctive shape; they respread in the late gastrula and neurula and remain alive until at least until the early tailbud stage. As noted above, the respreading is autonomous in culture and recapitulates the dorsal-to-ventral progression of bottle cell formation. In culture, apical constriction results in the cells turning their apices completely inward, forming a ball with a small hole on one side; respreading involves the ball turning inside out as the apices expand. It may be that this relaxation phase is only the first part of respreading, and subsequently, if these cells had access to an underlying mesodermal substrate as they do *in situ*, they would spread on it.

How is respreading related to the expansion of the archenteron? Although that part of the archenteron in which bottle cells are respreading increases in area and volume (Keller, 1981), it does not reach its definitive size until some time later. As mentioned above, it may be that bottle cells show some change in tissue affinity or spreading behavior *after* the initial apical relaxation. This should be evaluated experimentally.

Also, the mesoderm on which the bottle cells respread seems to vary with their previous dorsoventral (now anteroposterior) position. The lateral and ventral bottle cells respread to form a sizable area of the periphery of the archenteron lying near the lateral boundary of the somites (Keller, 1975, 1976, 1981). They never advance ahead of this point to reestablish connections with the mesoderm that was beneath them at the time of their formation; this was mesoderm that migrated ahead, forming the lateral and ventral leading edges of the mesodermal mantle. By contrast, the dorsal bottle cells respread anteriorly under the corresponding dorsal (head) mesodermal cells that were originally beneath them as they formed. Why do these dorsal bottle cells alone advance beyond what is typical for lateral and ventral ones? Are they somehow different in their behavior? Pharyngeal endodermal cells, lying ahead of the bottle cells, may enter the expanding anterior region of respreading bottle cells (see Nieuwkoop and Florschütz, 1950; Keller, 1975), which would explain the extra expansion there, but the fact that labeled bottle cell populations respread without insertion of unlabeled cells into their midsts argues against this hypothesis (Hardin and Keller, 1988).

What is the function of respreading? Does respreading simply undo the shape change accomplished during bottle cell formation, or does respreading have a function in tissue interactions, as suggested below? Last, how does it occur? Are there changes in behavior and protrusive activity beyond apical relaxation? Do respreading bottle cells have a specialized protrusive activity and contact behavior? Are bottle cells of *Xenopus* more invasive in the respreading phase than in the formative phase? The respreading potential of bottle cells with respect to various nonbottle cell epithelia could be compared using the brilliant technique of Holtfreter that allows epithelia to compete with one another for area on a tissue substrate (Holtfreter, 1943b).

8. Possible Function of Bottle Cells in Tissue Interactions

Do bottle cells have a function in pattern formation? As bottle cells form, the area of their deep surfaces facing the underlying mesoderm is greatly reduced. At about the same time, this mesoderm begins migration and subsequently leads the advance of the mesodermal mantle toward the animal cap, without being associated with an overlying epithelium until the bottle cells respread in late gastrulation–neurulation. In contrast, the posterior axial mesoderm maintains contact with the epithelium throughout gastrulation and converges and extends in asso-

2. Cellular Basis of Amphibian Gastrulation

ciation with this tissue. Contact of the deep mesodermal cells with the overlying epithelium early in gastrulation is necessary to produce convergence and extension of the IMZ (Wilson, 1990). The epithelium of the dorsal IMZ can induce ventral deep mesodermal cells to converge and extend and make somites, whereas they would do neither with their native epithelium (Shih and Keller, 1992a). Is one function of bottle cell formation to deprive the anterior leading edge mesoderm of epithelial contact and thus send it toward the migrating mesodermal phenotype while the posterior axial mesoderm remains under the influence of the epithelium and thus differentiates into converging and extending mesoderm? We are currently testing these possibilities.

B. Bottle Cell Ingression and Mesoderm Formation

Among amphibians studied in detail thus far, most bottle cells of urodeles and anurans other than *Xenopus* are of mesodermal rather than endodermal fate, and they ingress to deeper layers where they form notochord and contribute to somitic mesoderm, rather than remain on the surface.

1. Ingression of Somitic and Notochordal Mesoderm in Urodeles

Vogt (1929) showed that a large part of the somitic and lateroventral mesoderm of urodeles is located in the superficial layer of the marginal zone, lying between the notochord dorsally and the subblastoporal endoderm. This mesoderm disappears into the deep region at the corners of the blastopore during gastrulation (Vogt, 1929; Holtfreter, 1944; Smith and Malacinski, 1983; Lundmark, 1986) and so brings the subblastoporal endoderm up the the edges of the notochord, which remains in the superficial layer during gastrulation. During neurulation the notochord ingresses at the dorsal midline, and the subblastoporal endodermal regions on either side fuse in the midline, forming the archenteron, defined as a cavity lined with endoderm (see Vogt, 1929; Löfberg, 1974; Brun and Garson, 1984; Lundmark, 1986).

Only anterior pharyngeal endoderm is located in the superficial layer of urodeles; it occupies a crescent-shaped area at the middorsal line and covers the prospective prechordal plate mesoderm of the early gastrula (Vogt, 1929). These endodermal bottle cells apparently stay on the surface like those of *Xenopus*. All the remaining bottle cells in urodeles presumably represent ingressing, mesodermal bottle cells rather than invaginating endodermal ones.

2. Ingression in Anurans

In anurans other than *Xenopus*, both the notochordal and somitic mesoderm is represented in the superficial layer (King, 1903; Vogt, 1929; Purcell, 1990,

1991). In some anurans the pattern of ingression is different from urodeles studied thus far. In *Ceratophrys ornata*, the somitic and notochordal ingressions occur at the same time during neurulation, in three zones of ingression—a central one for notochord and two lateral ones for somitic mesoderm (Purcell, 1992). The same is true of *Bombina orientalis*, another anuran, but this species shows less morphological definition between zones of ingression (S. Purcell, personal communication, 1992).

3. Mechanism of Ingression

Little is known about the mechanism of ingression in amphibians. The cells lose their epithelial junctional complexes, but it is not known whether ingressing populations of bottle cells have junctions different from those of adjacent, non-ingressing cells, or perhaps they have weaker junctions or junctions specialized for disassembly. The loss of the epithelial apical–basal polarity must also occur, but we do not know when or how. The superficial cells must also adopt protrusive activity characteristic of deep cells after or as they leave the surface. These changes would presumably involve regulation of or retargeting proteins characteristic of apical and basal surfaces, including ion pumps, channels, and cell surface adhesion molecules. We do not know whether ingression requires interaction with other tissues, or if it is autonomous to the ingressing cell population at the onset of gastrulation. Studies on ingression of chick embryo cells in the primitive streak (Revel *et al.*, 1973; Sanders and Prasad, 1989; Stern *et al.*, 1990), and primary mesenchyme cell ingression in echinoderms (Gibbins *et al.*, 1969; Tilney and Gibbins, 1969; Fink and McClay, 1985), serve as models for what should be done in amphibians. This represents a large, important, and largely unexplored area of amphibian gastrulation.

4. The Function of Ingression

What is the mechanical effect of a large population of epithelial cells leaving the surface layer during gastrulation? In urodeles, ingression at the corners of the slit-shaped or horseshoe-shaped blastopore can be viewed as decreasing the circumference of a family of hoops that initially originate at the subblastoporal endoderm on both sides, and arc across the dorsal marginal zone. At the end of gastrulation, the corresponding hoops are much shorter, arcing transversely across the archenteron roof. Shortening these hoops by ingression of mesoderm would generate forces that would aid involution and blastopore closure equivalent to those generated by convergence of the IMZ in *Xenopus* (see Fig. 7, Keller, 1986; Lewis, 1952). However, we have no direct way of measuring such forces at present. If this proposed mechanism holds up, it means that embryos can generate the same constriction forces with two different cellular

2. Cellular Basis of Amphibian Gastrulation

behaviors—ingression in one case and cell intercalation in the other (see Section IV, below). In the anurans, where it appears that not much ingression occurs during gastrulation, this mechanism would not apply.

Whether ingression of somitic and notochordal cells occurs simultaneously, as in the anurans, or in sequence, as in the urodeles, the effect is to remove mesoderm from the roof of what is technically up to that point a gastrocoel, making it a true archenteron (a cavity lined with endoderm). Whether the ingressing mesoderm actually pulls the subblastoporal endoderm dorsally to the midline where it fuses, or whether other forces come into play, is not known. It is clear that the lateral endodermal crests, as they were once called, do not have a free edge that actively migrates dorsally, across the inner surfaces of the somitic and notochordal mesoderm as was once thought (see Keller, 1986). No protrusive activity is seen in this region, nor is a free edge seen (Purcell, 1992).

C. Patterning of Bottle Cell Formation

1. Origin of Bottle Cell Types

Why do bottle cells form? The vegetal (subblastoporal) endoderm induces formation of mesodermal tissues in the IMZ as shown by explant recombinant experiments in urodeles (Nieuwkoop, 1969a,b) and in *Xenopus* (Sudarwati and Nieuwkoop, 1971) and in whole embryos of the latter (Stewart, 1990). Is the other component of the IMZ, the overlying endodermal epithelium, including the bottle cells, also induced by subblastoporal endoderm? Nieuwkoop (1969a) and Sudarwati and Nieuwkoop (1971) considered that suprablastoporal endoderm was induced at the same time, but this problem has not been studied since the time of its discovery by Nieuwkoop and associates.

The differences in organization of the marginal zone among amphibians suggest that several types of signals or responses are at work. In *Xenopus*, mesoderm is induced in the deep layer and endoderm in the superficial layer of the IMZ (Sudarwati and Nieuwkoop, 1971). A subpopulation of endoderm must be set aside around the full dorsoventral extent of the IMZ, as prospective bottle cells. In urodeles and in anurans other than *Xenopus*, endodermal cells, including *Xenopus*-type bottle cells, must be induced in a small crescent-shaped middorsal area of the IMZ, which becomes pharyngeal endoderm. Prospective mesoderm must be induced in the remaining superficial layer and in the deep layer (Nieuwkoop, 1969a,b). Mesoderm induction has been treated mostly in terms of whether the induced mesoderm is "dorsal" or "ventral" or "anterior" or "posterior" in character, but it is clear that much more specific types of cellular morphogenetic behaviors are being induced, and these have not been considered since the pioneering work of Nieuwkoop and colleagues.

2. Dorsoventral Progression

The dorsoventral progression of bottle cell formation and respreading is apparently set up early in development and is not propagated from dorsal to ventral by tissue interactions in *Xenopus*. If the ventral side is isolated from the dorsal during the blastula or gastrula stage, the ventral cells form on schedule (V. Cannon, personal communication, 1992). In the Mexican axolotl, tissue isolation and recombination experiments suggest that blastopore formation (bottle cell formation) is not dependent on development of the mesoderm induced in the marginal zone but arises from a previous dorsoventral polarization event (Doucet-de Brune, 1973). *Xenopus* embryos ventralized by UV treatment are radially symmetrical, at least superficially, and their bottle cells form late and simultaneously around the entire embryo (see Malacinski *et al.*, 1974, 1977; Scharf and Gerhart, 1980). It appears unlikely that dorsoventral order of bottle cell formation has a function in itself: Black (1989) found normal development after stimulating precocious formation of bottle cells on the ventral side, before they formed on the dorsal side, with a temperature gradient.

D. Experimental Tests of Function during Gastrulation

1. Bottle Cell Removal

Removal of bottle cells after their formation does not stop gastrulation but only truncates the peripheral regions of the archenteron, as expected from the fate map (Keller, 1981). We do not know if this defect affects later development. The remaining epithelial endodermal cells heal over the site of missing bottle cells but do not regenerate bottle-shaped cells (Keller, 1981); however, these endodermal cells may regulate and compensate for the functions of the missing bottle cells in later development. These extirpation experiments argue against the idea that bottle cells migrate inward, towing the remaining epithelial cells after them during gastrulation. The endodermal roof of the archenteron tends to slip backward after bottle cell removal, suggesting that bottle cells may serve as a peripheral anchor point for the endodermal roof of the archenteron as it is towed inward by the underlying mesoderm (Keller, 1981).

2. Removal of Bottle Cells prior to Their Formation

The major role of *Xenopus* bottle cells in gastrulation appears to occur during their *formation*. To test this idea, we removed them microsurgically before they formed, but lack of immediate healing caused columnarization of underlying deep cells, changing the mechanics of the blastopore region and making interpretation uncertain (R. Keller, unpublished experiments, 1981). In preliminary experiments, UV irradiation of late blastulae prevented bottle cell formation, but blastoporal groove formation appeared to proceed as head mesoderm migration began, independent of bottle cell formation (V. Cannon, preliminary experiments, 1991). Thus it appears that head (leading edge) mesoderm can initiate involution without bottle cell formation, although these experiments are preliminary and the reliability of initiation under these conditions is not established.

E. Evolutionary Implications

It is important to realize that bottle-shaped cells in various amphibians differ in their behavior, their prospective germ layer, and their function in gastrulation. We do not yet know the full extent of differences between the ingressing and invaginating type bottle cells, but one important variable in the evolution of amphibian gastrulation is the presence and distribution of these two types of cells in the IMZ. This variable presumably reflects differences in the signal or in the response system, as marginal zone tissues are induced in the blastula by vegetal endoderm (see Nieuwkoop, 1969a,b; Sudarwati and Nieuwkoop, 1971). The fact that similar involutions and blastopore closures can occur among amphibians with such different cell behaviors in the marginal zone suggests that very different cell behaviors can be used to produce common patterns of mechanical force, depending on the geometry and context of their exercise. Far more of the mechanism of morphogenesis lies in the way cell behaviors are integrated into the mechanics of cell populations than realized heretofore, and this should be reflected in design and interpretation of experiments at lower levels of organization (molecular and genetic) and at higher levels of organization, in the evolution of morphogenetic mechanisms.

III. Mesodermal Cell Migration

A. Description of the Process

Before gastrulation the prospective mesoderm forms a ring surrounding the embryo in the IMZ below the equator. During gastrulation this ring involutes at the blastopore lip and, once apposed to the inner surface of the blastocoel roof (BCR), the mesoderm moves across this substrate toward the animal pole of the embryo (Vogt, 1929; Keller, 1976).

1. Involvement of Active Cell Migration in Mesoderm Movement

Active cell migration is involved in translocation of the mesoderm relative to the BCR. Light and electron microscopic studies on a number of anuran (*Bufo, Rana, Xenopus*) and urodele species (*Cynops, Ambystoma, Pleurodeles*) show that involuted mesoderm cells attach to the BCR surface and extend filiform, lamelliform, or sometimes loboform, cytoplasmic protrusions along the substrate that are presumably involved in active migration (Nakatsuji 1974, 1975a,b, 1976, 1984; Keller and Schoenwolf, 1977; Karfunkel, 1977; Kubota and Durston, 1978; Nakatsuji et al., 1982; Brick and Weinberger, 1984; Boucaut et al., 1985; Lundmark, 1986; Johnson, 1986; Johnson et al., 1990; Winklbauer et al., 1991). Filopodial and lamellipodial protrusions of migrating mesoderm cells are usually polarized and oriented in the direction of mesoderm movement (toward the animal pole) both in anurans (Nakatsuji and Johnson, 1982; Johnson, 1986; Winklbauer et al., 1991) and in urodeles (Holtfreter, 1944; Kubota and Durston, 1978; Nakatsuji et al., 1982; Nakatsuji, 1984). Oriented protrusions are not restricted to the cells of the leading edge of the advancing mesodermal mass, but are also found on submarginal cells, suggesting that all mesodermal cells in contact with the BCR, not just the ones at the leading edge, are migrating actively and directionally.

In addition to this indirect evidence for active cell migration, researchers have observed it directly, although in cultures of cells or explants rather than *in vivo*, because amphibian embryos are opaque. Isolated, single, mesoderm cells from anuran and urodele gastrulae are motile *in vitro* and migrate on artificial substrates, such as fibronectin (FN), laminin, collagen treated with serum, and extracellular matrix (ECM) (deposited by conditioning substrates with explants of BCR) (Weinberger and Brick, 1980; Nakatsuji and Johnson, 1982, 1983a, 1984a,b; Nakatsuji, 1984, 1986; Johnson, 1985; Johnson and Silver, 1986; Johnson et al., 1990; Winklbauer, 1990; Winklbauer et al., 1991). The velocities *in vitro* range from 1.5 μm/min (Weinberger and Brick, 1980; Winklbauer, 1990) to 5.3 μm/min (Nakatsuji and Johnson, 1982). This is similar to the rate of advance of the mesoderm on the BCR (2 μm/min) determined by time-lapse recording of the leading edge of the mesoderm through the blastocoel roof of *Xenopus* (R. Keller unpublished work, 1991). Not only single cells, but also whole pieces of explanted mesoderm move as coherent cell aggregates *in vitro* (Shi et al., 1989, 1990; Riou et al., 1990; Winklbauer, 1990; Winklbauer et al., 1991).

We have also observed migration under conditions closer to those *in vivo*. Keller and Hardin (1987), Winklbauer (1990), and Winklbauer et al. (1991) observed the migration of single mesoderm cells on explanted BCR in *Xenopus*, showing that these cells are motile on their *in situ* substrate. Kubota and Durston (1978) filmed the directed migration of the coherent mass of mesoderm toward the animal pole on the BCR of opened gastrulae of the Mexican axolotl, *Ambystoma mexicanum*. Single cells occasionally move ahead but soon stop until they are reintegrated into the advancing mesoderm. Likewise, in *Xenopus* directional migration of a coherent mass of mesoderm on the BCR occurs in explants, offering tissue arrangements similar to those in the embryo (Winklbauer, 1990). Again, single cells rarely migrate ahead of the main cell population, and the margin of the advancing mesoderm retains its integrity. The velocity of migration at the leading edge is 2 μm/min in *Xenopus* (Winklbauer, 1990) and 1 μm/min in the axolotl (Kubota and Durston, 1978).

2. Migration of the Mesoderm as a Coherent Stream of Cells

Not all moving mesodermal cells are in contact with the substrate. Histological sections and SEM of fractured embryos show that the involuted mesoderm is typically multilayered both in anurans (Nieuwkoop and Florschütz, 1950; Tarin, 1971; Nakatsuji, 1975b, 1976; Keller and Schoenwolf, 1977; Brick and Weinberger, 1984; Johnson, 1986; Johnson *et al.*, 1990) and in urodele embryos (Vogt, 1929; Nakatsuji, 1975a; Lundmark, 1986; Shi *et al.*, 1987, 1989, 1990). To move the mesodermal mass as a whole, the individual cells must adhere to each other.

This cohesion varies among amphibian species. When the BCR of urodeles is peeled back, a layer of loosely connected, migrating mesodermal cells usually remains attached to the BCR (Nakatsuji *et al.*, 1982; Nakatsuji, 1984). In the same manipulation of the anuran *Xenopus*, the BCR usually (e.g., Winklbauer *et al.*, 1991), but not always (see, e.g., Keller and Hardin, 1987), separates cleanly from the involuted mesoderm, which remains a coherent mass of cells. This holds true for living and fixed material, and presumably reflects a relatively greater mesoderm–BCR adhesion in urodeles, and a relatively greater mesoderm cell–cell cohesion in anurans. But these differences may be between species rather than orders in the amphibians, because the mesodermal cells of *Ceratophrys ornata*, an anuran, resemble those of urodeles in the above manipulations (S. Purcell and R. Keller, unpublished observations).

Regardless of these differences, the migrating mesoderm has sufficient cohesion to behave as a coherent cell stream (see Trinkaus, 1976, 1984) in all species examined. Involuted mesoderm explanted *in vitro* on an adhesive substrate spreads and moves as a coherent cell mass, with single cells rarely separating from the explant, both in anurans (Winklbauer, 1990; Winklbauer *et al.*, 1991) and in urodeles (Shi *et al.*, 1989, 1990; Riou *et al.*, 1990). Scanning electron microscopy and dissections rarely show single cells migrating ahead of the leading edge of the mesoderm in *Bufo* (Nakatsuji, 1976), in *Rana* (Brick and Weinberger, 1984; Johnson, 1986; Johnson *et al.*, 1990), in *Xenopus* (Keller and Schoenwolf, 1977; Nakatsuji, 1984; Winklbauer *et al.*, 1991), or in the urodele *Pleurodeles* (Shi *et al.*, 1987, 1989, 1990). Isolated, individual cells in front of the leading edge are found only after experimental manipulations (Kubota and Durston, 1978; Keller and Hardin, 1987). In *Pleurodeles*, migrating mesodermal cells show enough cohesion to carry their neighboring cells across a nonadhesive patch of BCR placed in their path (Boucaut *et al.*, 1984a). The same is true in *Rana* (Johnson, 1986) and *Xenopus* (R. Keller, unpublished results, 1991). Thus, besides the fact that individually migrating cells would pose mechanical problems for coordinated inward movement of the whole multilayered mesoderm (and the attached endoderm in *Xenopus*), empirical evidence suggests that the mesoderm in both urodele and anuran embryos moves as a coherent cell mass, despite the variation among species of relative strengths of mesodermal cohesion and adhesion to the BCR.

In summary, the movement of the involuted mesoderm relative to the BCR during amphibian gastrulation is brought about by, or at least involves, active cell migration, whereby cells move directionally as a coherent cell stream on a planar substrate.

In the following sections, the substrate of migration, the nature of the guiding cues that direct the mesoderm to its target region on the BCR, the motile behavior of mesoderm cells, and the role of cell–cell interaction in mesoderm cell migration will be analyzed in more detail.

B. Extracellular Matrix of the Blastocoel Roof and the Substrate of Mesodermal Cell Migration

1. Components and Structure of the Extracellular Matrix

Abundant ECM fills the blastocoel cavity and intercellular spaces in all regions of amphibian gastrulae (Johnson 1977a,b, 1984; Johnson *et al.*, 1979; Komazaki, 1982, 1985, 1986a,b). This matrix is preserved by special fixation procedures and is visualized in the interstices between blastomeres by toluidine blue or lanthanum staining of light and electron microscopic sections, suggesting that it is mainly mucopolysaccharide (Tarin, 1971; Johnson, 1977a,b). Preliminary biochemical characterization suggests that a main component of this matrix is a high molecular weight glycoconjugate whose polysaccharide moiety is similar, but not identical, to hyaluronic acid (Höglund and Løvtrup, 1976; Johnson, 1977c,d, 1978, 1984). So far, no one has identified a functional role for this ECM in gastrulation. The glycosaminoglycans chondroitin-4-sulfate, chondroitin-6-sulfate, heparan sulfate/heparin (Kosher and Searls, 1973; Höglund and Løvtrup, 1976), and hyaluronic acid (Höglund and Løvtrup, 1976), have been identified in early amphibian embryos, but we do not know much about their spatial distribution in the gastrula. Recently, Brickman (1990) localized chondroitin-6-sulfate to the involuting mesoderm of the *Xenopus* gastrula. Synthesis of collagen in the amphibian gastrula was inferred from the conversion of radioactively labeled proline to hydroxyproline (Green *et al.* 1968; Klose and Flickinger, 1971). Otte *et al.* (1990) suggested the presence of a collagen VI-like molecule near the outer surface of the *Xenopus* gastrula, and external application of a monoclonal antibody against this molecule appears to stop gastrulation. However, because of its location this collagen matrix is probably not involved in mesoderm migration. Altogether, too little is known about glycosaminoglycans or proteoglycans, and

2. Cellular Basis of Amphibian Gastrulation

collagen in the amphibian gastrula to allow for speculations about their role in gastrulation.

The best characterized component of the embryonic ECM of amphibians is a distinct network of fibrils found exclusively on the inner surface of the BCR in anuran (Nakatsuji and Johnson, 1983b, 1984a,b; Nakatsuji et al., 1985a) and urodele embryos (Nakatsuji et al., 1982; Nakatsuji and Johnson, 1983a,b; Nakatsuji, 1984; Nakatsuji et al., 1985a; Darribere et al., 1985; Boucaut et al., 1985). Its localization and appearance at the beginning of gastrulation suggest a role in mesodermal cell migration.

Scanning electron microscopy shows the BCR matrix as a two-dimensional network of extracellular fibrils, about 0.1 μm in diameter, on the basal surface of the innermost BCR cells. The network varies in density among species, being apparently less dense in anuran gastrulae (Nakatsuji et al., 1982; Nakatsuji and Johnson, 1983b; Nakatsuji et al., 1985a,b; Darribere et al., 1985; Boucaut et al., 1985). In *Xenopus*, the area of the basal surface of the BCR covered by these matrix fibrils increases from only 1% at the initial gastrula stage to 4% at the middle gastrula (calculated from figures in Nakatsuji and Johnson, 1983b). Thus, these fibrils do not form a continuous, occluding barrier between cells of the BCR and migrating mesoderm. Instead, direct cell–cell contact between these two cell populations is possible and is observed in electron micrographs (Nakatsuji, 1975a, 1976). Thus the BCR matrix does not preclude the *cells* of the BCR from also serving as substrate for mesodermal cell migration.

One component of these fibrils is the well-characterized adhesive glycoprotein fibronectin (FN). Each of the two nearly identical subunits of this dimeric protein consists of a linear series of domains, each endowed with specific binding properties that mediate interactions with other ECM molecules and with cellular receptors (see Hynes, 1985, for a review). A major cell binding site is defined by an Arg-Gly-Asp (RGD) amino acid sequence (Pierschbacher et al., 1981; Pierschbacher and Ruoslahti, 1984; Yamada and Kennedy, 1984; Hynes, 1987), but other potential cell-binding sites have been described (Humphries et al., 1986, 1987; McCarthy et al., 1988; Saunders and Bernfield, 1988). Slightly different isoforms of FN subunits are generated by alternate splicing (Hynes, 1985). Interestingly, some possible cell-binding sites are present on alternatively spliced regions of the molecule (Humphries et al., 1986, 1987), and FN isoforms show characteristic tissue distributions in a variety of systems (ffrench-Constant and Hynes, 1988, 1989; ffrench-Constant et al., 1989).

The RGD and other cell-binding sites on the FN molecule are recognized by cellular FN receptors belonging to the integrin family of integral membrane cell surface receptors. The integrins are dimeric transmembrane glycoproteins, each consisting of an α and a β subunit. Different combinations of subunits differ in ligand specificity, with the main RGD cell-binding site of FN being recognized primarily by integrin $\alpha_5\beta_1$ (Hynes, 1987; Buck and Horwitz, 1987; Akiyama et al., 1990; Humphries, 1990). However, other cell–FN interactions are possible; for example, the binding of the integral membrane proteoglycan syndecan to a heparin-binding site of FN (Saunders and Bernfield, 1988).

In *Xenopus* and in *Pleurodeles*, FN is present already at early cleavage stages, but the amount per embryo increases markedly after the midblastula stage, due at least in part to an increased rate of synthesis of this protein from stored maternal mRNA made available for translation at the midblastula stage (Lee et al., 1984; Darribere et al., 1984). After its synthesis, FN is found in the network of fibrils of the BCR matrix described above. Very little FN is detected outside the fibrils or on migrating mesoderm cells (Boucaut and Darribere, 1983a,b; Boucaut et al., 1985; Darribere et al., 1985; Nakatsuji et al., 1985a).

There is limited information on the presence of FN receptors in the amphibian embryo. In *Xenopus*, two closely related integrin β_1 subunits have been cloned and respective mRNAs are first detected at the onset of gastrulation (DeSimone and Hynes, 1988). In *Pleurodeles*, an antigen recognized by antiserum against an avian integrin receptor complex is present in cleavage, blastula, and gastrula stages. It is located on the periphery of all cells, except on the outer surface of the embryo, and it is concentrated on the inner surface of the BCR (Darribere et al., 1988).

Laminin is a 900-kDa glycoprotein characteristic of the basal lamina. The best characterized laminin, from the mouse Engelbreth-Holm-Swarm (EHS) tumor, consists of a large A chain and two smaller B chains that are covalently linked by disulfide bridges to form a cross-shaped complex (Martin and Timpl, 1987). Several antibodies against mouse laminin cross-react with laminin-related polypeptides from *Pleurodeles* embryos in immunoblotting experiments (Darribere et al., 1986), and are localized by immunohistochemical procedures in *Pleurodeles* and in *Cynops* embryos (Nakatsuji et al., 1985b; Darribere et al., 1986). The pattern of laminin synthesis in *Pleurodeles* is similar to that of FN in this species and in *Xenopus*. Laminin is synthesized at a low level already in oocytes, during cleavage, and in the early blastula, apparently from a stored maternal mRNA. At the midblastula stage, the rate of synthesis increases to a higher level, which is maintained throughout gastrulation (Riou et al., 1987). In the *Xenopus* gastrula laminin has not yet been demonstrated. An antiserum against a cloned and expressed fragment of a *Xenopus* laminin B chain first detects laminin around the forming notochord toward the end of gastrulation. Fibrils on the BCR are not stained by this antiserum (Fey and Hausen, 1990). It is not known whether other ECM components, like collagens and proteoglycans, are associated with these fibrils in urodeles or anurans. The extracellular matrix glycoprotein tenascin is not present on the BCR in *Pleurodeles*, but is incorporated into the preexisting FN matrix fibrils when added exogenously (Riou et al., 1990).

In *Xenopus* (Nakatsuji and Johnson, 1983b) and in *Rana* (Johnson et al., 1990) the FN fibril network forms rapidly with the onset of gastrulation, whereas it appears earlier and more gradually in *Cynops* (Komazaki, 1988) and in

Pleurodeles (Darribere *et al.*, 1990). In *Cynops* and in *Pleurodeles* gastrulae, laminin seems colocalized with FN in the fibrils of the BCR matrix (Nakatsuji *et al.*, 1985b; Darribere *et al.*, 1986). In *Xenopus*, the ability of BCR cells to adhere to the RGD cell-binding site of FN increases in parallel with FN fibril formation, suggesting an involvement of an FN receptor in fibrillogenesis (Winklbauer, 1988). In fact, RGD peptides or antibodies against the integrin β_1 chain inhibit FN fibril formation on the BCR in *Pleurodeles* (Darribere *et al.*, 1990) and in *Xenopus* (Winklbauer *et al.*, 1991). Also, antibodies to the cytoplasmic domain of the integrin β_1 subunit inhibit fibril formation when injected into blastomeres of the *Pleurodeles* embryo (Darribere *et al.*, 1990), and disruption of microfilaments with cytochalasin B inhibits fibrillogenesis in *Xenopus* (Winklbauer *et al.*, 1991), suggesting an involvement of the cytoskeleton in FN fibril assembly.

2. Substrate of Mesoderm Cell Migration

The substrate for the migrating mesoderm is the relatively sparse BCR matrix network and the surface of the BCR cells exposed between matrix fibrils. The roles of these two substrates in mesoderm cell migration must be defined experimentally.

An important discovery in this field has been that a substrate conditioned by explanted BCR supports mesoderm cell attachment, spreading, and locomotion. A piece of BCR cultured with its inner surface facing downward on an inert substrate deposits a network of fibrils similar to that observed by SEM on the BCR (Nakatsuji and Johnson, 1983a, 1984a,b). As on the BCR, these fibrils contain FN (Shi *et al.*, 1989; Winklbauer *et al.*, 1991). Mesoderm cells seeded onto this substrate spread, assume a morphology resembling that found *in situ*, and move around (Nakatsuji and Johnson, 1983a, 1984a,b; Nakatsuji, 1984; Johnson *et al.*, 1990). Explanted pieces of the dorsal marginal zone are also able to spread and move on this conditioned substrate (Shi *et al.*, 1989; Riou *et al.*, 1990; Winklbauer *et al.*, 1991).

The FN in the matrix fibrils mediates mesoderm cell adhesion and locomotion on the BCR. Antibodies against FN and against the β_1 subunit of a putative integrin-type FN receptor block mesodermal cell spreading and migration on conditioned substrate in *Pleurodeles* (Shi *et al.*, 1989). An RGD-containing peptide, mimicking the RGD cell-binding site of FN, and competitively inhibiting the cellular FN receptor, has the same effect (Shi *et al.*, 1989), suggesting that mesodermal cells specifically interact with the RGD cell-binding site of FN on the conditioned substrate. Interaction of mesodermal cells with laminin, which is present in BCR matrix fibrils of *Pleurodeles*, but not of *Xenopus*, could not be demonstrated (Shi *et al.*, 1989).

As expected from these results, amphibian mesodermal cells attach to, spread, and migrate on an artificial substrate of purified FN (Johnson, 1985; Johnson and

Silver, 1986; Nakatsuji, 1986; Komazaki, 1988; Darribere *et al.*, 1988; Shi *et al.*, 1989, 1990; Riou *et al.*, 1990; Johnson *et al.*, 1990; Smith *et al.*, 1990; Winklbauer, 1990; Winklbauer *et al.*, 1991). These interactions are also inhibited by RGD peptides, again demonstrating that mesoderm cells recognize the RGD cell-binding site of FN (Smith *et al.*, 1990; Winklbauer, 1990). This suggests that an integrin-type FN receptor is expressed on migrating mesodermal cells, as on BCR cells.

BCR cells apparently interact with FN by assembling it into fibrils, whereas mesodermal cells use these FN fibrils as a substrate for migration. Both cell types recognize the RGD cell-binding site of FN, but much higher FN concentrations are required to mediate BCR cell adhesion than mesoderm cell binding to FN (Winklbauer, 1988). When BCR cells are induced *in vitro* to become mesoderm, they spread on substrates made with low FN concentrations like normal mesodermal cells. This change in cell behavior is not paralleled by an increase in integrin β_1 subunit synthesis, indicating a more subtle regulation of cell–FN interaction in the amphibian embryo (Smith *et al.*, 1990).

Its promotion of adhesion and migration of mesodermal cells *in vitro*, and its presence on the inner surface of the BCR, suggest that FN plays a role in mesoderm cell migration in the embryo. Antibodies to FN or to the β_1 subunit of integrin, or RGD-containing peptides injected into the blastocoel of *Pleurodeles*, result in an arrest of gastrulation in this urodele embryo (Boucaut *et al.*, 1984a,b; Darribere *et al.*, 1988). Also, intracellular injection of antibodies against the cytoplasmic domain of the β_1 chain of integrin, which prevents FN fibril formation on the BCR, and injection of tenascin into the blastocoel (which becomes incorporated into FN fibrils and interferes with cell–FN interaction) both lead to an arrest of gastrulation in *Pleurodeles* (Darribere *et al.*, 1990; Riou *et al.*, 1990). Taken together, these experiments suggest that cell–FN interaction is essential for the gastrulation process, and mesodermal cell migration is a likely target of the applied inhibitors.

To define more precisely the role of FN in migration, investigators have experimentally separated the effects of FN on cell adhesion, morphology, and behavior, as expressed by cell spreading and locomotion on FN substrates. In the gastrula, migrating mesoderm cells are in direct, close contact with the basal surface of BCR cells (Nakatsuji, 1975a, 1976). As expected from this observation, when cell–FN interaction is prevented (e.g., by the presence of an RGD-containing peptide in the incubation medium), mesodermal cells still attach to the BCR *in vitro*, but remain globular and do not spread and extend cytoplasmic protrusions along the substrate (Winklbauer, 1990; Winklbauer *et al.*, 1991). This FN-independent attachment to the BCR shows that adhesion of mesoderm cells to a substrate does not by itself lead to cell spreading and the formation of protrusions. Instead, these latter processes seem to depend specifically on cell–FN interaction. Indirect evidence from comparison of cell spreading on artificial substrates, including FN, suggests that the formation of lamellipodia is specifi-

2. Cellular Basis of Amphibian Gastrulation

cally induced by FN in mesoderm cells of *Xenopus* (Johnson and Silver, 1986; Winklbauer *et al.*, 1991). When their interaction with FN is inhibited, mesodermal cells show some random movement on the explanted BCR, but a more efficient translocation is observed when FN-dependent protrusion formation and spreading on the BCR are allowed (Winklbauer, 1990; Winklbauer *et al.*, 1991).

Thus FN may contribute to mesoderm cell adhesion to the BCR in proportion to the FN fibril density, but its more specific role in mesoderm migration may be induction of protrusive activity of mesoderm cells. This function seems especially appropriate in situations where FN is present not as a continuous layer but as a discrete network of fibrils.

3. Guiding Cues of the Extracellular Matrix in Mesodermal Cell Migration

Amphibian mesodermal cells move directionally toward a region near the animal pole. They are oriented, with their locomotory protrusions pointing in the direction of advance. The evidence shows that both an intrinsic tissue polarity in the mesoderm and external cues in the BCR may contribute to this directionality.

A tissue polarity that could serve this function was demonstrated for the mesoderm of *Pleurodeles* (Shi *et al.*, 1989) and *Xenopus* (Winklbauer, 1990). In *Xenopus* a stable gradient of adhesiveness to FN extends along the anteroposterior axis of the dorsal mesoderm, that is, along the axis of its movement (Winklbauer, 1990). Moreover, in *Pleurodeles* (Shi *et al.*, 1989) and in *Xenopus* (Winklbauer, 1990) a strip of prospective mesoderm migrates on FN *in vitro* in the direction of its anterior end, as it would in the embryo. This demonstrates a mesodermal polarity able to determine the direction of movement, but the relevance of this phenomenon for mesoderm movement *in situ* is not clear. It may be that in the mechanical context of the whole gastrulating embryo, mesodermal tissue polarity is sufficient to ensure movement of the mesoderm in the proper direction. In fact, when the dorsal BCR is rotated 180° in the embryo at the onset of gastrulation, development nevertheless proceeds normally, implying a substrate-independent determination of the direction of mesoderm translocation (Cooke, 1972).

Guiding cues directing the movement of mesodermal cells also reside in the ECM of the BCR. In *Ambystoma*, single cells migrating on a substrate conditioned by explanted BCR show a slight preference for moving toward the animal pole position on this substrate (Nakatsuji and Johnson, 1983a; Nakatsuji, 1984), an effect that is much more pronounced when the movements of whole mesodermal explants are studied. In *Pleurodeles* (Shi *et al.*, 1989; Riou *et al.*, 1990) and in *Xenopus* (Winklbauer *et al.*, 1991), explanted mesoderm moves as a coherent cell aggregate toward the animal pole position on conditioned substrata. Thus, substrates bearing ECM transferred from the BCR have a directionality sufficient to guide the mesoderm toward the animal pole.

The nature of these guidance cues and their mechanism of action are unknown. One possibility is haptotaxis, where cells move up a gradient of adhesiveness of the substrate (Carter, 1965), but so far there is no evidence for this mechanism. Fibronectin seems to be the only matrix component mediating mesodermal cell adhesion to conditioned substrate (Shi *et al.*, 1989), but a global gradient in FN density along the blastopore lip–animal pole axis of the BCR could not be found in *Xenopus* (Nakatsuji *et al.* 1985a; R. Winklbauer, unpublished results, 1991). Furthermore, immunogold labeling of the FN matrix with antibodies did not reveal a detectable gradient of FN content along individual FN fibrils of *Xenopus* (Nakatsuji *et al.*, 1985a) or *Pleurodeles* (Darribere *et al.*, 1985; Boucaut *et al.*, 1985). Thus if there is a FN gradient, it is not detectable by current means. This is not conclusive evidence against a gradient of *adhesiveness*, because this quality could arise from a uniform distribution of FN that is differentially modified along the axis of mesoderm movement.

Another argument against haptotaxis in this system comes from inhibition of FN *fibril formation* during substrate conditioning by an RGD peptide or by cytochalasin B. Fibronectin is still deposited on the substrate, and mesoderm cells or explants can still attach to it, spread, and migrate, but the directionality of migration is lost (Winklbauer *et al.*, 1991). Apparently, an intact FN fibril network is necessary for directional cell migration. This is difficult to explain on the basis of a gradient of adhesiveness, whether it is due to a spatial variation in FN concentration or to spatially regulated FN modifications. So there is considerable but not yet conclusive evidence against haptotaxis operating in substrate-dependent mesoderm guidance.

The dependence of directionality of migration on FN fibril formation suggests that the relevant cues may be related to some property of these fibrils, for example, their spatial arrangement or molecular structure. In *Ambystoma*, Nakatsuji and others reported a weak alignment of FN fibrils along the blastopore lip–animal pole axis on the BCR (Nakatsuji *et al.*, 1982; Nakatsuji, 1984); when transferred to an *in vitro* substrate, this slight orientation of fibrils yields some alignment of the trails of mesodermal cells migrating on them (Nakatsuji and Johnson, 1983a; Nakatsuji, 1984). When the fibril network is artificially aligned perpendicular to the original orientation, cells again migrate preferentially along the axis of fibril orientation (Nakatsuji and Johnson, 1984a).

However, a general role for fibril alignment in mesoderm cell migration *in situ* is questionable. First, no obvious FN fibril alignment was found in *Xenopus* (R. Winklbauer, unpublished results, 1991), *Pleurodeles* (Darribere *et al.*, 1985), or *Rana* (Johnson *et al.*, 1990), although conditioned substrate nevertheless directs cell migration in the first two species. Second, although mechanisms by which fibril alignment *orients* cell migration have been established (Dunn and Heath, 1976; Dunn, 1982), it is not easy to understood how this alignment could give *directionality* to the movement of the mesoderm, unless one invokes desmotaxis. Desmotaxis is directional movement of cells on aligned fibrils that are all attached at only one and the same end; thus the cell moves when it attempts to pull the free end directly *away from the anchor point*, but collapses the fibril when it attempts to pull the free end *back toward the anchor point* (Boocock, 1989; see Newgreen, 1990). However, there is no evidence indicating FN fibrils are attached at one end. Even if they were, it is hard to conceive how this property could be preserved during substrate conditioning.

Nevertheless, the striking property of the BCR fibril network is its stimulation of polarized, directional migration toward the animal pole rather than movement in both directions along the blastopore lip–animal pole axis. It could be argued that contact inhibition of cells prevents movement of mesoderm cells back toward the blastopore lip, and that the combined effects of substrate alignment and contact inhibition would lead to directional translocation (Nakatsuji *et al.*, 1982). However, this explanation obviously does not apply to single cells (Nakatsuji and Johnson, 1983a; Nakatsuji, 1984) or to relatively small isolated pieces of mesoderm, which have free edges at both ends but move directionally on conditioned substrate (Winklbauer *et al.*, 1991).

This draws our attention to the exciting possibility, first expressed by Nakatsuji and Johnson (1983a), that the ECM substrate is polarized in the architecture of its fibrillar constituents. For example, FN molecules could arrange themselves in an orderly, polar manner within these fibrils. Fibril alignment would increase the efficiency of such a mechanism but not be necessary as long as the net polarity of the randomly oriented fibrils points to the animal pole.

In summary, it has become clear now that the ECM of the BCR serves a dual role in the movement of the mesoderm during amphibian gastrulation. First, it supports cell migration by regulating the protrusive activity of mesoderm cells and by contributing to the adhesion of these cells to their BCR substrate, and second it contains guidance cues that direct the movement of the mesoderm toward the animal pole.

C. Motile Activity and Behavior of Migratory Mesoderm Cells

1. Mesodermal Cell Motility

In situ, migratory mesoderm cells are typically rounded and only moderately spread on the BCR substrate, with filiform and lamelliform protrusions extending from the globular cell body along the substrate or to neighboring cells (Nakatsuji, 1975a,b, 1976, 1984; Keller and Schoenwolf, 1977; Kubota and Durston, 1978; Nakatsuji *et al.*, 1982; Boucaut *et al.*, 1985; Johnson, 1986; Johnson *et al.*, 1990; Winklbauer *et al.*, 1991). *In vitro*, filiform protrusions extend spontaneously, independent of interaction with the substrate in unattached migratory mesoderm cells (Nakatsuji and Johnson, 1984b; Winklbauer *et al.*, 1991) whereas lamellipodia formation requires interaction with a substrate. As mentioned above, FN substrate induces mesoderm cell spreading and the extension of lamelliform protrusions *in vitro*. These protrusions appear simultaneously, and usually at opposite ends of a cell, giving it a bipolar morphology, although Winklbauer and others (1991) also observed multipolar cell spreading. Spreading of the cell body is moderate and, with the presence of lamellipodia and filopodia, the morphology of mesoderm cells on FN *in vitro* is reasonably close to their shape *in situ* (Nakatsuji, 1986).

Detailed analysis of the migration of mesoderm cells on FN *in vitro* reveals the basic events governing the migratory behavior of mesoderm cells are (1) the formation and the retraction of lamellae, (2) the splitting of a single lamella in two, and (3) the lateral movement of lamellae along the cell circumference. With the predominantly bipolar structure of the cells, this leads to a stepwise, intermittent mode of translocation. Typically a cell is stretched by the centrifugal movement of lamellae at opposite ends of the cell. On retraction of one lamella, the cell body moves suddenly forward to the position of the remaining lamella, which then often splits in two, leading to a respreading of the cell in a direction perpendicular to the original one. As this cycle is repeated, movement is not only intermittent but also accompanied by frequent changes in direction. Rarely does an isolated cell maintain a unipolar structure with a single leading lamella; this allows a short interval of persistent, efficient translocation (Winklbauer *et al.*, 1991).

The bipolar cell morphology and the nonpersistent, intermittent mode of movement are not artifacts of the *in vitro* FN substrate but occur as well on collagen substrate treated with serum (Nakatsuji and Johnson, 1982), on BCR-conditioned substrate (Nakatsuji and Johnson, 1983a, 1984a,b; Nakatsuji, 1984; Winklbauer *et al.*, 1991), and on the BCR itself (Winklbauer, 1990; Winklbauer *et al.*, 1991). This mode of translocation is not well suited for efficient, directional migration in the embryo. However, mesoderm cells do not move *in situ* as isolated cells but as part of a coherent cell stream in which cell–cell interaction modifies their migratory behavior (see below). Thus analysis of isolated cells yields only a partial understanding of the motile behavior underlying mesodermal cell migration.

2. Changes at Onset of Migration

Regardless of when they are explanted, prospective mesodermal cells begin migration on FN *in vitro* when gastrulation begins in control embryos, suggesting that the onset of migration is controlled by changes within mesodermal cells rather than by the appearance of FN on the BCR (Shi *et al.*, 1989). In fact, the FN fibril network develops before the onset of gastrulation and mesoderm movement in *Pleurodeles* (Darribere *et al.*, 1990) and in *Cynops* (Komazaki, 1988); but the adhesiveness of gastrula cells to FN increases (Johnson, 1985; Johnson and Silver, 1986), and mesodermal cells acquire the ability to spread on FN at the onset of gastrulation (Komazaki, 1988).

It is not known what changes within the mesodermal cells initiate migration. Appearance of an FN receptor on migratory mesoderm cells at the beginning of gastrulation could explain the above observations, but *Xenopus* BCR cells artificially induced to become mesoderm show no detectable increased rate of synthesis of the integrin β_1 subunit, as they adopt the normal mesodermal spreading behavior on FN (Smith *et al.*, 1990). It follows that the ability to spread on FN is probably not solely regulated by synthesis of the β_1 subunit of the FN receptor. Furthermore, uninvoluted mesoderm cells of *Xenopus*, which are still part of the BCR, are completely immobile when spread on an FN substrate, whereas after involution they are motile on FN under the same conditions (Winklbauer, 1990). It would seem that acquiring the ability to interact with FN, by whatever mechanism, may not be sufficient to initiate mesodermal cell migration. Apparently a change in the motile apparatus of the cells is required.

D. Role of Cell Interactions in Mesoderm Cell Migration

The mesodermal cells move as a coherent cell stream, and thus cell interactions may play a significant role in their migration. Contact inhibition is one possible way of regulating contact behavior. When lamellipodia of two isolated mesoderm cells on a substrate *in vitro* come in contact, they stop immediately and retract (Johnson, 1976a; Nakatsuji and Johnson, 1982). This is a local reaction, because other lamellipodia of these cells are not affected. The reaction is triggered only when lamellipodia meet lamellipodia on other cells and not when they contact cell bodies (Johnson, 1976a; Nakatsuji and Johnson, 1982; Winklbauer and Selchow, 1992). This fits the observation that cytoplasmic protrusions extend from one cell to the cell body of a neighboring mesoderm cell in the embryo. Because mesoderm movement occurs on a planar substrate, and lamelliform protrusions are preferentially extended along this substrate, contact inhibition of protrusive activity could well play a role *in situ*, although at present no direct evidence supports this notion.

Contact interactions also mediate the cohesion of the mesodermal population. When a lamellipodium of a mesoderm cell contacts the cell body of another cell, the two often adhere and remain together while moving. In this way, isolated cells form motile aggregates *in vitro*, and move much like cultured epithelial cell clusters (DiPasquale, 1975; Kolega, 1981). Cells of an aggregate glide past each other and may exchange neighbors, but rarely manage to break loose from the aggregate (Winklbauer *et al.*, 1991). Aggregate formation depends on the presence of Ca^{2+} in the medium and, conversely, pieces of mesoderm can be dissociated into single cells in a Ca^{2+}-free buffer (Nomura *et al.*,

1986). A novel member of the cadherin family of Ca^{2+}-dependent cell adhesion molecules, U-cadherin, has been identified on the surface of migrating mesoderm cells in *Xenopus* (Angres et al., 1991; Winklbauer et al., 1991). In the presence of antibodies to U-cadherin, a mesoderm aggregate migrating on FN *in vitro* disintegrates, and the individual cells attached to FN disperse over the substrate (Winklbauer et al., 1991). This suggests that U-cadherin mediates the Ca^{2+}-dependent mutual adhesion of migrating mesoderm cells, and is responsible for the cohesive behavior and morphology of mesoderm explants.

We do not know in detail how cell interactions modify the behavior of mesoderm cells, but the importance of aggregate formation and the attendant cell interactions for mesoderm movement is apparent from the finding that single cells migrate directionally on conditioned substrate or on the BCR much less reliably than whole mesoderm explants. Isolated *Ambystoma* mesodermal cells show a tendency for directional migration on conditioned substrate (Nakatsuji and Johnson, 1983a; Nakatsuji, 1984), but cells from *Xenopus* (Nakatsuji and Johnson, 1983a) and *Rana* (Nakatsuji and Johnson, 1984b) do not move directionally under similar conditions. Also, isolated mesoderm cells cultured on the BCR of *Xenopus* (Keller and Hardin, 1987; Winklbauer, 1990; Winklbauer et al., 1991), and of *Ambystoma* (Kubota and Durston, 1978) do not move directionally. In contrast, movement of mesodermal explants is directional on conditioned substrate in *Pleurodeles* (Shi et al., 1989; Riou et al., 1990) and in *Xenopus* (Winklbauer et al., 1991) and on the BCR in *Ambystoma* (Kubota and Durston, 1978) and in *Xenopus* (Winklbauer, 1990). Individual cells migrate with relatively low persistence on FN, on conditioned substrate, and on explanted BCR, apparently as a consequence of their bi- or multipolar structure, which entails frequent, endogenous changes in the direction of protrusive activity (Nakatsuji and Johnson, 1982; Winklbauer et al., 1991). In contrast, migration becomes highly persistent, with cells moving on straight pathways, in aggregates of cells moving on conditioned substrate or on the BCR (Winklbauer, 1990; Winklbauer et al., 1991). Apparently, being part of an aggregate stabilizes protrusive activity: persistent and directional migration in aggregates occurs in conjunction with a unipolar cell shape and with locomotory protrusions of individual cells pointing in the direction of movement of the whole mesodermal cell mass (Winklbauer et al., 1991).

Development of polarized behavior and morphology requires interaction with the substrate and among mesodermal cells. As shown above, attachment to its *in vivo* substrate is not sufficient to induce a unipolar shape in a mesodermal cell. Also, a unipolar structure does not simply result from cell aggregate formation alone. Instead, cells must be part of an aggregate and be attached to the ECM of the BCR to assume a unipolar morphology, which is probably a prerequisite for persistent and directional migration (Winklbauer et al., 1991). Thus mesoderm cell–cell interaction is an essential, integrated feature of their directional migration.

E. Function of Mesoderm Cell Migration in Gastrulation

Polarized traction of the mesodermal cells on the roof of the blastocoel probably generates forces that move the mesoderm toward the animal pole; looked at another way, it pulls the blastocoel roof down over the mesoderm. Such tissue movements should also contribute to involution of the IMZ. Moreover, the migrating mesoderm probably accounts for some of the behavior often attributed to bottle cells. As noted above, the mesoderm in all amphibians has enough cohesion that the cells migrating on the BCR substratum carry along those not bounding the BCR; this argument extends as well to the bottle cells and overlying endodermal epithelium, where such endoderm exists in various species. As noted above, the endodermal epithelium, strongly attached by the bottle cells, seems to ride along on the underlying mesoderm.

As discussed below, the mesoderm is multifunctional; it also shows convergence and extension movements while migrating. It would be interesting to know what migration, on the one hand, and convergence and extension, on the other, contribute to gastrulation. The evidence suggests that amphibian species vary in the strength, autonomy, and timing of convergence and extension, some making great use of these movements in gastrulation and others depending mostly, if not solely, on migration. Gastrulation in embryos with a strong autonomous convergent extension machinery may appear independent of the process of mesoderm cell migration, and experimental interference with cell migration might have no spectacular effects on gastrulation as a whole.

Xenopus has such an embryo. Even high concentrations of an RGD peptide do not block gastrulation in *Xenopus* (Winklbauer, 1989; R. Winklbauer and R. Keller, unpublished results, 1991), although lower concentrations have a specific effect on mesoderm cell morphology and migration on the BCR (Winklbauer, 1990; Winklbauer *et al.*, 1991). Convergence and extension are completely autonomous and independent of external substrata, including the BCR (see Section IV, below), and presumably these movements are sufficient to drive much of the gastrulation movements in *Xenopus*. One would not expect this substrate-independent convergent extension to be inhibited by RGD peptides in *Xenopus*, and indeed it is not (Smith et al., 1990). Another way to test the function of mesoderm cell migration in gastrulation is to remove the substrate of migration, the BCR, from the embryo. This operation does not prevent involution of the mesoderm and endoderm, or closure of the blastopore, or their dorsal convergence and extension in *Hyla* (Holtfreter, 1933) and *Xenopus* (Keller *et al*., 1985a,b; Keller and Jansa, 1992).

This does not mean that migration of mesoderm is unimportant in *Xenopus*. The evidence suggests that the mesoderm of *Xenopus* can be divided into anterior leading edge mesoderm that aggressively migrates (Winklbauer, 1990) but does not converge and extend (Keller and Tibbetts, 1989). Behind this mesoderm is the posterior, axial, and paraxial (prospective notochordal and somitic) mesoderm that aggressively converges and extends (Keller et al., 1985a,b; Keller and Danilchik, 1988), but also participates in migration as demonstrated by the moving aggregates of mesodermal cells described above (Winklbauer, 1990). We will discuss below how axial and paraxial mesoderm can both migrate and converge and extend. Although convergence and extension are powerful enough to involute the IMZ, constrict the blastopore, and push the migrating mesodermal cells ahead to where the animal cap would have been, the migrating mesoderm does not distribute itself normally in embryos lacking a BCR (Keller and Jansa, 1992). Moreover, without the mesoderm–BCR interaction, all the mesoderm sinks into the endoderm during convergence and extension (Keller et al., 1985a,b; Keller and Jansa, 1992).

In contrast, in *Pleurodeles*, an autonomous convergent extension mechanism could not be demonstrated, at least until late gastrulation (Shi et al., 1987). Accordingly, injection into the blastocoel of a number of probes inhibiting cell–FN interaction, and thus mesodermal cell migration, results in complete arrest of gastrulation (Boucaut et al., 1984a,b; Darribere et al., 1988, 1990; Riou et al., 1990). So it appears that embryos of this type are dependent on mesodermal cell migration to bring about involution, movement of the mesoderm into place, and blastopore closure.

IV. Convergence and Extension Movements

The marginal zone of amphibians undergoes dramatic narrowing (convergence) and lengthening (extension) during gastrulation and neurulation (Vogt, 1929; Keller, 1975, 1976). Spemann, Vogt, Mangold, Holtfreter, Schechtman, and others (see Spemann, 1938, Schechtman, 1942; Keller, 1986) showed by explantation and microsurgical manipulations that convergence and extension are autonomous properties of the dorsal marginal zone. Despite their importance, these processes were ignored until recently, for several reasons (reviewed in Keller et al., 1991b), including a failure to understand their function in gastrulation. However, convergence and extension are major morphogenetic movements of the vertebrate organizer (see Spemann, 1938) and thus are fundamental in early vertebrate embryogenesis beyond their function in gastrulation. They establish the elongate shape and the dorsal position of the notochord, the somitic mesoderm, and the neural plate (Keller, 1986; Jacobson and Gordon, 1976; Jacobson, 1981), and thus physically define the anteroposterior axis of the vertebrate embryo (Gerhart and Keller, 1986). Both the function of convergence and extension in these morphogenetic events and the cellular behavior underlying these movements are discussed elsewhere (Wilson et al., 1989; Keller et al., 1989a,b; Wilson, 1990; Keller et al., 1991a,b; Keller et al., 1992c). Here we will focus on the function and cellular basis of these movements in gastrulation.

A. Convergence and Extension in *Xenopus:* Demonstration of Two Regions of Convergence and Extension in Sandwich Explants

Vital dye mapping (Keller, 1975, 1976) and time-lapse cinemicrography (Keller, 1978) show that the dorsal axial tissues of notochord, somite, and posterior neural plate of *Xenopus* show convergence and extension movements similar to those described in the classical literature. To learn whether these movements are passive responses to forces generated elsewhere or autonomous and independent, we made explants of the gastrula to see if they would converge and extend independently of potential mechanical or signaling influences from surrounding tissues. "Sandwich" explants were made by apposing the inner surfaces of two marginal zones of the early gastrula stage and allowing them to heal (Keller *et al.*, 1985a,b; Keller and Danilchik, 1988). These explants show not *one* but *two* regions of convergence and extension: one in the dorsal *involuting marginal zone* (IMZ), which differentiates into a central notochord with somites on both sides, and another in the dorsal *noninvoluting marginal zone* (NIMZ), which differentiates into neural tissue (Keller and Danilchik, 1988). These movements are autonomous and specific to the dorsal marginal zone, and explants of lateral and ventral regions show progressively fewer of these movements (Keller and Danilchik, 1988). But in the embryo, lateral and ventral sectors of the marginal zone are recruited to participate in convergence and extension to a greater degree than reflected in isolates of these sectors (Keller, 1975, 1976; Wilson *et al.*, 1989; Keller and Danilchik, 1988). Because both the recruitment process and the underlying cellular motors of extension and convergence in these two regions may be different, the regions were treated separately, and were named the IMZ and NIMZ, respectively (Keller and Danilchik, 1988). In *Xenopus*, convergence and extension of the NIMZ and IMZ are substantially different in amount of initial thinning, in the amount of extension relative to original length, in timing, and in tissue interactions controlling their expression (see Keller and Danilchik, 1988; Keller *et al.*, 1992a,b,c). It is not clear whether many of the extensions reported in the classical literature for species other than *Xenopus* (mostly urodeles) were of the IMZ or NIMZ types. In *Xenopus*, convergence and extension of the NIMZ are induced by planar signals from the adjacent IMZ (Keller *et al.*, 1992b).

B. Convergence and Extension in Other Amphibians

Autonomous convergence and extension were demonstrated in the classical period in several species of urodeles and anurans (see Spemann, 1938; Schechtman, 1942). However, in many cases it was not clear whether these movements occurred during gastrulation or later. In the case of urodeles, autonomous convergence and extension occur, at best, in the late gastrula stage of *Pleurodeles*, and thus their role in gastrulation of these embryos is probably minimal (Shi *et al.*, 1987). Convergence and extension are autonomous and have a major function in neurulation in urodeles, specifically in *Taricha torosa*, the California newt (Jacobson and Gordon, 1976). Convergence and extension also appear to function in gastrulation of *Hyla regilla*, the California tree frog (Schechtman, 1942), and in *Ceratophrys ornata*, another anuran (S. Purcell, preliminary experiments, 1991).

The lateness of convergence and extension in urodeles is perhaps a reflection of the fact that they have a short notochord and neural plate at the beginning of neurulation and do much of their extension in the neurula stage (Jacobson and Gordon, 1976; Jacobson, 1981). In contrast, *Xenopus* converges and extends earlier in the midgastrula stage. Convergence and extension must be examined in more species in all orders of amphibia to learn if precocious (gastrula stage) convergence and extension are characteristic of just anurans or if they are expressed this way in some urodeles and caecilians as well.

C. Roles of the Epithelial Cells and the Deep Mesodermal Cells in Convergence and Extension

The marginal zone of the *Xenopus* gastrula has a superficial epithelium of prospective endoderm and an underlying deep region, consisting of four or five layers of mesenchymal, prospective mesodermal cells (Keller, 1975, 1976). The epithelial layer appears to be stretched passively by forces produced by the deep mesodermal cells. Replacement of the epithelial layer of the dorsal marginal zone of the early gastrula with that from the animal cap does not disrupt convergence and extension in whole embryos (Keller, 1981) or explants (Keller and Danilchik, 1988), whereas the corresponding manipulation of the deep layer does block these movements (Keller, 1981, 1984). Thus most evidence suggests that the deep mesodermal cells produce the forces for convergence and extension and that the epithelial layer of cells follows along passively. However, the epithelial layer does appear to organize deep cell behavior and bias their axis of extension (Shih and Keller, 1992a; see Keller *et al.*, 1992c).

D. Convergence and Extension Occur by Radial and Mediolateral Intercalation

1. Epithelial Endodermal Cell Behavior

Time-lapse recordings reveal that superficial epithelial cells remain within the plane of the epithelium and intercalate between one another along the medi- convergence and extension, as in the IMZ (Keller *et al.*, 1992a). Superficial cells of the NIMZ intercalate mediolaterally during convergence and extension (Keller, 1978). We do not know much beyond this about cell intercalation in the

2. Cellular Basis of Amphibian Gastrulation

olateral axis (*mediolateral intercalation*) to form a longer but narrower array (Keller, 1978). Epithelial cells commonly rearrange during morphogenesis (reviewed in Keller, 1987; Fristrom, 1988), even in systems where a high-resistance physiological barrier is maintained throughout the rearrangement (Keller and Trinkaus, 1987).

2. Deep Mesodermal Cell Behavior

Using a coherent block of labeled deep cells to monitor intercalation in the embryo, we found that in the posterior converging and extending parts of the mesodermal mantle, the prospective notochordal and somitic mesodermal cells accomplished these movements by mediolateral intercalation, whereas in the migrating leading edge region of the mesodermal mantle, the prospective head mesodermal cells did not intercalate but spread coherently on the BCR (Keller and Tibbetts, 1989). Because the deep cells appear to generate much of the force for convergence and extension, it is important to understand their protrusive activity. So we developed the open-faced explant system that allows normal development and morphogenesis of the deep cells without the protection of an opaque epithelium on all sides (Keller et al., 1985a,b). This system has been used with high-resolution time-lapse videomicrography to reveal the cell behavior underlying convergence and extension during gastrulation (Wilson, 1990; Wilson and Keller, 1991; Shih and Keller, 1992b,c) and neurulation (Wilson et al., 1989; Keller et al., 1989a; Wilson, 1990).

These two principal cell behaviors, *radial intercalation* and *mediolateral intercalation*, are each driven by specific types of cell protrusive activity. In a given region, they occur sequentially, perhaps with some overlap. This sequence of behaviors begins in the dorsal midline at the vegetal end of the IMZ, which is prospective anterior notochord, and spreads progressively in the animal (prospective posterior) direction and also laterally and ventrally within the IMZ. Finally, mediolateral intercalation becomes specialized in the notochord and somitic mesoderm as the boundary forms between these tissues in the late midgastrula, occurring by new cell behaviors and producing different results in each tissue. The patterning and progression of these behaviors has been discussed elsewhere for the neurula stage (Wilson et al., 1989; Keller et al., 1991b) and the gastrula stage (Wilson and Keller, 1991; Keller, et al., 1991a,b; Keller et al., 1992c; Shih and Keller, 1992c) and will be summarized here.

From the onset of gastrulation through the midgastrula stage, the IMZ thins and extends but converges very little, as deep cells intercalate between one another, along the radial aspect of the embryo, forming fewer layers of greater area (*radial intercalation*). This increased area is channeled into extension along the animal–vegetal (future anteroposterior) axis, rather than in all directions, for reasons we do not understand (see Wilson and Keller, 1991; Keller et al., 1991a,b). At the midgastrula stage, radial intercalation has stopped in the vegetal (prospective anterior) end of the IMZ but continues at the animal (prospective posterior) end. At this point, the explant converges and extends as the deep cells at the vegetal end begin *mediolateral intercalation*, moving between one another along the mediolateral axis to produce a narrower but longer array. Note that in the early gastrula, radial intercalation produces extension as a result of thinning, whereas in the middle and late gastrula, mediolateral intercalation produces extension as a result of convergence. When convergence and extension are paired, they are referred to as "convergent extension." We have named these cell behaviors "intercalations" (Keller *et al.*, 1985a,b), meaning "to insert between or among existing elements" (*Webster's Third International Dictionary*), which is more accurate than the previously used "interdigitation" (Keller, 1984) and more specific and descriptive than the general term "cell rearrangement" (see Keller *et al.*, 1991b).

3. Convergence and Extension of the Noninvoluting Marginal Zone By Radial and Mediolateral Intercalation

Labeling deep cell populations shows that they undergo sequential radial and mediolateral intercalations, as in the IMZ. Radial intercalation first thins the NIMZ without much convergence and then mediolateral intercalation produces convergence and extension, as in the IMZ (Keller *et al.*, 1992a). Superficial cells of the NIMZ intercalate mediolaterally during convergence and extension (Keller, 1978). We do not know much beyond this about cell intercalation in the NIMZ because we have not yet been able to approach this problem directly by videorecording cells in the NIMZ.

E. Cell Behavior during Mediolateral Intercalation

What motile activities are responsible for cell intercalation? We will focus on the IMZ because it is the only region in which we have been able to observe deep cell behavior directly, and on mediolateral intercalation, because we still know little about radial intercalation.

1. Protrusive Activity during Mediolateral Intercalation

We developed a method of observing the deep cells next to the overlying epithelium, using "shaved" explants, in which the deepest cell layers of the IMZ are removed until just one or two layers of deep cells remain on the inner surface of the epithelium (Shih and Keller, 1992b). There are two reasons for using this type of explant. First, the epithelium contributes to the organization of the mediolateral intercalation process (Shih and Keller, 1992a) and thus the deep cells next to the epithelium are most likely the ones showing the most powerful and

2. Cellular Basis of Amphibian Gastrulation

well-organized mediolateral intercalation. Second, the optics are better and analysis is simpler in this explant, because radial intercalation is minimal, with only one or two layers of deep cells at most.

The following protrusive activity occurs during convergence and extension (see Shih and Keller, 1992b). When convergence and extension begin at the midgastrula stage, the cells elongate, orient mediolaterally, and align parallel to one another in a transverse band located at the vegetal (prospective anterior) end of the dorsal IMZ. During their elongation and alignment the protrusive activity of these cells, visualized by low-light, time-lapse fluorescence microscopy, is restricted progressively to their medial and lateral ends. The cells apply these protrusions to the surfaces of adjacent deep cells, appear to exert traction on them, and appear to move the cells between one another along the mediolateral aspect, producing mediolateral intercalation. The anterior and posterior margins of the cells do not advance across adjacent cells. Scanning electron microscopy of the same cells in whole embryos (Keller et al., 1989b) shows numerous filiform protrusions on the anterior and posterior sides of fixed cells; these protrusions probably represent focal points of contact that are turned into retraction fibers by shrinkage, because they are not seen in this number and length in recordings of living cells. These protrusions probably hold the cells in a parallel array but allow shearing between cells, as mediolateral intercalation occurs.

2. Cell Interactions: Organizers of Protrusive Activity of Mediolateral Intercalation

During the early period of radial intercalation, the intercalating mesodermal cells first appear to be transiently bipolar in explants, similar to the mesodermal cells in culture (see above and Nakatsuji and Johnson, 1982; Winklbauer, 1990). But as they elongate, align, and begin mediolateral intercalation, the mesodermal cells become progressively more bipolar in protrusive activity. Cell interactions within the IMZ must be responsible for organized protrusive activity, because this never occurs among individual cells.

The epithelium might bias and organize this protrusive activity. The dorsal deep cells require contact with the epithelium in early gastrulation to produce convergence and extension but later become independent of the epithelium (Wilson, 1990; Shih and Keller, 1992a). This contact may aid the development of the bipolar, aligned array of cells. The epithelium might reinforce the bipolar state by directly altering the protrusive activity of the deep cells, and in turn reinforce alignment. Such an effect might be carried out by epithelial cues *orienting* cell behavior in the mediolateral direction, or the epithelium might bias the *direction* of deep cell movement with substrate cues, perhaps by haptotaxis. To decide among these possibilities we must observe deep cell behavior on the inner surface of the epithelium.

Contact interactions between mesodermal cells may reinforce their bipolarity and bring them into parallel alignment. The morphology and behavior of these cells resemble that of fibroblasts in culture that show contact-mediated self-organization into parallel arrays of elongate cells (Elsdale and Wasoff, 1976; Erickson, 1978a,b). Numerous small, lateral protrusions inhibit underlapping by approaching cells, and thus collisions are minimized and parallel movements and alignment are favored (see Erickson, 1978a; Elsdale and Wasoff, 1976). Transformed cells lacking the lateral protrusions do not align as well and criss-cross readily (Erickson, 1978b). This type of contact interaction may align and polarize mesodermal cells (Keller et al., 1989a). We know that the large medial and lateral protrusions move along the surfaces of adjacent cell bodies without contact inhibition and with little criss-crossing (Shih and Keller, 1992b). The small contacts at the anterior and posterior surfaces may be arranged so that they channel the advance of the large protrusions. To test this idea, we must learn more about how the large protrusions interact with one another and with the small anterior–posterior contacts in the explant.

The main protrusive activity is directed medially and laterally, parallel to the axis of intercalation, and we believe that this is the primary force-generating event. It is not clear exactly *how* protrusive activity organized in this fashion produces intercalation. Specifically, if the cells move equally well in medial and lateral directions, they should be able to change places without net extension and convergence. The cells may be biased either in the frequency, the timing, or in the effectiveness of their protrusive activity such that they crowd in one direction or the other. However, we do not yet have conclusive data on these parameters in the early stages of mediolateral intercalation before notochord formation.

Leaving aside for a moment the issue of mediolateral bias, our current hypothesis is that the mediolaterally directed traction pulls the cells between one another, generating a wedging action that pushes neighboring cells apart and extends the tissue at the expense of width. Deep cells should be under tension in the mediolateral direction and under compression in the anteroposterior direction if this is so. Tension could be a self-reinforcing property by increasing polarization of protrusive activity. In some cell types, protrusive activity is inhibited at cell margins under tension (see Kolega, 1986). If this applies to intercalating mesodermal cells, protrusive activity at the narrow ends would increase tension on the long sides, lessening the probability of protrusive activity there, and putting a positive feedback system into effect.

3. Boundary Polarization of Protrusive Activity

The invasive protrusive activity of intercalating cells is inhibited on contact with the notochord–somite boundary: when one end of the previously bipolar deep cell contacts the boundary, it becomes monopolar and directs all protrusive activity inward, away from the boundary (Keller et al., 1989a). Inwardly directed traction tends to pull other cells to the boundary, where they too would be

2. Cellular Basis of Amphibian Gastrulation

"captured" by its inhibitory effect. Accumulation of cells at the boundary would expand its area, in this case, an increase channeled primarily into length (Keller et al., 1989a). This idea of capture of cells at a boundary has been invoked to explain why the notoplate–neural plate boundary is essential for extension of the neural anlagen of the newt (Jacobson et al., 1986). In the late neurula, invasive protrusive activity spreads to the top and bottom edges of notochord cells, and this functions in the spreading of the cells circumferentially to form the pizza-slice shape of the cells. This shape and pattern of connections between cells channels the forces generated by the subsequent vacuolation into continued extension, straightening, and stiffening of the notochord in the tailbud stages (Koehl et al., 1990).

Computer simulations of cell behavior (Weliky and Oster, 1991; Weliky et al., 1991) have been used to evaluate the mechanical consequences of the cell behavior and cell interactions seen in video recordings. Simulations show that inhibition of protrusive activity at the notochord boundary will produce elongation of the simulated notochord and appropriate behavior of the boundary cells, but not the expected elongation and alignment of internal notochord cells. When intrinsic polarization is added, the internal cells elongate, as seen in real notochords, but do not align. Finally, addition of contact inhibition of protrusive activity to the previous rules produces all major features of notochord morphogenesis. These investigations support the idea that cumulative action of several observed intrinsic motile properties, and cell interactions known to occur among other cell types, can account for the observed mediolateral intercalation of cells (Weliky and Oster, 1991; Weliky et al., 1991). Whether or not these contact interactions occur *in vivo* must be answered experimentally.

4. Simultaneous Migration and Mediolateral Intercalation of Mesodermal Cells

The converging and extending axial and paraxial mesodermal cell population may be able to participate in mediolateral intercalation and migration at the same time. The migratory mesoderm at the leading edge of the mesodermal mantle does not converge or extend and does not show mediolateral intercalation (Keller and Tibbetts, 1989). However, the posterior, axial, and paraxial mesodermal cells participating in convergence and extension also participate in migration on the BCR, although they show somewhat different behavior (Winklbauer and Nagel, 1991). How can they do two things at once? The answer may lie in the observation that the most powerful participants in mediolateral intercalation are the cells near the surface, immediately beneath the endodermal epithelium, whose role in organizing deep cells for convergence and extension we described above (Shih and Keller, 1992a,b). In contrast, the cells that would migrate on the BCR are farthest from the epithelium, several cell layers away.

The deepest of these cells, the ones next to the overlying BCR, appear to participate in migration. They form protrusions that attach to the BCR (Nakatsuji, 1975b; Keller and Schoenwolf, 1977; Winklbauer et al., 1991), and they migrate as individual cells on the BCR (Keller and Hardin, 1987) and in culture, although with less intrinsic persistence than the anterior migrating mesodermal cells (Winklbauer, 1990). Thus the deepest mesodermal cells, farthest from the influence of the superficial layer, may be free to engage in active migration while those next to the overlying endodermal epithelium, and most strongly under its influence, participate most strongly in active mediolateral intercalation. As pointed out above, in the section on bottle cells, the leading edge mesoderm is not associated with the endodermal epithelium at all.

F. Function of Convergence and Extension in Gastrulation

1. Function of Convergence and Extension in *Xenopus* Gastrulation

The geometry and mechanics of how convergence and extension produce involution and blastopore closure in the spherical geometry of the gastrula have been discussed elsewhere in detail (Keller, 1986; Wilson and Keller, 1991; Keller et al., 1991a), and are briefly summarized here. In cultured sandwich explants, convergence and extension are displayed serially, first in the IMZ and then in the NIMZ. *In vivo*, they act in parallel—the IMZ converges and extends primarily in the postinvolution position and the NIMZ converges and extends on the outside. At the outset, the vegetal edge of the IMZ is is turned inward, initiating involution, probably by a combination of forces mentioned above. The formation of bottle cells early in gastrulation probably applies a bending force and may aid movement of the leading edge mesoderm upward, although this has not yet been tested directly (see above and Hardin and Keller, 1988). Second, the head mesoderm probably gets a purchase on the blastocoel roof, and pulls on the leading edge of the converging and extending mesoderm. Likewise, this is consistent with the appearance of cell behaviors and movements but has not been tested directly.

After initiation of involution, radial intercalation begins in the converging and extending mesoderm, producing extension in the IMZ. This extension aids initial involution (Wilson and Keller, 1991). Only after it has extended somewhat and its leading edge has involuted does the IMZ show mediolateral intercalation (also described below), which produces convergence and *extension* just at the lip and in the postinvolution position. The increased length of the axial mesoderm is taken up in two ways: it pushes backward across the vegetal region and anteriorly toward the animal pole, where its movement is aided by the migration of the head mesodermal cells in the same direction. Meanwhile, preinvolution IMZ and all the NIMZ are continuing to extend, by virtue of their continued radial intercalation, which tends to push the remaining uninvoluted IMZ over the blastoporal lip. As the latter passes over the lip, it begins mediolateral intercalation and thus converges and extends. Convergence narrows the circumference of the marginal zone, putting a constriction force on the marginal zone just outside of the blastoporal lip, aiding involution and squeezing the blastopore shut (see Keller, 1986; Wilson and Keller, 1991). Thus, reducing the circumference of transverse hoops across the dorsal IMZ would contribute to its involution, and the contribution of this constriction to involution must be large. In embryos lacking all the animal cap and the NIMZ, the IMZ can still involute, at least as well or better than controls, as it constricts by convergence (Keller and Jansa, 1992). However, if the converging tissue has no lateral continuity, the dorsal sector should break free and extend into free space without involuting. In explants, the convergence movements show such lateral continuity: the dorsal sector pulls itself toward the ventral side (Keller and Danilchik, 1988). When lateral continuity is disrupted in whole embryos, the dorsal sector extends into space and involution does not occur (Keller, 1984; Schechtman, 1942).

2. Gastrulation in *Xenopus* without Bottle Cells and without Blastocoel Roofs

The function of convergence and extension in *Xenopus* gastrulation is uniquely fundamental: interdiction of all the other identified regional processes does not stop the major movements of gastrulation in *Xenopus*. Removal of the bottle cells truncated the periphery of the archenteron, but gastrulation continues (Keller, 1981) Removal of the entire blastocoel roof (animal cap) stops both migration of mesodermal cells on the roof and the epiboly of the animal cap itself, but most of gastrulation continues, including involution of the marginal zone, closure of the blastopore, and extension of the axial mesodermal tissues in their proper pattern (Keller et al., 1985a,b; Keller and Jansa, 1992; also see Holtfreter, 1933). In contrast, manipulation of the marginal zone, particularly the deep mesodermal region, results in failure of all movements, including involution, blastopore closure, and elongation and narrowing of the axial tissues (Keller, 1981, 1984). The marginal zone plays an important role in accomplishing all these movements. This role does not depend on the presence of either the blastocoel roof, its matrix, or the bottle cells. Accordingly, it must involve only the autonomous convergence and extension movements of the IMZ.

3. Function of Convergence and Extension in Other Amphibians

Convergence and extension movements contribute to the gastrulation of *Hyla regilla* (Schechtman, 1942) and *Ceratophrys ornata* (S. Purcell, preliminary results, 1991). But in other amphibians, particularly urodeles, there are no well-characterized examples of autonomous convergence and extension in the gastrula. Explants of *Pleurodeles* failed to show gastrula extension (Shi *et al.*,

1987), although considerable elongation and narrowing of vital dye marks occurs during urodele gastrulation (Vogt, 1929). If autonomous convergence and extension do not account for these movements, what does?

V. The Ignored Movement: Epiboly

A. Epiboly in *Xenopus*

In *Xenopus*, the animal cap spreads monotonically throughout blastulation and gastrulation (Keller, 1978). The superficial epithelial cells divide and spread and the deep cells intercalate along the radius of the embryo (radial intercalation), forming a thinner array of greater area. These behaviors have been quantitated by time-lapse cinemicrography and by morphometrics of scanning electron microscopic data (Keller, 1978, 1980). During the whole process there is no intercalation of deep cells into the superficial epithelial layer, and epithelial cells do not crawl into the deep layers.

B. Epiboly in Urodeles

In the urodeles that have been studied, epiboly is thought to occur by a more extreme version of the above process, in which deep cells also intercalate into the outer epithelial layer. This idea was set forth by Holtfreter (1943b) and is consistent with the observation that many of these amphibians have a single-layered animal cap by the early neurula stage, and that their animal cap pigmentation lightens in "salt and pepper" fashion during epiboly, presumably by intercalation of unpigmented deep cells into the surface layer (see Keller, 1986). However, no one has directly observed and quantitated the intercalation of deep cells into the overlying epithelium. If this egression does happen, it raises interesting questions about how mesenchymal cells invade epithelial cell sheets and finally become epithelial themselves. These putative egressions should be documented and, if they exist, be investigated further.

C. Evidence for Autonomy of Epiboly

Evidence for the autonomy of animal cap epiboly is rather weak. Epiboly was thought to be autonomous because isolated explants of animal caps are corrugated, an appearance that suggests that the epithelial sheet has expanded and thrown itself into folds lacking any process, such as involution, to absorb the excess (see Schechtman, 1942). There is some other evidence that epiboly is autonomous, but it is indirect (Keller, 1980). The putatively expanding animal caps of isolates should be recorded in time-lapse mode with sufficient resolution to allow cell tracing and quantitative morphometrics (Keller, 1978), to determine whether or not these corrugated structures represent expansion or, perhaps more likely, collapse and thickening. In a number of such explants of *Xenopus*, video recordings show contraction of the cell apices and thickening of the explant rather than expansion and thinning (R. Keller, unpublished results, 1985). If epiboly is autonomous, it may be driven by active intercalation of deep cells between one another, perhaps by stabilizing and thus maximizing deep contact with the undersurface of the epithelial layer (Keller, 1980).

VI. Involution

In a mechanical sense, involution appears to be a product of other behaviors and is a process different from migration, convergence/extension, and invagination. Several forces seem to be at work in producing involution. Early on, bottle cell formation could contribute by bending the marginal zone. Migration of the leading edge mesoderm probably contributes as much or more than the bottle cells, and the shearing the mesoderm produces at its boundary with the BCR, by the directional migration described above, would aid involution. Constriction forces generated by convergence probably play an increasingly strong role as gastrulation proceeds, by the hoop stress mechanism outline above, and this alone is sufficient to produce involution in animals lacking the entire BCR, including the NIMZ (Keller *et al.*, 1985a,b; Keller and Jansa, 1992). Last, the relative rates of extension of the NIMZ and IMZ probably contribute to regulating the rate of involution.

However, specific changes in cell behavior may occur in relation to the sequential involution of the mesoderm, and these changes may function in transformation of the mesoderm from occupying a static position in the BCR to a cell stream migrating across the BCR. Reintegration of involuted mesodermal cells back into the BCR must be prevented. That is, a stable BCR–mesoderm interface must be maintained. When a piece of preinvolution mesoderm is placed on the inner surface of the BCR, it readily reintegrates into the BCR cell layer. But involuted mesoderm under the same conditions does not mix with BCR cells; rather it stays on top of the BCR explant (R. Keller, unpublished results, 1979). The ECM of the BCR does not isolate the BCR and mesodermal cell populations (R. Winklbauer and R. Keller, unpublished results, 1991). We do not know the cell behavior underlying this sorting out phenomenon. Another interesting observation is that involuted mesoderm cells, but not preinvolution mesoderm cells that are still part of the the BCR, are motile on FN *in vitro* (Winklbauer, 1990; see above). This implies a change in the motile behavior of mesodermal cells during involution. It is hard to conceive how the mechanical forces and conditions related to involution, as discussed above, and the changes in cell behavior expected to accompany involution, are coordinated to move the mesoderm over the blastopore lip.

VII. Some Other Issues to Be Explored

A. Evolution and the Design of Gastrulation: Variation among Amphibians

As Ballard (1976) has pointed out, gastrulation is not a conservative process that is difficult to modify. The above analysis of amphibian gastrulation shows it to involve similar gross movements but to be variable in the sense that different combinations of a small number of basic cell behaviors are used in each species. These may not vary according to order. Indeed, thus far *Xenopus* seems to be the odd species.

1. Determining the Number of Basic Patterns

We should explore more amphibians to determine how many basic patterns of gastrulation there are. Are there any other amphibians sharing the *Xenopus* pattern of having only deep mesoderm? Is precocious, gastrula-stage convergence and extension widely represented among anurans or only in a few species? Is the simultaneous ingression of notochordal and somitic mesoderm through multiple zones of ingression found in some anurans also found in any urodeles, or do urodeles show only the sequential pattern of somitic mesoderm ingression followed by notochordal ingression. How common is the egression of deep cells into the epithelial layer that supposedly occurs in some urodeles? Because we know nearly nothing about the third order of amphibians, the caecilians, all the above questions apply to them. Rather than pursuing these issues along ordinal lines, it might be useful to pursue them along some other parameter, such as egg size, or time required for development, which may affect the design of gastrulation machinery.

2. Diversity of Cellular Behaviors

So far we have examples of cell migration, cell intercalation of two types (radial and mediolateral), change in cell shape among epithelial cells, and ingression of epithelial cells as primary cell behaviors in gastrulation. Are some of the behaviors related to others? For example, is bottle cell formation related to ingression and mesodermal cell migration to mediolateral intercalation, and if so, at what level? Are there amphibians with additional cellular behaviors in their repertoire? Does oriented cell division and local growth come into play in any system? Does cell shape change other than bottle cell formation come into play as it does in imaginal disk evagination (Condic et al., 1991)? Do secretion and swelling of extracellular matrix come into play in some species of amphibia, as they do in echinoderms (Lane, 1990, 1992)?

B. Significance of Cytomechanics and Cell Interactions in Populations

Morphogenetically specific behaviors emerge as the cells become part of a cell population, and their behavior is constrained by cell interactions. Examples include the effect of the epithelial cells on deep cell intercalation (Shih and Keller, 1992a), induction of convergence and extension of the NIMZ by planar contact with the IMZ (Keller et al., 1991c), and the modification of migrating mesodermal cell behavior by cell interactions (Winklbauer et al., 1991). Likewise, the functions of these specific cell behaviors are dependent on biomechanical and geometric context. Notable are the formation of bottle cells (Keller and Hardin, 1988), the bipolar protrusive activity of intercalating mesodermal cells directed perpendicular to the axis of extension and cell displacement (Shih and Keller, 1992b), and the forces developed by swelling of notochord cells being channeled by cell arrangement and by properties of the notochordal sheath (Adams et al., 1990; Koehl et al., 1990). This means that in morphogenesis of cell populations the function of cell behaviors and even the genes and molecules underlying these behaviors will make little sense analyzed outside the context of the mechanics of cell populations.

Fortunately, powerful new techniques are available to analyze cell movements in morphogenetically relevant contexts, in cell populations, including explantation and high-resolution videomicrography, using epiillumination or low light fluorescence microscopy, mentioned above. These methods negate one of the chief disadvantages of amphibian material (their opacity) and play to their chief strength—they can be explanted and grafted in ways unmatched by perhaps any other organism, and now visualized as well as any gastrulating system.

Acknowledgments

We thank John Shih, Paul Wilson, Amy Sater, Kurt Johnson, and Roger Pedersen for their insightful comments and useful suggestions.

References

Adams, D., Keller, R., Koehl, M. A. R. (1990). The mechanics of notochord elongation, straightening, and stiffening in the embryo of *Xenopus laevis*. *Development (Cambridge, UK)* **110**, 115–130.

Akiyama, S. K., Nagata, K., and Yamada, K. M. (1990). Cell surface receptors for extracellular matrix components. *Biochim. Biophys. Acta* **1031**, 91–110.
Angres, B., Muller, A. H. J., Kellermann, J., and Hausen, P. (1991). Differential expression of two cadherins in *Xenopus laevis*. *Development (Cambridge, UK)* **111**, 829–844.
Baker, P. (1965). Fine structure and morphogenetic movements in the gastrula of the treefrog, *Hyla regilla*. *J. Cell Biol.* **24**, 95–116.
Baker, P. and Schroeder, T. E. (1967). Cytoplasmic filaments and morphogenetic movement in the amphibian neural tube. *Dev. Biol.* **15**, 432–450.
Ballard, W. (1976). Problems of gastrulation: Real and verbal. *BioSciences* **26**, 36–39.
Betchaku, T., and Trinkaus, J. P. (1986). Programmed endocytosis during epiboly of *Fundulus heteroclitus*. *Am. Zool.* **26**, 193–199.
Black, S. (1989). Experimental reversal of the normal dorsal-ventral timing of blastopore formation does not reverse axis polarity in *Xenopus laevis* embryos. *Dev. Biol.*, **134**, 376–381.
Boocock, C. A. (1989). Unidirectional displacement of cells in fibrillar matrices. *Development (Cambridge, UK)* **107**, 881–890.
Boucaut, J.-C., and Darribere, T. (1983a). Presence of fibronectin during early embryogenesis in the amphibian *Pleurodeles waltlii*. *Cell Differ.* **12**, 77–83.
Boucaut, J.-C., and Darribere, T. (1983b). Fibronectin in early amphibian embryos: Migrating mesodermal cells contact fibronectin established prior to gastrulation. *Cell Tissue Res.* **234**, 135–145.
Boucaut, J.-C., Darrribere, T., Boulekbache, H., and Thiery, J.-P. (1984a). Prevention of gastrulation but not neurulation by antibodies to fibronectin in amphibian embryos. *Nature (London)* **307**, 364–367.
Boucaut, J.-C., Darribere, T., Poole, T.J., Aoyama, H., Yamada, K.M., and Thiery, J.-P. (1984b). Biologically active synthetic peptides as probes of embryonic development: A competitive peptide inhibitor of fibronectin function inhibits gastrulation in amphibian embryos and neural crest cell migration in avian embryos. *J. Cell Biol.* **99**, 1822–1830.
Boucaut, J.-C., Darribere, T., Shi, D.-L., Boulekbache, H., Yamada, K.M., and Thiery, J.-P. (1985). Evidence for the role of fibronectin in amphibian gastrulation. *J. Embryol. Exp. Morphol.* **89**(Suppl.), 211–227.
Brady, R. C., and Hilfer, S. R. (1982). Optic cup formation: A calcium regulated process. *Proc. Natl. Acad. Sci. U.S.A.* **79**, 5587–5591.
Brick, I., and Weinberger, C. (1984). Electrophoretic properties, cell surface morphology and calcium in amphibian gastrulation. *Am. Zool.* **24**, 629–647.
Brickman, M. (1990). Isolation and characterization of glycosaminoglycans from *Xenopus laevis* gastrula stage embryos. *J. Cell Biol.* **111**, 484a.
Brun, R. B., and Garson, J. A. (1984). Notochord formation in the Mexican salamander (*Ambystoma mexicanum*) is different from notochord formation in *Xenopus laevis*. *J. Exp. Zool.* **229**, 235–240.
Burnside, B. (1971). Microtubules and microfilaments in newt neurulation. *Dev. Biol.* **26**, 416–441.
Burnside, B. (1973). Microtubules and microfilaments in amphibian neurulation. *Am. Zool.* **13**, 989–1006.
Buck, C. A., and Horwitz, A. F. (1987). Cell surface receptors for extracellular matrix molecules. *Annu. Rev. Cell Biol.* **3**, 179–205.
Carter, S. B. (1965). Principles of cell motility: The direction of cell movement and cancer invasion. *Nature (London)* **208**, 1183–1187.
Condic, M., Fristrom, D., and Fristrom, J. W. (1991). Apical cell shape changes during *Drosophila* imaginal leg disc elongation: A novel morphogenetic mechanism. *Development (Cambridge, UK)* **111**, 23–33.
Cooke, J. (1972). Properties of the primary organization field in the embryo of *Xenopus*. III.

Retention of polarity in cell groups excised from the region of the early organizer. *J. Embryol. Exp. Morphol.* **28,** 47–56.

Darribere, T., Bouchet, D., Lacroix, J.-C., and Boucaut, J.-C. (1984). Fibronectin synthesis in the early embryogenesis of *Pleurodeles waltlii*. *Cell Differ.* **49,** 171–177.

Darribere, T., Boulekbache, H., Shi, D.-L., and Boucaut, J.-C. (1985). Immunoelectron microscopic study of fibronectin in gastrulating amphibian embryos. *Cell Tissue Res.* **239,** 79–80.

Darribere, T., Riou, J.-F., Shi, D.-L., Delarue, M., and Boucaut, J.-C. (1986). Synthesis and distribution of laminin-related polypeptides in early amphibian embryos. *Cell Tissue Res.* **246,** 45–51.

Darribere, T., Yamada, K. M., Johnson, K. E., and Boucaut, J.-C. (1988). The 140-kDa fibronectin receptor complex is required for mesodermal cell adhesion during gastrulation in the amphibian *Pleurodeles waltlii*. *Dev. Biol.* **126,** 182–194.

Darribere, T., Guida, K., Larjava, H., Johnson, K. E., Yamada, K. M., Thiery, J.-P., and Boucaut, J.-C. (1990). *In vivo* analyses of integrin b1 subunit function in fibronectin matrix assembly. *J. Cell Biol.* **110,** 1813–1823.

DeSimone, D. W., and Hynes, R. O. (1988). *Xenopus laevis* integrins. Structural conservation and evolutionary divergence of b subunits. *J. Biol. Chem.* **263,** 5333–5340.

DiPasquale, A. (1975). Locomotory activity of epithelial cells in culture. *Exp. Cell Res.* **94,** 191–215.

Doucet-de Bruine, M. H. M. (1973). Blastopore formation in *Ambystoma mexicanum*. *Wilhelm Roux Arch. Entwicklungsmech. Org.* **173,** 136–163.

Dunn, G. (1982). Contact guidance of cultured tissue cells: A survey of potentially relevant properties of the substratum. *In* "Cell Behavior: A Tribute to Michael Abercrombie" (R. Bellairs, A. Curtis, and G. Dunn, eds.), pp. 247–280, Cambridge Univ. Press, London.

Dunn, G., and Heath, J. (1976). A new hypothesis of contact guidance in tissue cells. *Exp. Cell Res.* **101,** 1–14.

Elsdale, T., and Wasoff, F. (1976). Fibroblast cultures and dermatoglyphs. The topology of two planar patterns. *Wilhelm Roux's Arch. Dev. Biol.* **180,** 121–147.

Erickson, C. A. (1978a). Analysis of the formation of parallel arrays of BHK cells *in vitro*. *Exp. Cell Res.* **115,** 303–315.

Erickson, C. A. (1978b). Contact behavior and pattern formation of BHK and polyoma virus-transformed BHK fibroblasts in culture. *J. Cell Sci.* **33,** 53–84.

Ettensohn, C. (1985). Gastrulation in the sea urchin is accompanied by the rearrangement of invaginating epithelial cells. *Dev. Biol.* **112,** 383–390.

Fey, J., and Hausen, P. (1990). Appearance and distribution of laminin during development of *Xenopus laevis*. *Differentiation (Berlin)* **42,** 144–152.

ffrench-Constant, C., and Hynes, R. O. (1988). Patterns of fibronectin gene expression and splicing during cell migration in chicken embryos. *Development (Cambridge, UK)* **104,** 369–382.

ffrench-Constant, C., and Hynes, R. O. (1989). Alternative splicing of fibronectin is temporally and spatially regulated in the chicken embryo. *Development (Cambridge, UK)* **106,** 375–388.

ffrench-Constant, C., Van De Water, L., Dvorak, H. F., and and Hynes, R. O. (1989). Reappearance of an embryonic pattern of fibronectin splicing during wound healing in the adult rat. *J. Cell Biol.* **109,** 903–914.

Fink, R., and McClay, D. (1985). Three cell recognition changes accompany the ingression of sea urchin primary mesenchyme cells. *Dev. Biol.* **107,** 66–74.

Fristrom, D. (1988). The cellular basis of epithelial morphogenesis. A review. *Tissue Cell* **20,** 645–690.

Gerhart, J., and Keller, R. K. (1986). Region-specific cell activities in amphibian gastrulation. *Annu. Rev. Cell Biol.* **2,** 201–229.

Gibbins, J. R., Tilney, L. G., and Porter, K. R. (1969). Microtubules in the formation and development of the primary mesenchyme in *Arbacia punctulata*. I. The distribution of microtubules. *J. Cell Biol.* **41**, 201–226.

Green, H., Goldberg, B., Schwartz, M., and Brown, D. D. (1968). The synthesis of collagen during the development of *Xenopus laevis*. *Dev. Biol.* **18**, 391–400.

Hardin, J. (1990). Context-sensitive cell behaviors during gastrulation. *In* "Control of Morphogenesis by Specific Cell Behaviors" (R. Keller and D. Fristrom, eds.). Seminars in Developmental Biology **1**, 335–345.

Hardin, J., and Keller, R. (1988). The behavior and function of bottle cells during gastrulation of *Xenopus laevis*. *Development (Cambridge, UK)* **103**, 211–230.

Hayman, E. G., Pierschbacher, M.D., and Rouslahti, E. (1985). Detachment of cells from culture substrate by soluble fibronectin peptides. *J. Cell Biol.* **100**, 1948–1954.

Höglund, L.R., and Låvtrup, S. (1976). Changes in acid mucopolysaccharides during development of the frog *Rana temporaria*. *Acta Embryol. Exp.* **1**, 63–79.

Holtfreter, J. (1933). Die totale Exogastrulation, eine Selbstablösung des Ektoderms vom Entomesoderm. *Wilhelm Roux Arch. Entwicklungsmech. Org.* **129**, 669–793.

Holtfreter, J. (1943a). Properties and function of the surface coat in amphibian embryos. *J. Exp. Zool.* **93**, 251–323.

Holtfreter, J. (1943b). A study of the mechanics of gastrulation. Part I. *J. Exp. Zool.* **94**, 261–318.

Holtfreter, J. (1944). A study of the mechanics of gastrulation. Part II. *J. Exp. Zool.* **95**, 171–212.

Holtfreter, J. (1946). Structure, motility and locomotion in isolated embryonic amphibian cells. *J. Morphol.* **72**, 27–62.

Humphries, M. J. (1990). The molecular basis and specificity of integrin–ligand interactions. *J. Cell Sci.* **97**, 585–592.

Humphries, M. J., Akiyama, S. K., Komoriya, A., Olden, K. and Yamada, K. M. (1986). Identification of an alternatively spliced site in human plasma fibronectin that mediates cell-type specific adhesion. *J. Cell Biol.* **103**, 2637–2647.

Humphries, M. J., Komoriya, A., Akiyama, S. K., Olden, K., and Yamada, K. M. (1987). Identification of two distinct regions of the type III connecting segment of human plasma fibronectin that promote cell type-specific adhesion. *J. Biol. Chem.* **262**, 6886–6892.

Hynes, R. O. (1985). Molecular biology of fibronectin. *Annu. Rev. Cell Biol.* **1**, 67–90.

Hynes, R. O. (1987). Integrins: A family of cell surface receptors. *Cell (Cambridge, Mass.)* **48**, 549–554.

Jacobson, A. (1981). Morphogenesis of the neural plate and tube. *In* "Morphogenesis and Pattern Formation" (T. G. Connolly, L. Brinkley, and B. Carlson, eds.), pp. 223–263. Raven, New York.

Jacobson, A., and Gordon, R. (1976). Changes in the shape of the developing vertebrate nervous system analyzed experimentally, mathematically, and by computer simulation. *J. Exp Zool.* **197**, 191–246.

Jacobson, A., Oster, G., Odell G., and Cheng, L. (1986). Neurulation and the cortical tractor model for epithelial folding. *J. Embryol. Exp. Morphol.* **96**, 19–49.

Johnson, K. E. (1970). The role of changes in cell contact behavior in amphibian gastrulation. *J. Exp. Zool.* **175**, 391–428.

Johnson, K. E. (1972). The extent of cell contact and the relative frequency of small and large gaps between presumptive mesodermal cells in normal gastrulae of *Rana pipiens* and the arrested gastrulae of the *Rana pipiens* × *Rana catesbeiana* hybrid. *J. Exp. Zool.* **179**, 227–238.

Johnson, K. E. (1976a). Ruffling and locomotion in *Rana pipiens* gastrula cells. *Exp. Cell Res.* **101**, 71–77.

Johnson, K. E. (1976b). Circus movements and blebbing locomotion in dissociated embryonic cells of an amphibian, *Xenopus laevis*. *J. Cell Sci.* **22**, 575–583.

Johnson, K. E. (1977a). Changes in the cell coat at the onset of gastrulation in *Xenopus laevis* embryos. *J. Exp. Zool.* **199**, 137–142.

Johnson, K. E. (1977b). Extracellular matrix synthesis in blastula and gastrula stages of normal and hybrid frog embryos. I. Toluidine blue and lanthanum staining. *J. Cell Sci.* **25**, 313–322.

Johnson, K. E. (1977c). Extracellular matrix synthesis in blastula and gastrula stages of normal and hybrid frog embryos: II. Autoradiographic observations on the sites of synthesis and mode of transport of galactose- and glucosamine-labelled materials. *J. Cell Sci.* **25**, 323–334.

Johnson, K. E. (1977d). Extracellular matrix synthesis in blastula and gastrula stages of normal and hybrid frog embryos: III. Characterization of galactose- and glucosamine-labelled materials. *J. Cell Sci.* **25**, 335–356.,

Johnson, K. E. (1978). Extracellular matrix synthesis in blastula and gastrula stages of normal and hybrid frog embryos. IV. Biochemical and autoradiographic observations on fucose-, glucose, and mannose-labelled materials. *J. Cell Sci.* **32**, 109–136.

Johnson, K. E. (1984). Glycoconjugate synthesis during gastrulation in *Xenopus laevis*. *Am. Zool.* **24**, 605–614.

Johnson, K. E. (1985). Frog gastrula cells adhere to fibronectin-sepharose beads. In "Molecular Determinants of Animal Form" (G. M. Edelman, ed.) pp. 271–292. Alan R. Liss, New York

Johnson, K. E. (1986). Transplantation studies to investigate mesoderm-ectoderm adhesive cell interactions during gastrulation. *J. Cell Sci.* **82**, 99–117.

Johnson, K. E. and Silver, M. H. (1986). Cells from *Xenopus laevis* gastrulae adhere to fibronectin-sepharose beads and other lectin coated beads. *Scanning Electron Microsc.* **2**, 671–678.

Johnson, K. E., Silver, M. H., and Kelly, R. O. (1979). Scanning electron microscopy of changes in cell shape and extracellular matrix in normal and interspecific hybrid frog embryos. *Scanning Electron Microsc.* **3**, 517–526, 536.

Johnson, K. E., Darribere, T., and Boucaut, J.-C. (1990). Cell adhesion to extracellular matrix in normal *Rana pipiens* gastrulae and in arrested hybrid gastrulae *Rana pipiens* (m.) and *Rana esculenta* (f.). *Dev. Biol.* **138**, 86–99.

Karfunkel, P. (1971). The role of microtubules and microfilaments in neurulation in *Xenopus*. *Devel. Biol.* **25**, 30–56.

Karfunkel, P. (1977). SEM analysis of amphibian mesodermal migration. *Wilhelm Roux's Arch. Dev. Biol.* **181**, 31–40.

Keller, R. E. (1975). Vital dye mapping of the gastrula and neurula of *Xenopus laevis*. I. Prospective areas and morphogenetic movements of the superficial layer. *Dev. Biol.* **42**, 222–241.

Keller, R. E. (1976). Vital dye mapping of the gastrula and neurula of *Xenopus laevis*. II. Prospective areas and morphogenetic movements of the deep layer. *Dev. Biol.* **51**, 118–137.

Keller, R. E. (1978). Time-lapse cinemicrographic analysis of superficial cell behavior during and prior to gastrulation in *Xenopus laevis*. *J. Morphol.* **157**, 223–248.

Keller, R. E. (1980). The cellular basis of epiboly: An SEM study of deep cell rearrangement during gastrulation in *Xenopus laevis*. *J. Embryol. Exp. Morph.* **60**, 201–234.

Keller, R. E. (1981). An experimental analysis of the role of bottle cells and the deep marginal zone in gastrulation of *Xenopus laevis*. *J. Exp. Zool.* **216**, 81–101.

Keller, R. E .(1984). The cellular basis of gastrulation in *Xenopus laevis*: Active post-involution convergence and extension by medio-lateral interdigitation. *Am. Zool.* **24**, 589–603.

Keller, R. E. (1986). The cellular basis of amphibian gastrulation. In "Developmental Biology: A Comprehensive Synthesis Volume 2: *The Cellular Basis of Morphogenesis*" (L. Browder, ed.), pp. 241–327. Plenum, New York.

Keller, R. E. (1987). Cell rearrangement in morphogenesis. *Zool. Sci.* **4**, 763–779.

Keller, R. E., and Danilchik, M. (1988). Regional expression, pattern and timing of convergence and extension during gastrulation of *Xenopus laevis*. *Development (Cambridge, UK)* **103**, 193–210.

Keller, R. E., and Hardin, J. (1987). Cell behavior during active cell rearrangement: evidence and speculation. *J. Cell Sci.* **8**(Suppl.) 369–393.

Keller, R. E. and Jansa, S. (1992). Gastrulation in *Xenopus* embryos without the blastocoel roof. submitted.

Keller, R. E., and Schoenwolf, G. C. (1977). An SEM study of cellular morphology, contact, and arrangement, as related to gastrulation in *Xenopus laevis*. *Wilhelm Roux's Arch. Dev. Biol.* **181**, 165–182.

Keller, R. E., and Tibbetts, P. (1989). Mediolateral cell intercalation is a property of the dorsal, axial mesoderm of *Xenopus laevis*. *Dev. Biol.* **131**, 539–549.

Keller, R. E., and Trinkaus, J.P. (1987). Rearrangement of enveloping layer cells without disruption of the epithelial permeability barrier as a factor in *Fundulus* epiboly. *Dev. Biol.* **120**, 12–24.

Keller, R., and Winklbauer, R. (1990). The role of the extracellular matrix in amphibian gastrulation. *Semin. Dev. Biol.* **1**, 25–33.

Keller, R. E., Danilchik, M., Gimlich, R., and Shih, J. (1985a). Convergent extension by cell intercalation during gastrulation of *Xenopus laevis*. In "Molecular Determinants of Animal Form" (G. M. Edelman, ed.), pp. 111–141. Alan R. Liss, New York.

Keller, R. E., Danilchik, M., Gimlich, R., and Shih, J. (1985b). The function of convergent extension during gastrulation of *Xenopus laevis*. *J. Embryol. Exp. Morphol.* **89**,(Suppl.) 185–209.

Keller, R., Cooper, M. S., Danilchik, M., Tibbetts, P. and Wilson, P. A. (1989a). Cell intercalation during notochord development in *Xenopus laevis*. *J. Exp. Zool.* **251**, 134–154.

Keller, R. E., Shih, J., and Wilson, P. A. (1989b). Morphological polarity of intercalating deep mesodermal cells in the organizer of *Xenopus laevis* gastrulae, "Proceedings of the 47th Annual Meeting of the Electron Microscopy Society of America," p. 840. San Francisco Press, San Francisco.

Keller, R., Shih, J., and Wilson, P. A. (1991a). Cell motility, control and function of convergence and extension during gastrulation of *Xenopus*. In "Gastrulation: Movements, Patterns, and Molecules" (R. Keller, W. Clark, and F. Griffin, eds.), pp. 101–119. Plenum, New York.

Keller, R., Shih, J., Wilson, P. A., and Sater, A. (1991b). Cell motility and cell interactions during convergence and extension. In "Cell–Cell Interactions in Early Development" (J. C. Gerhart, ed.) (49th Symp. Soc. Dev. Biol.), pp. 31–62. Academic Press, New York.

Keller, R. E., Shih, J., and Sater, A. (1992a). The cellular basis of convergence and extension of the *Xenopus* neural plate. *Dev. Dynam.* **193**, 199–217.

Keller, R. E., Shih, J., Sater, A., and Moreno, C. (1992b). Planar induction of convergence and extension of the neural plate by the organizer of *Xenopus*. *Dev. Dynam.* **193**, 218–234.

Keller, R. (1992c). The patterning and functioning of protrusion activity during convergence and extrusion of the *Xenopus* organiser. *Development (Cambridge, U.K.)*, in press.

King, H. D. (1903). The formation of the notochord in the amphibia. *Biol. Bull.* **4**, 287–300.

Klose, J., and Flickinger, R. A. (1971). Collagen synthesis in frog embryo endodermal cells. *Biochim. Biophys. Acta* **232**. 207–211.

Koehl, M. A. R., Adams, D., and Keller, R. (1990). Mechanical development of the notochord. In *Xenopus* early tail-bud embryos. In "Biomechanics of Active Movement and Deformation of Cells, NATO ASI Series. Volume H 42" (N. Akkas, ed.), pp. 471–485. Springer-Verlag, Berlin.

Kolega, J. (1981). The movement of cell clusters *in vitro*: Morphology and directionality. *J. Cell Sci.*, 15–32.

Kolega, J. (1986). Effects of mechanical tension on protrusive activity and microfilament and intermediate filament organization in an epidermal epithelium moving in culture. *J. Cell Biol.* **102**, 1400–1411.

Komazaki, S. (1982). Morphological changes of the inner surface of the blastocoelic wall before and during gastrulation in the newt, *Cynops pyrrhogaster*. *Dev. Growth Differ.* **24**, 491–499.

Komazaki, S. (1985). Scanning electron microscopy of the extracellular matrix of amphibian gastrulae by freeze-drying. *Dev. Growth Differ.* **27**, 57–62.

Komazaki, S. (1986a). Accumulation and distribution of extracellular matrix as revealed by scanning electron microscopy in freeze-dried newt embryos before and during gastrulation. *Dev. Growth Differ.* **28**, 285–292.

Komazaki, S. (1986b). Extracellular matrix in the blastocoel of newt gastrula: Its effects on dissociated embryonic cells and some aspects of its biochemical nature. *Dev. Growth Differ.* **28**, 293–301.

Komazaki, S. (1988). Factors related to the initiation of cell migration along the inner surface of the blastocoelic wall during gastrulation. *Cell Differ.* **24**, 25–32.

Kosher, R. A., and Searls, R. L. (1973). Sulfated mucopolysaccharide synthesis during the development of *Rana pipiens*. *Dev. Biol.* **32**, 50–68.

Kubota, H. Y., and Durston, A. J. (1978). Cinematographical study of cell migration in the opened gastrula of *Ambystoma mexicanum*. *J. Embryol. Exp. Morphol.* **44**, 71–80.

Lane, C. (1990). Precocious invagination of an epithelium is accompanied by region-specific secretion of extracellular matrix. *J. Cell Biol.* **111**, 19a.

Lane, C. (1992). Apically-directed secretion of extracellular matrix drives epithelial invagination during gastrulation. submitted.

Lee, G., Hynes, R., and Kirschner, M. (1984). Temporal and spatial regulation of fibronectin in early *Xenopus* development. *Cell (Cambridge, Mass.)* **36**, 729–740.

Lee, H.-Y., and Nagele, R. G. (1985). Studies on the mechanism of neurulation in the chick: Interrelationship of contractile proteins, microfilaments, and the shape of neuroepithelial cells. *J. Exp. Zool.* **235**, 205–215.

Lee, H., Nagele, R., and Karasanyi, N. (1977). Inhibition of neural tube closure by ionophore A23187 in chick embryos. *Experientia* **34**, 518–520.

Lee, H. Y., Kosciuk, M. C., Nagele, R. G., and Rosen, F. J. (1983). Studies on the mechanism of neurulation in the chick: Possible involvement of myosin in elevation of neural folds. *J. Exp. Zool.* **225**, 449–457.

Lewis, W. (1947). Mechanics of invagination. *Anat. Rec.* **97**, 139–156.

Lewis, W. (1952). Gastrulation of *Ambystoma puntatum*. *Anat. Rec.* **112**, 473.

Löfberg, J. (1974). Apical surface topography of invaginating and non-invaginating cells. A scanning-transmission study of amphibian neurulae. *Dev. Biol.* **36**, 311–329.

Löfberg, J., and Jacobson, C.O. (1974). Effects of vinblastine sulphate, cholchicine, and guanosine phosphate on cell morphogenesis during amphbian neurulation. *Zoon* **2**, 85–98.

Lundmark, C. (1986). Role of bilateral zones of ingressing superficial cells during gastrulation of *Ambystoma mexicanum*. *J. Embryol. Exp. Morphol.* **97**, 47–62.

McCarthy, J. B., Chelberg, M. K., Mickelson, D. J., and Furcht, L. T. (1988). Localization and chemical synthesis of fibronectin peptides with melanoma adhesion and heparin binding activities. *Biochemistry* **27**, 1380–1388.

Malacinski, G., Allis, C. D., and Chung, H. M. (1974). Correction of developmental abnormalities resulting from localized ultraviolet irradiation of an amphibian egg. *J. Exp. Zool.* **189**, 249–254.

Malacinski, G. M., Brothers, A. J., and Chung, H.-M. (1977). Destruction of components of the neural induction system of the amphibian egg with ultraviolet irradiation. *Dev. Biol.* **56**, 24–39.

Martin, G. R., and Timpl, R. (1987). Laminin and other basement membrane components. *Annu. Rev. Cell Biol.* **3**, 57–85.

Nagele, R. G., and Lee, H.-Y. (1980). Studies on the mechanism of neurulation in the chick:

Microfilament-mediated changes in cell shape during uplifting of neural folds. *J. Exp. Zool.* **213,** 391–398.

Nagele, R. G., Pietrolungo, J. F., and Lee, H. (1981). Studies on the mechanisms of neurulation in the chick: The intracellular distribution of Ca^{2+}. *Experientia* **37,** 304–306.

Nakatsuji, N. (1974). Studies on the gastrulation of amphibian embryos: Pseudopodia in the gastrula of *Bufo bufo japonicus* and their significance to gastrulation. *J. Embryol. Exp. Morphol.* **32,** 795–804.

Nakatsuji, N. (1975a). Studies on the gastrulation of amphibian embryos: Light and electron microscopic observations of a urodele *Cynops pyrrhogaster*. *J. Embryol. Exp. Morphol.* **34,** 669–685.

Nakatsuji, N. (1975b). Studies on the gastrulation of amphibian embryos: Cell movement during gastrulation in *Xenopus laevis* embryos. *Wilhelm. Roux's Arch. Dev. Biol.* **178,** 1–14.

Nakatsuji, N. (1976). Studies on the gastrulation of amphibian embryos: Ultrastructure of the migrating cells of anurans. *Wilhelm Roux's Arch. Dev. Biol.* **180,** 229–240.

Nakatsuji, N. (1984). Cell locomotion and contact guidance in amphibian gastrulation. *Am. Zool.* **24,** 615–627.

Nakatsuji, N. (1986). Presumptive mesoderm cells from *Xenopus laevis* gastrulae attach to and migrate on substrata coated with fibronectin or laminin. *J. Cell Sci.* **86,** 109–118.

Nakatsuji, N., and Johnson, K. E. (1982). Cell locomotion *in vitro* by *Xenopus laevis* gastrula mesodermal cells. *Cell Motil.* **2,** 149–161.

Nakatsuji, N., and Johnson, K. E. (1983a). Conditioning of a culture substratum by the ectodermal layer promotes attachment and oriented locomotion by amphibian gastrula mesodermal cells. *J. Cell Sci.* **59,** 43–60.

Nakatsuji, N., and Johnson, K. E. (1983b). Comparative study of extracellular fibrils on the ectodermal layer in gastrulae of five amphibian species. *J. Cell Sci.* **59,** 61–70.

Nakatsuji, N., and Johnson, K. E. (1984a). Experimental manipulation of a contact guidance system in amphibian gastrulation by mechanical tension. *Nature (London)* **307,** 453–455.

Nakatsuji, N., and Johnson, K. E. (1984b). Ectodermal fragments from normal frog gastrulae condition substrata to support normal and hybrid mesodermal cell migration *in vitro*. *J. Cell Sci.* **68,** 49–67.

Nakatsuji, N., Gould, A. C., and Johnson, K. E. (1982). Movement and guidance of migrating mesodermal cells in *Ambystoma maculatum* gastrulae. *J. Cell Sci.* **56,** 207–222.

Nakatsuji, N., Smolira, M. A., and Wylie, C. C. (1985a). Fibronectin visualized by scanning electron microscopy immunocytochemistry on the substratum for cell migration in *Xenopus laevis*. *Dev. Biol.* **107,** 264–268.

Nakatsuji, N., Hashimoto, K., Hayashi, M. (1985b). Laminin fibrils in newt gastrulae visualized by the immunofluorescent staining. *Dev. Growth Differ.* **27,** 639–643.

Newgreen, D. (1990). Control of the directional migration of mesenchyme cells and neurites. *Semin. Dev. Biol.* **1,** 301–311.

Nieuwkoop, P. (1969a). The formation of the mesoderm in urodelean amphibians. I. Induction by the endoderm. *Wilhelm Roux Arch. Entwicklungsmech. Org.* **162,** 341–373.

Nieuwkoop, P. (1969b). The formation of the mesoderm in urodelean amphibians. II. The origin of the dorsoventral polarity of the mesoderm. *Wilhelm Roux Arch. Entwicklungsmech. Org.* **163,** 298–315.

Nieuwkoop, P, and Florschütz, P. (1950). Quelques caracteres speciaux de la gastrulation et de la neurulation de l'oeuf de *Xenopus laevis*, Daud. et de quelques autres anoures. 1ere partie. Etude descriptive. *Arch. Biol. (1880–1985)* **61,** 113–150.

Nomura, K., Uchida, M., Kageura, H., Shiokawa, K., and Yamana, K. (1986). Cell to cell adhesion systems in *Xenopus laevis*, the South African clawed frog I. Detection of Ca^{++} dependent and independent adhesion systems in adult and embryonic cells. *Dev. Growth Differ.* **28,** 311–319.

Odell, G. M., Oster, G., Alberch, P., and Burnside, B. (1981). The mechanical basis of morphogenesis. I. Epithelial folding and invagination. *Dev. Biol.* **85**, 446–462.
Oshima, H., and Kubota, T. (1985). Cell surface changes during cleavage of newt eggs: Scanning electron microscopic studies. *J. Embryol. Exp. Morphol.* **85**, 21–31.
Otte, A. P., Roy, D., Siemerink, M., Koster, C. H., Hochstenbach, F., Timmermans, A., and Durston, A. J. (1990). Characterization of a maternal type VI collagen in *Xenopus* embryos suggests a role for collagen in gastrulation. *J. Cell Biol.* **111**, 271–278.
Owaribe, K., and Masuda, H. (1982). Isolation and characterization of circumferential microfilament bundles from retinal pigmented epithelial cells. *J. Cell. Biol.* **95**, 310–315.
Perry, M.M. (1975). Microfilaments in the external surface layer of the early amphibian embryo. *J. Embryol. Exp. Morphol.* **33**, 127–146.
Perry, M. M., and Waddington, C.H. (1966). Ultrastructure of the blastopore cells in the newt. *J. Embryol. Exp. Morphol.* **15**, 317–330.
Pierschbacher, M. D., and Ruoslahti, E. (1984). Cell attachment activity of fibronectin can be duplicated by small synthetic fragments of the molecule. *Nature (London)* **309**, 30–33.
Pierschbacher, M. D., Hayman, E. G., and Ruoslahti, E. (1981). Location of the cell attachment site in fibronectin with monoclonal antibodies and proteolytic fragments of the molecule. *Cell (Cambridge, Mass.)* **26**, 259–267.
Purcell, S. (1990). A different type of anuran gastrulation and morphogenesis as seen in *Ceratophrys ornata*. *Am. Zool.* **29**, 85a.
Purcell, S. (1992). A different type of anuran gastrulation and dorsal mesoderm formation in *Ceratophrys ornata*, submitted.
Revel, J.-P., Yip, P., and Chang, L. L. (1973). Cell junctions in the early chick embryo—A freeze etch study. *Dev. Biol.* **35**, 302–317.
Rhumber, L. (1902). Zur Mechanik des Gastrulationsvorganges, insbesondere der Invagination. Eine entwicklungsmechanische Studie. *Arch. Entwicklungsmech. Org.* **14**, 401–476.
Riou, J.-F., Darribere, T., Shi, D.-L., Richoux, V., and Boucaut, J.-C. (1987). Synthesis of laminin-related polypeptides in oocytes, eggs and early embryos of the amphibian *Pleurodeles waltlii*. *Roux's Arch. Dev. Biol.* **196**, 328–332.
Riou, J.-F., Shi, D.-L., Chiquet, M., and Boucaut, J.-C. (1990). Exogenous tenascin inhibits mesodermal cell migration during amphibian gastrulation. *Dev. Biol.* **137**, 305–317.
Sanders, E. J., and Prasad, S. (1989). Invasion of a basement membrane matrix by chick embryo primitive streak cells *in vitro*. *J. Cell Sci.* **92**, 497–504.
Satoh, N., Kageyama, T., and Sirakami, K. I. (1976). Motility of dissociated embryonic cells in *Xenopus laevis*: Its significance to morphogenetic movements. *Dev. Growth Differ.* **18**, 55–67.
Saunders, S., and Bernfield, M. (1988). Cell surface proteoglycan binds mouse mammary epithelial cells to fibronectin and behaves as a receptor for interstitial matrix. *J. Cell Biol.* **106**, 423–430.
Scharf, S. R., and Gerhart, J. C. (1980). Determination of the dorsal–ventral axis in eggs of *Xenopus laevis*: Complete rescue of uv-impaired eggs by oblique orientation before first cleavage. *Dev. Biol.* **79**, 181–198.
Schechtman, A. M. (1942). The mechanics of amphibian gastrulation. I. Gastrulation-producing interactions between various regions of an anuran egg (*Hyla regilla*). *Univ. Calif. Publ. Zool.* **51**, 1–39.
Schoenwolf, G. C., and Powers, M. L. (1987). Shaping of the chick neuroepithelium during primary and secondary neurulation: Role of cell elongation. *Anat. Rec.* **218**, 182–195.
Schoenwolf, G. C., and Smith, J. L. (1990). Mechanisms of neurulation: Traditional viewpoint and recent advances. *Development (Cambridge, UK)* **109**, 243–270.
Schoenwolf, G. C., Folsom, D., and Moe, A. (1988). A reexamination of the role of microfilaments in neurulation in the chick embryo. *Anat. Rec.* **220**, 87–102.

Schroeder, T. (1973). Cell constriction: Contractile role of microfilaments in division and development. *Am Zool.* **13,** 949–960.

Shi, D.-L., Delarue, M., Darribère, T., Riou, J.-F., and Boucaut, J.-C. (1987). Experimental analysis of the extension of the dorsal marginal zone in *Pleurodeles waltl* gastrulae. *Development (Cambridge, UK)* **100,** 147–161.

Shi, D.-L., Darribere, T., Johnson, K. E., and Boucaut, J.-C. (1989). Initiation of mesodermal cell migration and spreading relative to gastrulation in the urodele amphibian *Pleurodeles waltl*. *Development (Cambridge, UK)* **105,** 351–363.

Shi, D.-L., Beetschen, J.-C., Delarue, M., Riou, J.-F., Daguzan, C., and Boucaut, J.-C. (1990). Lithium induces dorsal-type migration of mesodermal cells in the entire marginal zone of urodele amphibian embryos. *Roux's Arch. Dev. Biol.* **199,** 1–13.

Shih, J., and Keller, R. E. (1992a). The epithelium of the dorsal marginal zone of *Xenopus* has organiser properties. submitted.

Shih, J., and Keller, R. E. (1992b). Cell motility driving mediolateral intercalation in explants of *Xenopus*. submitted.

Shih, J., and Keller, R. E. (1992c). Patterns of cell motility in the organiser and dorsal mesoderm of *Xenopus*. submitted.

Smith, J., and Malacinski, G. (1983). The origin of the mesoderm in an anuran, *Xenopus laevis*, and a urodele, *Ambystoma mexicanum*. *Dev. Biol.* **98,** 250–254.

Smith, J. L., and Schoenwolf, G. C. (1988). Role of cell cycle in regulating neuroepithelial cell shape during bending of the chick neural plate. *Cell Tissue Res.* **252,** 491–500.

Smith, J., and Schoenwolf, G. (1989). Notochordal induction of cell wedging in the chick neural plate and its role in neural tube formation. *J. Exp. Zool.* **250,** 49–62.

Smith, J. C., Symes, K., Hynes, R. O., and DeSimone, D. (1990). Mesoderm induction and the control of gastrulation in *Xenopus laevis*: The role of fibronectin and integrins. *Development (Cambridge, UK)* **108,** 229–238.

Spemann, H. (1938). "Embryonic Development and Induction." Yale Univ. Press, New York.

Stern, C., Ireland, G., Herrick, S. E., Gherardi, E., Gray, J., Perryman, M., and Stoker, M. (1990). Epithelial scatter factor and development of the chick embryonic axis. *Development (Cambridge, UK)* **110,** 1271–1284.

Stewart, R. (1990). "The Active Inducing Center of the Embryonic Body Axis in *Xenopus*." Ph.D. Dissertation, University of California, Berkeley.

Sudarwati, S., and Nieuwkoop, P. (1971). Mesoderm formation in the Anuran *Xenopus laevis* (Daudin). *Wilhelm Roux Arch. Entwicklungsmech. Org.* **166,** 189–204.

Takeichi, M. (1988). The cadherins: Cell–cell adhesion molecules controlling animal morphogenesis. *Development (Cambridge, UK)* **102,** 639–655.

Tarin, D. (1971). Histological features of neural induction in *Xenopus laevis*. *J. Embryol. Exp. Morphol.* **26,** 543–570.

Tilney, L. G., and Gibbins, J. R. (1969). Microtubules in the formation and development of the primary mesenchyme in *Arbacia punctulata*. II. An experimental analysis of their role in development and maintainance of cell shape. *J. Cell Biol.* **41,** 227–250.

Trinkaus, J. P. (1976). On the mechanism of metazoan cell movements. In "The Cell Surface in Animal Embryogenesis and Development" (G. Poste and G. Nicolson, eds.), pp. 225–329. North-Holland Publ., New York.

Trinkaus, J. P. (1984). "Cells into Organs; Forces that Shape the Embryo," 2nd Ed. Prentice-Hall, Englewood Cliffs, New Jersey.

Ubbels, G. A., and Hengst, R. T. M. (1978). A cytochemical study of the distribution of glycogen and mucosubstances in the early embryo of *Ambystoma mexicanum*. *Differentiation (Berlin)* **10,** 109–121.

Vogt, W. (1929). Gestaltungsanalyse am Amphibienkeim mit örtlicher Vitalfärbung. II. Teil. Gastrulation und Mesodermbildung bei Urodelen und Anuren. *Wilhelm Roux Arch. Entwicklungsmech. Org.* **120,** 384–706.

Weinberger, C. P., and Brick, I. (1980). Locomotion and adhesion of amphibian gastrula and neurula cells cultured on substrata of varied surface charge. *Exp. Cell Res.* **130,** 251–263.

Weiss, P. (1961). Guiding principles in cell locomotion and cell aggregation. *Exp. Cell Res.*, **8**(Suppl.) 260–281.

Weliky, M., and Oster, G. (1991). Dynamical models for cell rearrangement during morphogenesis. *In* "Gastrulation: Movements, Patterns, and Molecules" (R. Keller, W. Clark, and F. Griffin, eds.), pp. 135–146. Plenum, New York.

Weliky, M., Minsuk, S., Keller, R., and Oster, G. (1991). Notochordmorphogenesis in *Xenopus laevis*: Simulation of cell behavior underlying tissue convergence and extension. *Development* **113,** 1231–1244.

Wilson, P. A. (1990). "The Development of the Axial Mesoderm in *Xenopus laevis*." Ph.D. dissertation, University of California, Berkeley.

Wilson, P. A., and Keller, R. E. (1991). Cell rearrangement during gastrulation of *Xenopus*: Direct observation of cultured explants. *Development (Cambridge, UK)* **112,** 289–305.

Wilson, P. A., Oster, G. and Keller, R. E. (1989). Cell rearrangement and segmentation in *Xenopus*: Direct observation of cultured explants. *Development (Cambridge, UK)* **105,** 155–166.

Winklbauer, R. (1988). Differential interaction of *Xenopus* embryonic cells with fibronectin *in vitro*. *Dev. Biol.* **130,** 175–183.

Winklbauer, R. (1989). A reinvestigation of the role of fibronectin in amphibian gastrulation. *J. Cell Biol.* **107,** 815a.

Winklbauer, R. (1990). Mesodermal cell migration during *Xenopus* gastrulation. *Dev. Biol.* **142,** 155–168.

Winklbauer, R., and Nagel, M. (1991). Directional mesodermal cell migration in the *Xenopus* gastrula. *Dev. Biol.* **148,** 573–589.

Winklbauer, R., and Selchow, A. (1992). Motile behavior and protrusive activity of migratory mesoderm cells from the *Xenopus* gastrula. *Dev. Biol.* **150,** 335–351.

Winklbauer, R., Selchow, A., Nagel, M., Stoltz, C., and Angres, B. (1991). Mesodermal cell migration in the *Xenopus* gastrula. *In* "Gastrulation: Movements, Patterns and Molecules" (R. Keller, W. Clark, and F. Griffin, eds.), pp. 147–168. Plenum, New York.

Yamada, K. M., and Kennedy, D. W. (1984). Dualistic nature of adhesive protein function: Fibronectin and its biologically active peptide fragments can autoinhibit fibronectin function. *J. Cell Biol.* **99,** 29–36.

3
Role of the Extracellular Matrix in Amphibian Gastrulation

Kurt E. Johnson
Department of Anatomy
The George Washington University Medical Center
Washington, D.C. 20037

Jean-Claude Boucaut
Laboratoire de Biologie Experimentale
Université Pierre et Marie Curie (Paris VI)
75005 Paris, France

Douglas W. DeSimone
Department of Anatomy and Cell Biology and
The Molecular Biology Institute
University of Virginia Medical Center
Charlottesville, Virginia 22408

    I. Introduction
       A. Basic Mechanisms Controlling Gastrulation
       B. Problem Areas
   II. The Fibrillar Extracellular Matrix
       A. Discovery and Distribution
       B. Composition
       C. Orientation
       D. Differences among Species
  III. Extracellular Matrix Gene Expression during Development
       A. Fibronectin
       B. Integrin
       C. Integrin Expression
  IV. Synthesis and Distribution of Other Extracellular Matrix and Cell Surface Glycoconjugates
       A. Glycoconjugates in Extracellular Matrix
       B. Cell Surface Glycoconjugates
   V. Experimental Evidence for the Role of the Fibrillar Extracellular Matrix
       A. Temporal Pattern of Fibrillar Extracellular Matrix
       B. Spatial Pattern of Fibrillar Extracellular Matrix
       C. Manipulation of Oriented Fibrils
       D. Contact Guidance by an Artificial Alignment of Fibrils Produced by Mechanical Tension
       E. Inversion of Blastocoel Roof
       F. Cell Adhesion to Fibronectin-Coated Substrata

VI. Probes to Disrupt Mesodermal Cell–Fibrillar Matrix Interaction
  A. Fab' Fragments of Anti-Fibronectin IgG
  B. Fab' Fragments of Anti-Integrin β_1 IgG
  C. RGD Peptides
  D. Perturbation of Gastrulation with Tenascin
  E. Perturbation of Gastrulation with Heparin
  F. Disruption of Extracellular Matrix in Arrested Hybrid Embryos
  G. Disruption of Extracelular Matrix by Mutation
VII. Future Directions
  A. An Experiment for the Future
  B. A Molecular Understanding of Morphogenesis
  References

I. Introduction

A. Basic Mechanisms Controlling Gastrulation

1. Fundamental Differences between Urodele and Anuran Fate Maps

Experimental studies (Vogt, 1929; Smith and Malacinski, 1983; Lundmark, 1986; Keller, 1986) reveal that the early amphibian gastrula consists of four main regions. We will use the terminology of Keller (see Keller, 1986, for terms and illustration) here for convenience and because it facilitates comparisons between urodeles and anurans.

1. The animal cap region (AC) forms ventral ectoderm at the end of gastrulation.
2. The noninvoluting marginal zone (NIMZ) forms dorsal ectoderm between the margin of the AC derivatives and the blastopore.
3. The involuting marginal zone (IMZ) has superficial and deep components. The superficial portion of the IMZ consists of suprablastoporal endoderm and notochord as well as somitic mesoderm and lateral plate mesoderm arrayed from dorsal to ventral, respectively. The prospective somitic and lateral plate mesoderm of the superficial IMZ leave the superficial layer by ingression and become incorporated into the mesodermal mantle formed by deep IMZ cells. The suprablastoporal endoderm and notochord involute and form the roof of the archenteron. Later the notochord separates from the roof of the archenteron. The deep portion of the IMZ also contains notochord, somitic mesoderm, and lateral plate mesoderm arrayed just as it is in the superficial portion of the IMZ.
4. The subblastoporal endoderm (SBE) forms the floor of the archenteron and a yolky endodermal mass inside the late gastrula.

3. Role of ECM in Amphibian Gastrulation 93

Experimental studies (Keller, 1975, 1976; Smith and Malacinski, 1983; Lundmark, 1986; Shi et al., 1987; Delarue et al., 1992) reveal that there are many similarities but also fundamental differences in the fate maps of urodeles and *Xenopus laevis*. The AC and NIMZ expand until they occupy the entire superficial layer of the late gastrula. The AC is a proportionally smaller region in urodeles and, consequently, epiboly of this region is more extensive than in *Xenopus laevis*. The deep layer of the IMZ in *Xenopus laevis* consists of a torus of notochord, somitic mesoderm, and lateral plate mesoderm that turns inside out to form the mesodermal mantle. In contrast, the superficial layer of the IMZ contains prospective endodermal cells that turn inside out during gastrulation and become the roof of the archenteron. The SBE forms the floor of the archenteron and the endodermal mass of the late gastrula. The chief difference between urodeles and *Xenopus laevis* is that the superficial IMZ in the former contains numerous cells with mesodermal fates while in the latter, it contains only cells with endodermal fates.

2. Tissue Affinities

The experiments of Holtfreter (1939, 1943, 1944) revealed that cells in embryos were driven in their morphogenetic rearrangements by changes in tissue affinities. Holtfreter believed that the inherent associative properties of tissues could account for some of the morphogenetic movements of gastrulation. For example, it was shown that isolated dorsal lip of the blastopore could invaginate into an endodermal explant and form a minute invagination site. Also, when fragments of ectoderm and endoderm were cultured together, it was found that they had an inherent tendency to segregate from one another. When some mesoderm was added to the ectoderm and endoderm, Holtfreter found that the ectoderm and endoderm would still self-isolate but now they were held together by an intermediate layer of mesoderm, much like in the intact gastrula. Townes and Holtfreter (1955) showed similar behavior when dissociated groups of cells were formed into mixed aggregates. Type-specific segregation of cells within mixed aggregates resulted in the formation of a final equilibrium configuration similar to that found when whole explanted fragments of embryos were placed in contact and cultured. Boucaut (1974) transplanted radioactively labeled cells into the blastocoel of living embryos and showed that labeled endodermal cells ended up in endoderm and labeled ectodermal cells ended up in ectoderm. These studies revealed that preferential associative behavior of different cell types could serve as an important morphogenetic mechanism. More recently, Turner et al. (1989) studied the behavior of dissociated and reaggregated *X. laevis* blastula and early gastrula cells. They found regional identity that was expressed as specific recognition properties as early as the blastula stage.

3. Autonomous and Correlative Processes

Schechtman (1942) used the embryos of *Hyla regilla* to study the behavior of isolated fragments in culture and to study the behavior of embryos with transplanted portions added to them. Schechtman wanted to determine the relative importance of autonomous movements and correlative movements for the overall process of gastrulation. The experimental results suggested that there are certain processes in gastrulation that are self-driven and independent (autonomous movements) and others that are the result of interaction between different parts of the embryo and therefore are interdependent (correlative movements). For example, explanted dorsal lip of the blastopore showed autonomous elongation movements. Embryos where the dorsal lip of the blastopore had been removed became ring embryos with a disruption of invagination and epiboly. Schechtman interpreted these results to indicate that convergent extension was an autonomous movement but that invagination and involution were correlative movements. It was possible to demonstrate the delicacy of these correlative movements. For example, when thin strips of the presumptive ectoderm from the blastocoel roof of an early gastrula were transplanted bilaterally into the lateral lips of the early blastopore, the autonomous convergent extension of the dorsal lip still occurred but its invagination and involution were strikingly disrupted, so that there was a conspicuous proboscis projecting from the dorsal lip region. In addition, epiboly was disrupted. As a result, a ring embryo with a large exposed endodermal mass was produced. When strips of cellophane were placed in the lateral lips of the blastopore, again convergent extension occurred but the overall process of gastrulation was striking disrupted. Unfortunately, Schechtman's results are difficult to interpret because of the failure to report the results of controls such as sham operations.

4. Changes in Cellular Adhesiveness

Johnson (1969, 1970, 1972) showed that changes in cellular adhesiveness occur during gastrulation in amphibian embryos. By direct observation in the dissecting microscope, one gets the impression that in the blastula stage, cells are more or less round and relatively loosely associated with one another. In contrast, once gastrulation has begun, cells within the embryo become more compacted and intimately associated with one another. This impression of increasing strength of intercellular adhesion was most clearly demonstrated by making quantitative measurements of the rate of dissociation of whole, bisected embryos in Ca^{2+},Mg^{2+}-free saline (Johnson, 1970). It was found that there was a progressive increase in the resistance to dissociation when one compared the late blastula, midgastrula, and late gastrula stage of *Rana pipiens* embryos. We have also found an increasing affinity for fibronectin-coated Sepharose beads during gastrulation in *R. pipiens* (Johnson, 1985; Johnson and Silver, 1989) and *X. laevis* (Johnson and Silver, 1986) gastrulae.

5. Changes in Cell Motility

Holtfreter (1946) observed a peculiar kind of locomotory activity in dissociated cells from *Ambystoma maculatum* gastrulae, called circus movement. In circus movements, rounded dissociated cells formed hyaline bulges on the cell surface. These bulges rotate around the cell with or without endoplasmic flow into the hyaline bulges. Satoh *et al.* (1976) and Johnson (1976) studied circus movements in dissociated *Xenopus laevis* cells taken from different developmental stages. They found that there was a striking increase in circus movements following a period of intense cell division just prior to the onset of gastrulation in dissociated cells taken from the AC and NIMZ. In addition, Johnson (1976) was able to show that circus movements were most pronounced in dissociated cells from the IMZ of the early gastrula. Similar results have also been observed with dissociated cells from *R. pipiens* gastrulae (Johnson and Adelman, 1981, 1984). Circus movements are suppressed as isolated cells reaggregate but masses of cells continue to show active writhing movements. All in all, these results show that there are striking increases in the kinetic activity of embryonic cells during gastrulation in both urodeles and anurans.

6. Epiboly and Convergent Extension: Driven by Cellular Rearrangements

During epiboly, there is a striking increase in the area occupied by the AC and NIMZ. Detailed analysis using scanning electron microscopy and time-lapse cinemicrography (Keller, 1978; 1980) revealed that epiboly in *X. laevis* is accomplished by an expansion in the area of superficial cells and by intercalation of deep cells. The superficial layer remains one cell layer thick but the deep layers thin from two or more layers into a single cell layer. Cells from the superficial layer do not become intercalated into the deep layer nor do deep cells become part of the superficial layer. The prospective ectoderm is two cell layers thick at the end of anuran gastrulation. In contrast, there is clear evidence that deep cells become incorporated into the superficial layer during the more extensive epiboly of AC and NIMZ characteristic of urodele embryos (Goodale, 1911; Holtfreter, 1943; Smith and Malacinski, 1983; Shi *et al.*, 1987). The prospective ectoderm is only one cell layer thick at the end of urodele gastrulation. Cellular rearrangements that convert a wide and short cellular array into a narrow and long cellular array are also important for driving convergent extension (Keller, 1984), a powerful morphogenetic force during gastrulation (see Keller, 1986 and chapter 2, this volume, for review). Keller and Tibbetts (1989) have found that there are two populations of cells contributing to the mesodermal mantle in *X. laevis* gastrulae. The mesoderm that involutes first and forms the leading edge of the mesodermal mantle (head, heart, and blood island mesoderm) spreads across the basal surface of the blastocoel roof without intercalation of constituent cells. In contrast, the mesoderm that involutes later and forms the posterior mesoderm (notochord and somite mesoderm) is driven by mediolateral intercalation, which leads to convergent extension. These rearrangements by intercalation are taking place in a collection of cells surrounded by secreted extracellular matrix glycoconjugates (Johnson, 1984). At present, we do not know what contribution, if any, this extracellular matrix makes to the regulation of intercalation. The interesting observation by Keller and Tibbetts (1989) that ". . . the pattern of cell intercalation is irregular and produces a spacing of labeled and unlabeled cells along the axis of extension that is different in every embryo" suggests that there is a general rather than local, precise spatiotemporal control of cell intercalation. We may discover one day that specific cell adhesion molecules and extracellular matrix molecules are involved in the control of these processes. Perhaps the extracellular matrix is synthesized and secreted by mesodermal cells and merely serves a passive role to push cells apart so that they can intermingle more or less helter skelter. Or, we may discover one day that a specific topographical pattern of extracellular matrix molecules produces boundaries in the embryo to prevent intercalation across boundaries.

B. Problem Areas

1. Initiation of Mesodermal Cell Migration

It is possible that a sudden change in the affinity of cells for the fibronectin (FN)-rich fibrillar extracellular matrix (ECM) initiates cell locomotion at the dorsal lip of the blastopore. Komazaki (1988) studied *Cynops pyrrhogaster* gastrulae, finding that the dorsal involuted mesoderm and endoderm cells formed a loose collection of cells with large intercellular gaps while their ventral counterparts remained tightly coherent to one another with small spaces between cells. Komazaki also demonstrated quite clearly that dissociated dorsal mesodermal and endodermal cells had a much higher affinity for FN-coated substrata when compared to their ventral counterparts. Winklbauer (1990) studied regional variations in cell affinity to FN-coated substrata in *X. laevis* gastrulae and found the highest affinity in anterior head mesoderm and decreasing affinity as one moved posteriorly through the involuting marginal zone and noninvoluting marginal zone. Both of these results could be due to regional variation in integrin expression. We have not observed such an anisotropic distribution of integrins using antibodies against the integrin β_1 subunit in either urodeles or anurans (K. E. Johnson, T. Darribère, and J.-C. Boucaut, unpublished observations, 1991).

Perhaps the initiation of movement occurs first on the dorsal side because of regional heterogeneity in the localization and expression of ECM components and/or cellular receptors involved in adhesion.

2. Directing of Cell Movements

We have seen that there are striking changes in the inherent kinetic properties of embryonic cells, and that groups of cells can undergo rearrangements with clear importance in generating substantial forces to drive gastrulation. What is much less clear, however, is how these forces are coordinated in such a way that the push produced by convergent extension leads to involution of cells rather than exogastrulation. Substantial evidence suggests that the composition of the ECM and the functional expression of cell surface receptors for ECM components play a central role in guiding the "raw forces" of morphogenetic change that take place during gastrulation.

3. The Molecular Composition of Extracellular Matrix in Amphibian Gastrulae

a. Fibronectin. The inner surface of the blastocoel roof at the early gastrula stage is covered by an anastomosing network of extracellular fibrils. These fibrils contain the adhesive glycoprotein fibronectin (FN) (see Section II for details).

b. Laminin. These extracellular fibrils also contain laminin (Nakatsuji *et al.*, 1985a), which codistributes with fibronectin (Darribère *et al.*, 1986).

c. Collagen. Until recently, it was thought that collagen played little role in early amphibian development. For example, new collagen synthesis is slight during gastrulation in *X. laevis* but begins to climb steadily after gastrulation (Green *et al.*, 1968). The methods used in this study detected *de novo* collagen synthesis after fertilization but did not take into account collagen produced during oogenesis and stored in eggs. Otte *et al.* (1990) found a collagen similar to type VI mammalian collagen and, by use of a monoclonal antibody, showed that the antigen was uniformly distributed in unfertilized *Xenopus* eggs. In addition, they found that this antigen became concentrated in the periphery of the egg after fertilization and remained concentrated in the superficial cells derived from egg cortex. Strong antigen signal was detected on the surface of blastomeres and in the intercellular spaces between blastomeres, and lining the blastopore groove and archenteron during gastrulation. They found that Fab' fragments of the antibody blocked gastrulation when applied to the outside of embryos without a vitelline membrane but had no effect when the vitelline membrane was intact or when antibody was injected into the blastocoel. These observations suggest a role for type VI collagen that is distinct from fibronectin in gastrulation.

d. Lectins. Johnson (1981) dissociated whole *R. pipiens* gastrulae with ethylenediaminetetraacetic acid (EDTA) and found a high molecular weight, papain-sensitive glycoconjugate released into the dissociation medium. This glycoconjugate could be coupled to CNBr-Sepharose beads and promoted adhesion of dissociated ectodermal cells to coated beads. Roberson and Barondes (1982, 1983) discovered a galactoside-binding lectin with a molecular weight range of 43,000–45,000 in oocytes and blastulae of *X. laevis*. This lectin was found in the intercellular spaces in blastulae. In gastrulae, the lectin was localized on the inner surface of the blastocoel roof and in the invagination groove formed by the blastopore and was suggested to play an important role in gastrulation (Outenreath *et al.*, 1988).

4. Factors Controlling Extracellular Matrix Synthesis and Assembly

We have some information about the nature and distribution of the components of the extracellular matrix in gastrulating amphibian embryos. We are much less informed, however, concerning the molecular mechanisms underlying the control of this complex assembly of the extracellular matrix. In addition, there must be a diverse set of cell surface receptors to regulate the binding and assembly of different components of the extracellular matrix found in amphibian gastrulae. In the future, one of the most significant challenges will be to identify and characterize these cell surface molecules and their receptors and examine their role in gastrulation.

II. Fibrillar Extracellular Matrix

A. Discovery and Distribution

The fibrillar extracellular matrix on the basal surface of the blastocoel roof was first discovered by Nakatsuji *et al.* (1982) in gastrulae of *A. maculatum* using scanning electron microscopy. Nakatsuji and Johnson (1983b) showed that this fibrillar material was present in the urodeles *A. maculatum, Ambystoma mexicanum*, and *C. pyrrhogaster* and in the anurans *X. laevis* and *R. pipiens*. Boucaut and Darribère (1983a) observed fibrillar extracellular matrix on the basal surface of the blastocoel roof in *Pleurodeles waltlii* gastrulae. Delarue *et al.* (1985) showed that these fibrils are present in *Bufo bufo* and *Bufo calamita*. *Rana sylvatica* gastrulae also have a fibrillar ECM on the basal surface of the blastocoel roof (K. E. Johnson, unpublished observation, 1987). Komazaki (1988) demonstrated that there is a dramatic increase in the abundance of the fibrillar ECM on the basal surface of the blastocoel roof beginning at gastrulation in *C. pyrrhogaster*.

B. Composition

Boucaut and Darribère (1983a) discovered that this fibrillar ECM in *Pleurodeles waltlii* cross-reacted with antibodies against *A. mexicanum* plasma FN. Double-label fluorescent immunocytochemistry demonstrated that each fibril contains both FN and laminin (LM) (Darribère *et al.*, 1986). Nakatsuji *et al.* (1985a) showed that the fibrils in *C. pyrrhogaster* gastrulae contain LM. In contrast, the fibrils in *X. laevis* gastrulae also contain FN (Lee *et al.*, 1984; Nakatsuji *et al.*, 1985b) but not LM (Fey and Hausen, 1990). Furthermore, Wedlich *et al.* (1989) found that LM immunofluorescence was not present in *Xenopus* embryos until somitigenesis. Fey and Hausen (1990) showed that LM transcripts are not detectable in *Xenopus* until the midgastrula stage (stage 11) and LM immunofluorescence is absent until the end of gastrulation (stage 12.5). These observations suggest that there may be a significant difference in the composition of the fibrillar extracellular matrix on the basal surface of the blastocoel roof in urodeles and *X. laevis*. Careful comparative studies between urodeles and anurans are necessary before generalizations can be made.

C. Orientation

Nakatsuji *et al.* (1982) performed a morphometric analysis of the orientation of migrating mesodermal cells and the fibrillar extracellular matrix in *A. maculatum* gastrulae. They showed that migrating mesodermal cells were significantly aligned in that their lamellipodia and filopodia were directed toward the animal pole. Furthermore, a close association between locomotory organelles and the fibrils was observed. Finally, there was a significant alignment of fibrils parallel to the dorsal lip–animal pole axis. These observations led to the hypothesis that oriented fibrils are responsible for directing migrating mesodermal cells away from the blastopore toward the animal pole.

D. Differences among Species

1. Urodeles

Fibronectin-rich ECM in urodeles is first detected by immunocytochemistry in early blastulae (Darribère *et al.*, 1984; 1985). We have studied *A. mexicanum, P. waltlii.* (Boucaut and Darribère, 1983b; Darribère *et al.*, 1990), and *A. mac-*

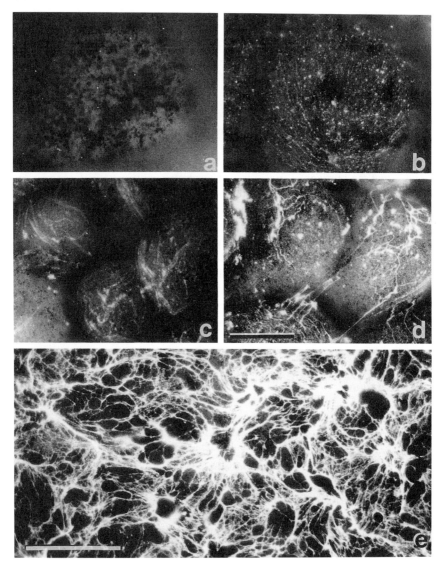

Fig. 1 *In situ* distribution of FN in whole-mount specimens of the blastocoel roof of *Ambystoma maculatum* embryos. (a) Early blastula (stage 7), animal pole. Note absence of FN staining. (b) Early blastula (stage 7), marginal zone. Fibronectin staining is present as specks and thin fibrils in linear arrays on the cell surface. (c) Late blastula (stage 9), animal pole. Note the FN fibrils on cell surfaces. (d) Late blastula (stage 9), marginal zone. This is the earliest stage when FN fibrils extend across cell boundaries (arrows). (e) Early gastrula (stage 10.5), near dorsal lip. Bar in (d) = 10 μm for (a–d). Bar in (e) = 50 μm for (e).

ulatum (Johnson *et al.*, 1992) and found similar patterns. At first, FN-positive materials are restricted to the blastomeres on the periphery of the animal cap near the marginal zone. They are absent from the animal pole region. They first appear as thin punctate fibrils on the cell surface. By the late blastula stage, these thin fibrils are present over the basal surface of animal cap blastomeres and across intercellular boundaries. By the early gastrula stage, a striking increase in the complexity of the fibrillar ECM is observed. At this stage, thin fibrils are present in an apparently random array. Thick, multifibrillar strands are also present and appear to have significant alignment parallel to the dorsal lip–animal pole axis. All throughout gastrulation, the complexity of the fibrillar ECM increases so that by the end of gastrulation, the basal surface of the blastocoel roof has a thick mat of fibrillar material (Fig. 1). The FN fibrils are restricted to the basal surface of the blastocoel roof. Mesodermal cells and endodermal cells have no FN fibrils on their surface.

2. Anurans

Prior to the early gastrula stage there are no FN fibrils detectable in *R. pipiens* (Johnson *et al.*, 1990a) or *X. laevis* (K. E. Johnson, T. Darribère, and J.-C. Boucaut, unpublished observations, 1988) embryos. Beginning at the early gastrula stage (stage 10), however, minute fibrils of FN begin to accumulate on the inner surface of the roof of the blastocoel in both species. During gastrulation in *R. pipiens*, there is a dramatic increase in the complexity of the fibrillar ECM. Fibronectin fibrils become organized into an anastomosing network that radiates away from the center of the cells and stretches across cell boundaries. The most dramatic increase in amount of FN fibrils on the blastocoel roof occurs between stages 11 and 11.5, the time when mesodermal cell migration is most dramatic. The FN fibrils in both anuran and urodele gastrulae are restricted to the basal surface of the blastocoel roof. Mesodermal cells and endodermal cells have no FN fibrils on their surfaces.

III. Extracellular Matrix Gene Expression during Development

A. Fibronectin

Fibronectin is first synthesized from maternal mRNAs at stage 8 in *Xenopus* (Lee *et al.*, 1984), coincident with the onset of zygotic transcription in these embryos at the midblastula transition (MBT; Newport and Kirschner, 1982). Although FN synthesis begins at the MBT, it is not dependent on the expression of new gene products at this stage because FN synthesis will occur on time in parthenogenetically activated eggs, which fail to undergo cytokinesis or to initiate the MBT (Lee *et al.*, 1984). This interesting pattern of regulation suggests that the timing of maternal FN expression is controlled by a separate "cytoplasmic clock." New synthesis of FN mRNA does not occur until mid- to late gastrula stages in *Xenopus*, which indicates that the FN mRNA translated during early gastrulation is entirely of maternal origin (Lee *et al.*, 1984; DeSimone *et al.*, 1992a). Understanding the translational and transcriptional regulatory events involved in FN synthesis is likely to be an important first step in determining how the initiation of gastrulation is controlled.

The FN protein is composed of a series of well-characterized functional domains (for a review, see Hynes, 1990). Multiple forms of the FN protein have been identified and shown to arise via alternative splicing of a common transcript. Studies done on avian and mammalian FNs have determined that this structural heterogeneity involves the complete or partial exclusion of three different exons. Two of these exons are situated near the central cell-binding domain of the protein. These exons are generally included in the cellular forms of FN (cFN) but are both absent from the plasma form of FN (pFN). It is unclear what functional differences, if any, may be attributed to these two major forms of FN. It should be noted, however, that the cellular form of FN predominates in early embryos (ffrench-Constant and Hynes, 1988, 1989). The third region of alternative splicing occurs in the so-called "V" or "type III connecting segment" (III CS) of FN. Additional cell-binding sites have been identified recently in the V region (Humphries *et al.*, 1987; Wayner *et al.*, 1989; Mould *et al.*, 1990). This suggests that alternative splicing may provide one mechanism by which cell binding to FN might be modulated. Because V-region variants of FN are likely to have different functional activities it is important to determine the precise spatial and temporal distribution of FN splice variants during embryogenesis.

cDNAs encoding *Xenopus* FNs have recently been isolated and the sequence determined (DeSimone *et al.*, 1992a). *Xenopus* FN shares approximately 70% amino acid identity with mammalian FNs. The RGD sequence is conserved along with an additional cell-binding site (EILDV) located within the V region. The temporal pattern of expression for each alternatively spliced form of the FN mRNA has been determined throughout early development. As expected, cFN is the predominant form expressed in the early embryo. The V region, which contains a putative non-RGD-dependent cell attachment site, is also included at these stages. This suggests that multiple integrin receptors for FN may be functioning in early morphogenesis.

The *in vitro* mesodermal cell adhesion assays described earlier have all been undertaken with exogenously supplied pFN of mammalian origin. Aside from the obvious species differences, it is worth noting that pFN is structurally distinct from the FN expressed in the embryo (cFN). Caution should be exercised, therefore, in the interpretation of cell behavior on exogenous sources of FN.

B. Integrins

Substantial progress has been made recently in the identification of cellular receptors for many ECM components, including FN and LM (for reviews, see Albelda and Buck, 1990; Hynes, 1990). These receptors have been termed *integrins* to denote their involvement in maintaining a transmembrane linkage between the cytoskeleton and the ECM. Integrins comprise a large family of heterodimeric receptor complexes made up of noncovalently associated α and β subunits. At least 11 different α subunits have thus far been identified in humans, each of which is able to associate with a smaller number of β subunits (from a total of at least 6) to generate a complex array of possible combinations. Different αβ combinations differ in their ligand-binding specificities. Recent studies in *Xenopus* indicate that the diversity of integrin structure observed in mammals is also conserved among lower vertebrates (DeSimone and Hynes, 1988; Smith *et al.*, 1990; Ransom and DeSimone, 1990; DeSimone *et al.*, 1992a,b).

In amphibians, initial attention has focused on the β_1 family of integrins, which includes several well-characterized receptors for FNs and LMs. In both urodeles (Darribère *et al.*, 1988) and anurans (DeSimone and Hynes, 1988; Smith *et al.*, 1990) integrin β_1 synthesis begins just prior to gastrulation. Metabolic labeling studies of *Xenopus* embryo fragments suggest that the β_1 subunit is synthesized in all regions of the early embryo and that the pattern of $\alpha\beta_1$ heterodimers may differ slightly in both the animal and vegetal hemispheres. However, more recent studies in *Xenopus* (DeSimone *et al.*, 1992b; J. E. Howard, J. C. Smith and D. W. DeSimone, unpublished observations, 1991), indicate that most of the β_1 integrin expressed during gastrulation is the immature form of the protein, which is not expressed at the cell surface and, therefore, is not functional. Studies are underway to determine if the relatively small amount of $\alpha\beta_1$ heterodimers expressed on the cell surface is sufficient to account for the FN-dependent cell adhesion observed at these stages. Alternatively, it is also possible that non-β_1 integrin expression during gastrulation plays a more important role in supporting FN adhesion (see Hynes, 1990). In this regard, it is interesting to note that Ransom and DeSimone (1990) have demonstrated that β_3 integrin mRNAs are present in eggs and early cleavage stage *Xenopus* embryos.

To understand fully the roles played by FN in orchestrating cellular events at gastrulation, it is of course necessary to identify and characterize all of the receptors for FN that are expressed in the embryo. Integrin heterogeneity is likely to be important not only in supporting mesodermal cell migration (Darribère *et al.*, 1988) but also for directing the assembly of FN into the ECM lining the blastocoel roof (Darribère *et al.*, 1990). It has been suggested that localized expression of FN at gastrulation is due to localized expression of FN receptors in the cells that line the roof of the blastocoel (Lee *et al.*, 1984). The obvious prediction of this hypothesis would be evidence of integrin localization in this region of the embryo. The $\alpha_3\beta_1$ receptor is known to function as a low-affinity FN, LM, and collagen receptor in other systems (reviewed in Hynes, 1990). The distribution of "high-affinity" FN receptors (such as the $\alpha_5\beta_1$), which may influence FN assembly into the ECM (for a review, see McDonald, 1988), has not yet been determined in the amphibian embryo. The recent availability of cDNA clones for the *Xenopus* α_5 subunit (Ransom and DeSimone, 1990) should now make this possible.

C. Integrin Expression

1. Anurans

As pointed out earlier (Section II,E), it is clear that a number of distinct integrins are expressed in the early *Xenopus* embryo. Extensive sequence information obtained primarily from work done in humans has made it possible to isolate multiple integrin cDNAs from *Xenopus* embryos using degenerate oligodeoxynucleotide primers and polymerase chain reaction (PCR) methods (Ransom and DeSimone, 1990). Partial sequence analyses of all of the *Xenopus* integrins isolated thus far reveal greater than 70% deduced amino acid identities with their cross-species homologs. In addition to previously reported β_1 cDNAs (DeSimone and Hynes, 1988), PCR amplification experiments have yielded cDNAs encoding integrins β_2, β_3, α_3, α_5 and a novel β subunit designated β_x (Ransom and DeSimone, 1992).

The embryonic expression of these integrin α and β mRNAs has been determined by RNase protection analyses. Significant differences are noted in the pattern of expression for these integrins. Both the β_1 and β_3 integrins are encoded by maternal mRNAs. Zygotic expression of β_1 mRNA coincides with the midblastula transition whereas β_3 transcription does not begin until neurulation. It will be necessary to determine the precise spatiotemporal expression of integrins during embryogenesis to analyze the roles that they play in mediating cell adhesion in the embryo. Integrin heterogeneity may provide one mechanism by which positional information is specified during morphogenesis.

2. Urodeles

Recent cloning efforts have resulted in the isolation of cDNAs encoding FN and the integrin β_1 subunit from *P. waltlii* (Clavilier *et al.*, 1992). Not surprisingly, *Pleurodeles* FN and integrin share a high degree of amino acid identity with their cross-species homologs. For example, the alternatively spliced regions of *Xenopus* FN are conserved in *Pleurodeles*, including the non-RGD-dependent cell attachment site in the V region of the protein. The developmental expression of alternatively spliced FN and integrin mRNAs is now being determined for

3. Role of ECM in Amphibian Gastrulation

Pleurodeles. Comparisons of the patterns of expression for these molecules between urodeles and anurans may provide important clues regarding significant morphogenetic differences in these two classes of amphibia.

IV. Synthesis and Distribution of Other Extracellular Matrix and Cell Surface Glycoconjugates

A. Glycoconjugates in Extracellular Matrix

Johnson (1977a) studied the accumulation of ECM in normal *R. pipiens* embryos and the arrested hybrid embryos formed by fertilizing the eggs of *R. pipiens* with the sperm of *Rana catesbeiana*. Johnson showed that there was little stainable ECM in normal and hybrid blastulae; that ECM first accumulates near the dorsal lip of the blastopore in normal embryos, and accumulates throughout the embryo by the end of gastrulation. In contrast, the ECM in arrested hybrid embryos does not show the same dramatic pattern of regional accumulation. Pulse-chase autoradiographic experiments using [^{3}H]galactose and [^{3}H]glucosamine showed that these precursors were first incorporated into the Golgi apparatus and then transported via secretory vesicles to the extracellular space. The defective accumulation of ECM in arrested hybrids was related to a defect in the transport of ECM components from the cytoplasm to the extracellular compartment (Johnson, 1977b). In addition, it was shown that the amount of Golgi apparatus decreases strikingly in cells in arrested hybrid embryo (Johnson and Budge, 1981). Biochemical studies of ECM components (Johnson, 1977c, 1978) showed that the majority of high molecular weight materials labeled with neutral sugar or amino sugar precursors were complex polyanionic glycoconjugates containing galactose, mannose, fucose, and glucosamine. Similar results were obtained with *X. laevis* embryos (Johnson, 1984). These macromolecules are not collagenous because their electrophoretic mobility using sodium dodecyl sulfate-polyacrylamide gel electrophoresis (SDS-PAGE) is different from collagen. Furthermore, they are not glycosaminoglycans such as hyaluronic acid or chondroitin sulfate because their chromatographic behavior on DEAE-cellulose is different from hyaluronate and chondroitin sulfate and they are insensitive to broad-spectrum hyaluronidases.

B. Cell Surface Glycoconjugates

Johnson (1977a) showed that there were striking changes in the lanthanum stainability of the cell coat during gastrulation in *X. laevis*. Similar results were published by Komazaki and Hirakow (1982a,b). One set of experiments reveals the complexity of changes in the chemical composition of the cell surface during gastrulation. Riou *et al.* (1986) dissociated blastulae and gastrulae of *P. waltlii*. Dissociated cell suspensions were then treated with galactose oxidase followed by sodium borohydride reduction to label surface glycoproteins. Next, labeled cells were extracted and subjected to two-dimensional SDS-PAGE. They found 8 different labeled glycoproteins in late blastula or early gastrula cells but 23 different labeled glycoproteins in late gastrula cells. It is clear that there are exceedingly complex changes occurring in the cell surface during gastrulation. The remaining task is to sort out which of these changes have morphogenetic significance. Various approaches have been used to determine the function of these and other ECM components during amphibian gastrulation, including changes in morphogenetic behavior after injecting antibodies to these glycoproteins into living gastrulae to look for disruption of gastrulation.

V. Experimental Evidence for the Role of the Fibrillar Extracellular Matrix

A. Temporal Pattern of Fibrillar Extracellular Matrix

The temporal pattern of the appearance of the fibrillar ECM is compatible with the hypothesis that it is required for mesodermal cell adhesion and directing mesodermal cell migration. Migrating mesodermal cells arise near the dorsal lip of the blastopore when an invagination site is established. In urodeles, the fibrillar matrix appears as a sparse network in the late blastula, prior to formation of the blastopore. It appears somewhat later in anurans but is substantial by the time of invagination. In both urodeles and anurans the fibrillar matrix is in place on the basal surface of the roof of the blastocoel prior to the initiation of mesodermal cell migration.

B. Spatial Pattern of Fibrillar Extracellular Matrix

The fibrillar ECM is restricted to the basal surface of the blastocoel roof in both urodeles and anurans. In urodeles, it appears first on the portion of the animal cap nearest the marginal zone (early blastula) and then appears at the animal pole portion (late blastula). Migrating mesodermal cells and endodermal cells lack this fibrillar ECM. Thus, the spatial pattern of the fibrillar ECM is compatible with the hypothesis that it serves as a substratum for directing mesodermal cell migration. Kubota and Durston (1978) observed migrating mesodermal cells in opened *A. mexicanum* embryos with time-lapse cinemicrography. They showed directed movement of masses of mesodermal cells away from the endodermal mass, toward the animal pole, but showed that there was little migration of isolated cells. They interpreted these observations to indicate that interactions between moving migrating mesodermal cells were important for controlling their directed migration. They made no observations of the effects of their culture medium on the postulated contact guidance system of oriented fibrils on the inner surface of the blastocoel roof. Indeed, it is possible that these fibrils are lost under their culture conditions.

C. Manipulation of Oriented Fibrils

1. Observations in the Scanning Electron Microscope

Nakatsuji *et al.* (1982) show that filopodia and lamellipodia of migrating mesodermal cells often show close association with the fibrils, suggesting that the fibrils serve as a preferential adhesion site for locomotory organelles. Furthermore, the presence of the statistically significant alignment of such a fibrillar network along the blastopore–animal pole axis indicates an interesting possibility for an actual role *in vivo* for a long postulated contact guidance system (Weiss, 1945) by an aligned fibril network (Nakatsuji, 1984).

2. Substratum Conditioning Experiments

To test this hypothesis, Nakatsuji and Johnson (1984a) transferred the extracellular fibril network onto a coverslip surface (Fig. 2). A rectangular piece of the ectodermal layer was dissected from the dorsal part of early gastrulae of *A. maculatum* and explanted onto a plastic coverslip with the inner surface facing down. After 3–5 hr of culture, an outline of the explant and an arrow marking the direction toward the animal pole of the original embryo were scratched on the coverslip. After the explant was removed from the coverslip, dissociated mesoderm cells isolated from gastrulae were seeded on the conditioned surface. Cell movement was recorded by time-lapse cinemicrography, followed by fixation of the coverslip for scanning electron microscopy. Immunocytochemistry demonstrated that FN-containing fibrils were transferred onto the coverslip surface from the inner surface of the ectoderm layer. Mesodermal cells adhered rapidly to such conditioned surfaces (but not to unconditioned substrata) and migrated actively on the conditioned surface. Mesodermal cells produced large lamellipodia, and moved persistently in one direction for long periods.

Nakatsuji and Johnson (1983a) developed a computer program to analyze alignment of cell trails and fibrillar networks on coverslip surfaces. Curved or zigzag lines were divided into short segments. Displacement along the y axis (the blastopore–animal pole axis) (dy) and along the perpendicular x axis (dx) of each segment was measured. Their ratio was then calculated and designated r. To obtain a more symmetrical parameter suitable for statistical manipulation, we

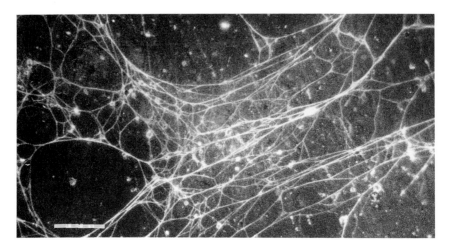

Fig. 2 Extracellular matrix fibrils deposited from the inner blastocoel roof of early *A. maculatum* gastrula (stage 10) onto plastic substrata and then stained for FN by immunocytochemistry. This portion of a larger conditioned area is located near the dorsal lip of the conditioning explant. The dorsal lip edge is on the left and the animal pole direction is toward the right. The long axis of the photograph is parallel to the dorsal lip–animal pole axis. Bar = 50 μm.

took $R = \log 2r$. In the case of random orientation (of fibrils or cell trails), R would be zero. An alignment along the y axis gave positive values of R, and along the x axis gave negative values of R.

Using this approach, Nakatsuji and Johnson (1983a) analyzed apparently random movement of the *Xenopus* mesodermal cells on a homogeneous surface coated with collagen and serum (Nakatsuji and Johnson, 1982). They obtained an R value of -0.03 ± 0.42 ($n = 58$), supporting the prediction that the mean value should be equal or close to zero. On the other hand, cell trails on the ECM-conditioned surface gave R values of $+0.22 \pm 0.66$ ($n = 65$) for *Ambystoma* cells, and $+0.25 \pm 0.45$ ($n = 63$) for *Xenopus* cells. In both cases, the shift of the mean to the positive value was statistically significant at a level of $P < 0.01$.

Analysis of the fibril network observed by SEM revealed that orientation of the cell trails coincided with alignment of the fibrils in the same area. For example, where cell trails appeared unoriented, the fibril network was not significantly aligned. On the other hand, where several cell trails were oriented strongly along the y axis, the fibril network was strongly aligned along the same axis.

We also discovered that isolated cells not only moved preferentially along lines parallel to the animal pole–dorsal lip axis of the conditioned substratum, but they also moved more frequently toward the animal pole portion of the conditioned substratum. These results suggest that isolated cells have the ability to respond in

3. Role of ECM in Amphibian Gastrulation

a polarized fashion with respect to conditioning fibrils. However, results by Kubota and Durston (1978) in *A. mexicanum* are contradictory to the above results. They made time-lapse films of opened gastrulae, showed that isolated mesodermal cells exhibited little migration on the basal surface of the blastocoel roof in their opened preparations and suggested that for truly effective directional migration to occur, migrating mesodermal cells needed interactions with neighboring cells. They argued that contact guidance was probably unimportant in guiding mesodermal cells because these cells overshot the animal pole and migrated all the way to the ventral marginal zone of their opened embryos. In an intact embryo, however, migrating cells also appeared in the ventral marginal zone, moved toward the animal pole, and (by contact inhibition of locomotion) prevented cells from the dorsal marginal zone from overshooting the animal pole region. However, the culture conditions used by Kubota and Durston may not have preserved the fibrillar extracellular matrix. In addition, their culture conditions may have been inadequate because cell migration occurred at a rate of about 1 μm/min, which is much slower than the rate observed *in vivo* (Ignat'eva, 1963) and the 3.4 to 3.7-μm/min rate of locomotion observed on conditioned substrata *in vitro* by Nakatsuji and Johnson (1983a).

Shi *et al.* (1989) showed that large multicellular explants of the dorsal marginal zone of *P. waltlii.* gastrulae would turn toward the animal pole region of substrata conditioned by a fragment of the blastocoel roof that extended from the dorsal lip of the blastopore to the animal pole. These results also suggest that there is some polarized information in the network of FN-containing fibrils deposited on the conditioned substratum. Similar results have been obtained with *A. maculatum* gastrulae (Fig. 3). Pesciotta Peters *et al.* (1990) studied the coassembly of plasma and cellular fibronectin into extracellular fibrils in human fibroblast cultures and found evidence for a model for linear arrangement of unfolded fibronectin dimers by end-to-end association at the amino terminals of dimers. Such an arrangement would create fibronectin fibrils with polarity.

D. Contact Guidance by an Artificial Alignment of Fibrils Produced by Mechanical Tension

Artificial mechanical tension on the conditioning ectodermal layer might be expected to realign the fibrillar matrix. If the matrix guides cell migration, one would expect that realignment of fibrils would cause alteration in the paths of migrating mesodermal cells. Nakatsuji and Johnson (1984a) applied mechanical tension continuously to the ectodermal layer during conditioning. A rectangular piece of the ectodermal layer was dissected from an early gastrula so that the long dimension of the rectangle was parallel to the animal pole–blastopore axis of the embryo. This explant was suspended across a thin strip of a coverslip so that the

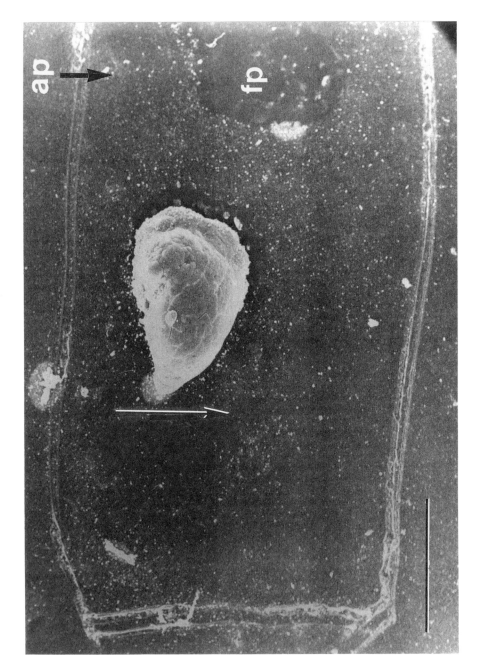

3. Role of ECM in Amphibian Gastrulation

long axis of the explant was parallel to the long axis of the coverslip. The free ends of the ectodermal explant dangled in the medium, so that their weight applied tension to the central part of the explant. This was done to create mechanical tension forces *perpendicular* to the normal animal pole–blastopore tension axis *in vivo*. After conditioning, the explant was removed from the coverslip surface and dissociated mesodermal cells were seeded on the conditioned surface. Scanning electron microscopy revealed that the transferred fibril network was indeed realigned along the axis of mechanical tension. Analysis of cell movement showed that cell trails were also strongly aligned along the axis of tension force. When the ectoderm explant from early gastrulae (stage 10) was put on the coverslip in an orientation so that its original animal pole–blastopore axis became perpendicular to the axis of tension, the R value of the cell trails with y axis adjusted to the tension axis was $+0.29 \pm 0.73$ (n=55). The mechanical tension and resulting stretching of the ectoderm layer apparently erased the original alignment along the blastopore–animal pole axis, and then produced alignment along the tension axis.

E. Inversion of Blastocoel Roof

Boucaut *et al.* (1984a) used *P. waltlii* gastrulae and showed that the apical surface of the blastocoel roof was devoid of FN fibrils. When they inverted a piece of the blastocoel roof from the dorsal lip region so that the apical surface now faced the blastocoel and the basal surface now faced the perivitelline space, they found that this inverted graft did not become coated with FN fibrils and that it did not serve as a substratum for mesodermal cells migration. However, they found that gastrulation proceeded normally because the migrating mesodermal cells were able to move around the inverted patches on areas of host embryos where the blastocoel roof was in its normal orientation and still make their way toward the animal pole. Johnson (1986) made inverted grafts of the central portion of the animal cap in *R. pipiens* gastrulae and found similar results, suggesting that in *P. waltlii*. and *R. pipiens* the fibrillar matrix plays a crucial role in gastrulation.

Fig. 3 Scanning electron micrograph of dorsal marginal zone (DMZ) explant on substratum conditioned by deposition of oriented fibrils from the inner blastocoel roof. The lines drawn in the plastic represent the boundary of the conditioning explant. The dorsal lip end is on the left and the animal pole (ap) region is on the right and marked by the thick black arrow. The thin black and white arrow represents the position of the DMZ explant at the beginning of incubation. Twenty-four hours later, there is a significant movement of the explant toward the AP region. The DMZ explant placed on the VMZ half of the conditioned area was lost during critical point drying, but the "footprint" (fp) of the lost DMZ explant overlaps the ap region of the conditioned area because the explant has moved extensively in that direction. Bar = 0.5 mm.

F. Cell Adhesion to Fibronectin-Coated Substrata

Nakatsuji (1986) examined the adhesion of *X. laevis* gastrula cells to FN- and LM-coated substrata, finding that mesodermal cells adhered avidly to substrata coated with low (5–10 µg/ml) concentrations of FN and LM but that ectodermal cells did not adhere at all at these concentrations. Mesodermal cells migrated more rapidly on FN-coated substrata (2.8 ± 1.1 µm/min) than on LM-coated substrata (1.8 ± 1.0 µm/min). Both mesodermal and ectodermal cells failed to adhere or move on substrata coated with type IV collagen or heparan sulfate. Winklbauer (1988) studied the adhesion of different cell populations from *X. laevis* gastrulae to FN-coated substrata and found that endodermal cells adhere to FN-coated substrata at low FN concentrations (2 µg/ml). Ectodermal cells will adhere but only at high concentrations of FN (200 µg/ml). Winklbauer also studied the adhesiveness of ectodermal cells as a function of embryonic age and found a dramatic increase in adhesiveness starting at the late blastula stage and continuing throughout gastrulation. Johnson and Silver (1986) showed similar developmental increases in ectodermal cell adhesiveness to FN-coated beads. These results correlate nicely with a morphometric study of the number of fibrils per unit area as a function of developmental stage in *X. laevis* (Nakatsuji and Johnson, 1983b). Darribère *et al.* (1988) showed that animal cap cells from *P. waltlii* embryos required a high FN coating concentration to show maximal adhesiveness. Komazaki (1988) studied cell adhesion to FN-coated substrata in *C. pyrrhogaster* gastrulae. Komazaki found stronger adhesiveness among presumptive mesodermal and endodermal cells located near the dorsal lip of the blastopore compared to presumptive mesodermal and endodermal cells isolated from the regions distant from the dorsal lip of the blastopore.

Winklbauer (1990) has published the most comprehensive study to date on mesodermal cell migration in *X. laevis*. Winklbauer dissected mesodermal cells from gastrulae and found a striking gradient of adhesiveness to FN-coated substrata, with anterior (head) mesoderm cells (first involuted) showing greater adhesiveness than posterior mesoderm cells (last involuted). Furthermore, when strips of the entire mesoderm were dissected, these strips migrated on FN-coated substrata and developed a spread-out, leading anterior margin and a compact, retracted posterior trailing margin. These explants had a head and a tail end and migrated with the head end leading the way. Dissociated head mesoderm cells migrated rapidly and randomly on the basal surface of the blastocoel roof. Coherent explants of migratory mesoderm, which showed directional migration on FN-coated substrata *in vitro*, failed to translocate on the blastocoel roof unless the explant was large and contained a part of the blastocoel roof, a dorsal blastoporal lip, and involuted mesoderm. The anterior margin of such large explants showed directional advance of the leading edge, even in the presence of inhibitory, RGD-containing peptides, although cells moved at a lower rate than when RGD-containing peptides were absent. Apparently some property of the

3. Role of ECM in Amphibian Gastrulation
113

large explant, mediated by FN-dependent cell adhesiveness, allowed it to migrate rapidly in a directed fashion across the inner surface of the blastocoel roof. These results also show convincingly that the fibrillar matrix on the basal surface of the blastocoel roof is not sufficient to direct isolated mesodermal cell locomotion in *X. laevis*.

Winklbauer's development of an *in vitro* model for studying cell migration on the basal surface of the blastocoel roof allows study of the control system regulating directional cell migration in developing amphibian embryos. Careful application of this model to both anuran and urodele embryos should be quite instructive regarding the role of the fibrillar ECM in guiding mesodermal cell migration during gastrulation.

VI. Probes to Disrupt Mesodermal Cell–Fibrillar Matrix Interaction

A. Fab' Fragments of Anti-Fibronectin IgG

Boucaut *et al.* (1984a; 1985) studied *P. waltlii* embryos and showed that injection of Fab' fragments of anti-fibronectin IgG into late blastulae or early gastrulae inhibited gastrulation in a dose-dependent manner. Inhibited embryos were strikingly similar to inhibited embryos injected with RGD peptides. Control embryos injected with buffer, bovine serum albumin (BSA), preimmune Fab', or Fab' anti-FN absorbed with FN all developed normally. Similar results have been obtained with *A. mexicanum*, *A. maculatum*, and *R. pipiens* (Fig. 4).

B. Fab' Fragments of Anti-Integrin β_1 IgG

Darribère *et al.* (1988) studied *P. waltlii* embryos and showed that injection into late blastulae or early gastrulae of Fab' fragments of anti-integrin β_1 IgG inhibited gastrulation in a dose-dependent manner. Inhibited embryos were strikingly similar to inhibited embryos injected with Fab' anti-FN IgG. Control embryos injected with preimmune Fab' or Fab' anti-integrin β_1 IgG absorbed with integrin β_1, developed normally. Similar results have been obtained with *A. maculatum*, and *R. pipiens* (Fig. 4).

Darribère *et al.* (1990) made an extensive study of the factors governing FN assembly into fibrils on the inner blastocoel roof in *P. waltlii* embryos. Native FN begins to assemble at the early blastula stage and progressively forms a complex extracellular matrix. Fluorescein isothiocyanate (FITC)-labeled bovine plasma FN injected into the blastocoel of living embryos was assembled into fibrils in the same spatiotemporal pattern as observed for endogenous FN. This suggests that cell surface receptor integrins rather than the supply of endogenous

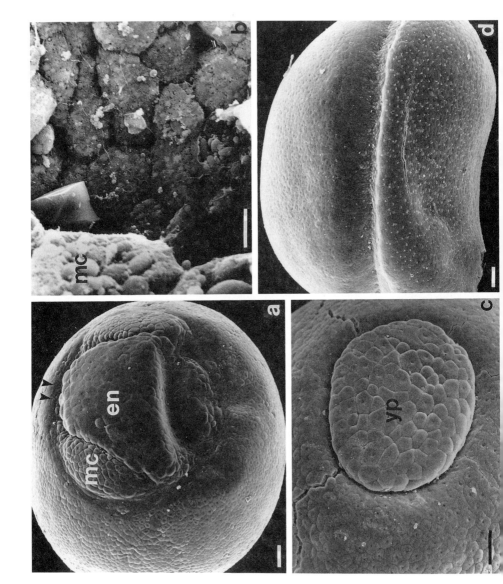

FN regulate fibrillogenesis. Fibrillogenesis of exogenous FITC-labeled FN is inhibited in a dose-dependent manner by both the GRGDS peptide and monospecific antibodies to amphibian integrin β_1 subunit. Injection of antibodies to the cytoplasmic domain of integrin β_1 subunit produces a reversible inhibition of FN fibril formation in progeny of injected blastomeres and causes delays in development. These results suggest that the integrin β_1 subunit and the RGD recognition signal are essential *in vivo* for the proper FN fibrillogenesis. They also suggest that normal gastrulation requires normal assembly of an FN-rich fibrillar extracellular matrix.

C. RGD Peptides

Many integrins recognize an Arg-Gly-Asp (RGD) peptide sequence found in many ECM glycoproteins. The cell-binding domain of FN has such a sequence in it. The addition of RGD-containing peptides to tissue culture media causes the detachment of cells spread on FN-coated substrata in many systems, including dissociated cells from amphibian gastrulae. These peptides are effective inhibitors of mesodermal cell migration during gastrulation in *P. waltlii*. (Boucaut *et al.*, 1984b), *A. mexicanum* (J.-C. Boucaut and T. Darribère, unpublished), and *X. laevis* mesodermal cell adhesion to FN-coated substrata (Winklbauer, 1990). When injected into the blastocoel of late blastulae or early gastrulae, these peptides effectively inhibit mesodermal cell migration and disrupt the entire ensemble of morphogenetic cell movements of gastrulation in a dose-dependent manner. Control peptides from the collagen-binding region of FN, adrenocorticotropic hormone, and buffer injection have no effect on gastrulation.

Scanning electron microscopy of Fab' antibody or peptide-blocked embryos have a characteristic morphology that includes three prominent features: (1) Conspicuous folding and wrinkling of the blastocoel roof, (2) a circular blastopore but little invagination and no archenteron formation; and (3) failure of involution of mesodermal and endodermal cells, so that a large yolky mass is exposed on the vegetal pole of the arrested embryo. Fractured embryos had a

---

Fig. 4 Scanning electron micrographs of *Rana pipiens* embryos incubated with probes to disrupt mesodermal cell–ECM interactions. (a and b) Fab' anti-integrin β_1 (10 mg/ml), incubated 12 hr after injection at stage 10, and then fixed and prepared for scanning electron microscopy. In (a), notice the partial blastopore (arrowheads), the large mass of exposed endodermal cells (en) flanked by presumptive mesodermal cells (mc) that have failed to involute. In (b), we are looking toward the basal surface of the cells of the blastocoel roof. Mesodermal cells (mc) have migrated very little on this substratum. (c) Fab' anti-FN (1 mg/ml), observed 24 hr after injection at stage 10. No neural induction has occurred and embryo has a large yolk plug (yp). (d) Fab' anti-FN + FN (10 mg/ml) (control). Normal neural folds present. Buffer controls were also negative. Bars in (a), (c), and (d) = 100 μm. Bar in (b) = 10 μm.

large blastocoel that was not collapsed. The outer surface of the roof of the blastocoel was extensively convoluted. In contrast, the inner surface was remarkably smooth. Migrating mesodermal cells formed a ringlike collection in the marginal zone but they failed to migrate across the inner surface of the roof of the blastocoel, presumably because they were unable to gain an appropriate foothold there. Migrating mesodermal cells could be seen to project small filopodia and lamellipodia toward the FN fibril-rich basal surface of the blastocoel roof but mesodermal cells failed to attach these locomotory organelles to the fibrillar matrix.

These results suggest that urodele mesodermal cell migration across the basal surface of the blastocoel roof is a FN-dependent process but that epiboly and bottle cell formation are FN-independent processes. Winklbauer (1990) showed that coherent explants of migratory mesoderm translocate on the blastocoel roof if the explant is large and contains a part of the blastocoel roof plus a dorsal blastoporal lip and involuted mesoderm. The anterior margin of such large explants shows directional advance of the leading edge, even in the presence of inhibitory, RGD-containing peptides, albeit at a lower rate than when RGD-containing peptides are absent. Apparently, some property of the large explant (not present in isolated cells), mediated by FN-independent cell adhesiveness, allows it to migrate rapidly in a directed fashion across the inner surface of the blastocoel roof. These results suggest that mesodermal cell migration in *X. laevis* may not require a FN-dependent system. More recent results by Winklbauer and Nagel (1991) suggest that ". . . different parts of the mesoderm . . . are . . . able to migrate independently on the inner surface of the blastocoel roof. The direction of mesoderm cell migration is determined by guidance clues in the [FN-rich] extracellular matrix of the blastocoel roof and by an intrinsic tissue polarity of the mesoderm." They were able to demonstrate that intact FN fibrils are required for directional mesodermal cell migration and were able to eliminate haptotaxis as a guidance mechanism. They have followed the lead of Nakatsuji and Johnson (1983a; Nakatsuji, 1984) in proposing that polarized extracellular matrix fibrils may play a role in guiding migrating mesodermal cells toward the animal pole.

D. Perturbation of Gastrulation with Tenascin

Tenascin (TN) is a noncollagenous glycoprotein found in the extracellular matrix. Tenascin was first identified by Chiquet and Fambrough (1984a,b) and was shown to show temporal and spatial regulation during development (Chiquet-Ehrismann *et al.*, 1986). Subsequently, it has been shown to modify integrin-mediated cell attachment to FN (Crossin *et al.*, 1986; Chiquet-Ehrismann *et al.*, 1988; Mackie *et al.*, 1988; Riou *et al.*, 1988; Boucaut *et al.*, 1991). Recently, we have shown that TN has striking effects on mesodermal cell migration in *Pleurodeles*. Mesodermal cells in explants of the dorsal marginal zone showed extensive outgrowth on FN-coated substrata. This outgrowth was inhibited by addition of TN to culture media. In addition, cell outgrowth was inhibited on substrata coated with FN and TN. When TN was injected into the blastocoel of living embryos, it colocalized with FN in the fibrillar ECM on the basal surface of the blastocoel roof. In addition, gastrulation was significantly inhibited in TN-injected embryos. The morphology of TN-inhibited embryos is reminiscent of embryos injected with other probes to disrupt mesodermal cell–extracellular matrix interaction. Finally, when a monoclonal antibody that masks the cell-binding site of TN is added to culture media or coinjected with TN, it reverses the effects of TN *in vitro* and *in vivo* (Riou *et al.*, 1990). Once again, in an experimental system where a probe disrupts mesodermal cell–FN interaction, gastrulation is specifically inhibited.

E. Perturbation of Gastrulation with Heparin

Mitani (1989) injected heparin or dextran sulfate into the blastocoel of *Xenopus* blastulae and found a dose-dependent retardation of gastrulation. After a retardation of 2–3 hr, gastrulation proceeded normally. No such effect was found with hyaluronic acid or chondroitin sulfate. We have confirmed these heparin effects in *X. laevis* (K. E. Johnson, unpublished observation, 1988).

F. Disruption of Extracellular Matrix in Arrested Hybrid Embryos

Delarue *et al.* (1985) produced nucleocytoplasmic hybrids by transplanting *B. bufo* blastula nuclei into enucleated *B. calamita* eggs. They found that most such hybrids showed developmental arrest at the early gastrula stage. In these arrested gastrulae, the FN-containing ECM on the basal surface of the blastocoel roof was either totally absent or sparse and punctate rather than organized into fibrils as observed in control normal embryos or in the few nucleocytoplasmic hybrids that were able to gastrulate.

Most interspecific hybrid embryos between various amphibian species stop their development at the early gastrula stage, and show almost no mesodermal cell migration (Moore, 1955; Johnson, 1970). In these embryos, the extracellular fibrils are absent or very much reduced in number on the inner surface of the ectoderm layer (Delarue *et al.*, 1985; Johnson *et al.*, 1990a). Using these hybrid embryos between *Rana* species, Nakatsuji and Johnson (1984b) did conditioning experiments with various combinations of the ectodermal layer and mesodermal cells. As expected, normal *R. pipiens* mesodermal cells attached to and actively migrated on surfaces conditioned by the ectodermal layer from normal embryos. Seeded mesodermal cells moved at a mean rate of 4.1 μm/min. The computer analysis of the cell trails, however, revealed very weak alignment along the blastopore–animal pole axis of the ectoderm layer ($R = 0.12 \pm 0.72$; $n = 84$). Such very weak alignment compared to the *Ambystoma* embryo may reflect the fact that the mesodermal cells in anuran embryos migrate as a packed mass of many cells, while those in urodele embryos migrate as individual cells, sometimes completely separated from other cells, thus perhaps needing more guidance for the oriented migration.

The ectodermal layer from arrested hybrid embryos deficient in fibrils *in vivo* had almost no conditioning effect on adhesion by mesodermal cells of either hybrid embryos or normal embryos. Another interesting finding was that if the mesodermal cells from arrested hybrid embryos were seeded on surfaces conditioned by normal ectodermal layer, they adhered to the surface and moved at moderate rates. For example, mesodermal cells from the hybrid, *R. pipiens* ♀ × *R. catesbeiana* ♂, moved at a mean rate of 1.8 μm/min on surfaces conditioned by ectodermal layer from *R. pipiens* normal embryos. This result suggests that the migratory deficiency in hybrid embryos is partially rescued by deposited extracellular fibrils from normal embryos.

Conspicuous convolution of the roof of the blastocoel occurs in three interspecific arrested hybrid embryos: *R. pipiens* ♀ × *Rana sylvatica* ♂ (Johnson, 1971), *R. pipiens* ♀ × *Rana temporaria* ♂ , and *R. pipiens* ♀ × *R. esculenta* ♂, but does not occur in other arrested hybrids (e.g., *R. pipiens* ♀ × *R. catesbeiana* ♂ or *R. pipiens* ♀ × *Rana clamitans* ♂ (Johnson, 1970). These observations, along with recent studies showing defects in FN function in arrested hybrid embryos (Johnson *et al.*, 1990a), suggest that arrested hybrid embryos share similarities with *P. waltlii* gastrulae probed with molecules capable of disrupting cell–FN interaction.

Elinson (1981) studied the hybrid *R. catesbeiana* ♀ × *R. clamitans* ♂ and found that they produced exogastrulae. Triploid embryos, produced by high-pressure suppression of expulsion of the second polar body, showed normal development (Elinson and Briedis, 1981). Gynogenetic diploid embryos also gastrulate normally. Their genetic analysis showed that the defect in arrested hybrids was due to the involvement of a single chromosome. Elinson and Briedis speculated that the interaction between an *R. clamitans* chromosome and the *R. catesbeiana* cytoplasm of the egg resulted in the lack of expression of genes for cell surface molecules required for mesodermal cell migration. It would be interesting to examine FN and integrin gene expression in this system.

G. Disruption of Extracellular Matrix by Mutation

Darribère *et al.* (1991) studied FN and INT in the *ac/ac* ("ascite caudale") maternal effect mutation in *P. waltlii*. Cleavage is apparently normal in progeny from homozygous *ac/ac* females but then all embryos show the "ectodermal syndrome" (Beetschen and Fernandez, 1979). Mutant progeny have deep furrows in the blastocoel roof. Epiboly and archenteron formation are also disturbed, usually leading to partial exogastrulation. The phenotype of severely arrested embryos is strikingly similar to the appearance of embryos injected with Fab' antibodies to FN or INT or injected with RGDS-containing peptides. In studying the extracellular matrix of *ac/ac* progeny, it was shown that the synthesis of FN and $\alpha_5\beta_1$ INT and mutant mesodermal cell adhesion to and locomotion on FN-coated substrata were comparable to wild-type embryos. In contrast, *ac/ac* progeny show a conspicuous defect in the assembly of either endogenous FN or exogenous injected FITC-labeled FN into a complex fibrillar ECM. The FN present in *ac/ac* mutant progeny occurs in small speckles rather than as an anastomosing fibrillar network, as found in wild-type embryos Although it is unclear why fibrillogenesis is defective in *ac/ac* progeny, the morphological similarity between *ac/ac* progeny and probed embryos once again suggests that a normal fibrillar ECM is required to support mesodermal cell migration and gastrulation in *P. waltlii*.

VII. Future Directions

A. An Experiment for the Future

1. Specific Blockage of Component Morphogenetic Movements of Gastrulation with Targeted Molecular Probes

One advantage of working with amphibian eggs and embryos is the possibility of utilizing so-called "reverse or pseudogenetic" approaches to analyze the structure and function of specific molecules. For example, proteins can be ectopically expressed in embryos following the injection of synthetic transcripts, synthesized *in vitro* from cloned cDNA templates, into fertilized eggs or individual early cleavage stage blastomeres. Alternatively, several investigators have taken advantage of antisense oligodeoxynucleotide (ODN) injections to specifically target and degrade maternal mRNAs in *Xenopus* oocytes, which have relatively high levels of endogenous RNase H activity (Dash *et al.*, 1987; Shuttleworth *et al.*, 1988; Woolf *et al.*, 1990).

The antisense ODN approach is likely to be feasible for analyzing FN function during gastrulation, because the FN synthesized just prior to this stage is encoded entirely by maternal mRNA. Greater than 95% of the total FN mRNA present in *Xenopus* oocytes can be eliminated following the injection of specific antisense ODNs into the cytoplasm (DeSimone *et al.*, 1992a). It is also possible to target specific variants of FN by injecting antisense ODNs that correspond to alternatively spliced exons. The difficulty with these experiments continues to be nonspecific toxic effects associated with ODN injections at concentrations necessary to eliminate mRNA (see Woolf *et al.*, 1990). However, Dagle *et al.* (1990)

have recently used modified ODNs that are effective at much lower concentrations for eliminating mRNAs and, correspondingly, their toxicity is greatly reduced. Preliminary results of FN antisense experiments in *Xenopus* embryos suggest that FN expression is necessary for mesodermal cell ingression but does not affect the timing of dorsal lip formation (D. W. DeSimone and Wylie, unpublished observations, 1991). These "FN-minus" embryos are arrested until the equivalent of stage 12 in control embryos, which coincides with the first zygotic expression of FN mRNAs. The arrested embryos are then able to continue gastrulating although involution of mesoderm appears incomplete. These experiments should be interpreted cautiously because of the RNA duplex unwinding activity present in *Xenopus* embryos (Rebagliati and Melton, 1987).

2. Reversal of Blockage

Experiments are in progress to experimentally "rescue" these delayed embryos by introducing full-length synthetic FN transcripts into ODN-treated eggs and, in related experiments, purified FN protein into the blastocoel of FN-minus embryos. Careful analyses of this type are necessary to validate the specificity of ODN induced effects in the embryo.

B. A Molecular Understanding of Morphogenesis

Why does a population of cells engage in a particular set of morphogenetic cell behaviors? Complex morphogenetic cell movements such as those observed during amphibian gastrulation must ultimately be understood by combining an analysis of cell behavior with an analysis of the molecular events underlying changes in cell behavior. Our goal is to understand the molecular events driving changes in cell behavior. Several criteria must be met for us to be sure that a particular behavior is caused by a particular molecular event. First, we must characterize the cell behavior. Then we must generate specific molecular probes that will alter this behavior in a nondestructive and specific way. Finally, we must be able to reverse the effects of a specific probe and restore normal cell behavior. Once normal cell behavior has been restored we should observe a concomitant resumption of normal morphogenesis. We are aware of an important caveat. If it turns out that there is some complex timing mechanism involved in the control of morphogenesis, such that once the process is set in motion an appropriate behavioral and molecular repertoire must unfold in an appropriate temporal and spatial arrangement, then the analytical task will become much more complex. Regardless of this potential difficulty, we are optimistic that we are on the threshold of a new era in studying gastrulation. Cell behaviorists have laid much of the groundwork for delineating those aspects of cell behavior that are amenable to a molecular analysis. Molecular biologists are in the process of defining proteins

3. Role of ECM in Amphibian Gastrulation

and genes that have particular morphogenetic significance. By combining behavioral and molecular analysis, we aim to definitively test the hypothesis that developmental regulation of gene expression results in the assembly of new combinations of extracellular matrix components and cell surface receptors, which in turn guide moving cells to appropriate destinations during gastrulation (Johnson *et al.*, 1990b; Boucaut *et al.*, 1990, 1991).

References

Albelda, S. M., and Buck, C. A. (1990). Integrins and other cell adhesion molecules. *FASEB J.* **4**, 2868–2880.

Beetschen, J. C., and Fernandez, M. (1979). Studies on the maternal effect of the semi-lethal factor *ac* in the salamander *Pleurodeles waltlii. In* "Maternal Effects in Development," (D. R. Newth and M. Balls, eds.), Br. Soc. Dev. Biol. Symp. 4, pp. 269–286. Cambridge Univ. Press, Cambridge.

Boucaut, J.-C. (1974). Ètude autoradiographique de la distribution de cellules embryonnaires isolées, transplantées dans le blastocoele chez *Pleurodeles waltlii* Michah (Amphibien, Urodele). *Ann. Embryol. Morphol.* **7**, 7–50.

Boucaut, J.-C., and Darribère, T. (1983a). Presence of fibronectin during early embryogenesis in the amphibian. *Pleurodeles waltlii. Cell Differ.* **12**, 77–83.

Boucaut, J.-C., and Darribère, T. (1983b). Fibronectin in early amphibian embryos: Migrating mesodermal cells are in contact with a fibronectin-rich fibrillar matrix established prior to gastrulation. *Cell Tissue Res.* **234**, 135–145.

Boucaut, J. C., Darribère, T., Boulekbache, H., and Thiery, J. P. (1984a). Prevention of gastrulation but not neurulation by antibody to fibronectin in amphibian embryos. *Nature (London)* **307**, 364–367.

Boucaut, J.-C., Darribère, T., Poole, T. J., Aoyama, H., Yamada, K. M., and Thiery, J. P. (1984b). Biologically active synthetic peptides as probes of embryonic development: A competitive peptide inhibitor of fibronectin function inhibits gastrulation in amphibian embryos and neural crest cell migration in avian embryos. *J. Cell Biol.* **99**, 1822–1830.

Boucaut, J.-C., Darribère, T., Shi, D.L., Boulekbache, H. Yamada, K.M., and Thiery, J.P. (1985). Evidence for the role of fibronectin in amphibian gastrulation. *J. Embryol. Exp. Morphol.* **89**(Suppl.), 211–227.

Boucaut, J.-C., Johnson, K.E., Darribère, T., Shi, D.-L., Riou, J.-F., Boulekbache, H., and Delarue, M. (1990). Fibronectin-rich fibrillar extracellular matrix controls cell migration during amphibian gastrulation. *Int. J. Dev. Biol.* **34**, 139–147.

Boucaut, J.-C., Darribère, T., Shi, D.-L., Riou, J.-F., Johnson, K. E., and Delarue, M. (1991). Amphibian gastrulation: The molecular bases of mesodermal cell migration in urodele embryos. *In* "Gastrulation: Movements, Patterns and Molecules" (R. Keller, W. Clark, and F. Griffin, eds.), pp. 169–184. Plenum, New York.

Chiquet, M., and Fambrough, D. M. (1984a). Chick myotendinous antigen. I A monoclonal as a marker for tendon and muscle morphogenesis. *J. Cell Biol.* **98**, 1926–1936.

Chiquet, M., and Fambrough, D. M. (1984b). Chick myotendinous antigen. I. A novel extracellular glycoprotein complex consisting of large disulfide linked subunits. *J. Cell Biol.* **98**, 1937–1947.

Chiquet-Ehrismann, R., Makie, E. J., Pearson, C. A., and Sakakura, T. (1986). Tenascin: An extracellular matrix protein involved in tissue interactions during fetal development and oncogenesis. *Cell (Cambridge, Mass.)* **47**, 131–139.

Chiquet-Ehrismann, R., Kalla, P., Pearson, C. A., Beck, K., and Chiquet, M. (1988). Tenascin interferes with fibronectin action. *Cell (Cambridge, Mass.)* **53**, 383–390.

Clavilier, L., Darrihère, T., Riou, J.-F., Shi, D. L., Boucaut, J.-C., and DeSimone, D. W. (1992). in preparation.

Crossin, K. L., Hoffman, S., Grumet, M., Thiery, J. P., and Edelman, G. M. (1986). Site-restricted expression of cytotactin during development of the chick embryo. *J. Cell Biol.* **102**, 1917–1930.

Dagle, J. M., Walder, J. A., and Weeks, D. L. (1990). Targeted degradation of mRNA in *Xenopus* oocytes and embryos directed by modified oligonucleotides: studies of An2 and cyclin in embryogenesis. *Nucleic Acids Res.* **18**, 4751–4757.

Darribère, T., Boucher, D., Lacroix, J. C., and Boucaut J.-C. (1984). Fibronectin synthesis during oogenesis and early development in the amphibian *Pleurodeles waltlii*. *Cell Differ.* **14**, 7–14.

Darribère, T., Boulekbache, H., Shi, D. L., and Boucaut, J.-C. (1985). Immunoelectron microscopic study of fibronectin in gastrulating amphibian embryos. *Cell Tissue Res.* **239**, 75–80.

Darribère, T., Riou, J.-F., Shi, D. L., Delarue, M., and Boucaut, J.-C. (1986). Synthesis and distribution of laminin-related polypeptides in early amphibian embryos. *Cell Tissue Res.* **246**, 45–51.

Darribère, T., Yamada, K.M., Johnson, K.E., and Boucaut, J.-C. (1988). The 140 kD fibronectin receptor complex is required for mesodermal cell adhesion during gastrulation in the amphibian *Pleurodeles waltlii*. *Dev. Biol.* **126**, 182–194.

Darribère, T., Guida, K., Larjava, H., Johnson, K. E., Yamada, K. M., Thiery, J.-P., and Boucaut, J.-C. (1990). In vivo analyses of integrin β_1 subunit function in fibronectin matrix assembly. *J. Cell Biol.* **110**, 1813–1823.

Darribère, T., Riou, J.-F., Guida, K., Duprat, A.-M., Boucaut, J.-C., and Beetschen, J.-C. (1991). A maternal-effect mutation disturbs extracellular matrix organization in the early *Pleurodeles waltl.* embryo. *Cell Tissue Res.* **263**, 507–514.

Dash, P., Lotan, I., Knapp, M., Kandel, E. R., and Goelet, P. (1987). Selective elimination of mRNAs *in vivo*: Complementary oligodeoxynucleotides promote mRNA degradation by an RNase H-like activity. *Proc. Natl. Acad. Sci. U.S.A.* **84**, 7896–7900.

Delarue, M., Darribère, T., Aimer, C., and Boucaut, J.-C. (1985). Bufonid nucleocytoplasmic hybrids arrested at the early gastrula stage lack a fibronectin-containing fibrillar extracellular matrix. *Wilhelm Roux's Arch. Dev. Biol.* **194**, 275–280.

Delarue, M., Sanchez, S., Johnson, K. E., Darribère, T., and Boucaut, J.-C. (1992). A fate map of superficial and deep circumblastoporal cells in the early gastrula of *Pleurodeles waltl*. *Development* (Cambridge, UK) **114**, 135–146.

DeSimone, D. W., and Hynes, R. O. (1988). *Xenopus laevis* integrins. Structural and evolutionary divergence of integrin β subunits. *J. Biol. Chem.* **263**, 5333–5340.

DeSimone, D. W., Norton, P. A., and Hynes, R. O. (1992a). Identification and characterization of alternatively spliced fibronectin mRNAs expressed in early *Xenopus* embryos. *Dev. Biol.* **149**, 357–369.

DeSimone, D. W., Smith, J. C., Howard, J. E., Ransom, D. G., and Symes, K. (1992b). The expression of fibronectins and integrins during mesodermal induction and gastrulation in Xenopus. *In* "Gastrulation: Movements, Patterns and Molecules" (R. Keller, W. Clark, and F. Griffin, eds.), pp. 185–198. Plenum, New York.

Elinson, R. P. (1981). Genetic analysis of developmental arrest in an amphibian hybrid (*Rana catesbeiana, Rana clamitans*). *Dev. Biol.* **81**, 167–176.

Elinson, R. P., and Briedis, A. (1981). Triploidy permits survival of an inviable amphibian hybrid. *Dev. Genet.* **2**, 357–367.

Fey, J., and Hausen, P. (1990). Appearance and distribution of laminin during development of *Xenopus laevis*. *Differentiation (Berlin)* **42**, 144–152.

3. Role of ECM in Amphibian Gastrulation

ffrench-Constant, C., and Hynes, R.O. (1988). Patterns of fibronectin gene expression and splicing during cell migration in chicken embryos. *Development (Cambridge, UK)* **104**, 369–382.

ffrench-Constant, C., and Hynes, R. O. (1989). Alternative splicing of fibronectin is temporally and spatially regulated in the chicken embryo. *Development (Cambridge, UK)* **106**, 375–388.

Goodale, H. D. (1911). The early development of *Spelerpes bilineatus* (Green). *Am. J. Anat.* **12**, 173–247.

Green, H., Goldberg, B., Schwartz, M., and Brown, D. D. (1968). The synthesis of collagen during the development of *Xenopus laevis*. *Dev. Biol.* **18**, 391–400.

Holtfreter, J. (1939). Gewebeaffinitat, ein mittel der embryonalen formbildung. *Arch. Exp. Zellforsch. Besonders Gewebezuecht.* **23**, 169–209. Available in translation *In* "Foundations of Experimental Embryology" (B. H. Willier and J. M. Oppenheimer, eds.), pp. 186–225. Prentice-Hall, Englewood Cliffs, New Jersey.

Holtfreter, J. (1943). A study of the mechanics of gastrulation, I. *J. Exp. Zool.* **94**, 261–318.

Holtfreter, J. (1944). A study of the mechanics of gastrulation, II. *J. Exp. Zool.* **95**, 171–212.

Holtfreter, J. (1946). Structure, motility and locomotion in isolated embryonic amphibian cells. *J. Morphol.* **79**, 27–61.

Humphries, M. J., Komoriya, A., Akiyama, S. K., Olden, K., and Yamada, K. M. (1987). Identification of two distinct regions of the type III connecting segment of human plasma fibronectin that promote cell-type specific adhesion. *J. Biol. Chem.* **262**, 6886–6892.

Hynes, R. O. (1990). "Fibronectins." Springer-Verlag, New York.

Ignat'eva, G. M. (1963). Comparison of the dynamics of the chordamesoderm in sturgeon and axolotl embryos. *Dokl. Akad. Nauk SSSR* **151**, 973–976.

Johnson, K. E. (1969). Altered contact behavior of presumptive mesodermal cells from hybrid amphibian embryos arrested at gastrulation. *J. Exp. Zool.* **170**, 325–332.

Johnson, K. E. (1970). The role of changes in cell contact behavior in amphibian gastrulation. *J. Exp. Zool.* **175**, 391–428.

Johnson, K. E. (1971). A biochemical and cytological investigation of differentiation in the interspecific hybrid amphibian embryo *Rana pipiens* ♀ × *Rana sylvatica* ♂. *J. Exp. Zool.* **177**, 191–206.

Johnson, K. E. (1972). The extent of cell contact and the relative frequency of small and large gaps between presumptive mesodermal cells in normal gastrulae of *Rana pipiens* and the arrested gastrulae of the *Rana pipiens* ♀ × *Rana catesbeiana* ♂ hybrid. *J. Exp. Zool.* **179**, 227–238.

Johnson, K. E. (1976). Circus movements and blebbing locomotion in dissociated embryonic cells of an amphibian, *Xenopus laevis*. *J. Cell Sci.* **22**, 575–583.

Johnson, K. E. (1977a). Extracellular matrix synthesis in blastula and gastrula stages of normal and hybrid embryos. I. Toluidine blue and lanthanum staining. *J. Cell Sci.* **25**, 313–332.

Johnson, K. E. (1977b). Extracellular matrix synthesis in blastula and gastrula stages of normal and hybrid embryos. II. Autoradiographic observations on the sites of synthesis and mode of transport of galactose- and glucosamine-labelled material. *J. Cell Sci.* **25**, 323–334.

Johnson, K. E. (1977c). Extracellular matrix in blastula and gastrula stages of normal and hybrid embryos. III. Characterization of galactose- and glucosamine-labelled material. *J. Cell Sci.* **25**, 335–354.

Johnson, K. E. (1978). Extracellular matrix synthesis in blastula and gastrula stages of normal and hybrid frog embryos. IV. Biochemical and autoradiographic observations on fucose-, glucose, and mannose-labelled materials. *J. Cell Sci.* **32**, 109–136.

Johnson, K. E. (1981). Normal frog gastrula extracellular materials serve as a substratum for normal and hybrid cell adhesion when covalently coupled with CNBr-activated sepharose beads. *Cell Differ.* **10**, 47–55.

Johnson, K. E. (1984). Glycoconjugate synthesis during gastrulation in *Xenopus laevis*. *Am. Zool.* **24**, 605–614.

Johnson, K. E. (1985). Frog gastrula cells adhere to fibronectin-Sepharose beads. *In* "Molecular

Determinants of Animal Form" (G. M. Edelman, ed.), pp. 271-292. Alan R. Liss, New York.
Johnson, K. E. (1986). Transplantation studies to investigate mesoderm-ectoderm adhesive cell interactions during gastrulation. *J. Cell Sci.* **82**, 99-117.
Johnson, K. E., and Adelman, M. R. (1981). Circus movements in dissociated cells in normal and hybrid frog embryos. *J. Cell Sci.* **49**, 205-216.
Johnson, K. E., and Adelman, M. R. (1984). Circus movements in dissociated cells from two new hybrid frog embryos. *J. Cell Sci.* **68**, 69-82.
Johnson, K. E., and Budge, J. H. (1981). The Golgi apparatus in blastula and gastrula stages of normal and hybrid amphibian embyros. *Cell Diff.* **10**, 219-227.
Johnson, K. E., and Silver, M. H. (1986). Cells from *Xenopus laevis* gastrulae adhere to fibronectin-Sepharose beads and other lectin coated beads. *Scanning Electron Microsc.* **2**, 671-678.
Johnson, K. E., and Silver, M. H. (1989). Cells from *Rana pipiens* gastrulae and arrested hybrid gastrulae show differences in adhesion to fibronectin-Sepharose beads. *J. Exp. Zool.* **251**, 155-166.
Johnson, K. E., Darribère, T., and Boucaut, J.-C. (1990a) Cell adhesion to extracellular matrix in normal *Rana pipiens* gastrulae and in arrested hybrid gastrulae *Rana pipiens* ♀ × *Rana esculenta* ♂. *Dev. Biol.* **137**, 86-99.
Johnson, K. E., Nakatsuji, N., and Boucaut, J.-C. (1990b). Extracellular matrix control of cell migration during amphibian gastrulation. *In* "Cytoplasmic Organization Systems" (G. Malacinski, ed.), pp. 349-374. McGraw-Hill, New York.
Johnson, K. E., Darribère, T., and Boucaut, J.-C. (1992). *Ambystoma maculatum* gastrula have an oriented fibronectin containing extracellular matrix. *J. Exp. Zool.* **261**, 458-471.
Keller, R. E. (1975). Vital dye mapping of the gastrula and neurula of *Xenopus laevis*. I. Prospective areas and morphogenetic movements in the superficial layer. *Dev. Biol.* **42**, 222-241.
Keller, R. E. (1976). Vital dye mapping of the gastrula and neurula of *Xenopus laevis*. II. Prospective areas and morphogenetic movements in the deep region. *Dev. Biol.* **51**, 118-137.
Keller, R. E. (1978). Time-lapse cinemicrographic analysis of superficial cell behavior during and prior to gastrulation in *Xenopus laevis*. *J. Morphol.* **157**, 223-248.
Keller, R. E. (1980). The cellular basis of epiboly: An SEM study of deep-cell rearrangement during gastrulation in *Xenopus laevis*. *J. Embryol. Exp. Morphol.* **60**, 201-234.
Keller, R. E. (1984). The cellular basis of gastrulation in *Xenopus laevis*: Postinvolutional convergence and extension. *Am. Zool.* **25**, 589-602.
Keller, R. E. (1986). The cellular basis of amphibian gastrulation. *In* "Developmental Biology, A Comprehensive Synthesis. Volume 2, The Cellular Basis of Morphogenesis" (L. W. Browder, ed.), pp. 241-327. Plenum, New York.
Keller, R. E., and Tibbetts, P. (1989). Mediolateral cell intercalation in the dorsal, axial mesoderm of *Xenopus laevis*. *Dev. Biol* **131**, 539-549.
Komazaki, S. (1988). Factors related to the initiation of cell migration along the inner surface of the blastocoelic wall during amphibian gastrulation. *Cell Differ.* **24**, 25-32.
Komazaki, S., and Hirakow, R. (1982a). Ultrastructural demonstration of extracellular matrix in *Xenopus laevis* gastrulae by using cationic dyes, with special reference to filamentous masses stained with alcian blue-lanthanum nitrate. *Proc. Jpn. Acad. Ser. B* **58**, 17-20.
Komazaki, S., and Hirakow, R. (1982b). An ultrastructural study of extracellular matrix components of gastrulating *Xenopus laevis* embryos. *Proc. Jpn. Acad. Ser. B* **58**, 131-134.
Kubota, H. Y., and Durston, A. J. (1978). Cinematographical study of cell migration in the opened gastrulae of *Ambystoma mexicanum*. *J. Embryol. Exp. Morphol.* **44**, 71-80.
Lee, G., Hynes, R., and Kirschner, M. (1984). Temporal and spatial regulation of fibronectin in early *Xenopus* development. *Cell (Cambridge, Mass.)* **36**, 729-740.

3. Role of ECM in Amphibian Gastrulation 125

Lundmark, C. (1986). Role of bilateral zones of ingressing superficial cells during gastrulation of *Ambystoma mexicanum*. *J. Embrol. Exp. Morphol.*. **97**, 47–62.
McDonald, J. A. (1988). Extracellular matrix assembly. *Annu. Rev. Cell Biol.* **4**, 183–207.
Mackie, E. J., Tucker, R. P., Halfter, W., Chiquet-Ehrismann, R., and Epperlein, H. H. (1988). The distribution of tenascin coincides with pathways of neural crest cell migration. *Development (Cambridge, UK)* **102**, 237–250.
Mitani, S. (1989). Retarded gastrulation and altered subsequent development of neural tissues in heparin-injected *Xenopus* embryos. *Development (Cambridge, UK)* **107**, 423–435.
Moore, J. A. (1955). Abnormal combinations of nuclear and cytoplasmic systems in frogs and toads. *Adv. Genet.* **7**, 139–182.
Mould, A. P., Wheldon, L. A., Komoriya, A., Wayner, E. A., Yamada, K. M., and Humphries, M. J. (1990). Affinity chromatographic isolation of the melanoma adhesion receptor for the III CS region of fibronectin and its identification as the integrin $\alpha_4\beta_1$. *J. Biol. Chem.* **265**, 4020–4024.
Nakatsuji, N. (1984). Cell locomotion and contact guidance in amphibian gastrulation. *Am. Zool.* **24**, 615–627.
Nakatsuji, N. (1986). Presumptive mesodermal cells from *Xenopus laevis* gastrulae attach to and migrate on substrata coated with fibronectin or laminin. *J. Cell Sci.* **86**, 109–118.
Nakatsuji, N., and Johnson, K. E. (1982). Cell locomotion *in vitro* by *Xenopus laevis* gastrula mesodermal cells. *Cell Motil.* **2**, 149–161.
Nakatsuji, N., and Johnson, K. E. (1983a). Conditioning of a culture substratum by the ectodermal layer promotes attachment and oriented locomotion by amphibian gastrula mesodermal cells. *J. Cell Sci.* **59**, 43–60.
Nakatsuji, N., and Johnson, K. E. (1983b). Comparative study of extracellular fibrils on the ectodermal layer in gastrulae of five amphibian species. *J. Cell Sci.* **59**, 61–70.
Nakatsuji, N., and Johnson, K. E. (1984a). Experimental manipulation of a contact guidance system in amphibian gastrulation by mechanical tension. *Nature (London)* **307**, 453–455.
Nakatsuji, N., and Johnson, K. E. (1984b). Substratum conditioning experiments using normal and hybrid frog embryos. *J. Cell Sci.* **68**, 49–67.
Nakatsuji, N., Gould, A., and Johnson, K. E. (1982). Movement and guidance of migrating mesodermal cells in *Ambystoma maculatum* gastrulae. *J. Cell Sci.* **56**, 207–222.
Nakatsuji, N., Smolira, M. A., and Wylie, C. C. (1985a). Fibronectin visualized by scanning electron microscopy immunocytochemistry on the substratum for cell migration in *Xenopus laevis* gastrula. *Dev. Biol.* **107**, 264–268.
Nakatsuji, N., Hashimoto, K., and Hayashi, M. (1985b). Laminin fibrils in newt gastrulae visualized by immunofluorescent staining. *Dev. Growth Differ.* **27**, 639–643.
Newport, J., and Kirschner, M. (1982). A major developmental transition in early *Xenopus* embryos: 1. Characterization and timing of cellular changes at the midblastula stage. *Cell (Cambridge, Mass.)* **30**, 675–686.
Otte, A. P., Debjani, R., Siemerink, M., Koster, C. H., Hochstenbach, F., Timmermans, A., and Durston, A. J. (1990). Characterization of a maternal type VI collagen in *Xenopus* embryos suggests a role for collagen in gastrulation. *J. Cell Biol.* **111**, 271–278.
Outenreath, R. L., Roberson, M. R., and Barondes, S. H. (1988). Endogenous lectin secretion into the extracellular matrix of early embryos of *Xenopus laevis*. *Dev. Biol.* **125**, 187–194.
Pesciotta Peters, D. M., Portz, L. M., Fullenwider, J., and Mosher, D. F. (1990). Co-assembly of plasma and cellular fibronectins into fibrils in human fibroblast cultures. *J. Cell Biol.* **111**, 249–256.
Ransom, D. G., and DeSimone, D. W. (1990). Cloning and characterization of multiple integrin α and β subunits expressed in *Xenopus* embryos. *J. Cell Biol.* **111**, 142a.
Ransom, D. G., and DeSimone, D. W. (1992). in preparation.

Rebagliati, M. R., and Melton, D. A. (1987). Antisense RNA injection in fertilized frog eggs reveals an RNA duplex unwinding activity. *Cell (Cambridge, Mass.)* **48**, 599–605.

Riou, J.-F., Darribère, T., and Boucaut, J.-C. (1986). Cell surface glycoproteins change during gastrulation in *Pleurodeles waltlii*. *J. Cell Sci.* **82**, 23–40.

Riou, J.-F., Shi, D.-L., Chiquet, M., and Boucaut, J.-C. (1988). Expression of tenascin in response to neural induction in amphibian embryos. *Development (Cambridge, UK)* **104**, 511–524.

Riou, J.-F., Shi, D.-L., Chiquet, M., and Boucaut, J.-C. (1990). Exogenous tenascin inhibits mesodermal cells migration during amphibian gastrulation. *Dev. Biol.* **137**, 305–317.

Roberson, M.M., and Barondes, S.H. (1982). Lectin from embryos and oocytes of *Xenopus laevis*. Purification and properties. *J. Biol. Chem.* **257**, 7520–7524.

Roberson, M. M., and Barondes, S. H. (1983). *Xenopus laevis* lectin is localized at several sites in *Xenopus* oocytes, eggs, and embryos. *J. Cell. Biol.* **97**, 1875–1881.

Satoh, N., Kageyama, T., and Sirakami, K.-I. (1976). I. Motility of dissociated embryonic cells in *Xenopus laevis*: Its significance to morphogenetic movements. *Dev. Growth Differ.* **18**, 55–67.

Schechtman, A. M. (1942). The mechanism of amphibian gastrulation. I. Gastrulation-promoting interactions between various regions of an anuran egg (*Hyla regilla*). *Univ. Calif. Publ. Zool.* **51**, 1–40.

Shi, D. L., Delarue, M., Darribère, T., Riou, J.-F., and Boucaut, J.-C. (1987). Experimental analysis of the extension of the dorsal marginal zone in *Pleurodeles waltlii* gastrulae. *Development (Cambridge, UK)* **100**, 147–161.

Shi, D. L., Darribère, T., Johnson, K. E., and Boucaut, J-C. (1989). Initiation of mesodermal cell migration and spreading relative to gastrulation in the urodele amphibian *Pleurodeles waltlii*. *Development (Cambridge, UK)* **105**, 351–363.

Shuttleworth, J., Matthews, G., Dale, L., Baker, C., and Colman, A. (1988). Antisense oligodeoxyribonucleotide-directed cleavage of maternal mRNA in *Xenopus* oocytes and embryos. *Gene* **72**, 267–275.

Smith, J. C., and Malacinski, G. M. (1983). The origin of the mesoderm in the anuran, *Xenopus laevis*, and a urodele, *Ambystoma mexicanum*. *Dev. Biol.* **98**, 250–254.

Smith, J. C., Symes, K., Hynes, R. O., and DeSimone, D. (1990). Mesoderm induction and the control of gastrulation in *Xenopus laevis*: The roles of fibronectin and integrins. *Development (Cambridge, UK)* **108**, 229–238.

Townes, P. L., and Holtfreter, J. (1955). Directed movements and selective adhesion of embryonic amphibian cells. *J. Exp. Zool.* **128**, 53–120.

Turner, A., Snape, A. M., Wylie, C. C., and Heasman, J. (1989). Regional identity is established before gastrulation in the *Xenopus* embryo. *J. Exp. Zool.* **251**, 245–252.

Vogt, W. (1929). Gestaltungsanalyse am Amphibienkeim mit örtlicher Vitalfärbung. II. Teil. Gastrulation and Mesodermbildung bei Urodelen und Anuren. *Wilhelm Roux Arch. Entwicklungsmech. Org.* **120**, 384–706.

Wayner, E. A., Garcia-Pardo, A., Humphries, M. J., McDonald, J. A., and Carter, W. G. (1989). Identification and characterization of the lymphocyte adhesion receptor for an alternative cell attachment domain (CS-1) in plasma fibronectin. *J. Cell Biol.* **109**, 1321–1330.

Wedlich, D., Hacke, H., and Klein, G. (1989). The distribution of fibronectin and laminin in the somitogenesis of *Xenopus laevis*. *Differentiation (Berlin)* **40**, 77–83.

Weiss, P. (1945). Experiments on cell and axon orientation *in vitro*: The role of colloidal exudates in tissue organization. *J. Exp. Zool.* **100**, 353–386.

Winklbauer, R. (1988). Differential interaction of *Xenopus* embryonic cells with fibronectin *in vitro*. *Dev. Biol.* **130**, 175–183.

Winklbauer, R. (1990). Mesodermal cell migration during *Xenopus* gastrulation. *Dev. Biol.* **142**, 155–168.

Winklbauer, R., and Nagel, M. (1991). Directional mesoderm cell migration in the *Xenopus* gastrula. *Dev. Biol.* **148,** 573–589.

Woolf, T. M., Jennings, C. G. B., Rebagliati, M. and Melton, D. (1990). The stability, toxicity and effectiveness of unmodified and phosphorothioate antisense oligodeoxynucleotides in *Xenopus* oocytes and embryos. *Nucleic Acids Res.* **18,** 1763–1769.

4
Role of Cell Rearrangement in Axial Morphogenesis

Gary C. Schoenwolf and Ignacio S. Alvarez
Department of Anatomy
University of Utah
School of Medicine
Salt Lake City, Utah 84132

I. Introduction
II. Cell Rearrangement during Neurulation
   A. Primary and Secondary Body Development
   B. Overview of Neurulation
   C. Cell Rearrangement Driving Shaping and Bending of the Neural Plate
III. Cell Rearrangement during Gastrulation
   A. Gastrulation in Sea Urchins: Archenteron Formation and Elongation
   B. Gastrulation in *Xenopus*: Epiboly and Convergent-Extension
   C. Gastrulation in Teleost Fishes: Epiboly
   D. Protrusive Activity in Cell Rearrangement during Gastrulation
IV. Summary and Model
   References

> *Aristotle among the ancients, and Hieronymus Fabricius of* Aquapendente *among the moderns, have written with so much accuracy on the generation and formation of the chick from the egg that little seems left for us to do.*
>
> William Harvey, 1578–1616, as quoted by Lillie (1919)

I. Introduction

In the 1990s, the statement, "little seems left for us to do," could not be further from the truth. We have made tremendous advances in developmental biology in the more than 2000 years since Aristotle first described formation of the chick embryo. Yet we have only scratched the surface. Each new advance reveals more complexity and highlights our vast state of ignorance concerning the ways in which the embryo goes about making itself.

The focus of this chapter is on the role of cell rearrangement in axial morphogenesis. Cell rearrangement is a process in which cells change positions by

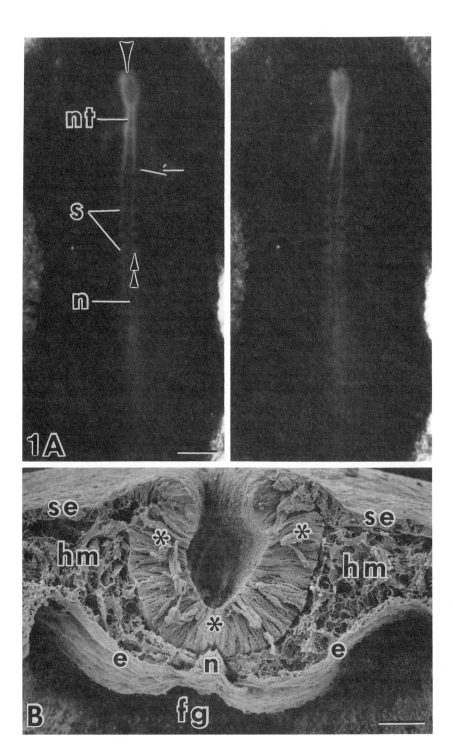

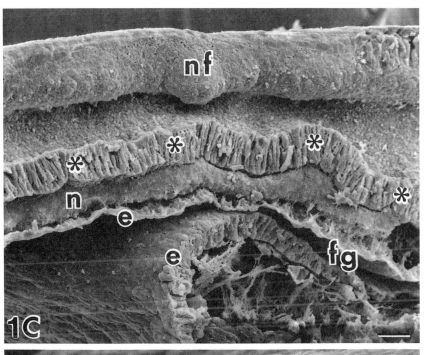
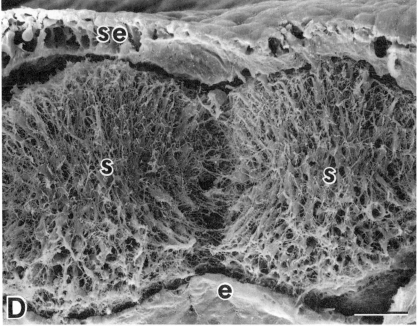

intermixing and, as a result, exchange neighbors. Cell rearrangement is one of a limited number of diverse, fundamental cell behaviors that drive the morphogenetic movements of gastrulation (i.e., formation of the three germ layers—the ectoderm, mesoderm, and endoderm) and neurulation (i.e., formation of the neural tube—the rudiment of the central nervous system). How cell rearrangement occurs is largely unknown, but it would seem that for cells to rearrange in an orderly way, they must have some "idea" of who they are, where they are residing before they start rearranging, where they are going as they are rearranging, and when they have arrived at the proper place and should stop rearranging. The coordination of cell rearrangement and the mechanisms underlying this process are certainly fascinating and intriguing problems to solve.

Through cell rearrangement, cells of early embryos change their locations and ultimately express a unique developmental fate, which is appropriate for their new position. The extensive cell rearrangements occurring during gastrulation and neurulation lead to the formation of the axial and associated rudiments (Fig. 1): the neural tube and notochord, and the neural crest (contributes to head mesenchyme), somites, somitomeres (another source of head mesenchyme), and gut (archenteron). Consequently, our discussion will center on those particular cell rearrangements that contribute to these important morphogenetic events. Cell rearrangement typically results in the process of convergent extension, in which developing rudiments concomitantly narrow in the transverse plane and lengthen in the longitudinal plane (Fig. 2). Understanding how convergent extension occurs lies at the center of our quest to determine the mechanisms of axial morphogenesis.

Below, we will first focus on cell rearrangement during neurulation. Our discussion will be based principally on avian embryos, the model system we have chosen for most of our studies. Then we will focus on cell rearrangement during gastrulation. Our discussion will involve three important aspects of gastrulation and will be based on echinoderms, amphibians, and teleost fishes, model systems that have been studied extensively. Although our discussion of cell rearrangement will be limited to a few systems, it is important to point out that cell rearrangement underlies a wide range of morphogenetic events, including imag-

---

Fig. 1 Views of the axial and associated rudiments of chick embryos during the second day of incubation. (A) Stereopair light micrograph of the dorsal surface; arrow indicates the transverse level shown in (B); large arrowhead indicates the midsagittal level shown in (C); double arrowhead indicates the parasagittal level shown in (D). (B) Scanning electron micrograph (SEM) of a transverse slice through the level indicated by the arrow in (A). (C) SEM of a midsagittal slice through the level indicated by the large arrowhead in (A) (rostral is toward the right). (D) SEM of a parasagittal slice through the level indicated by the double arrowhead in (A) (rostral is toward the right). e, Endoderm; fg, foregut; hm, head mesenchyme; n, notochord; nf, neural fold; nt, neural tube; s, somite; se, surface epithelium; asterisks, median and dorsolateral hinge points. Bars = 500 μm (A); 20 μm (B–D).

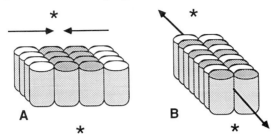

Fig. 2 Schematic drawings of a cluster of epithelial cells (A) rearranging to effect the process of convergent extension (B). During such rearrangement, cells intercalate by moving toward the midline (reducing the transverse extent of the cluster and causing convergence) and pile up rostrocaudally (enlarging the longitudinal extent of the cluster and causing extension). Arrows, directions of cell movements; asterisks, midline.

inal disc formation in *Drosophila* (Fristrom, 1976, 1988; Fristrom and Chihara, 1978; Held, 1979), regeneration in *Hydra* (Graf and Gierer, 1980; Bode and Bode, 1984), scale pattern formation in the moth wing (Nardi and MaGee-Adams, 1986), elongation of the pronephric duct in urodeles (Poole and Steinberg, 1981), wound healing (Honda *et al.*, 1982), Sertoli cell differentiation (Russell, 1979), and notochord formation in ascidians and amphibians (Miyamoto and Crowther, 1985; Keller *et al.*, 1989).

II. Cell Rearrangement during Neurulation

A. Primary and Secondary Body Development

Before discussing cell rearrangement during neurulation, it is necessary to provide some general information on the development of birds and mammals. In these organisms, the axial and associated rudiments are formed in two separate modes called primary and secondary body development (Holmdahl, 1925a,b; Schoenwolf, 1983, 1991a). Primary body development begins with gastrulation (Fig. 3). During gastrulation, cells migrate toward the caudal end of the epiblast (the dorsalmost layer of the blastoderm), where they accumulate as the initial primitive streak (Fig. 3A). Subsequent elongation (i.e., progression) of the primitive streak (Fig. 3B) transforms the formerly broad accumulation of cells into a narrow, longitudinal strip, the definitive primitive streak (Fig. 3C). During further development, the primitive streak shortens, owing principally to regression (caudal movement) of its rostral part (Fig. 3D and E). Movement of epiblast cells through the primitive streak (i.e., ingression) begins during streak progression and continues throughout its regression (Fig. 3F). Only prospective endodermal and mesodermal cells migrate through the primitive streak; cells remaining on

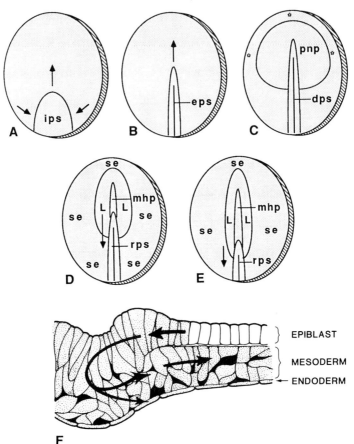

Fig. 3 Schematic drawings of the dorsal surface of the epiblast (A–E) or of a transverse section through the primitive streak (F) during avian gastrulation and neurulation. Arrows indicate directions of cell movement associated with formation of the primitive streak (A), progression of the primitive streak (A and B), regression of the primitive streak (D and E), and ingression of cells through the primitive streak (F). dps, Definitive primitive streak; eps, elongating primitive streak; ips, initial primitive streak; L, lateral cells of the neural plate; mhp, median hinge-point cells of the neural plate; pnp, prospective neural plate (prospective neurepithelium); rps, regressing primitive streak; se, surface epithelium; asterisks, prospective surface epithelium. In (F), ingressing and ingressed cells are shaded.

the surface form the ectoderm. As gastrulation is underway, primary development continues with the gradual assembly of distinct rudiments from each of the forming germ layers, namely, the neural tube and neural crest from the ectoderm, the notochord, somites, and somitomeres from the mesoderm, and the gut (fore-, mid-, and hindgut) from the endoderm. Thus during primary development, axial rudiments, such as the neural tube, progressively form just rostral to the primitive streak as the latter undergoes its regression (Fig. 3D and E).

Secondary development occurs independently of germ layer formation. Instead, during this process rudiments develop directly from the tail bud, a spherical accumulation of mesenchymal cells situated at the caudal end of the embryo and derived after the completion of gastrulation from remnants of the primitive streak (Schoenwolf, 1979). Mapping studies have revealed that the tail bud forms the caudal portions of the neural tube, neural crest, and somites but not the notochord (Schoenwolf, 1977; Schoenwolf and Nichols, 1984; Schoenwolf et al., 1985). Rather, the notochord arises during primary development and extends caudally from more cranial regions into areas undergoing secondary development (Schoenwolf, 1978).

B. Overview of Neurulation

The focus of this and the subsequent section will be on primary neurulation, a process that occurs in four distinct but temporally and spatially overlapping stages: (1) formation of the neural plate, (2) shaping of the neural plate, (3) bending of the neural plate, and (4) closure of the neural groove. To set the stage for our discussion of cell rearrangement during neurulation, each of these stages will be described briefly below (for additional details, see reviews by Karfunkel, 1974; Schoenwolf, 1982; Gordon, 1985; Schoenwolf and Smith, 1990).

During formation of the neural plate, the ectoderm becomes demarcated according to two developmental fates: neurepithelium and surface epithelium (Fig. 3C). The neural plate consists of neurepithelial cells, which became taller than adjacent surface epithelial cells and begin to express unique molecular markers (e.g., Keane et al., 1988). Specification of neurepithelial cell fate has been long known to be the result of induction (e.g., reviewed by Spemann, 1938; Gurdon, 1987; Yamada, 1990). The discovery of induction by Spemann and Mangold (see the historical account by Hamburger, 1988) was a much simpler step than deciphering how this process comes about. Although still not fully understood, it is known that during neural induction, signaling occurs—mediated by diffusible factor(s) and cognate receptor(s)—between the chordamesoderm (amphibians) or endoderm/mesoderm (birds and mammals) and overlying ectoderm, ultimately setting off a cascade of inductive events (reviewed by Jacobson and Sater, 1988).

Shaping of the neural plate begins shortly after neural plate formation and results in a change in the overall configuration of the nascent rudiment (cf. Fig. 3C–E). During shaping, the neural plate on the average thickens apicobasally (i.e., its cells get taller), narrows mediolaterally (i.e., its width or transverse extent decreases), and lengthens longitudinally (i.e., its length or craniocaudal extent increases) (Burnside and Jacobson, 1968; Jacobson and Gordon, 1976;

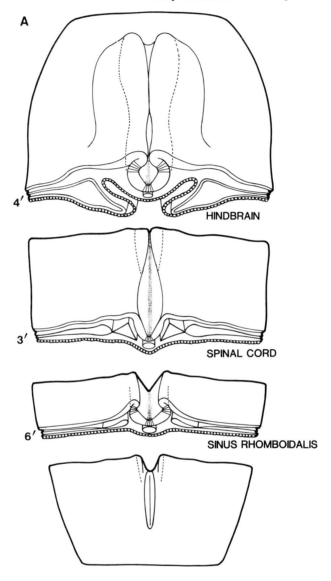

Fig. 4 Drawings of chick embryos undergoing neurulation. (A) Dorsal view at the midneurula stage showing representative transverse levels. Levels indicated by 3', 4', and 6' indicate, respectively, levels shown in parts 3, 4, and 6 of (B). Shading indicates the location of the median hinge point; dotted lines indicate the basal position of the dorsolateral hinge points. (B) Parts 1–4 are schematic drawings of transverse sections at the brain level: 1, flat neural plate [stage shown is younger than that in (A)]; 2, neural plate with furrowed median hinge point (M), consisting of wedge-shaped neurepithelial cells and notochord (N), and flanking spindle-shaped lateral neurepithelial cells (L) [stage shown is younger than that in (A)]; 3, neural plate undergoing folding with neural folds (asterisks)

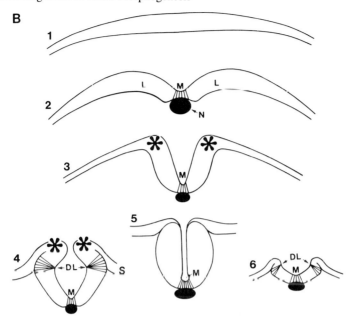

Fig. 4 (*Continued*)
elevating around the furrowed median hinge point (M) [stage shown is comparable to that at level 3' of (A)]; 4, neural plate undergoing folding with neural folds (asterisks) fully elevated around the furrowed median hinge point (M) and each converging around a furrowed dorsolateral hinge point (DL), which consists of wedge-shaped neurepithelial cells and adjacent surface epithelium (S) [stage shown is comparable to that at level 4' of (A)]; 5 and 6, schematic drawings of transverse sections at the rostral and caudal (sinus rhomboidalis) spinal cord levels, respectively: 5, neural plate undergoing folding with neural folds fully elevated around the furrowed median hinge point (M) [dorsolateral hinge points do not form at this level of the neuraxis; stage shown is older than that in (A)]; 6, neural plate undergoing folding with neural folds fully converged around the dorsolateral hinge points (DL) but with neural folds just beginning their elevation around the median hinge point (M) [stage shown is comparable to that at level 6' of (A)]. Parts 1–6 are adapted from Fig. 3 of Smith (1988). The median and dorsolateral hinge points contain wedge-shaped neurepithelial cells as indicated in transverse sections.

Jacobson and Tam, 1982; Morriss-Kay, 1981; Schoenwolf, 1985; Tuckett and Morriss-Kay, 1985).

Bending of the neural plate begins while its shaping is underway and involves two basic processes (Figs. 4–6): *furrowing* of the neural plate and *folding* of the neural plate (Schoenwolf, 1982, 1983). Furrowing of the neural plate, the formation in the originally flat neural plate of shallow longitudinal grooves (Fig. 5B) lined with wedge-shaped neurepithelial cells (Fig. 6A and B), occurs in areas called hinge points. Three hinge points form during bending of the neural plate (Fig. 4), one median (the MHP) and paired dorsolateral (the DLHPs), and each consists of a localized area of neural plate anchored to adjacent tissues

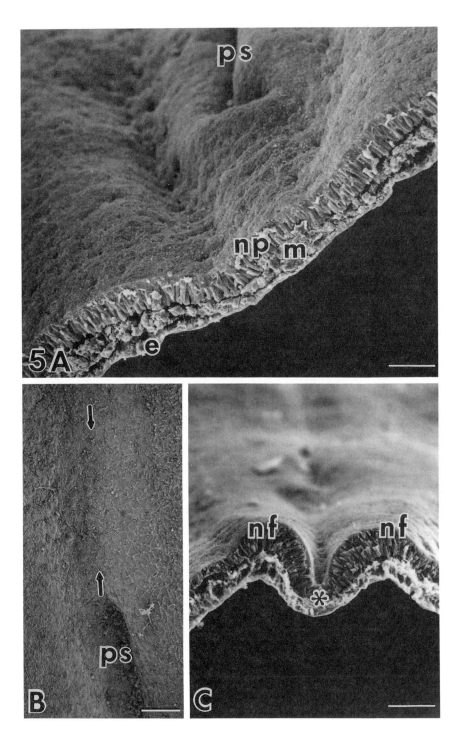

(notochord for the MHP and surface epithelium of the neural folds for the DLHPs). Furrowing of the neural plate within each hinge point is followed by folding of the neural plate around each hinge point. Folding involves two movements: *elevation* of the neural folds dorsally, and *convergence* of the neural folds medially (Fig. 4). Specifically, during elevation of the neural folds, the latter rotate around a longitudinal axis centered at the MHP and, during convergence of the neural folds, each of the latter rotate around a longitudinal axis centered at its corresponding DLHP. Furrowing and subsequent folding around the median hinge point convert the flat neural plate into a V-shaped neural groove (as viewed in cross-section; Fig. 4B, parts 2, 3, and 5), and furrowing and subsequent folding around each dorsolateral hinge point convert the V-shaped neural groove into a diamond-shaped structure (as viewed in cross-section; Fig. 4B, parts 4 and 6).

As a result of bending of the neural plate, the neural folds are brought into apposition in the dorsal midline, where they adhere and subsequently fuse. The stage of closure of the neural groove, which results in the formation of the roof plate of the neural tube and the associated neural crest and overlying surface epithelium, is poorly understood. For example, it is known that the neural folds contain a surface coat, which makes the initial contact across the midline during closure (Moran and Rice, 1975; Lee *et al.*, 1976a,b, 1978; Rice and Moran, 1977; Mak, 1978; Sadler, 1978; Silver and Kerns, 1978; O'Shea and Kaufman, 1980; Rovasio and Monis, 1981; McLone *et al.*, 1983; McLone and Knepper, 1985/86; Smits-van Prooije *et al.*, 1986a; Takahashi, 1988; Takahashi and Howes, 1986), but its composition and role are largely unknown.

C. Cell Rearrangement Driving Shaping and Bending of the Neural Plate

Two aspects of primary neurulation are driven partially by cell rearrangement: shaping and bending of the neural plate. Consequently, only these two processes will be discussed further.

At least three populations of ectodermal cells undergo extensive rearrangement during shaping and bending of the neural plate (Fig. 7): surface epithelial (SE) cells, lateral neuroepithelial (L) cells, and median hinge-point neuroepithelial (MHP) cells. The movements of these cells have been followed in either chick embryos, injected with heritable cell markers, or in quail/chick transplantation chimeras in which plugs of chick epiblast are removed and replaced with com-

---

Fig. 5 Scanning electron micrographs of chick embryos during the first day of incubation. (A) Transverse slice through the neural plate. (B) Dorsal view of the midline furrow associated with the median hinge point. (C) Transverse slice through the neural groove. e, Endoderm; m, mesoderm; nf, neural fold; np, neural plate; ps, primitive streak; arrows, extent of midline furrow; asterisk, median hinge point. Bars = 40 μm (A and C) and 20 μm (B).

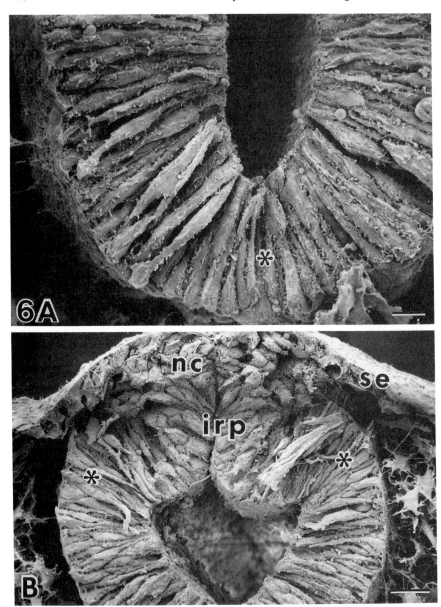

Fig. 6 Scanning electron micrographs of transverse slices through the incipient neural tube of chick embryos. Asterisks, median hinge point (A) and dorsolateral hinge points (B); irp, incipient roof plate; nc, neural crest; se, surface epithelium. Bars = 5 μm (A) and 10 μm (B).

parable plugs of quail epiblast. Recently, fluorescent vital dyes have been used to label the donor plugs and to follow the rearrangement of their cells in living embryos over time (Fig. 8). Such experiments, in combination with additional ones utilizing microsurgery and other techniques of experimental embryology (see below), have suggested that the movement of MHP and L cells within the neurepithelium contributes intrinsic forces for shaping of the neural plate (i.e., forces generated *within* the neural plate), whereas the movements of SE cells (i.e., ectodermal cells located lateral to the neurepithelium and contributing to the surface layer covering the body of the embryo) contribute extrinsic forces for bending of the neural plate (i.e., forces generated *outside* the neural plate in lateral tissue). We will first consider the evidence for intrinsic and extrinsic neurulation forces, and then describe patterns of cell rearrangement contributing to the shaping and bending of the avian neural plate. Subsequently, we will discuss briefly two other cell behaviors, cell division and change in cell shape, that assist cell rearrangement in the shaping and bending of the neural plate. Finally, we will conclude this section by suggesting that multiple aberrant cell behaviors can lead to the formation of neural tube defects.

1. Evidence of Intrinsic and Extrinsic Neurulation Forces

It has become clear over the last few years that shaping and bending of the neural plate are driven by the coordinated actions of both intrinsic and extrinsic neurulation forces, and that such forces are generated by a limited repertoire of common morphogenetic cell behaviors (Schoenwolf and Smith, 1990). Such behaviors include changes in cell shape, size, position, number, and cell–cell and cell–extracellular matrix associations.

Neurulation generally occurs over a relatively short time course, the length of which is species dependent. It occurs in concert with several other developmental events as diverse as gastrulation, subdivision of the mesoderm (including segmentation of the somites and somitomeres), formation of the heart, development of the body folds, and formation of the gut. So the question becomes, if we are interested in determining the mechanism of neurulation, and if neurulation is a coordinated part of the overall morphogenesis of the embryo, which cells are the most important ones to study? Logically enough, the traditional focus has been on the neurepithelial cells composing the neural plate.

Fig. 7 Scanning electron micrographs showing the morphology of three types of ectodermal cells in chick embryos. (A) Whole mount; arrow indicates the level of the transverse slice shown in (B). (B) Transverse slice from the level shown in (A); arrowheads indicate three levels enlarged in (C) through (E). (C–E) Enlargements, in medial to lateral order, of the areas indicated by arrowheads in (B). e, Endoderm; L, lateral neurepithelial cells; m, mesoderm; mhp, median hinge-point neurepithelial cells; n, notochord; nf, neural fold; ps, primitive streak; se, surface epithelium. Bars = 200 μm (A), 100 μm (B), and 10 μm (C–E).

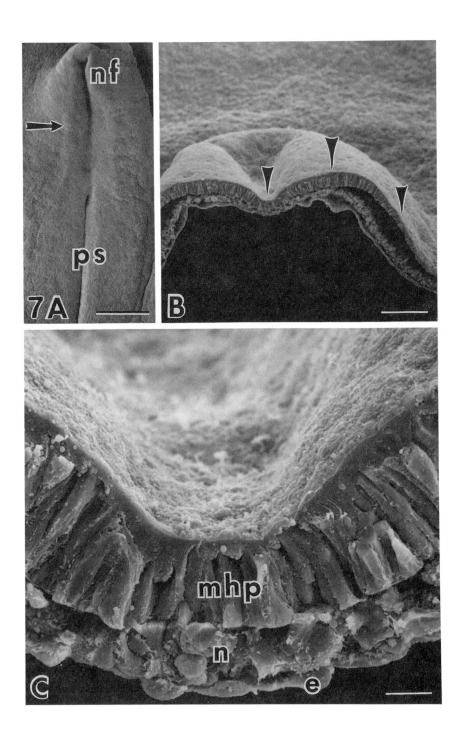

142

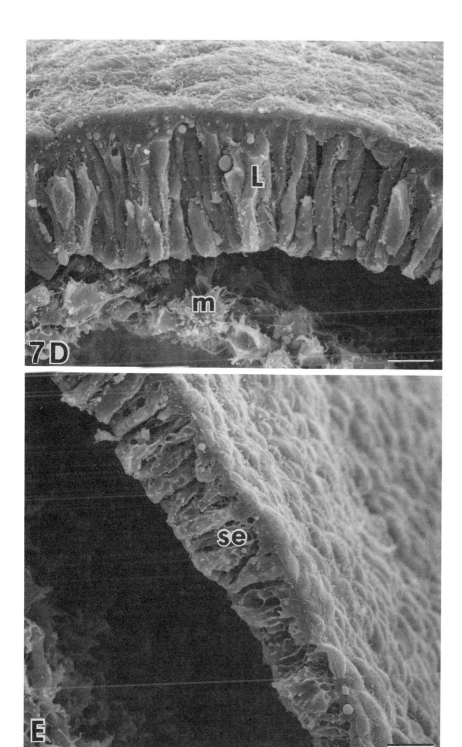

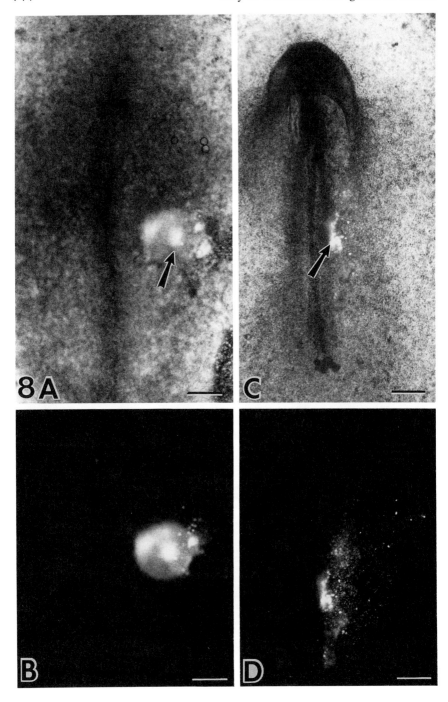

But the almost exclusive study of neural plate cells, which has continued over the last 100 years, has been done largely for the wrong reason—the belief that intrinsic neurulation forces by themselves are both necessary and sufficient to drive all of neurulation, or at least its most crucial aspects. This erroneous belief arose with the classical experiment of Roux (1885), in which the neural plate, when isolated *in vitro*, rolled up into a structure resembling a neural tube. More recently, it has been shown that epithelial sheets from early embryos quickly roll up when isolated in a fluid medium, but they consistently roll up inside out; that is, with their original basal side on the inside and their original apical side on the outside (Burnside, 1972; Jacobson, 1981; Vanroelen *et al.*, 1982; Stern *et al.*, 1985; Schoenwolf, 1991b). During normal neurulation, the original apical side of the neural plate ultimately lines the inside of the neural tube and the original basal side, the outside. Because isolated epithelial sheets curl in a direction exactly opposite to that purported to be driven by intrinsic neurulation forces, such experiments tell us how sheets act under artificial conditions, but they reveal little or nothing about mechanisms of neurulation occurring in the embryo.

Instead, our knowledge about the origin of neurulation forces derives largely from (1) microsurgical studies conducted on avian blastoderms, and (2) chemical perturbation studies conducted on both avian and mammalian embryos to disrupt the extracellular matrix.

The results of the first type of experiment suggest collectively that shaping of the neural plate is driven by principally intrinsic forces, whereas bending (i.e., furrowing and folding) is driven by both intrinsic and extrinsic forces. The MHP is essentially the only region where neurepithelial cell wedging occurs during *elevation* of the neural folds; traditionally, cell wedging has been believed to provide the major intrinsic force for neurulation (discussed in detail by Schoenwolf and Smith, 1990). In experiments of the first type, the MHP was either removed entirely or was manipulated so that cell wedging was prevented (Smith and Schoenwolf, 1989, 1991). Nevertheless, shaping and folding of the neural plate still occurred often with closure of the neural groove (in those experiments in which cell wedging was blocked, furrowing of the MHP failed to occur, providing evidence that wedging causes furrowing; see Schoenwolf and Smith, 1990, for additional evidence). Thus, this group of experiments (and particularly the MHP removal experiment; see discussion in Smith and Schoenwolf, 1991) demonstrates that at least some forces necessary for neurulation must arise outside the neural plate. This conclusion is supported further by similar experiments in amphibian embryos in which the entire neural plate was removed microsurgically or "inactivated" chemically (Jacobson and Jacobson, 1973; Brun and

---

Fig. 8 Light micrographs of dorsal views of chick embryos containing grafts of epiblast plugs labeled with a fluorescent vital dye. (A and B) Two hours postgrafting [(A) brightfield; (B) fluorescence]; (C and D) Fifteen hours postgrafting [(C) brightfield; (D) fluorescence]. Arrows, site of graft. Bars = 200 μm.

Garson, 1983); the lateral SE still migrated up toward the midline as "neural folds" and underwent fusion. In other experiments of the first type the MHP was left intact, but all nonneurepithelial tissue lateral to the neural plate was removed (Schoenwolf, 1988). Shaping of the neural plate and furrowing of the MHP still occurred (supporting a role for intrinsic forces), but folding of the neural plate was strongly inhibited (supporting a role for extrinsic forces). In cases in which the lateral tissues were removed from one side only, a half-neural tube formed on the intact side, demonstrating (when considered in conjunction with additional controls; see Schoenwolf and Smith, 1990, for discussion) that the trauma of microsurgery per se does not block neurulation. In the final experiment of the first type, mesodermal and endodermal tissues underlying and lateral to the neural plate were dissected away shortly after their formation but before elevation of the neural folds had begun (Alvarez and Schoenwolf, 1992). This experiment differed from the one just described, mainly in that the *surface epithelium* was left intact. Under these conditions, shaping and bending of the neural plate, sometimes with closure of the neural groove, still occurred. Thus, this experiment suggests that a principal extrinsic force for bending of the neural plate is generated by the surface epithelium (the only lateral tissue left intact).

Experiments examining the role of the extracellular matrix in providing extrinsic neurulation forces have been performed in both avian and mammalian embryos (Anderson and Meier, 1982; Morriss-Kay and Crutch, 1982; Morriss-Kay et al., 1986; Morriss-Kay and Tuckett, 1989; Schoenwolf and Fisher, 1983; Smits-van Prooije et al., 1986b; Tuckett and Morriss-Kay, 1989). The mesenchymal compartment underlying the neural folds is particularly rich in matrix. Depletion of the extracellular matrix, either by removing hyaluronic acid, chondroitin sulfate-proteoglycan, or heparan sulfate proteoglycan, or by inhibiting chondroitin sulfate-proteoglycan synthesis, delays or blocks neurulation. Unfortunately, there are two major problems with these experiments that cloud their interpretation: (1) most of the chemical agents used for matrix perturbation are embryotoxic, and (2) attempts to remove one specific component of the matrix can result in the dislodging of other spatially affiliated components. Nevertheless, these experiments at least support a role for the extracellular matrix in generating neurulation forces, and additional studies are encouraged. One avenue to explore, in view of the last microsurgical experiment described above (Alvarez and Schoenwolf, 1992), is the possible role of the surface epithelium-associated extracellular matrix (see Schoenwolf and Fisher, 1983, for plausible mechanisms remaining to be tested).

2. Patterns of Cell Rearrangement

Patterns of cell rearrangement during neurulation have been studied most extensively in avian embryos (Schoenwolf and Alvarez, 1989, 1991; Schoenwolf et al., 1989; Schoenwolf and Sheard, 1989, 1990; Alvarez and Schoenwolf, 1991), although data are also available for amphibian (Jacobson and Gordon, 1976;

4. Cell Rearrangement in Axial Morphogenesis

Suzuki and Harada, 1988) and mammalian (Morriss-Kay and Tuckett, 1987) embryos. Results from these three classes of vertebrates are in agreement, and here we will focus on avian embryos. Three patterns of cell rearrangement will be described. Each pattern is associated with a group of cells contributing to different regions of the ectoderm. Two patterns are generated within the neural plate, one for prospective MHP cells and one for prospective L cells; an additional pattern is generated within the surface epithelium, specifically, in its intraembryonic portion (i.e., the portion contributing to the outer surface of the embryo proper).

Prospective MHP and L cells arise within the portion of the epiblast surrounding the cranial part of the primitive streak. Median hinge point cells arise from a small, circumscribed, midline area just rostral to and overlapping Hensen's node (the cranial end of the primitive streak), whereas L cells arise from paired epiblast zones immediately flanking the cranial part of the primitive streak (Fig. 9; Schoenwolf and Alvarez, 1989; Schoenwolf et al., 1989; Schoenwolf and Sheard, 1990; also see Rosenquist, 1966, 1983; Nicolet, 1970). Median hinge point cells of the future *brain* originate principally cranial to Hensen's node, in contrast to those of the future *spinal cord*, which originate both cranial to as well as in Hensen's node (Smith and Schoenwolf, 1991; Schoenwolf et al., 1992).

The rearrangement of MHP and L cells closely parallels the regression of Hensen's node that occurs during gastrulation (Fig. 9). Hensen's node contains, in addition to some prospective MHP cells, the precursor cells of the notochord. Prospective MHP cells rearrange during the regression of Hensen's node and during the concomitant formation of the notochord and shaping and bending of the neural plate. They undergo as a group the process of convergent extension— that is, the width of the forming MHP area progressively decreases while its length increases, owing to the lateral-to-medial intercalation of its cells. Thus, the definitive MHP region eventually becomes a narrow strip of cells extending along the length of the ventral midline of the neural tube. Here, these cells collectively form the floor plate (Schoenwolf et al., 1989; Fraser et al., 1990), a specialized, apparently nonneuronal region, extending from the midbrain to caudal spinal cord levels. Each L cell strip also undergoes convergent extension in concert with that of the forming MHP. During this process, prospective L cells intercalate, so that the lateralmost cells on each side move toward the medialmost cells flanking the incipient MHP. Here, they collectively form the lateral walls of the brain and spinal cord.

Cells from the perinodal sites also move rostrally to contribute to the forebrain level of the neural tube (Fig. 9C; Schoenwolf and Alvarez, 1989). Such a "flow" of cells rostrally was described first for mammalian embryos (Morriss-Kay and Tuckett, 1987). Cells labeled with wheat germ agglutinin were injected into the neurepithelium of the future midbrain and hindbrain levels of rat embryos developing in whole-embryo culture. After an additional 24 hr of development, labeled cells were found more rostrally, indicating that they had moved forward with expansion of the forebrain region of the neural plate.

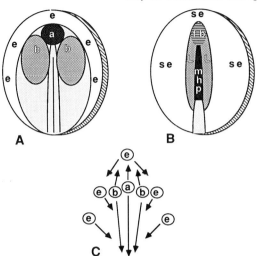

Fig. 9 Schematic drawings of dorsal views of the avian epiblast. Designation of areas in (A) by letters a, b, and e corresponds to the nomenclature established by our laboratory (Schoenwolf and Alvarez, 1989, 1991; Schoenwolf et al., 1989). (A) Stage comparable to that in Fig. 3C. (B) Stage comparable to that in Fig. 3E. (C) Arrows indicate the directions of cell movements between the stages in (A) and (B). a, Prenodal area that forms the median hinge-point neurepithelial cells (mhp); b, paranodal areas that form the lateral neurepithelial cells (L); e, peripheral areas that form the surface epithelium (se); fb, forebrain.

Prospective SE cells also undergo convergent extension, which is coordinated spatially and temporally with that occurring within the MHP and L regions (Fig. 9). These cells originate as far laterally in the epiblast as the area pellucida–area opaca border and then rearrange on each side in a lateral to medial direction. Consequently, the epithelial integrity of the epiblast is maintained during the extensive convergent extension process underlying the shaping and bending of the neural plate, as cells from more lateral regions (i.e., SE cells) move medially to replace those cells that have moved further medially (i.e., L and MHP cells). Thus, as the neural plate narrows transversely during its shaping, SE cells move medially to fill the positions occupied formerly by L cells (and L cells move medially to fill the positions occupied formerly by MHP cells). This process of medial movement of SE cells accelerates during bending of the neural plate and contributes to the elevation and convergence of the neural folds (Alvarez and Schoenwolf, 1992; Schoenwolf and Alvarez, 1991). It must be emphasized, however, that the medial spread of SE cells is not owing solely to cell rearrangement. At least two other cell behaviors play a role: division and change in shape. Furthermore, these same two additional cell behaviors also act in conjunction with cell intercalation *within the neural plate*, contributing to its convergent extension (as well as the expansion of its future forebrain level). The specific roles of cell division and change in cell shape in shaping and bending of the neural plate are discussed below.

3. Roles of Cell Division and Change in Cell Shape in Shaping and Bending of the Neural Plate

Cell division and change in cell shape are important behaviors underlying shaping and bending of the neural plate. In higher vertebrates (i.e., amniotes), including humans, the embryo grows during neurulation. Consequently, the volume of the embryo increases, and differential growth would be expected to generate morphogenetic forces. Little is known about change in *cell* size in embryos during the period of neurulation. By contrast, cell division has received substantial attention. In avian and mammalian embryos most, if not all, cells of the neural plate divide throughout neurulation (Langman *et al.*, 1966; Jacobson and Tam, 1982; Smith and Schoenwolf, 1988). This division represents true growth, that is, each daughter cell enlarges to the former size of the parental cell, thereby increasing the volume of the neural plate (Jacobson and Tam, 1982; Schoenwolf, 1985). The direction of placement of daughter cells within the plane of the epithelium also plays an important role in morphogenesis. The study of mitotic spindles has revealed that differences exist in their orientation (and presumably as a consequence, in the placement of daughter cells within the neurepithelium) depending on the area examined (Langman *et al.*, 1966; Martin, 1967; Jacobson and Tam, 1982; Tuckett and Morriss-Kay, 1985; Zieba *et al.*, 1986). Moreover, modeling studies have suggested also that daughter cells would have to be positioned differently at different levels of the neuraxis (Schoenwolf and Alvarez, 1989). For example, daughter cells would be placed within the transverse plane of the neural plate at the future forebrain level, thereby contributing to the widening that occurs at this level, and within the longitudinal plane further caudally, thereby contributing to the lengthening. Precisely how this positioning is effected, and whether it is dependent on the orientation of the mitotic spindle, the cleavage furrow, or some other mitotic or nonmitotic (i.e., population density) event, remains to be determined.

Neurepithelial cells of avian embryos undergo two to three cycles of division during shaping and bending of the neural plate (Smith and Schoenwolf, 1987; Schoenwolf and Alvarez, 1989). In addition, their cycles are regulated differentially during formation of the hinge points (Smith and Schoenwolf, 1987, 1988; van Straaten *et al.*, 1988). Such regulation has been postulated to lead to localized changes in neurepithelial cell shape (Schoenwolf and Franks, 1984). Because the neural plate of avian and mammalian embryos is a pseudostratified, columnar epithelium, its dividing cells characteristically exhibit the "to and fro" movements of interkinetic nuclear migration (Sauer, 1935; Watterson, 1965; Langman *et al.*, 1966); that is, cell nuclei migrate toward the apex of the neurepithelium to undergo mitosis and then each daughter nucleus migrates back toward the base of the neurepithelium. The widest portion of each neurepithelial cell generally contains the nucleus (Schoenwolf and Franks, 1984). Consequently, as the position of the nucleus changes during interkinetic nuclear migration, the overall shape of the cell changes; for example, from globular, when the cell has rounded up for division and resides at the apex of the neurepithelium; to inverted wedge shaped, when the daughters are each extending a process toward the base of the neurepithelium; to spindle shaped, when the nucleus has migrated from the apex of the neurepithelium but has not yet reached its base; to wedge shaped, when the nucleus resides at the base of the neurepithelium. With the onset of the next cell cycle, the cell changes from wedge shaped to spindle shaped, when its nucleus has migrated from the base of the neurepithelium and is moving toward the apex; to inverted wedge shaped, when the nucleus has reached the apex, but the basal process has not yet been fully withdrawn; to globular, when the cell has rounded up in preparation for its next division. Therefore, regulation of interkinetic nuclear migration, namely, changing its timing, could be a mechanism controlling cell shape. This hypothesis is supported by the observation that MHP cells, most of which are wedge shaped, have as a population longer cell-cycle lengths than do L cells, most of which are spindle shaped (Smith and Schoenwolf, 1987), and that the nucleus resides at the base of the neurepithelium in MHP cells (i.e., its characteristic position in wedge-shaped cells) during the prolonged phases of the cell cycle (Smith and Schoenwolf, 1988).

Alternatively, change in neurepithelial cell shape could be mediated by the apical constriction of neurepithelial cells. It has long been known that embryonic epithelial cells (including MHP, L, and SE cells) contain circumferentially arrayed, apical bands of microfilaments, and that the disruption of these bands with cytochalasins inhibits neurulation, but careful analysis has revealed that neurepithelial cells can still undergo interkinetic nuclear migration and become wedge shaped in the presence of cytochalasins and in the absence of apical bands of microfilaments (Schoenwolf et al., 1988). Moreover, the size of the apex of spindle-shaped neurepithelial cells appears similar to that of wedge-shaped cells (Schoenwolf and Franks, 1984). Thus, although microfilaments likely function in controlling neurepithelial cell shape, they do not seem to be involved in interkinetic nuclear migration, especially in the repositioning of nuclei at the base of the neurepithelium, nor are they necessary for cell wedging. Therefore, a basal expansion mechanism, rather than an apical constriction mechanism, seems to be the more important one for neurepithelial cell wedging.

4. Role of Abnormal Cell Behavior in Formation of Neural Tube Defects

Neural tube defects are considered to arise in the vast majority of instances by a failure of primary neurulation. As a result of such failure, the neural groove stays unclosed, and neural tissue remains exposed on the dorsal surface of the embryo (i.e., the neural tube exhibits dysraphism). How do such defects arise? The traditional answer has been based on the assumptions that neurulation forces are intrinsic to the neural plate and that they are generated by principally microfilament-mediated neurepithelial cell wedging. As emphasized above these assumptions are incorrect: neurulation is a multifactorial process involving several morphogenetic cell behaviors, and multiple mechanisms likely underlie each behavior. Moreover, change in cell behavior both within the neural plate and outside of this structure is required for normal neurulation. The fact that multiple cell behaviors contribute neurulation forces is consistent with the observation that a wide variety of teratogenic agents cause neural tube defects (reviewed by Copp et al., 1990).

So, then, what is the answer to the question of how neural tube defects arise? It seems most likely that there are several answers, that a range of aberrant behaviors and mechanisms are involved. Our purpose in stating this is not to leave our readers disconcerted; instead, we offer a plea for additional detailed studies of the cell behaviors that underlie not only normal development but also the formation of congenital anomalies such as neural tube defects. It is our belief that through such studies, our understanding of the complexity of normal and abnormal development will increase and that some of the most elusive secrets of the embryo will be revealed.

III. Cell Rearrangement during Gastrulation

Gastrulation involves several events in both invertebrate and vertebrate embryos. Such events include the formation and elongation of the archenteron (the primitive endodermal gut), the convergent extension process during formation of the mesoderm (the middle germ layer), and epiboly (the spreading of a layer or layers of cells over the yolk). Rather than discussing these processes exhaustively for each organism, we will provide a brief overview of each process in a model system where the role of cell rearrangement has been studied most thoroughly. Then we will describe the evidence for cell rearrangement in each system and its role in gastrulation. Finally, we will discuss briefly the postulated role of cell protrusive activity in cell rearrangement during gastrulation.

A. Gastrulation in Sea Urchins: Archenteron Formation and Elongation

1. Overview

Following total (holoblastic) cleavage, the sea urchin egg consists of a hollow blastula composed of a single layer of blastomeres enclosing the large blastocoel. Cells at the vegetal pole of the epithelium ingress (i.e., migrate as individual cells) into the blastocoel, marking the beginning of gastrulation (Fig. 10). Such

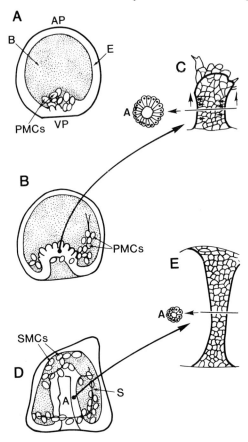

Fig. 10 Drawings illustrating sea urchin gastrulation. (A, B, and D) Highlights [adapted from Figs. 2–4, Trinkaus (1984)]. (C and E) Cell rearrangement (small arrows) within the archenteron during its elongation (large arrows) [adapted from Fig. 13, Hardin and Cheng (1986)]. A, Archenteron; AP, animal pole; B, blastocoel; E, epidermis; PMCs, primary mesenchymal cells; S, spicule; SMCs, secondary mesenchymal cells; VP, vegetal pole.

ingressing blastomeres constitute the primary mesenchymal cells (PMCs). Following their migration into the interior, the PMCs aggregate into a precise pattern and secrete the spicules or skeleton of the sea urchin embryo.

Gastrulation can be subdivided in two steps after ingression of the PMCs is completed: primary and secondary invagination. During primary invagination, the vegetal pole of the blastula flattens as it forms the vegetal plate, and shortly thereafter it bends inward (invaginates) into the blastocoel, initiating formation of the archenteron (primitive gut) (Fig. 10B). The vegetal plate, during primary invagination, extends between one-quarter and one-half the animal–vegetal length of the blastocoel and then arrests temporarily. After a short pause, secondary invagination begins. The archenteron elongates considerably (two to threefold) during secondary invagination, until it reaches the blastocoel roof (Fig. 10D). At the same time that secondary invagination is occurring, cells bulge from the tip of the elongating archenteron and extend filopodia. These cells are the secondary mesenchymal cells (SMCs); their filopodial extensions serve to attach the tip of the archenteron to the blastocoel roof. Later, they migrate into the blastocoel and form the mesodermal rudiments. The area of contact between the archenteron tip and the blastocoel roof marks the future position of the mouth. The blastomeres that remain on the surface of the blastula, rather than ingressing and invaginating into its interior, differentiate into the epidermis and its derivatives. Through this simple sequence of events, the basic body plan of the radially symmetric sea urchin embryo is established.

2. Evidence for Cell Rearrangement

Evidence for cell rearrangement during sea urchin gastrulation is of three types. First, morphometric analysis of the number of cells in sections through the archenteron of fixed material has revealed that cross-sectional cell number decreases while elongation progresses (cf. Fig. 10C and E; Ettensohn, 1985; Hardin and Cheng, 1986). This decrease has also been shown by following over time the distribution of living cells labeled fluorescently (Hardin, 1989). The decrease in the number of cells that span the circumference of the archenteron can be explained by postulating that archenteron cells intercalate in a circumferential (or spiral) direction, concomitantly with the narrowing and elongation of the tissue. Loss of cells from the archenteron, during formation of the SMCs, does not occur until the end of secondary invagination, and cell death apparently does not occur. Consequently, the decrease in the number of cells spanning cross-sectional levels of the archenteron must be attributed to cell rearrangement.

The second type of evidence for cell rearrangement during elongation of the archenteron comes from the observation that labeled clones of cells introduced into the archenteron during gastrulation collectively change shape as a unit in a manner similar to that of the archenteron as a whole (i.e., the clone narrows and elongates in an animal–vegetal direction). Moreover, such clones exhibit both jagged borders, where they interdigitate with unlabeled cells, and discontinuities formed by intervening unlabeled cells (Hörstadius, 1935; Cameron et al., 1987; Wray, 1987; Hardin, 1989), strongly suggesting that the marked cells intercalate with the unlabeled host population.

The final type of evidence for cell rearrangement during archenteron elongation comes from direct observation of living cells with differential interference contrast optics. By monitoring cells in this manner, Hardin (1989) found that cell position changed over time. Cells that were followed intercalated circumferentially, producing narrower and longer arrays as the archenteron elongated.

3. Role of Cell Rearrangement

The mechanisms postulated to underlie sea urchin gastrulation have been revised recently by Hardin and Cheng (1986). Although earlier studies explained the elongation of the archenteron as resulting from the traction of the SMC filopodia (reviewed by Gustafson and Wolpert, 1963, 1967), cell rearrangement seems to play a more significant role (Ettensohn, 1985; Hardin, 1988, 1989).

What is the evidence for this revised view? First, filopodial traction by the SMCs cannot explain the elongation of the archenteron in exogastrulae. In exogastrulae, the vegetal region of the blastula *evaginates* (i.e., bends outward) rather than invaginates (Dan and Okazaki, 1956; Dan and Inaba, 1968). In such cases, elongation of the archenteron is almost normal even though SMCs fail to form; however, cell rearrangement still occurs normally (Hardin and Cheng, 1986). Second, in species of sea urchin having SMCs that either lack filopodia or have misdirected filopodia, elongation of the archenteron still occurs normally (*Eucidaris tribuloides:* Schroeder, 1981; Trinkaus, 1984; *Lytechinus variegatus:* Morrill and Santos, 1985). Finally, the archenteron still elongates normally following laser ablation of secondary mesenchymal cells (Hardin, 1988). Together, these observations demonstrate that SMC traction is not a sufficient explanation for elongation of the archenteron.

B. Gastrulation in *Xenopus*: Epiboly and Convergent Extension

1. Overview

After total (holoblastic) cleavage of the *Xenopus* egg, different regions can be distinguished in the resulting blastula by fate mapping studies (see Keller, 1986, for a thorough description of *Xenopus* gastrulation and its blastula fate map) (Fig. 11). The animal cap (AC) is located at the animal pole and, therefore, constitutes the upper portion of the blastula. The marginal zone (MZ) is vegetal to the animal cap, lying at the equatorial level. The marginal zone can be subdivided into two regions: (1) the noninvoluting marginal zone (NIMZ) which, as its name suggests, does not involute (i.e., roll inward) during gastrulation and is in contact at its animal side with the AC, and (2) the involuting marginal zone (IMZ), which involutes through the blastopore during gastrulation. The more vegetal region of the blastula is occupied by a mass of yolky cells that undergoes little movement during gastrulation and is called the subblastoporal endoderm (SBE) because of its position with respect to the blastopore and its endodermal fate.

Each of the regions of the blastula is multilayered, consisting of superficial and deep cells. Both the superficial and deep layers in the AC and in the NIMZ are destined to form ectodermal tissues because, as already stated, they remain on the surface rather than involuting into the blastocoel during gastrulation. The IMZ as a whole (both its superficial and deep layers) undergoes involution and

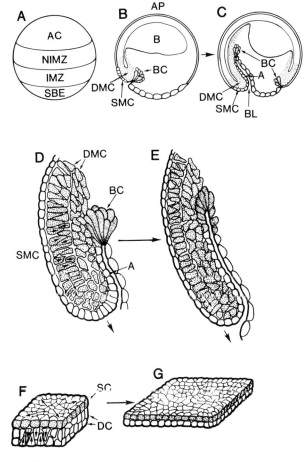

Fig. 11 Drawings illustrating *Xenopus* gastrulation. (A–C) Highlights [adapted from Fig. 9, Keller (1981)]. (D and E) Enlargements of the dorsal lip of the blastopore and the rearrangement of its deep cells (arrows) [adapted from Fig. 13, Keller (1981)]. (F and G) The blastocoel roof and the rearrangement of its cells (arrows) during epiboly [adapted from Fig. 3, Gerhart and Keller (1986)]. A, Archenteron; AC, animal cap; AP, animal pole; B, blastocoel; BC, bottle cells; BL, blastopore; DC, deep cells; DMC, deep marginal cells; IMZ, involuting marginal zone; NIMZ, noninvoluting marginal zone; SC, superficial cells; SBE, subblastoporal endodermal cells; SMC, superficial marginal cells.

gives rise to the archenteron (derived from the superficial layer) and the associated mesodermal structures (derived from the deep layer), including the head mesoderm, notochord, and somites. Some of the superficial cells of the IMZ change their shape and are called bottle cells. The bottle cells mark the initial site of blastopore formation (see Chapter 2, this volume).

As the IMZ involutes, its former position becomes occupied by the NIMZ, and the former position of the NIMZ in turn becomes occupied by the AC. This indicates that the NIMZ and AC expand or undergo epiboly during gastrulation. Expansion of the AC occurs uniformly and radially and it involves both of its layers (Fig. 11F and G). Similarly, expansion of the NIMZ involves both of its layers, but in contrast to the AC its expansion is asymmetric, with the NIMZ extending principally in the animal–vegetal direction (future anterior–posterior axis) and undergoing a contraction or convergence toward the midline of the embryo. In other words, the NIMZ undergoes convergent extension. Both layers of the IMZ also undergo convergent extension in conjunction with their involution. The IMZ elongates in a rostrocaudal direction and converges toward the midline in concert with the convergent extension of the NIMZ (the IMZ involutes in such a way as to maintain contact with the NIMZ; Fig. 11D and E).

During later gastrulation, the involuted mesoderm spreads over the inner surface of the blastocoel roof (Fig. 11C). This process is brought about by the migration of the leading cells of the IMZ, a process that apparently requires an extracellular substratum (Nakatsuji and Johnson, 1983, 1984; Nakatsuji, 1984; Keller and Hardin, 1987; Winklbauer, 1990; see Chapter 3, this volume). Although such migratory movements could assist in involution and extension of the IMZ by creating traction, it seems likely that cell rearrangement is the major contributor to these processes.

2. Evidence for Cell Rearrangement

Direct observation and vital staining have revealed the displacement of cells within the superficial layer of the AC of *Xenopus*, as well as the intercalation of these cells with one another as epiboly occurs (Keller, 1975, 1976, 1978). Superficial cells intercalate only with other superficial cells, not with deep cells. Thus, intercalation occurs only in a circumferential or tangential direction within the spherical gastrula and is referred to as circumferential or lateral intercalation. The deep layer of cells in the AC cannot be observed directly in intact embryos, but evidence for their rearrangement also exists. During epiboly, the number of cell rows in the deep layer of the blastocoel roof progressively decreases (Fig. 11F and G), as shown particularly by scanning electron microscope studies (Keller and Schoenwolf, 1977; Keller, 1980). Because deep cells do not migrate into the superficial layer or into the blastocoel underlying the AC, and because cell degeneration does not occur during amphibian gastrulation, such a reduction in the number of rows can be explained only by means of cell intercalation in a direction parallel to the radius of the egg. Such mixing among different layers of deep cells in a radial direction is called radial intercalation. In summary, cells in both the superficial and deep layers of the AC rearrange during epiboly, although the details of this process are different in each layer.

The importance of the convergent extension process in amphibian gastrulation was realized long ago by several investigators (Vogt, 1922a, b; Spemann, 1902; Mangold, 1920; Lehman, 1932; Schechtman, 1942). Over 50 years ago, it was proposed that convergent extension was driven by cell rearrangement (Waddington, 1940), but direct evidence supporting this possibility was not obtained until recently (see Keller *et al.*, 1985a,b; Keller, 1986, for reviews). Cells in the superficial layer of both the NIMZ and IMZ rearrange, as showed by direct observation in living embryos (Keller, 1978, 1981). Likewise, deep cells rearrange in both the NIMZ and IMZ (Fig. 11D and E). Morphometric data reveal that the number of rows of deep cells in the NIMZ decreases (Keller, 1980) in a manner similar to that described for the deep cells of the AC, which undergo radial intercalation. Furthermore, when cells in the NIMZ were marked with lineage tracers, extensive interdigitation occurred among marked and nonmarked cells, mainly in the mediolateral (circumferential) direction (Keller *et al.*, 1985a,b). Similar analyses have demonstrated cell intercalation in the IMZ as this stream of cells involutes and forms the mesodermal structures (Keller *et al.*, 1985a,b, 1989; Keller and Danilchik, 1988; Keller and Tibbetts, 1989). In additional studies, using an explant culture system, cell behavior in deep cells has been observed directly (Keller *et al.*, 1985a,b). Such direct observation in the so-called "sandwich" explants has confirmed the evidence presented above that cells actively rearrange during convergent extension and has provided information on differences that exist depending on the prospective fate of each group of deep cells examined (i.e., whether they form notochord, head mesoderm, or somites).

3. Role of Cell Rearrangement

Epiboly differs in the superficial and deep layers of the AC owing to the nature of its cells. The superficial layer is an epithelial sheet in which cells are joined tightly, whereas the deep layer consists of loosely packed cells (Keller and Schoenwolf, 1977; Keller, 1980). In addition to cell rearrangement, there is considerable cell division in the superficial layer of the AC, and the radial positioning of daughter cells after division seems to favor the expansion of this layer (Keller, 1978, 1980). Furthermore, changes in cell shape occur in both the superficial and the deep layers, which can account partially for expansion, especially of the superficial layer. Finally, during epiboly, the superficial layer of the AC is under tension, and this tension would again favor the expansion of this layer. Therefore, in addition to cell rearrangement, multiple factors are involved in the epiboly of the superficial layer of the AC during gastrulation.

The situation differs for cells within the deep layer of the AC. This tissue is not under as much tension during epiboly as the superficial layer, and few cell divisions occur while epiboly is underway. Moreover, changes in cell shape seem to be of small consequence and are transitory. Consequently, cell rearrangement seems to be the major process underlying expansion of the deep layer of the AC during epiboly (see Keller, 1986, for a more complete discussion). Additionally, determination of the amount of expansion that radial intercalation could produce in the deep layer of the AC (i.e., by decreasing the number of rows and, therefore, increasing the surface area of the sheet) reveals that cell rearrangement alone is sufficient to account for the amount of epiboly that occurs during gastrulation. Furthermore, such expansion may also play a role in convergent extension and involution of the IMZ, by pushing the deep layer of this region toward and through the blastopore.

Despite the importance of epiboly, convergent extension seems to be the major morphogenetic process contributing to *Xenopus* gastrulation (reviewed by Keller, 1986; Gerhart and Keller, 1986). Before the importance of cell rearrangement as a morphogenetic force was realized, most hypotheses explained gastrulation as a series of pulling forces and mechanical tensions generated within the gastrula as it underwent development. For example, several authors assigned major roles to bottle cells, largely because of their appearance at the beginning of the involution, but the proposed role for bottle cells in directing involution by mechanical traction has not been supported by experimental study. When bottle cells were extirpated, involution and convergent extension still occurred (Keller, 1981; Hardin and Keller, 1988).

Sandwich explants provide direct evidence that cell rearrangement is the major cell behavior involved in convergent extension of the MZ. As opposed to the AC, which does not exhibit normal morphogenetic movements when explanted (it forms a knoblike structure), the isolated MZ (both its NIMZ and IMZ portions) exhibits development that is highly reminiscent of development occurring *in vivo* (Keller *et al.*, 1985a,b). Furthermore, the MZ shows almost complete autonomy from more animal and vegetal regions when isolated in culture. Thus, the isolated *Xenopus* MZ presents an ideal proving ground to examine the cell behaviors involved in convergent extension in the absence of extrinsic forces. Such an explant system has provided strong evidence that cell rearrangement by itself can generate the convergent extension process (Keller *et al.*, 1985a,b; Keller, 1986; Keller and Danilchik, 1988).

C. Gastrulation in Teleost Fishes: Epiboly

1. Overview

During partial (meroblastic) cleavage in teleost fishes, the blastoderm is formed at the animal pole of the egg (Fig. 12). From this location, the blastoderm expands vegetally over the yolk and at the same time flattens considerably. This spreading is called epiboly and in many ways it resembles the epiboly described in the previous section. Through epiboly, the blastoderm expands uniformly over the large yolk mass until the latter is completely covered.

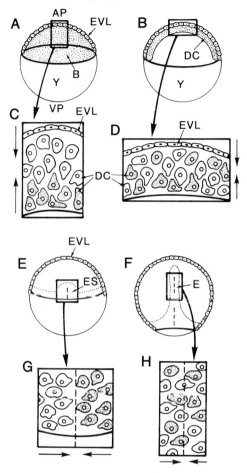

Fig. 12 Drawings illustrating teleost fish gastrulation. (A, B, E, and F) Highlights [adapted from Figs. 11 and 13, Warga and Kimmel (1990)]. (C and D) Cell rearrangement during epiboly; (G and H) convergent extension [adapted from Figs. 11 and 13, Warga and Kimmel (1990)]. AP, Animal pole; B, blastocoel; DC, deep cells; E, embryo; ES, embryonic shield; EVL, enveloping layer; VP, vegetal pole; Y, yolk.

The blastoderm during epiboly consists of two layers: a superficial layer of cells called the enveloping layer (EVL), and the deep layer of cells. The EVL consists of an epithelial sheet one cell thick. By contrast, the deep layer is initially a multicellular layer, about four cells thick, lying deep to the EVL and consisting of loosely packed, irregularly shaped cells called deep cells (DC). During epiboly, this layer thins to one or two cell layers (cf. Fig. 12C and D). The marginal cells of the blastoderm undergo extensive morphogenetic movements that on the whole result in a convergent extension. As this process occurs, the embryonic shield forms at the future posterior midline of the embryo (Fig. 12E–H). The embryonic shield elongates with continued convergent extension and ultimately forms the body of the embryo. In some species (i.e., zebrafish), formation of the embryo also involves involution; cells at the posterior margin of the blastoderm turn inward and migrate into the embryonic shield region. In other species (i.e., *Fundulus*), formation of the embryonic shield involves only convergent extension.

2. Evidence for Cell Rearrangement

Cells in the EVL, and more specifically in its marginal edge, undergo rearrangement during epiboly. Direct evidence for this rearrangement comes from examination of the number of cells spanning the marginal edge. During epiboly, the number of cells in the edge progressively decreases as the edge advances from the equator toward the vegetal pole. This has been documented by staining cell boundaries with silver nitrate (Kageyama, 1980, 1982; Keller and Trinkaus, 1987). Additional observations of the EVL of *Fundulus* failed to detect rearrangement in nonmarginal edge regions (Trinkaus and Lenzt, 1967; Trinkaus, 1976, 1984; Betchaku and Trinkaus, 1978).

In studies on zebrafish, microinjection has been used either to label individual blastomeres with a single fluorescent marker (Kimmel and Law, 1985a) or neighboring blastomeres with two different fluorescent markers (Warga and Kimmel, 1990). The pattern of distribution of marked cells during blastulation and gastrulation was then followed by the direct observation of living embryos over time, as well as by the examination of sectioned, fixed material. Using this approach, cell rearrangement has been observed in the deep layer during both epiboly and subsequent convergent extension. Deep cells intercalate radially during epiboly, resulting in the flattening of this layer through the reduction in the number of rows of cells present in the thickness of the blastoderm. Deep cells also rearrange in the medial-lateral direction during convergent extension (Kimmel and Law, 1985a,b; Warga and Kimmel, 1990; Kimmel *et al.*, 1990).

3. Role of Cell Rearrangement

The role of cell rearrangement in epiboly of the teleost fishes is very similar to that occurring during epiboly in *Xenopus*. The EVL (in both composition and movements) resembles the superficial layer of the *Xenopus* AC, and the DCs correspond to the deep cells of both the AC and NIMZ of *Xenopus*. As in the AC of *Xenopus*, intercalation of cells in the EVL of teleost fishes seems to play only a minor role in blastoderm expansion. There are three reasons for this. First, active cell rearrangement occurs principally in the marginal edge of this layer.

Second, during epiboly of the EVL there is oriented mitotic activity (Kageyama, 1987), which can account partially for EVL expansion during epiboly. Third, and most importantly, the EVL is under tension during epiboly, and such tension favors the vegetal expansion of the blastoderm (see Trinkaus, 1984, for review). Therefore, cell rearrangement in the EVL during epiboly is only one of a series of factors whose relative importance still lies within the realm of modeling and speculation (Weliky and Oster, 1990).

In a way reminiscent of the decisive role of radial intercalation in epiboly of the AC of *Xenopus*, intercalation of the DCs of zebrafish embryos seems to be an important mechanism involved in the expansion of the blastoderm. Alternative mechanisms are lacking currently, but perhaps this is owing to less extensive analyses of the DCs than has been done for other cells. Unfortunately, an explant culture system similar to the sandwich explants used in *Xenopus* is not available for teleost fishes. Thus, the direct observation of DCs in an *in vitro* system has not yet been possible.

D. Protrusive Activity in Cell Rearrangement during Gastrulation

As cells rearrange, they move within epithelial sheets or in more loosely organized mesenchymal tissues. Little is known about how migratory behavior is regulated. One possibility is that cells rearrange as a passive response to mechanical forces applied to them by other morphogenetic events. Although the existence of such "external tensions" has been shown to exist during epiboly in amphibians and teleost fishes (Trinkaus, 1951; Keller, 1980; Keller and Trinkaus, 1987), the requirement for mechanical pressure has not been demonstrated definitively. Furthermore, many of the tissues in which cell rearrangement is occurring are not under significant external tension and, therefore, most of the impetus for cell movement must be generated within the local environment of the cell.

Protrusive activity (also called pulsatile activity or blebbing) occurs during the rearrangement of many populations of gastrulating cells. For example, vigorous protrusive activity is widespread in the cells of the invaginating archenteron of sea urchins (Gustafson and Kinnander, 1956; Kinnander and Gustafson, 1960; Ettensohn, 1985), in the deep cells of the AC of *Xenopus* during the radial intercalation accompanying epiboly (Keller, 1980; Keller *et al.*, 1985a,b), in the deep cells of the NIMZ and IMZ, during intercalation accompanying convergent extension (Keller *et al.*, 1985a,b; Gerhart and Keller, 1986), and in the deep cells of the blastoderm of teleost fishes during epiboly (Trinkaus, 1973; Tickle and Trinkaus, 1973; Kageyama, 1987; Warga and Kimmel, 1990). Although protrusive activity is a characteristic feature of cells undergoing rearrangement, how it is generated and its precise role in cell movement is unknown. Mechanisms underlying cell rearrangement likely differ for epithelial and mesenchymal cells.

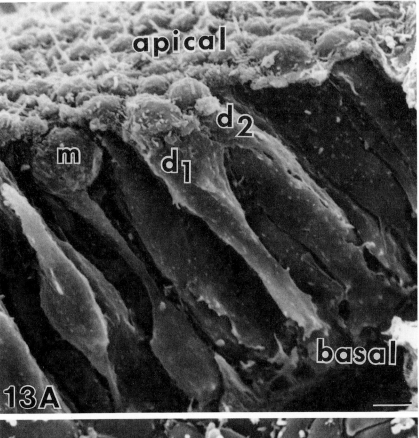

4. Cell Rearrangement in Axial Morphogenesis

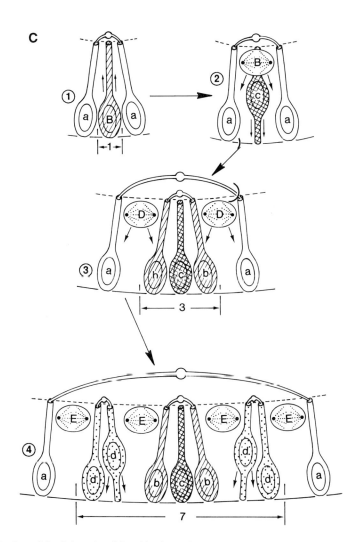

Fig. 13 A model of the role of interkinetic nuclear migration, change in cell shape, and cell division in cell rearrangement postulated from observations based largely on scanning electron microscopy. (A) SEM of a transverse slice. (B) Dorsal view of the apex of the neurepithelium. (C) Schematic model; see text for explanation. The top and bottom of each panel of (C) represent, respectively, the apical and basal sides of the neurepithelium; m, mitotic figure, d_1, d_2, presumed daughter cells; small arrows, cytokinetic bridges connecting cell apices. Bars = 2 μm (A and B).

In the former type, cells are interconnected tightly, whereas in the latter, cells are more loosely organized. Determining how cells move in the different layers of embryos remains a challenge for future studies.

IV. Summary and Model

We have discussed above the role of cell rearrangement in processes as diverse as chick neurulation, archenteron formation and elongation in sea urchin gastrulation, epiboly and convergent extension in *Xenopus* gastrulation, and epiboly in teleost fish gastrulation. It is important to point out that cell rearrangement is merely one of a number of fundamental cell behaviors having morphogenetic consequences. It is highly unlikely that any single, particular morphogenetic movement is driven by only one cell behavior. Rather, various behaviors typically act in concert.

To close this article, we wish to consider one possible interaction between three cell behaviors: change in cell shape, change in cell position, and change in cell number (i.e., cell division) during shaping and bending of the avian neural plate (Fig. 13). As discussed above, all three of these behaviors are involved in neurulation. Recall that the neural plate is a pseudostratified, columnar epithelium. Cells, especially their apices, are interconnected within this epithelium, by cell surface adhesive molecules (Thiery *et al.*, 1982; Edelman *et al.*, 1983; Crossin *et al.*, 1985; Keane *et al.*, 1988), intercellular junctions (see especially Revel *et al.*, 1973; Decker and Friend, 1974; Bellairs *et al.*, 1975; Revel and Brown, 1976; Geelen and Langman, 1979), and cytokinetic bridges (Backhouse, 1974; Bancroft and Bellairs, 1974, 1975; Bellairs and Bancroft, 1975; Waterman, 1976, 1979; Nagele and Lee, 1979, 1980; Schoenwolf, 1982, 1983; Everaert *et al.*, 1988). The presence of such long cytokinetic bridges, with intervening cell apices (Fig. 13B), has been suggested to provide morphological evidence of cell rearrangement (Alvarez and Navascués, 1990). Additionally, their presence seems to play a role in maintaining the integrity of the epithelial sheet during cell division and rearrangement (Sandig and Kalnins, 1990). Throughout neurulation, all neurepithelial cells are dividing and, as they do so, they exhibit interkinetic nuclear migration. We propose a model that links cell rearrangement with cell division and change in cell shape, the latter occurring as a consequence of interkinetic nuclear migration (see above, Section II,C,3). In this model (Fig. 13C) we assume (1) that as a nucleus moves from the base toward the apex in preparation for mitosis (Fig. 13C, step 1), space is vacated and that such space can be filled subsequently by another cell moving into this area (Fig. 13C, step 2) and (2) that after a cell divides and its two daughters extend toward the base, the space vacated at the apex can be filled by other mitotic figures (Fig. 13C, step 3).

Using these assumptions, we can follow the displacement (i.e., rearrangement) of two daughter cells (still connected by a cytokinetic bridge) labeled *a*, who are initially separated by one-cell diameter, namely, a parental cell, labeled B (Fig. 13C, step 1). As cell B rounds up to divide at the apex of the neurepithelium, space is vacated at the basal side of this layer, which allows cell c to bulge into it (Fig. 13C, step 2). This basally directed bulging can be considered as a type of protrusive activity. Parental cell B then divides, with each daughter expanding toward the base, thereby separating daughter cells a by three cell diameters (Fig. 13C, step 3). Two parental cells, labeled D, round up at the apex (Fig. 13C, step 3) and then divide, with each daughter expanding toward the base, thereby separating daughter cells a by seven cell diameters (Fig. 13, step 4). Four parental cells, labeled E, then round up at the apex prior to their division.

This model presents only one of multiple ways in which change in cell shape and cell division (particularly interkinetic nuclear migration) could act in concert to bring about cell rearrangement within an epithelial sheet. We present it here in the hope that it will provide a stimulus for future studies.

Acknowledgments

The superb technical assistance of Fahima Rahman, artistic assistance of Julian Maack, and secretarial assistance of Jennifer Parsons are acknowledged gratefully. The research described herein was supported by grants principally from the National Institutes of Health. I.S.A. was a Fulbright Scholar from the University of Extremadura during the tenure of this study.

References

Alvarez, I. S. and Navascués, J. (1990). Shaping, invagination and closure of the chick embryo otic vesicle. Scanning electron microscopic and quantitative study. *Anat. Rec.* **228,** 315–326.

Alvarez, I. S., and Schoenwolf, G. C. (1991). Patterns of neuroepithelial cell rearrangement during avian neurulation are established independently of notochordal inductive interactions. *Dev. Biol.* **143,** 78–92.

Alvarez, I. S., and Schoenwolf, G. C. (1992). Expansion of surface epithelium provides the major extrinsic force for bending of the neural plate. *J. Exp Zool.*, (in press).

Anderson, C. B., and Meier, S. (1982). Effect of hyaluronidase treatment on the distribution of cranial neural crest cells in the chick embryo. *J. Exp. Zool.* **221,** 329–335.

Backhouse, M. (1974). Observations on the development of the early chick embryo. *Scanning Electron Microsc.* **1974,** 25–532.

Bancroft, M., and Bellairs, R. (1974). The onset of differentiation in the epiblast of the chick blastoderm (SEM and TEM). *Cell Tissue Res.* **155,** 399–418.

Bancroft, M., and Bellairs, R. (1975). Differentiation of the neural plate and neural tube in the young chick embryo. *Anat. Embryol.* **147,** 309–335.

Bellairs, R., and Bancroft, M. (1975). Midbodies and beaded threads. *Am. J. Anat.* **143,** 393–398.

Bellairs, R., Breathnach, A. S., and Gross, M. (1975). Freeze-fracture replications of junctional complexes in unincubated and incubated chick embryos. *Cell Tissue. Res.* **162**, 235–252.

Betchaku, T., and Trinkaus, J. P. (1978). Contact relations, surface activity, and cortical microfilaments of marginal cells of the enveloping layer and of the yolk syncytial and yolk cytoplasmic layer of *Fundulus* before and during epiboly. *J. Exp. Zool.* **206**, 381–426.

Bode, P. M., and Bode, H. (1984). Formation of pattern in regenerating tissue pieces of *Hydra attenuata*. III. The shaping of the body column. *Dev. Biol.* **106**, 315–325.

Brun, R. B., and Garson, J. A. (1983). Neurulation in the Mexican salamander (*Ambystoma mexicanum*): A drug study and cell shape analysis of the epidermis and the neural plate. *J. Embryol. Exp. Morphol.* **74**, 275–295.

Burnside, B. (1972). Experimental induction of microfilament formation and contraction. *J. Cell Biol.* **55**, 33a.

Burnside, M. B., and Jacobson, A. G. (1968). Analysis of morphogenetic movements in the neural plate of the newt *Taricha torosa*. *Dev. Biol.* **18**, 537–552.

Cameron, R. A., Hough-Evans, B. R., Britten, R. J., and Davidson, E. H. (1987). Lineage and fate of each blastomere of the eight-cell sea urchin embryo. *Genes Dev.* **1**, 75–84.

Copp, A. J., Brook, R. A., Estibeiro, J. P., Shum, A. S. W., and Cockroft, D. L. (1990). The embryonic development of mammalian neural tube defects. *Prog. Neurobiol.* **35**, 363–403.

Crossin, K. L., Chuong, C.-M., and Edelman, G. M. (1985). Expression sequences of cell adhesion molecules. *Proc. Natl. Acad. Sci. U.S.A.* **82**, 6942–6946.

Dan, K., and Inaba, D. (1968). Echinoderma. *In* "Invertebrate Embryology" (M. Kumé and K. Dan, eds.), pp. 280–332. NOLIT Publ., Belgrade.

Dan, K., and Okazaki, K. (1956). Cyto-embryological studies of sea urchins. III. Role of secondary mesenchyme cells in the formation of the primitive gut in sea urchin larvae. *Biol. Bull.* **110**, 29–42.

Decker, R. S., and Friend, D. S. (1974). Assembly of gap junctions during amphibian neurulation. *J. Cell Biol.* **62**, 32–47.

Edelman, G. M., Gallin, W. J., Delouvée, A., Cunningham, B. A., and Thiery, J.-P. (1983). Early epochal maps of two different cell adhesion molecules. *Proc. Natl. Acad. Sci. U.S.A.* **80**, 4384–4388.

Ettensohn, C. A. (1985). Gastrulation in the sea urchin embryo is accompanied by the rearrangement of invaginating epithelial cells. *Dev. Biol.* **112**, 383–390.

Everaert, S., Espeel, M., Bortier, H., and Vakaet, L. (1988). Connecting cords and morphogenetic movements in the quail blastoderm. *Anat. Embryol.* **177**, 311–316.

Fraser, S., Keynes, R., and Lumsden, A. (1990). Segmentation in the chick embryo hindbrain is defined by cell lineage restrictions. *Nature (London)* **344**, 431–435.

Fristrom, D. (1976). The mechanism of evagination of imaginal discs of *Drosophila*. III. Evidence for cell rearrangement. *Dev. Biol.* **54**, 163–171.

Fristrom, D. (1988). The cellular basis of epithelial morphogenesis. A review. *Tissue. Cell* **20**, 645–690.

Fristrom, D., and Chihara, C. (1978). The mechanism of evagination of imaginal discs of *Drosophila melanogaster*. V. Evagination of disc fragments. *Dev. Biol.* **66**, 564–570.

Geelen, J. A. G., and Langman, J. (1979). Ultrastructural observations on closure of the neural tube in the mouse. *Anat. Embryol.* **156**, 73–88.

Gerhart, J., and Keller, R. (1986). Region-specific cell activities in amphibian gastrulation. *Annu. Rev. Cell Biol.* **2**, 201–229.

Gordon, R. (1985). A review of the theories of vertebrate neurulation and their relationship to the mechanics of neural tube birth defects. *J. Embryol. Exp. Morphol.* **89**(Suppl.), 229–255.

Graf, L., and Gierer, A. (1980). Size, shape and orientation of cells in budding hydra and regulation of regeneration in cell aggregates. *Wilhelm Roux's Arch. Dev. Biol.* **188**, 141–151.

4. Cell Rearrangement in Axial Morphogenesis

Gurdon, J. B. (1987). Embryonic induction—Molecular prospects. *Development (Cambridge, UK)* **99**, 285–306.

Gustafson, T., and Kinnander, H. (1956). Microaquaria for time-lapse cinematographic studies of morphogenesis in swimming larvae and observations on sea urchin gastrulation. *Exp. Cell Res.* **11**, 36–51.

Gustafson, T., and Wolpert, L. (1963). The cellular basis of morphogenesis and sea urchin development. *Int. Rev. Cytol.* **15**, 139–214.

Gustafson, T., and Wolpert, L. (1967). Cellular movement and contact in sea urchin morphogenesis. *Biol. Rev. Cambridge Philos. Soc.* **42**, 442–498.

Hamburger, V. (1988). "The Heritage of Experimental Embryology. Hans Spemann and the Organizer." Oxford Univ. Press, New York.

Hardin, J. (1988). The role of secondary mesenchyme cells during sea urchin gastrulation studied by laser ablation. *Development (Cambridge, UK)* **103**, 317–324.

Hardin, J. (1989). Local shifts in position and polarized motility drive cell rearrangement during sea urchin gastrulation. *Dev. Biol.* **136**, 430–445.

Hardin, J. D., and Cheng, L. Y. (1986). The mechanisms and mechanics of archenteron elongation during sea urchin gastrulation. *Dev. Biol.* **115**, 490–501.

Hardin, J., and Keller, R. (1988). The behaviour and function of bottle cells in gastrulation of *Xenopus laevis*. *Development (Cambridge, UK)* **103**, 211–230.

Held, L. I. (1979). The high-resolution morphogenetic map of the second-leg basitarsus in *Drosophila melanogaster*. *Wilhelm Roux's Arch. Dev. Biol.* **187**, 129–150.

Holmdahl, D. E. (1925a). Die erste Entwicklung des Körpers bei den Vögeln und Säugetieren, inkl. dem Menschen, besonders mit Rücksicht auf die Bildung des Ruckenmarks, des Zöloms und der entodermalen Kloake nebst einem Exkurs über die Entstehung der Spina bifida in der Lumbosakralregion. *Gegenbaurs Morph. Jahrb.* I **54**, 333–384.

Holmdahl, D. E. (1925b). Experimentelle Untersuchungen über die Lage der Grenze zwischen primärer und sekundärer Körperentwicklung beim Huhn. *Anat. Anz.* **59**, 393–396.

Honda, H., Ogita, Y., Higuchi, S., and Kani, K. (1982). Cell movements in a living mammalian tissue: Long-term observation of individual cells in wounded corneal endothelia of cats. *J. Morphol.* **174**, 25–39.

Hörstadius, S. (1935). Über die Determination im Verlaufe der Eiachse bei Seeigeln. *Pubbl. Stn. Zool. Napoli* **14**, 251–429.

Jacobson, A. G. (1981). Morphogenesis of the neural plate and tube. In "Morphogenesis and Pattern Formation" (T. G. Connelly, L. L. Brinkley and B. M. Carlson, eds.), pp. 233–263. Raven, New York.

Jacobson, A. G., and Gordon, R. (1976). Changes in the shape of the developing vertebrate nervous system analyzed experimentally, mathematically and by computer simulation. *J. Exp. Zool.* **197**, 191–246.

Jacobson, A. G., and Sater, A. K. (1988). Features of embryonic induction. *Development (Cambridge, UK)* **104**, 341–359.

Jacobson, A. G., and Tam, P. P. L. (1982). Cephalic neurulation in the mouse embryo analyzed by SEM and morphometry. *Anat. Rec.* **203**, 375–396.

Jacobson, C. O., and Jacobson, A. (1973). Studies on morphogenetic movements during neural tube closure in amphibia. *Zoon* **1**, 17–21.

Kageyama, T. (1980). Cellular basis of epiboly of the enveloping layer in the embryo of Medaka, *Oryzias latipes*. I. Cell architecture revealed by silver staining method. *Dev. Growth Differ.* **22**, 659–668.

Kageyama, T. (1982). Cellular basis of epiboly of the enveloping layer in the embryo of the Medaka, *Oryzias latipes*. II. Evidence for cell rearrangement. *J. Exp. Zool.* **219**, 241–256.

Kageyama, T. (1987). Mitotic behavior and pseudopodial activity of cells in the embryo of *Oryzias latipes* during blastula and gastrula stages. *J. Exp. Zool.* **244**, 243–252.

Karfunkel, P. (1974). The mechanisms of neural tube formation. *Int. Rev. Cytol.* **38**, 245–271.
Keane, R. W., Mehta, P. P., Rose, B., Honig, L. S., Loewenstein, W. R., and Rutishauser, U. (1988). Neural differentiation, NCAM-mediated adhesion, and gap junctional communication in neuroectoderm. A study *in vitro*. *J. Cell. Biol.* **106**, 1307–1319.
Keller, R. E. (1975). Vital dye mapping of the gastrula and neurula of *Xenopus laevis*. I. Prospective areas and morphogenetic movements of the superficial layer. *Dev. Biol.* **42**, 222–241.
Keller, R. E. (1976). Vital dye mapping of the gastrula and neurula of *Xenopus laevis*. II. Prospective areas and morphogenetic movements of the deep layer. *Dev. Biol.* **51**, 118–137.
Keller, R. E. (1978). Time-lapse cinemicrographic analysis of superficial cell behavior during and prior to gastrulation in *Xenopus laevis*. *J. Morphol.* **157**, 223–247.
Keller, R. E. (1980). The cellular basis of epiboly: An SEM study of deep cell rearrangement during gastrulation in *Xenopus laevis*. *J. Embryol. Exp. Morphol.* **60**, 201–234.
Keller, R. E. (1981). An experimental analysis of the role of bottle cells and the deep marginal zone in gastrulation of *Xenopus laevis*. *J. Exp. Zool.* **216**, 81–101.
Keller, R. E. (1986). The cellular basis of amphibian gastrulation. In "Developmental Biology: A Comprehensive Synthesis" (L. Browder, ed.), pp. 241–327. Plenum, New York.
Keller, R. E., and Danilchik, M. (1988). Regional expression, pattern, and timing of convergence and extension during gastrulation of *Xenopus laevis*. *Development (Cambridge, UK)* **103**, 193–209.
Keller, R. E., and Hardin, J. (1987). Cell behavior during active cell rearrangement: Evidence and speculation. *J. Cell Sci.* **8**(Suppl.), 369–393.
Keller, R. E., and Schoenwolf, G. C. (1977). An SEM study of cellular morphology, contact, and arrangement, as related to gastrulation in *Xenopus laevis*. *Wilhelm Roux's Arch. Dev. Biol.* **182**, 165–186.
Keller, R. E., and Tibbets, P. (1989). Mediolateral cell intercalation in the dorsal axial mesoderm of *Xenopus laevis*. *Dev. Biol.* **131**, 539–549.
Keller, R. E., and Trinkaus, J. P. (1987). Rearrangement of enveloping layer cells without disruption of the epithelial permeability barrier as a factor in *Fundulus* epiboly. *Dev. Biol.* **120**, 12–24.
Keller, R. E., Danilchik, M., Gimlich, R., and Shih, J. (1985a). Convergent extension by cell intercalation during gastrulation of *Xenopus laevis*. In "Molecular Determinants of Animal Form" (G. M. Edelman, ed.), UCLA Symposia on Molecular and Cellular Biology, New Series 31, pp. 111–141. Alan R. Liss, New York.
Keller, R. E., Danilchik, M., Gimlich, R., and Shih, J. (1985b). The function and mechanism of convergent extension during gastrulation of *Xenopus laevis*. *J. Embryol. Exp. Morphol.* **89**(Suppl.), 185–209.
Keller, R. E., Cooper, M. S., Danilchik, M., Tibbetts, P., and Wilson, P. A. (1989). Cell intercalation during notochord development in *Xenopus laevis*. *J. Exp. Zool.* **251**, 134–154.
Kimmel, C. B., and Law, R. D. (1985a). Cell lineage of zebrafish blastomeres. I. Cleavage pattern and cytoplasmic bridges between cells. *Dev. Biol.* **108**, 78–85.
Kimmel, C. B., and Law, R. D. (1985b). Cell lineage of zebrafish blastomeres. III. Clonal analysis of the blastula and gastrula stages. *Dev. Biol.* **108**, 94–101.
Kimmel, C. B., Warga, R. M., and Schilling, T. F. (1990). Origin and organization of the zebrafish fate map. *Development (Cambridge, UK)* **108**, 581–594.
Kinnander, H., and Gustafson, T. (1960). Further studies on the cellular basis of gastrulation in the sea urchin larva. *Exp. Cell Res.* **19**, 278–290.
Langman, J., Guerrant, R. L., and Freeman, B. G. (1966). Behavior of neuro-epithelial cells during closure of the neural tube. *J. Comp. Neurol.* **127**, 399–412.
Lee, H.-Y., Nagele, R. G., and Kalmus, G. W. (1976a). Further studies on neural tube defects caused by concanavalin A in early chick embryos. *Experientia* **32**, 1050–1052.

Lee, H.-Y., Sheffield, J. B., Nagele, R. G., and Kalmus, G. W. (1976b). The role of extracellular material in chick neurulation. I. Effects of concanavalin A. *J. Exp. Zool.* **198**, 261–266.

Lee, H.-Y., Sheffield, J. B., and Nagele, R. G. (1978). The role of extracellular material in chick neurulation. II. Surface morphology of neuroepithelial cells during neural fold fusion. *J. Exp. Zool.* **204**, 137–154.

Lehman, F. E. (1932). Die Beteiligung von Implantats- und Wirtsgewebe bei der Gastrulation und Neurulation induzierter Embryonalanlagen. *Wilhelm Roux Arch. Entwicklungsmech. Org.* **125**, 566.

Lillie, F. R. (1919). "Problems of Fertilization." Chicago Univ. Press, Chicago.

McLone, D. G., and Knepper, P. A. (1985/86). Role of complex carbohydrates and neurulation. *Pediatr. Neurosci.* **12**, 2–9.

McLone, D. G., Suwa, J., Collins, J. A., Poznanski, S., and Knepper, P. A. (1983). Neurulation: Biochemical and morphological studies on primary and secondary neural tube defects. *Concepts Pediatr. Neurosurg.* **4**, 15–29.

Mak, L. L. (1978). Ultrastructural studies of amphibian neural fold fusion. *Dev. Biol.* **65**, 435–446.

Mangold, O. (1920). Fragen der regulation und Determination an umgeordneten Furchungsstadien und verschmolzenen Keimen von Triton. *Wilhelm Roux Arch. Entwicklungsmech. Org.* **47**, 250–301.

Martin, A. H. (1967). Significance of mitotic spindle fibre orientation in the neural tube. *Nature (London)* **216**, 1133–1134.

Miyamoto, D. M., and Crowther, R. (1985). Formation of the notochord in living ascidian embryos. *J. Embryol. Exp. Morphol.* **86**, 1–17.

Moran, D., and Rice, R. W. (1975). An ultrastructural examination of the role of cell membrane surface coat material during neurulation. *J. Cell Biol.* **64**, 172–181.

Morrill, J. B., and Santos, L. L. (1985). A scanning electron microscopical overview of cellular and extracellular patterns during blastulation and gastrulation in the sea urchin, *Lytechinus variegatus*. *In* "The Cellular and Molecular Biology of Invertebrate Development" (R. H. Sawyer and R. M. Showman, eds.), pp. 3–33. South Carolina Univ Press, Columbia, South Carolina.

Morriss-Kay, G. M. (1981). Growth and development of pattern in the cranial neural epithelium of rat embryos during neurulation. *J. Embryol. Exp. Morphol.* **65**(Suppl.), 225–241.

Morriss-Kay, G. M., and Crutch, B. (1982). Culture of rat embryos with β-D-xyloside: Evidence of a role for proteoglycans in neurulation. *J. Anat.* **134**, 491–506.

Morriss-Kay, G., and Tuckett, F. (1987). Fluidity of the neural epithelium during forebrain formation in rat embryos. *J. Cell Sci.* **8**(Suppl.), 433–449.

Morriss-Kay, G. M., and Tuckett, F. (1989). Immunohistochemical localisation of chondroitin sulphate proteoglycans and the effects of chondroitinase ABC in 9- to 11-day rat embryos. *Development (Cambridge, UK)* **106**, 787–798.

Morriss-Kay, G. M., Tuckett, F., and Solursh, M. (1986). The effects of *Streptomyces* hyaluronidase on tissue organization and cell cycle time in rat embryos. *J. Embryol. Exp. Morphol.* **98**, 59–70.

Nagele, R. G., and Lee, H. Y. (1979). Ultrastructural changes in cells associated with interkinetic nuclear migration in the developing chick neuroepithelium. *J. Exp. Zool.* **210**, 89–106.

Nagele, R. G., and Lee, H. Y. (1980). A transmission and scanning electron microscopic study of cytoplasmic threads of dividing neuroepithelial cells in early chick embryos. *Experientia* **36**, 338–340.

Nakatsuji, N. (1984). Cell locomotion and contact guidance in amphibian gastrulation. *Am. Zool.* **24**, 615–627.

Nakatsuji, N., and Johnson, K. (1983). Comparative study of extracellular fibrilla on the ectodermal layer in gastrulae of five amphibian species. *J. Cell Sci.* **59**, 61–70.

Nakatsuji, N., and Johnson, K. (1984). Experimental manipulation of a contact guidance system in amphibian gastrulation by mechanical tension. *Nature (London)* **307**, 453–455.

Nardi, J., and MaGee-Adams, S. M. (1986). Formation of scale spacing patterns in a moth wing. I. Epithelial feet may mediate cell rearrangement. *Dev. Biol.* **116**, 278–290.

Nicolet, G. (1970). Analyse autoradiographique de la localisation des différentes ébauches présomptives dans la ligne primitive de l'embryon de Poulet. *J. Embryol. Exp. Morphol.* **23**, 79–108.

O'Shea, K. S., and Kaufman, M. H. (1980). Phospholipase C-induced neural tube defects in the mouse embryo. *Experientia* **36**, 1217–1219.

Poole, T., and Steinberg, M. (1981). Amphibian pronephric duct morphogenesis: Segregation, cell rearrangement and directed migration of the *Ambystoma* duct rudiment. *J. Embryol. Exp. Morphol.* **63**, 1–16.

Revel, J.-P., and Brown, S. S. (1976). Cell junctions in development, with particular reference to the neural tube. *Cold Spring Harbor Symp. Quant. Biol.* **40**, 443–455.

Revel, J.-P., Yip, P. and Chang, L. L. (1973). Cell junctions in the early chick embryo. A freeze etch study. *Dev. Biol.* **35**, 302–317.

Rice, R. W., and Moran, D. J. (1977). A scanning electron microscope and X-ray microanalytic study of cell surface material during amphibian neurulation. *J. Exp. Zool.* **201**, 471–478.

Rosenquist, G. C. (1966). A radioautographic study of labeled grafts in the chick blastoderm. Development from primitive-streak stages to stage 12. *Carnegie. Contrib. Embryol. No. 262* **38**, 31–110.

Rosenquist, G. C. (1983). The chorda center in Hensen's node of the chick embryo. *Anat. Rec.* **207**, 349–355.

Roux, W. (1885). Beiträge zur entwicklungsmechanik des embryos. *Z. Biol. (Munich)* **21**, 411–524.

Rovasio, R. A., and Monis, B. (1981). Ultrastructure (TEM and SEM) of the glycocalyx of the neural folds of normal and carrageenan-injected chick embryos. *Biol. Cell* **42**, 173–180.

Russell, L. D. (1979). Observations on the interrelationships of Sertoli cells at the level of the blood-testis barrier: Evidence for formation and resorption of Sertoli-Sertoli tubulobulbar complexes during the spermatogenic cycle of the rat. *Am. J. Anat.* **155**, 259–279.

Sadler, T. W. (1978). Distribution of surface coat material on fusing neural folds of mouse embryos during neurulation. *Anat. Rec.* **191**, 345–350.

Sandig, M., and Kalnins, V. I. (1990). Reorganization of circumferential microfilament bundles in retinal epithelial cells during mitosis. *Cell Motil. Cytoskeleton* **17**, 133–141.

Sauer, F. C. (1935). Mitosis in the neural tube. *J. Comp. Neurol.* **62**, 377–405.

Schechtman, A. (1942). The mechanism of amphibian gastrulation. I. Gastrulation-promoting interactions between various regions of an anuran egg (*Hyla regilla*). *Univ. Calif. Publ. Zool.* **51**, 1–39.

Schoenwolf, G. C. (1977). Tail (end) bud contributions to the posterior region of the chick embryo. *J. Exp. Zool.* **201**, 227–246.

Schoenwolf, G. C. (1978). Effects of complete tail bud extirpation on early development of the posterior region of the chick embryo. *Anat. Rec.* **192**, 289–296.

Schoenwolf, G. C. (1979). Histological and ultrastructural observations of tail bud formation in the chick embryo. *Anat. Rec.* **193**, 131–147.

Schoenwolf, G. C. (1982). On the morphogenesis of the early rudiments of the developing central nervous system. *Scanning Electron Microsc.* 1982/I, 289–308.

Schoenwolf, G. C. (1983). The chick epiblast: A model for examining epithelial morphogenesis. *Scanning Electron Microsc.* 1983/III, 1371–1385.

Schoenwolf, G. C. (1985). Shaping and bending of the avian neuroepithelium: Morphometric analyses. *Dev. Biol.* **109**, 127–139.

Schoenwolf, G. C. (1988). Microsurgical analyses of avian neurulation: Separation of medial and lateral tissues. *J. Comp. Neurol.* **276**, 498–507.

Schoenwolf, G. C. (1991a). Cell movements in the epiblast during gastrulation and neurulation in avian embryos. In "Gastrulation: Movement, Patterns and Molecules" (R. E. Keller, J. W. Clark, and F. Griffin, eds.), pp. 1–28. Plenum, New York.

Schoenwolf, G. C. (1991b). Cell movements driving neurulation in avian embryos. *Development (Cambridge, UK)* (1991 Suppl. 2), 157–168.

Schoenwolf, G. C., Garcia-Martinez, V., and Dias, M. (1992). Mesoderm movement and fate during avian gastrulation and neurulation, *Dev. Dynamics,* in press.

Schoenwolf, G. C., and Alvarez, I. S. (1989). Roles of neuroepithelial cell rearrangement and division in shaping of the avian neural plate. *Development (Cambridge, UK)* **106,** 427–439.

Schoenwolf, G. C., and Alvarez, I. S. (1991). Specification of neuroepithelium and surface epithelium in avian transplantation chimeras. *Development (Cambridge, UK)* **112,** 713–722.

Schoenwolf, G. C., and Fisher, M. (1983). Analysis of the effects of *Streptomyces* hyaluronidase on formation of the neural tube. *J. Embryol. Exp. Morphol.* **73,** 1–15.

Schoenwolf, G. C., and Franks, M. V. (1984). Quantitative analyses of changes in cell shapes during bending of the avian neural plate. *Dev. Biol.* **105,** 257–272.

Schoenwolf, G. C., and Nichols, D. H. (1984). Histological and ultrastructural studies on the origin of caudal neural crest cells in mouse embryos. *J. Comp. Neurol.* **222,** 496–505.

Schoenwolf, G. C., and Sheard, P. (1989). Shaping and bending of the avian neural plate as analysed with a fluorescent-histochemical marker. *Development (Cambridge, UK)* **105,** 17–25.

Schoenwolf, G. C., and Sheard, P. (1990). Fate mapping the avian epiblast with focal injections of a fluorescent-histochemical marker: Ectodermal derivatives. *J. Exp. Zool.* **255,** 323–339.

Schoenwolf, G. C., and Smith, J. L. (1990). Mechanisms of neurulation: Traditional viewpoint and recent advances. *Development (Cambridge, UK)* **109,** 243–270.

Schoenwolf, G. C., Chandler, N. B., and Smith, J. (1985). Analysis of the origins and early fates of neural crest cells in caudal regions of avian embryos. *Dev. Biol.* **110,** 467–479.

Schoenwolf, G. C., Folsom, D., and Moe, A. (1988). A reexamination of the role of microfilaments in neurulation in the chick embryo. *Anat. Rec.* **220,** 87–102.

Schoenwolf, G. C., Bortier, H., and Vakaet, L. (1989). Fate mapping the avian neural plate with quail/chick chimeras: Origin of prospective median wedge cells. *J. Exp. Zool.* **249,** 271–278.

Schroeder, T. (1981). Development of a "primitive" sea urchin (*Eucidaris tribuloides*): Irregularities in the hyaline layer, micromeres, and primary mesenchyme. *Biol. Bull.* **161,** 141–151.

Silver, M. H., and Kerns, J. M. (1978). Ultrastructure of neural fold fusion in chick embryos. *Scanning Electron Microsc.* 1978/II, 209–215.

Smith, J. L. (1988). "Regulation of Neuroepithelial Cell Shape During Bending of the Chick Neural Plate." Ph.D. Dissertation, University of Utah, Salt Lake City.

Smith, J. L., and Schoenwolf, G. C. (1987). Cell cycle and neuroepithelial cell shape during bending of the chick neural plate. *Anat. Rec.* **218,** 196–206.

Smith, J. L., and Schoenwolf, G. C. (1988). Role of cell-cycle in regulating neuroepithelial cell shape during bending of the chick neural plate. *Cell Tissue Res.* **252,** 491–500.

Smith, J. L., and Schoenwolf, G. C. (1989). Notochordal induction of cell wedging in the chick neural plate and its role in neural tube formation. *J. Exp. Zool.* **250,** 49–62.

Smith, J. L., and Schoenwolf, G. C. (1991). Further evidence of extrinsic forces in bending of the neural plate. *J. Comp. Neurol.* **307,** 225–236.

Smits-van Prooije, A. E., Poelmann, R. E., Gesink, A. F., Groeningen, M. J. v., and Vermeij-Keers, C. (1986a). The cell surface coat in neurulating mouse and rat embryos, studied with lectins. *Anat. Embryol.* **175,** 111–117.

Smits-van Prooije, A., Poelmann, R., Dubbeldam, J., Mentink, M., and Vermeij-Keers, C. (1986b). The formation of the neural tube in rat embryos, cultured *in vitro,* studied with teratogens. *Acta Histochem.* **32,** 41–45.

Spemann, H. (1902). Entwicklungsphysiologische Studien am *Triton*-Ei II. *Wilhelm Roux's Arch. Entwicklungsmech. Org.* **15,** 448–534.

Spemann, H. (1938). "Embryonic Development and Induction." Yale Univ. Press, New Haven, Conn.
Stern, C. D., Manning, S., and Gillespie, J. I. (1985). Fluid transport across the epiblast of the early chick embryo. *J. Embryol. Exp. Morphol.* **88,** 365–384.
Straaten, H. W. M., van, Hekking, J. W. M., Wiertz-Hoessels, E. J. L. M., Thors, F., and Drukker, J. (1988). Effect of the notochord on the differentiation of a floor plate area in the neural tube of the chick embryo. *Anat. Embryol.* **177,** 317–324.
Suzuki, A. S., and Harada, K. (1988). Prospective neural areas and their morphogenetic movements during neural plate formation in the *Xenopus* embryo. II. Disposition of transplanted ectoderm pieces of *X. borealis* animal cap in prospective neural areas of albino *X. laevis* gastrulae. *Dev. Growth Differ.* **30,** 391–400.
Takahashi, H. (1988). Changes in peanut lectin binding sites on the neuroectoderm during neural tube formation in the bantam chick embryo. *Anat. Embryol.* **178,** 353–358.
Takahashi, H., and Howes, R. I. (1986). Binding pattern of ferritin-labeled lectins (RCA_1 and WGA) during neural tube closure in the bantam embryo. *Anat. Embryol.* **174,** 283–288.
Thiery, J.-P., Duband, J.-L., Rutishauser, U., and Edelman, G. M. (1982). Cell adhesion molecules in early chicken embryogenesis. *Proc. Natl. Acad. Sci. U.S.A.* **79,** 6737–6741.
Tickle, C., and Trinkaus, J. (1973). Change in surface extensibility of *Fundulus* deep cells during early development. *J. Cell Sci.* **13,** 721–726.
Trinkaus, J. P. (1951). A study of the mechanism of epiboly in the egg of *Fundulus heteroclitus*. *J. Exp. Zool.* **118,** 269–320.
Trinkaus, J. (1973). Surface activity and locomotion of *Fundulus* deep cells during blastula and gastrula stages. *Dev. Biol.* **30,** 68–103.
Trinkaus, J. (1976). On the mechanism of metazoan cell movements. *In* "The Cell Surface in Animal Embryogenesis and Development" (G. Poste and G. L. Nicolson, eds.), pp. 225–329. North-Holland, Amsterdam.
Trinkaus, J. P. (1984). "Cells into Organs, the Forces That Shape the Embryo," 2nd Ed. Prentice-Hall, Englewood Cliffs, New Jersey.
Trinkaus, J. P., and Lentz, T. (1967). A fine structural study of cytodifferentiation during cleavage, blastula and gastrula stages of *Fundulus heteroclitus*. *J. Cell. Biol.* **32,** 121–138.
Tuckett, F., and Morriss-Kay, G. M. (1985). The kinetic behaviour of the cranial neural epithelium during neurulation in the the rat. *J. Embryol. Exp. Morphol.* **85,** 111–119.
Tuckett, F., and Morriss-Kay, G. M. (1989). Heparitinase treatment of rat embryos during cranial neurulation. *Anat. Embryol.* **180,** 393–400.
Vanroelen, C., Verplanken, P., and Vakaet, L. C. A. (1982). The effects of partial hypoblast removal on the cell morphology of the epiblast in the chick blastoderm. *J. Embryol. Exp. Morphol.* **70,** 189–196.
Vogt, W. (1922a). Die Einrollung und Streckung der Urmundlippen bei Triton nach Versuchen mit einer neuer Methode embryonaler transplantation. *Verh. Zool. Bot. Ges. Wein* **27,** 49–51.
Vogt, W. (1922b). Opertiv bewirkte "Exogastrulation" bei *Triton* und ihre Bedeutung fur die Theorie dur Wirbeltiergastrulation. *Verh. Anat. Ges.* **55,** 53–64.
Waddington, C. H. (1940). "Organizers and Genes." Cambridge Univ. Press, Cambridge, England.
Warga, R. M., and Kimmel, C. B. (1990). Cell movements during epiboly and gastrulation in zebrafish. *Development (Cambridge, UK)* **108,** 569–580.
Waterman, R. E. (1976). Topographical changes along the neural fold associated with neurulation in the hamster and mouse. *Am. J. Anat.* **148,** 151–172.
Waterman, R. E. (1979). Scanning electron microscope studies of central nervous system development. *Birth Defects, Orig. Artic. Ser.* **15,** 55–77.
Watterson, R. L. (1965). Structure and mitotic behavior of the early neural tube. *In* "Organogenesis" (R. L. DeHaan and H. Ursprung, eds.), pp. 129–159. Holt, Rinehart, and Winston, New York.

Weliky, M., and Oster, G. (1990). The mechanical basis of cell rearrangement. I. Epithelial morphogenesis during *Fundulus* epiboly. *Development (Cambridge, UK)* **109,** 373–386.

Winklbauer, R. (1990). Mesodermal cell migration during *Xenopus* gastrulation. *Dev. Biol.* **142,** 155–168.

Wray, G. A. (1987). "Heterochrony and Homology in the Evolution of Echinoid Development." Ph.D. Dissertation, Duke Univ., Durham, North Carolina.

Yamada, T. (1990). Regulations in the induction of the organized neural system in amphibian embryos. *Development (Cambridge, UK)* **110,** 653–659.

Zieba, P., Strojny, P., and Lamprecht, J. (1986). Positioning and stability of mitotic spindle orientation in the neuroepithelial cell. *Cell Biol. Int. Rep.* **10,** 91–100.

5
Mechanisms Underlying the Development of Pattern in Marsupial Embryos

Lynne Selwood
Department of Zoology
La Trobe University
Bundoora, 3083 Victoria, Australia

I. Introduction
II. Initial Polarity
   A. Apolar State in Oocytes
   B. Polar State in Oocytes
   C. Vitellogenesis and Yolk
   D. Ovulation
III. Definitive Polarity
   A. Fertilization
   B. Egg Envelopes
   C. Cleavage
IV. Disappearance and Renewal of Polarity
   A. Unilaminar Blastocyst
   B. Bilaminar Blastocyst
   C. Trilaminar Blastocyst
V. Discussion and Conclusions
   A. Polarity and the Investments
   B. Cleavage Results in Two Cell Populations
   C. Capacity to Regulate
   D. Pattern and Order of Division in Specification of Cell Fate
   E. Axis Formation
   F. Marsupials in Context
   References

I. Introduction

Since the development of the inside–outside hypothesis (Tarkowski and Wroblewska, 1967) and its subsequent experimental analysis (for review, see Denker, 1976; Johnson, 1981; Gardner, 1983; Pedersen, 1986) of how the cells of a mammalian embryo could receive and respond to positional signals (reviewed by Johnson and Maro, 1986) the marsupial embryo has presented a paradox to developmental biologists because of the external position of the embryoblast (Hill, 1910; Hartman, 1916; McCrady, 1938; Selwood and Young, 1983; Selwood, 1986a,b). The embryoblast of the marsupial appears in a blastocyst in which all cells are superficial and apparently identical. This caused McCrady (1938) to call the blastocyst epithelium a "protoderm" because of the apparent totipotency of the cells. Marsupial embryos are not unique in possessing a unilaminar blastocyst with no inner cell mass. Such a structure is also found in a number of other mammals including the elephant shrew, *Elephantulus* (van der Horst, 1942), and the tenrec, *Hemicentetes* (Bluntschli, 1938; Goetz, 1939; see Wimsatt, 1975, for review). Of course, in the monotremes *Ornithorynchus* and *Tachyglossus*, which have meroblastic cleavage (see Griffiths, 1978, for review) embryonic and extraembryonic cells are also generated from an initially unilaminar structure, the blastoderm.

Eutherian mammalian embryos such as the mouse have considerable capacity to regulate for loss or addition of blastomeres (Tarkowski, 1961; Mintz, 1965). Specification of cell fate is due largely, if not entirely, to positional effects (Pedersen, 1986; Johnson and Maro, 1986) and maternal determinants are thought to play a minimal role (Johnson et al., 1986). These positional effects are derived from environmental stimuli, not so much from diffusable components unique to an inside or outside location, but because inside or outside cells have particular relationships with their neighboring cells (Pedersen and Spindle, 1980). In eutherian mammals, in which some cells (the inner cell mass, ICM) are enclosed by others (the trophectoderm, TE), the cellular positional signals act in a three-dimensional framework within the morula. In marsupials, cellular positional signals may act instead in the two-dimensional framework of an epithelial sheet because of the unilaminar structure of the blastocyst. Any three-dimensional signals acting on early marsupial differentiation would have to be derived from the outside environment (the investments and factors in the uterine lumen) and the inside environment (the blastocoel) acting in concert with cell–cell interactions. It is difficult, however, to envision how outside and inside information could lead to specialization of some blastocyst epithelial cells at the embryonic pole (destined to form the primary endoderm cells) because all blastocyst epithelial cells appear to be exposed to the same external and internal environments.

This review provides an outline of marsupial embryonic development during preimplantation stages up to the trilaminar blastocyst stage. Because the developing polarity of the marsupial embryo is initiated during oogenesis, the review also covers oogenesis and fertilization in marsupials. It does not deal with reproductive physiology of marsupials, which has been comprehensively reviewed in recent years (Tyndale-Biscoe and Renfree, 1987). It also does not deal with embryonic development past the trilaminar blastocyst stage, and so does not include organogenesis. [In most marsupials the embryo is implanted after neural tube and somite formation have begun (Selwood, 1989a) when the embryo has about 15–25 somites.]

Accordingly, the major theme of this review will be to explore the idea that the marsupial embryo acquires a polarized state that is initially related to the position of the nucleus and the pattern of distribution of cytoplasmic organelles, especially yolky storage products, in oocytes and zygotes. The particular nature of this polarized state generates a particular and highly specific pattern of cleavage. This in turn ensures that blastomeres during cleavage end up occupying a particular part of the blastocyst wall. The potential role of positional signals, the order of cell division and maternal determinants operating at a two-dimensional level in specifying cell fate, and early steps in determination will be examined.

II. Initial Polarity

The first signs of polarization of the marsupial egg occur in the ovary. Primordial germ cells and oogonia are apparently apolar. Polarization is initially related to an eccentric position of the germinal vesicle in primary oocytes and subsequently also to the distribution of yolky storage products. These two factors play a variable and sometimes combined role in the expression of oocyte polarity (Fig. 1). The eccentric position of the nucleus has been described in stages as early as primary oocytes in leptotene or later, after the onset of vitellogenesis. Polarization of cytoplasmic organelles is most frequently described in primary and secondary oocytes and is usually related to the formation or distribution of yolky storage products.

In this review the use of such terms as "animal pole" and "vegetal pole" will be avoided because their use is misleading in this discussion. In follicular oocytes, whether the distribution of yolk is in a radial or polar pattern, the

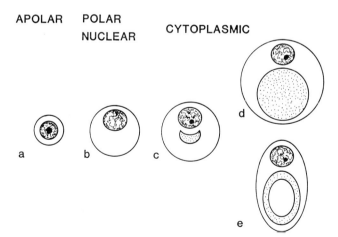

Fig. 1 Development of the polarized state during oogenesis in marsupials. The apolar state exists in the primordial germ cell (a). Nuclear polarization is found in the primary oocyte in leptotene (b). Cytoplasmic polarization is found in the primary oocyte (c) in primordial follicles because of the location of the paranuclear complex (stippled). During vitellogenesis, yolk (stippled) is located centrally (d) or radially (e). Not drawn to scale.

nucleus is peripheral and the polar body lies nearby, making identification of an "animal pole" relatively simple (Figs. 1 and 2). In fertilized eggs, however, the pronuclei move to a central position in eggs with a radial distribution of yolk or into the approximate centre of the crescent of less yolky cytoplasm in eggs with polar distribution. The polar bodies, when visible, are found at any point on the less yolky hemisphere and even, occasionally, in the yolky hemisphere. Furthermore, once blastocyst formation commences, blastomeres flatten on the zona peripheral to the yolk, in other words at the equivalent of the "vegetal pole" of follicular oocytes. Therefore, marsupial eggs are not readily understood in animal–vegetal terms, and their polarity will be described here according to the relative position of their conspicuous organelles, the nucleus and the yolk.

A. Apolar State in Oocytes

Oogenesis, from the appearance of the primordial germ cells to ovulation, has not been fully described in any marsupial (see Selwood, 1982, for review). Primordial germ cells and oogonia have been described only in bandicoots (Ullman, 1981) and the tammar *Macropus eugenii* (Alcorn and Robinson, 1983). X chromosome activity during these stages and the primary oocyte stage have been analyzed in the euro, *Macropus robustus*, by Robinson *et al.* (1977). Migrating primordial germ cells have been located in 9.5-day-old embryos of *Isoodon macrourus*, the short-nosed bandicoot, and *Perameles nasuta*, the long-nosed bandicoot, in mesodermal and endodermal elements of the hind gut and proximal levels of the yolk sac (Ullman, 1981). They were identified on the basis of their large size, distinctive nuclear cytology, and low affinity for eosin. In a careful study of limited material, Ullman (1981) concluded that the origin and migration of the germ cells is similar to what has been found in eutherians. Thus primordial germ cells in marsupials arise at the interface between embryonic and extraembryonic endoderm and migrate via the hindgut into the adjacent splanchnic mesoderm. They then invade the dorsal mesentery migrating out into the lateral mesenchyme. From there they move into the dorsal edge of the gonadal ridges. Invasion of the gonadal ridges begins on day 11 of embryonic development in bandicoots and peaks during the perinatal period (days 12–13). Ultrastructurally, the germ cells contain few ribosomes and some electron-dense vesicles of unknown function, so that they are relatively less electron dense than their neighboring cells. Their light microscope appearance is similar to the primordial germ cells of the tammar (Alcorn and Robinson, 1983). Fraser (1919) identified, but did not describe, primordial germ cells in the genital ridge in an embryo (greater length, 13 mm) of *Trichosurus vulpecula*, the brush possum. They have also been identified in neonatal stages of the Virginian opossum (McCrady, 1938; Morgan, 1943) and *Dasyurus viverrinus*, the native cat (Ullman, 1984). The primordial germ cells and oogonia show no obvious polarity, but as study of these stages is very restricted, further examination may reveal an underlying polarity expressed, for example, in the cytoplasmic skeletal proteins or in relationship to migratory activity of the primordial germ cells.

B. Polar State in Oocytes

1. Nuclear Polarity

The first sign of oocyte polarity is detectable in the primary oocyte stage (Fig. 1) and is marked by the development of an eccentric location of the nucleus, coincident with the onset of oocyte growth in leptotene oocytes of the tammar 24–30 days after birth (Alcorn and Robinson, 1983) and paternal X chromosome inactivation in pachytene and diplotene stages in the euro (Robinson et al., 1977). The eccentric, peripheral position of the nucleus, at least by the time of meiotic maturation of the oocyte, appears to be a common feature of marsupial oocytes. This applies to small as well as larger oocytes and is independent of the type of distribution of cytoplasmic organelles (Table I). It is presumably a device that ensures that the meiotic divisions are not hampered by the presence of large numbers of yolky particles. The relationship between nuclear position and the

Table I Development of Polarity Based on Position of Nucleus and Yolk during Oogenesis and Fertilization in Marsupials

Animal	Stage	Diameter (μm)	Positions of Nucleus	Positions of Yolk	Ref.
Virginian opossum	Quiescent primordial oocyte	—	Polar	Polar	Hartman (1916, 1919); Guraya (1968); McGrady (1938)
	oocyte	—	Polar	Radial	
	Secondary oocyte	135 × 165	Polar	Radial	
	Tubal zygote	104 × 122	Polar	Radial	
	Uterine zygote	107 × 111	Polar, then central	Radial	
Brown antechinus	Primary oocyte	115 × 131	Polar	Central	Selwood (1982)
	Secondary oocyte	116 × 145	Polar	Central-polar	
	Tubal zygote	93 × 118	Polar	Polar	
	Uterine zygote	136 × 140	Polar	Polar	
Native cat	Primary oocyte	240	Polar	Central	Hill (1910)
	Secondary oocyte	—	Polar	Polar	
	Uterine zygote	245	Polar	Polar	

underlying cytoplasmic architecture as represented by cytoplasmic actin, tubulin, and microtubule-organizing centers (MTOCs) has yet to be investigated.

2. Cytoplasmic Polarity

The polarization conferred on the oocyte by the eccentric position of the nuclear material is further emphasized in some species by a polarization of cytoplasmic organelles associated with early vitellogenesis and accumulation of yolky storage products (Fig. 1). The earliest stage reported showing such polarization of cytoplasmic organelles is the quiescent primordial oocyte of the Virginian opossum (Guraya, 1968) and the short-nosed bandicoot (Ullman, 1978). In these oocytes, the large nucleus occupies most of one hemisphere of the oocyte and a "yolk nucleus" (Guraya, 1968) or "paranuclear complex" (Ullman, 1978) lies near the nucleus in the other hemisphere. In addition, the cytoplasm of this hemisphere contains a vesicle–microtubule complex and an aggregate of tubular cisternae in the bandicoot and lipid bodies in the opossum. With further oocyte growth and dispersal of the yolk nucleus in the opossum, the polarized cytoplasmic distribution of organelles is replaced by a mainly radial pattern of distribution (Hartman, 1916, 1919; Guraya, 1968). Some cytoplasmic polarity persists in the shape of the primary (Fig. 1e) and secondary oocytes of the opossum, which are elliptical, with the nuclear material usually located at one pole of the egg (Hartman, 1916; McCrady, 1938; Martinez-Esteve, 1942; Guraya, 1968). In secondary oocytes, the metaphase chromosomes lie in an area of relatively yolk-free cytoplasm, similar to that found in the central regions of the oocyte. Secondary oocytes with peripheral meiotic chromosomes and radially distributed yolky storage products similar to those of the Virginian opossum have also been described in one other didelphid, *Didelphis aurita*, the common opossum (Hill, 1918).

Where information is available on oocyte structure during development of the primary oocyte to formation of the secondary oocyte and ovulation, two patterns of distribution of cytoplasmic organelles and yolky storage products emerge (Fig. 1). In one, already described in the Virginian opossum (Hartman, 1916; McCrady, 1938; Guraya, 1968), organelles are distributed in a radial pattern that arises during vitellogenesis and persists in the secondary oocyte and during ovulation. The other pattern is found in the dasyurids, the native cat (Sandes, 1903; Hill, 1910), brown antechinus, *Antechinus stuartii* (Selwood, 1982), and in *Sminthopsis crassicaudata*, the fat-tailed dunnart (Breed and Leigh, 1990); and in the didelphid, *Monodelphis domestica*, the gray short-tailed opossum (Baggott and Moore, 1990). In this group, the yolky storage products are centrally located in primary (Fig. 1d) and secondary oocytes but become increasingly polarized in location during the progress of the maturation divisions (Table I) so that eventually they lie as a mass at the opposite pole to the nuclear material in the secondary oocyte. This process is very marked in the larger,

~240-μm secondary oocytes of the native cat and the fat-tailed dunnart, but less marked in the smaller oocytes of the brown antechinus (Table I). Microvilli are found all around the circumference of the primary oocyte in the fat-tailed dunnart, but disappear from the plasma membrane overlying the female chromatin in secondary oocytes (Breed & Leigh, 1990).

Oocytes at various stages late in vitellogenesis, especially primary oocytes in the dictyate stage of meiosis I, have been described or illustrated in a number of studies on folliculogenesis or reproductive cycles. The oocyte (as pictured or described) has an eccentrically placed germinal vesicle or meiotic chromosomes in the peripheral cytoplasm and a central accumulation of yolky storage products. Primary oocytes with these characteristics are found in *Pseudocheirus perigrinus*, the common ring-tail possum (Hughes et al., 1965), *Sarcophilus harrisii*, the Tasmanian devil (Hughes, 1982), the tammar wallaby (Panyaniti et al., 1985), and secondary oocytes in *Setonix brachyurus*, the quokka (Sharman, 1955a), the common ring-tail possum (Hughes et al., 1965), *Sminthopsis macroura* (= *larapinta*), the stripe-faced dunnart (Godfrey, 1969), and the brushtailed possum (Pilton and Sharman, 1962; Hughes, 1977). Lintern-Moore et al. (1976), in a study of oocyte and follicular growth in marsupials, found a biphasic pattern, with growth of the oocyte ceasing once antrum formation begins. In some dasyurids, however, the oocyte appears to increase in diameter during antrum formation (Godfrey, 1969; Selwood, 1982).

In summary then, an initial polarity is created by the eccentric location of the nucleus in early primary oocytes. This is reinforced by the location of the yolk nucleus in the opposite hemisphere at the onset of vitellogenesis. As oocyte growth progresses, the pattern of distribution of yolky storage products fall into two classes (Fig. 1): a radial pattern as shown in some didelphids, and a centralized accumulation that becomes polarized opposite the chromosomes as maturation progresses, as found in the dasyurids and other didelphids. As can be seen from Table I, the different patterns of yolk assembly are not simply a function of the size of the egg. Small eggs, such as those of the Virginian opossum, have a radial and nonpolar localization, while those of the antechinus initially have a central, then a polarized, distribution.

C. Vitellogenesis and Yolk

Some workers find difficulties in the use of the term "yolk" to describe the storage products in the marsupial egg because the yolk mass that is segregated from the remainder of the cytoplasm contains a variety of organelles as well as yolk particles (Lyne & Hollis, 1976; Hughes and Hall, 1984). Its use for marsupials is, however, quite consistent with its use in embryological literature (Balinsky, 1981). In addition to being used to refer to lipid and protein storage products in the ooplasm, "yolk" is also used to refer to that part of the ooplasm containing mainly yolk, as in the yolk of the hen's egg or other telolecithal or mesolecithal eggs. The chemical composition, the morphological form of the yolk bodies, and the origin of the yolk, either from oocyte or maternal tissues or both, vary from species to species (Balinsky, 1981). Yolk is usually divided into fatty yolk and protein yolk and both have been identified in ultrastructural studies of marsupial eggs (Lyne and Hollis, 1976; Selwood and Sathananthan, 1988) and in a histochemical study (Guraya, 1968). Nevertheless, much remains to be clarified on the yolk of marsupials. Vitellogenesis has not been studied nor have the mechanisms of yolk elimination and its utilization during development. In addition, proper identification of cytoplasmic components is needed to clarify which subcellular bodies are yolk bodies and which contribute the electron-dense particulate material to the perivitelline space that has been located under the zona pellucida (Talbot and Dicarlantonio, 1984; Krause and Cutts, 1983; Selwood and Sathananthan, 1988; Baggott and Moore, 1990) and between blastomeres (Selwood and Sathananthan, 1988). For the purposes of this article "yolk bodies" will be used to refer both to vesicles of protein yolk and droplets of lipid yolk that are found in the cytoplasm and are released individually into the pervitelline space in didelphids (Hartman, 1916; Hill, 1918; Baggott and Moore, 1990). "Yolk mass" will be used to refer to the membrane-bounded accumulation of yolk bodies and other cytoplasmic components that is segregated from the remainder of the cytoplasm during the first division in dasyurid marsupials, as in Fig. 2 (Hill, 1910; Selwood, 1980; Selwood and Young, 1983; Selwood and Sathananthan, 1988; Breed and Leigh, 1990). "Yolk" will be used as a general term to describe storage products in the cytoplasm.

Ultrastructural studies have shown that the yolk bodies of marsupials consist of lipid droplets and relatively simple protein yolk vesicles (Lyne and Hollis, 1977; Selwood and Sathananthan, 1988; Breed and Leigh, 1990). When in a yolk mass, they are associated with a variety of cytoplasmic organelles including mitochondria, membranous structures, and a finely granular cytoplasmic matrix (Fig. 3). Fibrous arrays, showing linear periodicity, are also found in the yolk mass and cytoplasm of ova from the brown antechinus (Selwood and Sathananthan, 1988) (Fig. 4).

D. Ovulation

Examination of a number of species from two different marsupial families, Dasyuridae (Hill, 1910; Godfrey, 1969; Selwood, 1982; Selwood and Woolley, 1991) and Didelphidae (Hartman, 1919, 1928; Rodger and Bedford, 1982b; Baggott and Moore, 1990), reveals the general features of ovulation in marsupials. Cumulus cells detach from the oocyte shortly before ovulation so that the oocyte is naked when shed. The shedding of cumulus cells occurs close to ovulation and is apparently related in the stripe-faced dunnart to secretory gran-

5. Development of Pattern in Marsupial Embryos 183

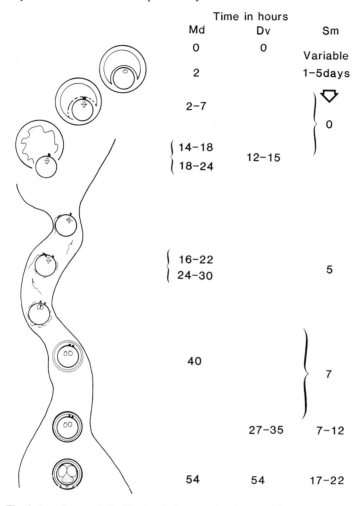

Fig. 2 The timing of events in fertilization in three species of marsupials: *Monodelphis domestica* (Md) (Baggott et al., 1987; Baggott and Moore, 1990), *Didelphis virginiana* (Dv) (Hartman, 1928; McCrady, 1938; Rodger and Bedford, 1982a), and *Sminthopsis macroura* (Sm) (Selwood, 1987; Selwood and Woolley, 1991). Time in hours is shown opposite the events depicted in the ovary, oviduct, and uterus. Time zero is the time of observed copulation for *M. domestica* and *D. virginiana* and time of ovulation in *S. macroura*, in which time from coitus to ovulation is variable.

ules that accumulate between zona and cumulus cells prior to and during cumulus dispersal (Selwood and Woolley, 1991). These granules may be enzymatic in nature and may be secreted by either the oocyte or the cumulus cells. Shedding of cumulus cells has also been reported in a number of other species from different families, including the ring-tail possum (Hughes *et al.*, 1965) and the brush

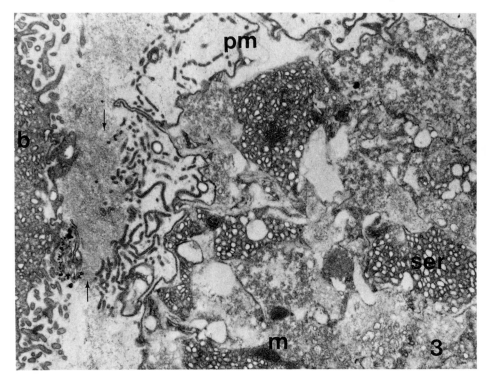

Fig. 3 Portion of a plasma membrane-bounded (pm) yolk mass of a two-cell embryo of the brown antechinus (right half of figure), containing a number of organelles, including profiles of smooth endoplasmic reticulum (ser) and mitochondria (m) as well as protein and lipid yolk material. It has recently separated (arrows) from a blastomere (b) lying on the left. ×9100. [From Selwood and Sathananthan (1988) with permission.]

possum (Hughes, 1977). The first polar body is given off in the ovary and the egg is ovulated at metaphase II of meiosis. No obvious changes have been noted in distribution of cytoplasmic organelles during these later stages of oogenesis. The hormonal profiles at ovulation have been described for a number of marsupials (see reviews by Tyndale-Biscoe and Renfree, 1987; Tyndale-Biscoe and Hinds, 1989). Ovulation is spontaneous in marsupials with the exception of the gray opossum (Fadem and Rayve, 1985; Baggott et al., 1987). The timing of these events has been estimated in some species (Fig. 2). The particular polarized state of the egg is retained during ovulation but the shape of the egg may be slightly distorted immediately postovulation and during tubal transport. An as yet unexplained decrease in egg size is found following ovulation in all species where this has been examined (Hartman, 1919; McCrady, 1938; Godfrey, 1969; Selwood, 1982; Hughes and Hall, 1984).

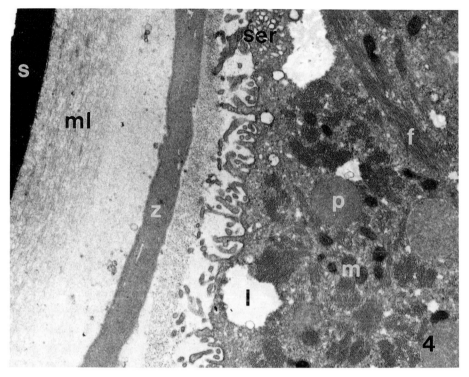

Fig. 4 Peripheral yolk-rich cytoplasm from one of the two-cell blastomeres of the brown antechinus, containing many fibrous arrays (f), protein (p), and lipid (l) yolk bodies. Mitochondria (m) and smooth endoplasmic reticulum (ser) are visible. The embryo is surrounded by shell (s), mucoid layer (ml), and zona (z) envelopes. Microvilli from the egg surface penetrate the electron-dense layer under the zona pellucida. ×9100. [From Selwood and Sathananthan (1988) with permission.]

III. Definitive Polarity

A. Fertilization

1. Timing and Transport

Rapid tubal transport is a characteristic of all marsupials examined so far and is one feature that distinguishes marsupials from eutherian mammals, in which tubal transport usually takes 2–4 days (McLaren, 1982). The timing of events of fertilization has been studied in most detail in the gray opossum (Baggott *et al.*, 1987; Baggott and Moore, 1990) but timing of some stages is also available for the Virginian opossum (Rodger and Bedford, 1982a) and the stripe-faced dunnart

(Selwood and Woolley, 1991) (Fig. 2). Hartman (1928), on the basis of experiments performed using nematode eggs as markers, suggested that transport through the oviduct is extremely rapid in the opossum and takes not more than 24 hr. Subsequent studies in a variety of other marsupials have confirmed that transport through the oviduct is rapid, such as 24 hr or less for the quokka (Sharman, 1955a), brush possum (Hughes and Hall, 1984; Rodger and Mate, 1988), native cat (Hill, 1910), and stripe-faced dunnart (Godfrey, 1969) or 12 hr or less for the brown antechinus (Selwood, 1980; Selwood and Smith, 1990) and the stripe-faced dunnart (Selwood and Woolley, 1991). Such rapid transport implies that synchronization of the events of fertilization with ovulation is essential to successful fertilization. Hartman (1916) suggested that in the opossum the timing of mating and the transport of spermatozoa up the oviduct must be sufficiently synchronized with ovulation to ensure that spermatozoa reach the oocyte before mucin deposition occurs. During the relatively short period of tubal transport, the initial polarity of the egg is emphasized, especially during the activation changes.

Fertilization in marsupials was described only recently *in vivo* in the brown antechinus (Selwood, 1982), the fat-tailed dunnart (Breed and Leigh, 1988, 1990; Breed *et al.*, 1989), and the gray opossum (Baggott and Moore, 1990) and *in vitro* in the opossum (Rodger and Bedford, 1982a,b). These studies confirm Hartman's suggestion that timing of events is critical. All authors also agree with Hartman that mucin deposition may contribute to a barrier to polyspermy. So far, a low incidence of polyspermy has been reported only in the Dasyuridae (Hill, 1910; Selwood, 1982; Breed and Leigh, 1990). Fertilization occurs in the ampulla in the Virginian opossum (Hartman, 1916; Rodger and Bedford, 1982b), the brown antechinus (Selwood, 1982), the fat-tailed dunnart (Breed and Leigh, 1988), and the gray opossum (Baggott and Moore, 1990) (Fig. 2).

2. Sperm–Egg Interactions

Usually, spermatozoa approach and penetrate the zona by an oblique path so that the inner face of the acrosomal membrane lies in contact with the surface of the zona. Rodger and Bedford (1982b) suggest that penetration is achieved mainly by enzymatic activity in the opossum. In contrast, Breed and Leigh (1988) suggest that penetration in the dunnart is achieved by a combination of enzymatic activity and spermatozoan motility because the spermatozoan penetrates the zona via a very narrow slit, and bulging of zona material occurs at the point of spermatozoan entry. Decondensation of the nucleus occurs almost immediately after entry in the opossum but is more delayed in the dunnart. Rotation of the head of the spermatozoan occurs around the time of decondensation of the head in both species. Spermatozoan membranes, possibly nuclear in origin, are found associated with the nucleus of the dunnart spermatozoan, but not that of the opossum. Some spermatozoan plasma membrane is incorporated into the ooplasm in the opossum, but in the dunnart most of the spermatozoan plasma membrane is shed. A small portion to the level of the midpiece in the dunnart is fused with the egg plasma membrane (Breed and Leigh, 1988) and may contribute to polarization of the egg plasma membrane. In the opossum, the egg plasma membrane fuses with the inner acrosomal membrane of the spermatozoan (Roger and Bedford, 1982b). The site of spermatozoan entry has not been examined. It seems likely that the less yolky hemisphere would be the major site for spermatozoan entry.

3. Effects of Activation on Polarity

a. Nuclear Polarity. In the Virginian opossum, pronuclei initially lie to one side in an area of yolk-free cytoplasm and then move to lie in a central region of relatively yolk-free cytoplasm (Fig. 5i–l). In species where the cytoplasm is polarized into yolky and less yolky hemispheres, the pronuclei lie in the less yolky hemisphere (Fig. 5a–d) (Hill, 1910; Selwood and Young, 1983; Breed and

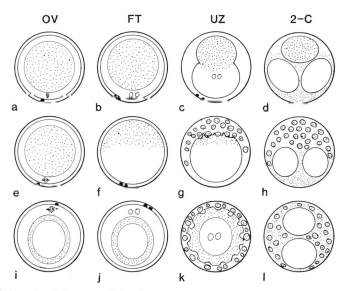

Fig. 5 Polar and radial patterns during fertilization and cleavage in marsupials. A diagrammatic representation of the location of the nucleus and yolky cytoplasm (coarse stippling) in follicular eggs (OV), fertilized tubal eggs (FT), and uterine zygotes (UZ) and during the emissions of cytoplasmic material as yolk (coarse stippling) or extracellular matrix (fine stippling) in two-cell stages (2-C). Circles enclosing coarse stippling represent a yolk mass in (d) or yolk vesicles in (g), (h), (k), and (l). Polarized yolk mass pattern (a–d); polarized-particulate pattern (e–h); radial pattern (i–l). The zona pellucida, but not mucoid and shell envelopes, is shown. In the polarized-particulate pattern, the pronuclei are not visible in the fertilized tubal egg (f) or uterine zygote (g).

Leigh, 1990). The position of the pronuclei in the fertilized egg of the gray opossum is not clear because they are obscured by yolky vesicles (Fig. 5e–h).

The polarity expressed by the oocyte in distribution and shape of nuclear material, yolky storage products, and surface features is thus initiated during oogenesis and enhanced during folliculogenesis and following fertilization. It is unknown whether the initial oocyte polarity determines the site of entry of spermatozoa, although one would speculate that spermatozoa would enter via the less yolky hemisphere, because this is commonly found in animal species with a polarized distribution of yolk (Balinsky, 1981). Presumably, the presence of numerous yolk bodies would obstruct the migration of the spermatozoan nucleus after penetration.

b. Cytoplasmic Polarity. Following spermatozoan entry, the pattern of distribution of yolky cytoplasm found in the secondary oocyte in the ovary is enhanced during the activation process (Fig. 5). Activation-induced changes can be identified in the ampulla and are completed in the uterus, at the pronuclear stage. Two patterns of distribution of yolk, polarized and radial (Fig. 5), are found. In the polarized pattern, yolky cytoplasm is accumulated as a spherical mass at the pole opposite the pronuclei (Fig. 5b and c) as in the koala (Caldwell, 1887), native cat (Hill, 1910), quokka (Waring et al., 1955), brown antechinus (Selwood, 1980; Selwood and Young, 1983), stripe-faced dunnart (Selwood, 1987), and the fat-tailed dunnart (Selwood, 1987; Breed and Leigh, 1990). In the brown antechinus, the ultrastructure of the cytoplasm in the two hemispheres is different with respect to relative proportions of organelles (Selwood and Sathananthan, 1988) (Fig. 6). Alternatively, yellow, refractive yolky cytoplasm may accumulate as a cap at one pole and a more amorphous material at the other pole (Fig. 5f and g), as in the gray opossum (Baggott and Moore, 1990). The radial pattern of distribution (Fig. 5j and k) is shown by the Virginian opossum (Hartman, 1916; McCrady, 1938) and the common opossum (Hill, 1918). In recently ovulated oocytes in these forms, the yolky cytoplasm is distributed in a submarginal zone but following fertilization this yolky material moves to the cortical regions.

Cortical granule release is a part of the activation process and has been shown to occur in the Virginian opossum (Rodger and Bedford, 1982b), the gray opossum (Baggott and Moore, 1990), and the fat-tailed dunnart (Breed and Leigh, 1990) but the pattern of release has not been described. The distribution of microvilli over the surface of the egg has been described only in the fat-tailed dunnart (Breed and Leigh, 1990), in which the microvilli are lost at the point of spermatozoan penetration, where a small fertilization and incorporation cone also forms. In the brown antechinus (Fig. 4) (Selwood and Sathananthan, 1988), the fat-tailed dunnart (Breed and Leigh, 1990), and the gray opossum (Baggott and Moore, 1990) a large amount of electron-dense material is found in the perivitelline space of fertilized eggs. In primary and secondary oocytes this material is absent in the gray opossum and present in only small quantities in the fat-tailed dunnart, so possibly it comes from cortical granules or other vesicles releases at this time.

B. Egg Envelopes

An unfortunate dependence on the terminology of egg envelopes as defined by Boyd and Hamilton (1952) has generated some ambiguities in descriptions of the marsupial egg plasma membrane and envelopes. The analysis by Boyd and Hamilton was made without the benefits of electron microscopy and has since been replaced by a more appropriate classification (Dumond and Brummett, 1985; Dietl, 1989), so that effective comparisons can be made at the ultrastructural level between widely separated phylogenetic groups, including invertebrates (Balinsky, 1981). Like every egg, the limiting membrane of the marsupial egg is the plasma membrane. In this article, the modern practice of referring to any investment outside the plasma membrane as an "envelope" will be adopted. It is misleading to refer to the plasma membrane as the vitelline membrane because the "vitelline membrane" of lower vertebrates and invertebrates lies *outside* the plasma membrane and is regarded as being equivalent to the zona pellucida of mammals (Balinsky, 1981). In some species, such as the opossum (Hartman, 1916, 1919), expansion of the blastocyst within the investments is eccentric and is related to embryo polarity (see Section IV).

1. Zona Pellucida

The origin of the zona pellucida or vitelline envelope (Fig. 4) of marsupials is uncertain but it first appears during oocyte growth in the Virginian opossum (Guraya, 1968) and in the brush possum (Hughes, 1977). Interdigitating microvillous processes from both the oocyte and granulosa cells penetrate the zona in the opossum and in bandicoots (Lyne and Hollis, 1983). An additional extracellular matrix component of hyaluronic acid-rich filaments and proteinaceous granules is secreted by the granulosa cells into the perivitelline space (Talbot and Dicarlantonio, 1984). This layer disappears in fertilized eggs in opossums. The zona pellucida is relatively narrow and varies in width from 1 μm in the opossum (Hartman, 1916) to 6.3 ± 1.4 μm in the tammar (Renfree and Tyndale-Biscoe, 1978).

2. Mucoid Envelope

Deposition of the mucoid envelope (Figs. 2 and 4) begins in the ampulla during penetration of the zona by the spermatozoan in the didelphids (Rodger and Bedford, 1982a; Baggott and Moore, 1990) and dasyurids (Selwood, 1982;

Breed and Leigh, 1988). The homologies of this and other egg envelopes with those of other mammals and vertebrates have been examined and reviewed by Hughes (1977). The mucoid material is an acid glycoprotein (Hughes, 1977), periodic acid–Schiff stain (PAS)-positive substance (Rodger and Bedford, 1982a) secreted by the nonciliated cells of the oviduct by a process of exocytosis (Breed and Leigh, 1988; Baggott and Moore, 1990). Deposition of the mucoid envelope is asymmetrical on eggs in the ampulla of the gray opossum (Baggott and Moore, 1990). Like the zona pellucida, the mucoid envelope is compressed and obliterated during blastocyst expansion (Lyne and Hollis, 1977). Hartman (1916, 1919) speculated that the mucoid envelope plays a role in nutrition of the embryo. It varies widely in width from 6.7 μm in the long-nosed bandicoot (Lyne and Hollis, 1976) to 140–230 μm in the Virginian opossum (Hartman, 1916; McCrady, 1938).

3. The Shell

The outermost envelope is the shell (Figs. 2 and 4) that encloses all marsupial embryos from the zygote stage until just prior to implantation, usually at early stages in organogenesis (Selwood, 1989a). The shell is ovokeratinous in nature (Hughes, 1977), but its composition has not been analyzed in any detail. There may be species differences in composition, because the shell of the brush-tailed possum is insoluble in trypsin and pepsin but is disrupted by pronase or α-chymotrypsin at 37°C (Hughes, 1977). In contrast, the shell of the brown antechinus was reduced in width and disrupted after 2.5 hr in 2.5% pancreatin at 35°C and after 4 hr in 0.05% trypsin at 35°C. In addition, 0.1% pronase appeared to penetrate the shell and digest the zona pellucida, causing disruption of the embryo and release of blastomeres, but left a relatively intact shell following incubation at 35°C of embryos of both the brown antechinus and stripe-faced dunnart (Selwood, 1989b). The shell varies in thickness from 1.0 μm in the Virginian opossum (McCrady, 1938) and the long-nosed bandicoot (Lyne and Hollis, 1976) to 5.9 μm in the tammar (Tyndale-Biscoe, and Renfree, 1987). The difference in digestibility of the shell may be due partly to differences in thickness, because the shell of the brush-tailed possum (4.8 μm) (Hughes and Hall, 1984) is thicker than the shell of the brown antechinus (2.2 μm) (Selwood and Young, 1983).

Various sites have been given for the secretion of the shell material. The lower oviduct has been implicated in shell secretion in the Virginian opossum (Hartman, 1916; McCrady, 1938), native cat (Hill, 1910), brown antechinus (Selwood, 1980), stripe-faced dunnart (Selwood, 1987), and fat-tailed dunnart (Breed *et al.*, 1989). Uterine endometrial glands have been implicated in shell secretion in the brush-tailed possum (Hughes and Hall, 1984), the gray opossum (Baggott and Moore, 1990), and possibly the fat-tailed dunnart (Breed and Leigh, 1990). Evidence of increase in width of the shell during the first few cleavages does not eliminate the glandular cells in the oviduct as a source of the shell in dasyurids, because secretions from the isthmus may pass into the upper half of the uterus near the uterotubal junction where the zygotes initially are found. Eggs found at the uterotubal junction are surrounded by a cloud of material including spermatozoa and cell debris and the shell is sticky (Selwood and Smith, 1990). In addition, the change in environment from oviduct to uterus may alter the width of the shell, by hydration, for example.

The shell is initially relatively thin, increases in thickness during cleavage to formation of the unilaminar blastocyst, then decreases in thickness as the blastocyst expands (Hill, 1910; Hartman, 1916, 1919; Selwood, 1980; Selwood and Young, 1983; Lyne and Hollis, 1976, 1977; Hollis and Lyne, 1977; Hughes and Hall, 1984). It is lost just prior to implantation at early stages in organogenesis (Hughes, 1977; Selwood, 1989a). In the tammar (Denker and Tyndale-Biscoe, 1986), loss of the shell membrane is associated with glycogenolytic enzymatic secretion beginning in the nonvascular yolk sac. Hughes and Shorey (1973) have shown that the shell is porous to ferritin and horseradish peroxidase and would not present a barrier to passage of embryonic nutrients and waste. Sharman (1963) has suggested that the shell may play an immunoprotective role but this has yet to be established. Hartman (1920) suggested that the shell prevents fusion of embryonic membranes of adjacent embryos. Fusion of membranes does occur between embryos after shell loss in philander opossum (Enders and Enders, 1969), bandicoots, antechinus, and dunnarts (L. Selwood, unpublished observations, 1991). The structural integrity of the investments has also been implicated in blastocyst formation (Selwood, 1989b), owing to the requirement for blastomere zona adhesion during late cleavage when cell cell adhesion first develops.

C. Cleavage

The polarization of the embryo and its blastomeres during cleavage can be related to the polarized distribution of pronuclei, cytoplasmic organelles, and yolky storage products in the zygote. Thus, the pattern established during oogenesis, ovulation, and fertilization affects, in turn, the mode of cytoplasmic emissions during cleavage and eventually the cleavage pattern.

1. Cytoplasmic Emissions

Marsupial early cleavage is characterized by the elimination of a variety of cytoplasmic organelles into the perivitelline space. These cytoplasmic emissions can be subdivided into two categories: yolk elimination and extracellular matrix emission.

a. Yolk Elimination. Elimination of yolk (deutoplasmolysis) occurs in three major patterns: polarized yolk mass, polarized particulate, and radial (Fig. 5). The type of pattern does not appear to be determined by the size of the egg, as the brown antechinus and Virginian opossum have eggs of a similar size (Table 1) but elimination is polar in antechinus and radial in the opossum.

A polar pattern of yolk elimination is found in the native cat (Hill, 1910), brown antechinus (Selwood, 1980), stripe-faced dunnart (Selwood, 1987), fat-tailed dunnart (Selwood, 1987; Breed and Leigh, 1990), and the gray short-tailed opossum (Baggot and Moore, 1990). In these forms, yolky material is accumulated centrally during oogenesis and is markedly polarized by the end of fertilization. In species showing a polarized yolk mass pattern (Fig. 5a–d), including the native cat, brown antechinus, and the dunnarts, a yolk mass is found at the pole opposite the pronuclei following fertilization. In uterine zygotes, the yolk mass is further segregated to one pole and is eventually extruded as a membrane-bound yolk mass from the two blastomeres during the first division. Further material is eliminated from the surface of the blastomeres as separate particles during the first and second divisions. In the species showing the polarized particulate pattern, the gray opossum (Baggot and Moore, 1990) (Fig. 5e–h), clear lipid-like yolk material forms a cap at one pole of the egg in tubal zygotes. This material is eliminated into the perivitelline space as many separate vesicles of varying size, prior to and during the first division. Additional material is eliminated from the blastomeres at the two-cell stage and during the four-cell stage.

The radial pattern (Fig. 5i–l) is found in the Virginian opossum (Hartman, 1916, 1919; McCrady, 1938) and common opossum (Hill, 1918). In these forms, a radial distribution of yolk vesicles, at first submarginal and then cortical, is developed during oogenesis and fertilization. Their uterine zygotes have a cortical layer of vesicular yolk. Elimination of yolky material into the perivitelline space begins prior to the first cleavage division in the Virginian opossum and continues during the second cleavage division until the four blastomeres round up. Elimination of yolk vesicles in the common opossum occurs during the two- and four-cell stages but a fertilized uterine egg has not been described.

Because these stages of fertilization and early cleavage have not been fully described in other marsupials, it is not possible from the descriptions to fit those species into these patterns. Exceptions are the Tasmanian devil (Hughes, 1982), which appears to have a polarized yolk mass pattern, and bandicoots, which appear to have a polarized particulate pattern (Lyne and Hollis, 1976, 1977). No mention is ever made of a single yolk mass in descriptions of macropodid development (Tyndale-Biscoe and Renfree, 1987), so presumably they fit into either the radial pattern or the polarized particulate pattern.

b. Extracellular Matrix Emission. There is some evidence from light microscope studies (Hill, 1910, 1918; Hartman, 1916, 1919; McCrady, 1983) and even more convincing evidence from electron microscope studies (Selwood and Sathananthan, 1988; Breed and Leigh, 1990; Baggott and Moore, 1990) that there is a qualitative difference in the nature and sites of cytoplasmic emissions from the eggs of marsupials. The emissions consist of yolk material either in the form of separate vesicles or a single yolk mass and an electron-dense matrix consisting of fibrous material and occasional granules. Talbot and Dicarlantonio (1984) and Breed and Leigh (1988), using ruthenium red to enhance preservation of extracellular matrix material, demonstrated a fibrous and granular matrix lining the zona pellucida of follicular oocytes in the opossum and fat-tailed dunnart, respectively. Material with a similar ultrastructure but not stained with ruthenium red has been reported in the perivitelline space of oocytes of the brush possum (Hughes and Hall, 1984) and in increasing quantities during early cleavage in the brown antechinus (Figs. 3 and 4) (Selwood and Sathananthan, 1988), fat-tailed dunnart (Breed and Leigh, 1990), and the gray opossum (Baggott and Moore, 1990). Further work is necessary to establish whether the material surrounding preovulatory oocytes is the same as that found in fertilized eggs and early cleavage stages. Selwood and Sathananthan (1988) showed that in the brown antechinus, from the fertilized egg to the four cell stage, this fibrous material originates from vesicles that discharge their contents into the perivitelline space. It surrounds the blastomeres and forms a lining under the zona pellucida (Fig. 4).

Time-lapse studies (Selwood and Smith, 1990) suggest that this material is also polarized in distribution because vesicles are not discharged from the yolk mass. Thus the zona above the yolk mass is relatively free of it while that of the other hemisphere would contain a greater amount. This needs to be confirmed by ultrastructural observations but if confirmed could provide a possible mechanism for the blastomeres initially, preferentially populating the zona above and around the yolk mass in the native cat (Hill, 1910), brown antechinus (Selwood and Young, 1983), stripe-faced and fat-tailed dunnart (Selwood, 1987), and above the polarized yolky vesicles in the gray opossum (Baggott and Moore, 1990). Because it is found between blastomeres (Selwood and Sathananthan, 1989), it may prevent blastomere–blastomere adhesion during early cleavage stages. Breed and Leigh (1990) and Baggott and Moore (1990) suggest that it may assist in blastomere–zona adhesion. This material may have an alternative or additional role as the gellike matrix that holds blastomeres and yolk mass together (Selwood, 1989b). If the extracellular matrix is capable of hydrating under different osmotic conditions as suggested by Selwood and Sathananthan (1988) it would explain how the space inside the zona pellucida expands in early cleavage before the blastocyst wall is complete (Selwood and Young, 1983). Insufficient studies have been done on this extracellular matrix material but these few tantalizing observations merit further study.

2. Site of Cleavage

With the exception of one animal, the brush-tailed possum, in which cleavage stages have been located in the oviduct (Hughes and Hall, 1984) as well as in the uterus (Sharman, 1961), cleavage stages have been found only in the uterus (Fig. 2). The zygotes usually enter the uterus at the pronuclear stage. Retrograde movement of eggs can occur. In the brown antechinus, a single two-cell stage embryo with all egg envelopes was located at the fimbria (L. Selwood, unpublished observations, 1991) while all other two-cell embryos were found in the uterus. *In vitro* studies (Selwood and Young, 1983; Selwood, 1987; Baggott and Moore, 1990) have established that a complex culture medium is required to support marsupial cleavage. The simpler medium used for culture of eutherian cleavage stages does not support marsupial cleavage in the species examined so far (Selwood, 1987).

3. Cleavage Patterns

While a number of different cleavage patterns have been identified in marsupials, they all have the following characteristics in common:

1. The manner of cytoplasmic emissions influences cleavage pattern.
2. The cleavage pattern is stylized and characteristic of the genus or even more inclusive taxa.
3. Blastomere–zona adhesion precedes blastomere–blastomere adhesion, which occurs in late cleavage, so that no morula is formed.
4. Two cell populations are established as a result of cleavage.
5. The order of cell division contributes to the cleavage pattern.

a. Cytoplasmic Emissions. Yolk elimination and elimination of extracellular matrix material have been dealt with in Section III,C,1. Particular modes of yolk elimination are associated with particular cleavage patterns. It seems possible that where yolk material is polarized, the elimination of extracellular matrix material is also polarized as in some dasyurids (Selwood and Smith, 1990), and both could presumably affect the cleavage pattern.

b. Types of Patterns. A major cleavage pattern, the polar type (Fig. 7a–d), has been described in dasyurid marsupials (Hill, 1910; Selwood, 1980, 1986b, 1989b; Selwood and Young, 1983). In these animals, the zygote is polarized and a polar yolk mass is eliminated into the perivitelline space during the first division. The first three divisions are meridional with respect to the polar yolk mass and with some rotation of blastomeres in the second division (Selwood and Smith, 1990) in one species. The fourth division is latitudinal and unequal and

Fig. 6 Polarized distribution of organelles in the cytoplasm and polarized shape of the nucleus (n) in a blastomere at the four-cell stage in the brown antechinus. The cytoplasm at the top of the figure lies nearest the yolk mass and contains much lipid (l) and protein (p) yolk as well as fibrous arrays (f). The cytoplasm at the bottom of the plate contains mainly mitochondria (m) and vesicular bodies (vb). no, Nucleolus. ×9100. [From Selwood and Sathananthan (1988) with permission.]

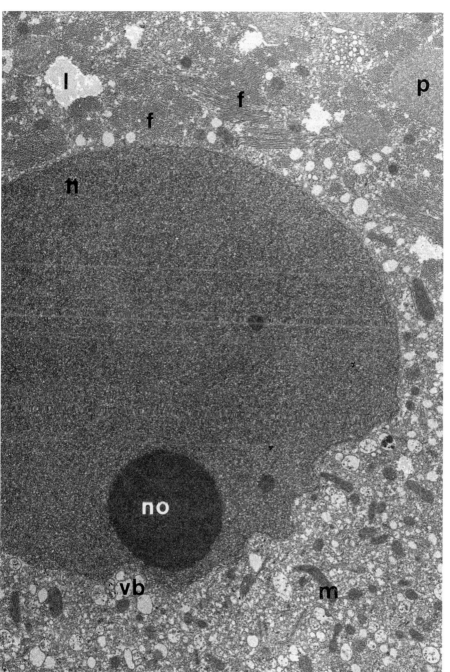

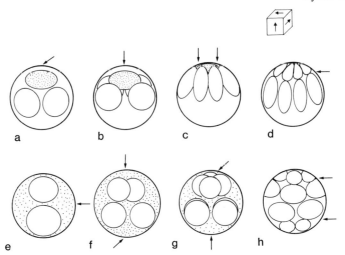

Fig. 7 A diagrammatic representation of the two major cleavage patterns found in marsupials from the 2-cell to the 16-cell stage. (a–d) The polar pattern, found in the dasyurids; (e–h) the radial pattern, found in the Virginian and common opossum. The cleavage planes are represented by arrows. Egg envelopes outsize the zona pellucida are not shown.

creates a tier of eight smaller, more rounded cells near the yolk mass and a tier of eight larger blastomeres. Subsequent divisions are obliquely latitudinal, the plane of division being dependent on the position of the blastomere in the embryo. Selwood and Sathananthan (1988) have established by electron microscopy that, in zygotes and blastomeres to the four-cell stage of the brown antechinus, the distribution of cytoplasmic organelles is polarized with respect to the yolk mass (Fig. 6). This has also been shown in histological studies of the brown antechinus (Selwood and Young, 1983) and the native cat (Hill, 1910), in which polarization of cell contents to the eight-cell stage precedes subdivision into two tiers of histologically distinct cells.

A variation on the polar type of pattern has been identified in the embryos of the gray opossum (Baggott and Moore, 1990). The zygotes of the gray opossum are polarized, and polarized-particulate yolk emission occurs. Description of cleavage in this species is incomplete because only a small number of specimens were collected at some stages, and in particular the yolky vesicles obscure much of the detail in cleavages after the two-cell stage. Baggott and Moore (1990) reported that the first division is meridional to the yolk pole and the second division is latitudinal, in contrast to the meridional second division of other polar species. However, because movement or rotation of blastomeres at the second division is a feature of didelphid embryos, this observation should be investigated further for verification. The subsequent two divisions are both meridional as far as could be determined.

5. Development of Pattern in Marsupial Embryos   197

Another pattern, the radial type (Fig. 7e–h), is found in the Virginian opossum (Hartman, 1916, 1919; McCrady, 1938) and the common opossum (Hill, 1918). These two species are characterized by a basically apolar egg after fertilization and by radial elimination of separate yolky vesicles. Polarity is created at the unequal, latitudinal first division. Each blastomere then divides meridionally and at right angles to each other so that the blastomeres lie in the typical eutherian cross-shaped formation. Hartman (1916), believed that this happens because of rotation of blastomeres but McCrady (1938) felt that the division planes are a result of blastomeres accommodating to the available space inside the egg investments. The third division is meridional and results in a tier of four small blastomeres lying adjacent to a tier of four large blastomeres. During the next two divisions the blastomeres divided by oblique to latitudinal division planes dependent on the position of the blastomeres and thus complete the wall of the blastocyst.

Cleavage has been described in only a few families of marsupials. Two families in the polyprotodonts, the didelphids (Hartman, 1916, 1919; Spurgeon and Brooks, 1916; Hill, 1918; McCrady, 1938; Baggott and Moore, 1990) and the dasyurids (Hill, 1910; Selwood, 1980, 1987, 1989a,b; Selwood and Young, 1983), have received the most attention, and cleavage patterns can be decided for most of the animals studied. Cleavage patterns are unknown for most other marsupials, because in most cases too few stages are described or they are described in insufficient detail. Because of blastomere rotation during early divisions, comments on blastomere size need to be supported by measurements of blastomeres, in the fresh state if possible. Determination of cleavage planes requires large numbers of embryos and analysis by time-lapse cinematography or reconstructions from sections. The pictures published of the normal embryos of the Tasmanian devil (Hughes, 1982), a dasyurid, suggest that it probably fits into the pattern of other dasyurids. In the long-nosed bandicoot, a peramelid polyprotodont, descriptions of a possibly abnormal 4-cell stage, an 8-cell stage (Lyne and Hollis, 1976), and a 30- and 32-cell embryo (Hill, 1910) did not identify any difference between the cells or confirm whether a single yolk mass was present.

In the diprotodonts cleavage stages of some families have been investigated: a 2-cell stage with equal-sized blastomeres in the greater glider, *Schoinobates volans* (Bancroft, 1973) in the Petauridae, the 4- and 16-cell stages of the brush-tailed possum (Sharman, 1961; Hughes and Hall, 1984) (Phalangeridae), a 4-cell stage surrounded by separate yolk vesicles in the feather-tail glider, *Acrobates pygmaeus* (Burramyidae) (Ward and Renfree, 1988), a 4-cell stage of *Macropus giganteus*, the eastern gray kangaroo, and early cleavage stages of the tammar (Macropodidae) (Tyndale-Biscoe and Renfree, 1987). Without examination of sectioned material and measurement of blastomeres it is impossible to comment further on the cleavage patterns in these animals. Caldwell (1887) claimed that cleavage in the koala is meroblastic. Hill's assertion (1910) that this claim is incorrect is well founded, because Caldwell mistakenly identified a uterine-zygote as an early cleavage stage—what he enterpreted as "segmentation nuclei" appear to be polar bodies in the pervitelline space.

c. Blastomere–Zona Adhesion. Blastomere–zona adhesion appears very early in marsupial cleavage during the two-cell to eight-cell stages (Fig. 8) and is essential to the formation of the marsupial blastocyst (Selwood, 1989b). The appearance of blastomere–zona adhesion may be earlier than stated in Fig. 8 in some species. In the stripe-faced dunnart, for instance, time-lapse cinematography (Selwood and Smith, 1990) established that temporary processes from the blastomeres first attach to the zona at the two-cell stage, not the four-cell stage (Selwood, 1987, 1989b). Blastomere–zona adhesion precedes blastomere–blastomere adhesion by one or two divisions (Fig. 8). In embryos with a polarized emission of yolk (Fig. 7), blastomeres first adhere to the portion of the zona lying nearest the yolk material. In the stripe-faced dunnart, blastomeres at the two- and four-cell stages attach to a protruding rim of zona that encircles the embryo between the yolk mass and the blastomeres (Selwood and Smith, 1990) (Fig. 9d). In embryos with a radial emission of yolk, blastomeres appear to adhere to any part of the zona but an ultrastructural study could reveal subtle differences not detected in light microscope studies. The attachment of the blastomeres to particular sites on the zona, which is followed by stretching of the blastomeres between those sites, may in turn influence the cleavage plane, which is at right angles to the long axis of the blastomere (Selwood and Young, 1983).

During the early cleavage divisions, after blastomere–zona adhesion is initiated, blastomeres attach and flatten on the zona immediately prior to and following each division but round up between divisions (Selwood and Young, 1983). In early stages (Figs. 9 and 10a) blastomeres round up and separate after the division, then detach from the zona. Later, blastomeres retain more contact with the zona and round up less following the divisions (Figs. 9 and 10b and c), until eventually at the end of cleavage, when the unilaminar blastocyst is almost complete, blastomeres remain flattened on the zona (Figs. 9 and 10d). In cultured embryos, rounding up of blastomeres in late cleavage is the first obvious sign of embryonic failure (Selwood and Young, 1983; Selwood, 1989b). Persistence of the flattened state is associated with increasing contact between blastomeres (Figs. 9 and 10) and eventually leads to a complete epithelium lining the zona pellucida. Blastomeres isolated from their investments show similar characteristics at each cleavage stage as they show in intact embryos (Selwood, 1989b) except that attachment is to the culture substrate. They also become migratory at the four- and eight-cell stages. In intact embryos at these stages, blastomeres attach to the zona nearest to them. At the two- and four-cell stages in the stripe-faced dunnart (Selwood and Smith, 1990), blastomeres extend processes and attach to the projection of the zona or rim formed between yolk mass and blastomeres around the latitude of the egg (Fig. 9). At the eight-cell stage blastomeres extend along the meridians of the egg from above the yolk mass to

5. Development of Pattern in Marsupial Embryos

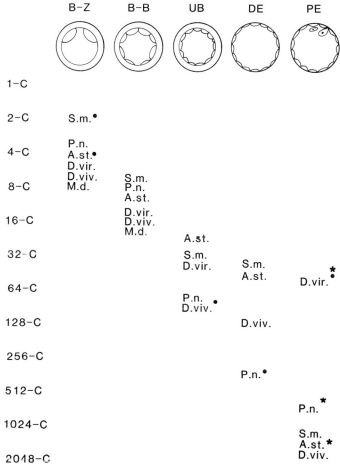

Fig. 8 Stages in relationship between cell number and adhesive behavior during cleavage, blastocyst formation, and primary endoderm formation in marsupials. The stages are shown on the horizontal axis: blastomere–zona adhesion (B-Z), blastomere–blastomere adhesion (B-B), complete unilaminar blastocyst (UB), definitive expansion (DE), and primary endoderm cell formation (PE). The number of cells are shown on the vertical axis on the left. The time of preliminary expansion is marked with a closed circle (●) and growth with an asterisk (*). The species represented are *Sminthopsis macroura* (S.m.), *Perameles nasuta* (P.n.), *Antechinus stuartii* (A.st.), *Didelphis virginiana* (D.vir.), *Dasyurus viverrinus* (D.viv.), and *Monodelphis domestica* (M.d.).

about the equator (Fig. 9). Possible interpretations of this are that the zona pellicuda itself has a polarized nature, or acquires polarity as a result of cytoplasmic emissions, so that the zona in the yolky hemisphere is the preferred site for initial attachment. In the gray opossum, with polarized, particulate yolk emission, blastomeres also adhere to the zona initially in the yolky hemisphere.

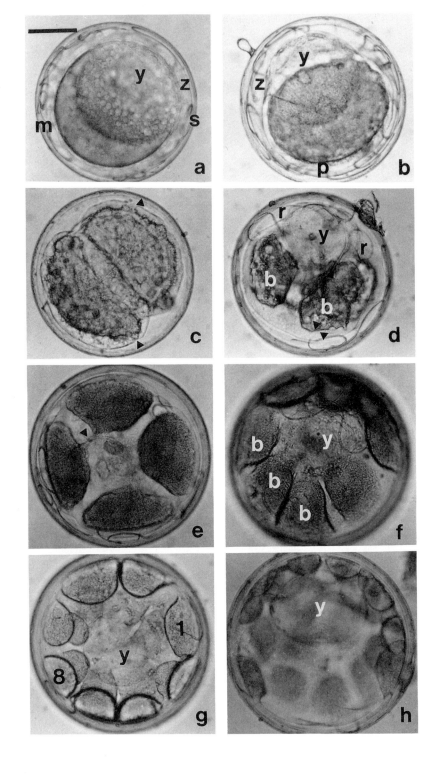

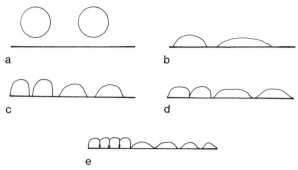

Fig. 10 Diagrammatic model of stages in formation of the blastocyst epithelium in marsupials. During early cleavage blastomeres lie separately within the zona (a). Blastomeres then flatten on the zona (b) prior to each division but round up slightly and pull apart after the division (c). After flattening in preparation for the next division (d) blastomeres divide but remain flattened with developing cell–cell junctions after the division (e). As blastomeres have less space in which to flatten (on the left), they remain more rounded and form contacts with neighbors sooner.

d. Blastomere–Blastomere Adhesion. Blastomere–blastomere adhesion is not characteristic of early cleavage in marsupials and blastomeres instead lie separately in the cleavage cavity (Figs. 7 and 8). The degree of separation has not been clearly established and would require serial sectioning possibly at the ultrastructural level to eliminate the possibility of persistent midbodies. In the brown antechinus at the two-cell stage, for instance, while the blastomeres lie far apart from each other and from the yolk mass in the cleavage cavity they remain connected to each other until just before the next division by a midbody and to the yolk by a narrow cytoplasmic bridge (Selwood and Sathananthan, 1988). Nor

Fig. 9 Embryos of the stripe-faced dunnart during cleavage and early blastocyst formation. [Bar = 100 μm for (a–h).] (a) Fertilized egg showing polarized yolk (y) at the top right pole and surrounded by zona (z), mucoid (m), and shell envelopes (s). (b) Fertilized egg showing yolk mass (y) being eliminated into the perivitelline space in the hemisphere with a thick zona (z). Microvilli and particle release into the perivitelline space (p) are associated with a thinner zona at the other pole. (c) Early two-cell stage viewed from the nonyolky pole showing blastomeres emitting particles into the perivitelline space (arrowheads). (d) Blastomeres lying near the yolk mass (y). Particle emission (arrowhead) into the perivitelline space continues from the blastomeres (b). The zona projection forming a rim (r) between yolk mass and blastomeres and a wider zona are visible in the yolky hemisphere. (e) Separation of blastomeres at the four-cell stage except for a persistent midbody between blastomeres on the left (arrowhead). The zona extends out into the space between blastomeres. (f) Blastomeres flattening on the zona around the yolk mass (y) at the eight-cell stage. Contacts are visible between some blastomeres (b). (g) Sixteen-cell stage viewed from the nonyolky pole; the first (1) and last (8) blastomere to divide at the fourth division lie opposite each other. More cell–cell contacts are found between cells lying over the yolk mass (y) and around the first cell to divide. (h) Thirty-two-cell stage viewed from the nonyolky hemisphere. Blastomeres are flattening and spreading on the zona over the top and to the sides of the yolk mass (y). The cells of the upper tiers lying over the yolk mass are closer together than cells of the lower tiers. [From Selwood and Smith (1990) with permission.]

do blastomeres engulf or spread over other blastomeres as do the blastomeres of eutherian mammals during the process of compaction as described by Ducibella and Anderson (1975). Marsupial embryos therefore do not form a morula. This separation of blastomeres in early cleavage stages and absence of a morula stage has been observed in all studies on marsupial cleavage whatever the cleavage pattern, and is found in blastomeres in intact embryos as well as blastomeres isolated from their investments (Selwood, 1989b). Eventual blastomere–blastomere adhesion is stage specific both in investment-free and enclosed blastomeres and begins to appear at the eight-cell stage in the brown antechinus and stripe-faced dunnart (Selwood, 1989b). Blastomere–blastomere adhesion follows blastomere–zona adhesion (Fig. 8) by one or two cell divisions. Lyne and Hollis (1976), in an electron microscope study, have shown that at the time of blastomere–blastomere adhesion in the bandicoot, small zones of attachment are found between blastomeres and in the 75-cell unilaminar blastocyst junctional complexes have developed. Baggott and Moore (1990) did not find junctional complexes at the 8-cell stage in the gray opossum but found tight junctions at the 16-cell stage.

e. Cell Populations during Cleavage. During cleavage in all marsupials examined so far, two distinct cell populations develop (Table II). These two populations are called the "embryoblast" and "trophoblast" for reasons outlined in Sections IV,A and IV,B,4. In marsupials with polarized yolk emission, two cell lineages are established at the 16-cell stage (Table II), one lying as a tier of cells nearest the yolk and the other in the less yolky hemisphere (Fig. 11). As in the dasyurids, two cell populations are distinct at the 16-cell stage (Table II) in the gray opossum (Baggott and Moore, 1990) with one lineage populating the hemisphere containing most of the yolk vesicles and the other the remainder of the zona pellucida. In the brush possum (Hughes and Hall, 1984), two populations distinguished by size are detectable until blastocysts have 60–80 cells. In the dasyurids, two lineages are created by a latitudinal fourth division (Figs. 7 and 9). These populations of cells differ in histological appearance, size, and position relative to the yolk mass (Hill, 1910; Selwood, 1980; Selwood and Young, 1983). Two populations remain detectable in whole mounts until blastocysts have between 32 and 64 cells in the brown antechinus (Selwood, 1980; Selwood and Young, 1983), about 130 cells in the native cat (Hill, 1910), and about 200 cells in the stripe-faced dunnart (Selwood and Woolley, 1991). They have also been detected in sectioned blastocysts with 75 cells (Lyne and Hollis, 1976) and 179 cells (Selwood, 1986a) in bandicoots. As well as these histological differences, the cells of the two lineages in the stripe-faced dunnart eventually differ in cell cycle time (Selwood and Woolley, 1991). Once the cell layer of the unilaminar blastocyst is complete, the cells furthest from the yolk mass (trophectoderm) retain a cell cycle time of about 8 hr and those nearest the yolk mass attain a cell cycle time of 24 hr. This situation is reversed several

Table II Stages at Which Two Cell Populations Are Detectable during Cleavage in Marsupials and the Basis of That Detection[a]

Animal	Stage	Basis	Ref.
Virginian opossum	Two-cell	Cell size, order of division, time of yolk and extracellular matrix elimination	Hartman (1916, 1919); McGrady (1938)
Gray short-tailed opossum	16-cell	Size, position	Baggott and Moore (1990)
Native cat	16-cell	Size, histology, position	Hill (1910)
Brown antechinus	16-cell	Size, histology, position	Selwood and Young (1983)
Stripe-faced dunnart	16-cell	Size, histology, position, cell cycle time	Selwood (1989b); Selwood and Woolley (1991)
Fat-tailed dunnart	16-cell	Size, histology, position	Selwood (1987)

[a] Only animals in which cleavage has been described from the zygote to the unilaminar blastocyst stage are included.

cycles later just prior to primary endoderm formation, when the trophectoderm cells cell cycle becomes 24 hr, and the other cells attain an 8-hr cycle. Other studies of cell cycle time have not been made, but the cell counts by Hartman (1919) on the opossum imply that there might be changes in cell cycle time before primary endoderm cells appear.

Two cell populations are initiated as early as the two-cell stage in animals with radial yolk emission, such as the Virginian opossum (Table II) and probably also the common opossum, although only one specimen of the two-cell stage was described by Hill (1918) for the latter species. During the radial emission of yolk particles, emissions at one pole are slightly in advance of those at the other, one cell is larger than the other, and one blastomere divides ahead of the other at the second division. These features of early cleavage in didelphids merit further analysis using modern lineage tracing techniques and time-lapse cinematography because of discrepancies in the descriptions. In the Virginian opossum, McCrady (1938) believed that the larger of the two blastomeres lay at the pole where yolk is first eliminated and that it was the first to divide. Hartman (1916, 1919) in the Virginian opossum and Hill (1918) in the common opossum believed that the smaller of the two blastomeres was more advanced than the other. This is an important point because the initial polarity after fertilization assigned to these opossum embryos is based on whether one blastomere is in advance of the other, not on any obvious morphological differences. Exact cell lineages have not been

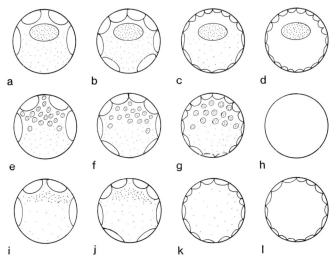

Fig. 11 Types of blastocyst formation in marsupials related to the pattern of yolk elimination (after Selwood, 1986a, with permission). Where a polarized yolk mass (a–d) or polarized particulate yolk (e–h) is eliminated into the blastocoel, blastomeres form a continuous epithelium in the hemisphere containing the yolk before the epithelium in the less yolky hemisphere is completed. The blastocyst is completed after further cell divisions and blastomere flattening in the dasyurids (d). The complete blastocyst has not been described in *Monodelphis*, so (h) is not completed. Where yolk is eliminated in a particulate radial pattern (i–l) the process is more random, except that the blastomeres that are descendants of the fastest dividing blastomere tend to completely line the zona first.

analyzed in opossums but two cell populations are distinct until shortly after the 16-cell stage and then disappear until the appearance of the primary endoderm cells distinguishes one part of the blastocyst epithelium as the embryonic area.

f. Order of Division. Cell divisions during cleavage in marsupials are asynchronous (Hill, 1910; Hartman, 1916; Selwood and Young, 1983), and this lack of synchrony appears to contribute to the development of distinct cell populations. Unfortunately, this has been examined only in two species and with time-lapse cinematography only in one. In the Virginian and common opossum, Hartman (1916, 1919), Hill (1918), and McCrady (1938) were all convinced that order of division segregated the cells into two groups: a faster and a slower dividing population. McCrady (1938) also suggested that the faster population was at the end of the embryo that had 60% of the yolk (Fig. 11), so even the opossum may have polarized yolk emissions albeit in a less conspicuous way.

In the stripe-faced dunnart, the order of division creates two foci in the cleaving embryo (Fig. 12) (Selwood and Smith, 1990). Descendants of the first cell to divide at the 4-cell stage are the first to divide at 8- and 16-cell stages. The descendants of the cell opposite the first cell to divide at the 4-cell stage are the last to divide at the 8- and 16-cell stages (Fig. 12). This precedence in division

5. Development of Pattern in Marsupial Embryos

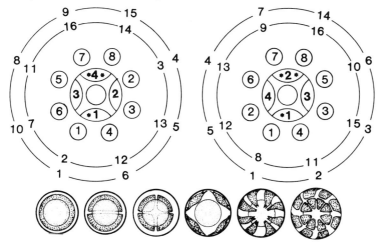

Fig. 12 Diagrammatic representation of the order of division from the four-cell stage onward in two embryos of the stripe-faced dunnart. The cells of each stage are numbered in their order of division to the subsequent stage. The yolk mass is represented by the innermost circle. The cells of the four-cell stage lie inside of the next circle (moving outward), and the cells of the eight-cell stage are represented by enclosed numbers outside this circle. The order of division at the 16-cell stage is shown on the 2 discontinuous outer circles. Of these, the inner circle represents the cells of the tier near the yolk mass and the outer circle represents the cells of the tier away from the yolk mass. Two foci are created at the four-cell stage around the first cell to divide at this and subsequent stages (*) and its opposite cell (**), the descendents of which contribute the cell that is last to divide at the two subsequent divisions. The diagrams along the bottom of the figure (Selwood and Young, 1983) represent cleavage from the zygote of the 16-cell stage viewed from the pole containing the yolk mass. [From Selwood and Smith (1990) with permission.]

order creates two foci (early versus late) in the embryo and the basis for a posterior–anterior axis between them. In addition, cell cycle length of the descendants of the first cell to divide at the four-cell stage are shorter than those of intermediate or opposite blastomeres at the fourth and fifth division. The difference between cell cycle length of the groups of blastomeres increases with successive divisions (Selwood and Smith, 1990). This process may start at the two-cell stage but continuous time-lapse observations were not available for the two- to four-cell stages. This precedence in division order could have significant consequences in the creation of pattern at the blastocyst stage (see Section V).

IV. Disappearance and Renewal of Polarity

A. Unilaminar Blastocyst

The unilaminar blastocyst, with all cells apparently identical in nature (Table III), is a feature of marsupial development and has been found in all species examined (see Selwood, 1986a,b, 1989a for review). McCrady (1938, 1944) described the

Table III Families of Marsupials in Which Unilaminar Blastocysts with a Uniform Population of Cells Have Been Recorded[a]

Family	Animal	Size (mm)	Cell number	Ref.
Didelphidae	Virginian opossum	0.11–0.34	32–50	Hartman (1916, 1919); Minot, (1911); McGrady, (1938)
Dasyuridae	Native cat	0.39–<3.5	>130–>1000	Hill, (1910)
	Brown antechinus	0.22–<1.3	>64–1000	Selwood, (1980); Selwood and Young (1983)
	Stripe-faced dunnart	0.34–<1.3	>200–2200	Selwood and Woolley (1991)
Peramelidae	Long-nosed bandicoot	0.23–<1.0	>75–1000	Hill (1910); Hollis and Lyne (1976, 1977)
Phalangeridae	Brush-tailed possum	>0.26–<0.75	>60–?	Hughes and Hall, (1984); Selwood, (1986a)
Petauridae	Greater glider	?–1.5	<50–643	Bancroft (1973)
Burramyidae	Feather-tail glider	0.4–2.0	?–2000	Ward and Renfree (1988)
	Pigmy possum	0.68	276–340	Clarke (1967)
Macropodidae	Red kangaroo	0.24	65–101	Clarke (1966)
	Red-neck wallaby (Bennett's wallaby)	0.21–<0.35	<62–~200	Walker and Rose (1981); Selwood (1986a)
	Tammar wallaby	0.21–0.30	~50–?	Renfree and Tyndale-Biscoe (1973)
Potoroinae	Tasmanian bettong	0.18–0.22	?–<80	Kerr (1935)

[a] Only animals in which information is reasonably complete are included.

blastocyst epithelium as a protoderm because it was considered capable of generating both embryonic and extraembryonic cell lineages. This now seems unlikely for a number of reasons. In all marsupials examined during cleavage, two populations of cells appear and persist into early stages of blastocyst formation and expansion. They become temporarily indistinguishable with further blastocyst expansion, but reappear as two populations, one of which, the embryoblast, generates the primary endoderm cells. When they reappear in the native cat (Hill, 1910), the brown antechinus (Selwood, 1980), and the stripe-faced dunnart (Selwood and Woolley, 1991) they reappear as two distinct populations already in existence and separated by a boundary or sutural line. It seems more likely that the two populations created during cleavage persist and give rise to the embryonic versus trophectoderm lineages, but are indistinguishable during blastocyst expansion. It is not so surprising that the two cell types are indistinguishable because all the cells become extremely flattened to 1 to 3 μm in width during blastocyst expansion (see Lyne and Hollis, 1976). Flattened epithelial cells from quite different tissues look similar. Also, in some dasyurids, especially the stripe-faced dunnart (Selwood and Woolley, 1991), the yolk mass persists until reappearance of the embryonic area or embryoblast. The yolk mass is associated with the embryoblast or the hemisphere containing it and acts as a natural marker for the embryoblast cells in these species. The physiology of blastocyst expansion and maintenance and the role of the two cell populations have not been examined at all.

The extraembryonic part of the blastocyst epithelium is called the trophectoderm or trophoblast in this article because these cells constitute the epithelium involved in implantation (Sharman, 1961; Tyndale-Biscoe and Renfree, 1987) and hence nourishment of the embryo. The relatively numerous perinuclear granule in trophectoderm cells suggest that these cells may also be embryotrophic during the preimplantation period. The trophectoderm of marsupials, like that of some other mammals such as the goat and pig (McLaren, 1982), but unlike that of mammals such as the mouse, does not enclose the embryoblast cells.

1. Modes of Blastocyst Formation

Certain events of cleavage, yolk elimination, and blastomere–zona followed by blastomere–blastomere adhesion are critical to formation of the unilaminar blastocyst. The type of yolk elimination influences the way in which the unilaminar blastocyst is formed (Selwood, 1986a). When yolk elimination is polarized, blastomeres of the upper tier lining the zona in the yolky hemisphere initially have relatively more contacts and more extensive areas of contact with neighboring cells than cells of the lower tier because they are populating a smaller area (Fig. 11). This continues for a further one or two divisions until the blastocyst wall is complete (Fig. 9). The cells of the upper tier spread over about one-third of the inner surface of the zona, while the cells of the lower tier spread over two-thirds (Fig. 11b and f). Blastomere–zona adhesion before blastomere–blastomere adhesion ensures that blastomeres do not clump together and form a morula but form an epithelium instead. Differences in size and the order of division between cells of the two lineages in the Virginian and common opossum, which have radial yolk elimination, ensure that the cells of one tier are crowded together on the zona and form the center of the developing blastocyst epithelium at one pole of the embryo (Fig. 11j).

2. Structure of the Unilaminar Blastocyst

The blastocyst is formed when the zona is completely lined by an epithelium. Embryos at this stage have between 22 and 32 cells in the brown antechinus (Selwood, 1980; Selwood and Young, 1983) or up to 108–130 cells in the native cat (Hill, 1910) (Fig. 8). The blastocyst epithelium is usually completed during the fourth to seventh cleavage division (Selwood, 1989a), with most being completed at about 32 cells. The stage at completion seems to be related to the diameter of the space inside the zona and to the relative size of the zygote itself and of the developing blastomeres. Two populations of cells are distinct (the embryoblast and trophoblast) for one or two divisions following blastocyst formation in the native cat (Hill, 1910), brown antechinus (Selwood, 1980), stripe-faced dunnart (Selwood and Woolley, 1991), bandicoot (Lyne and Hollis, 1976), and the brush possum (Hughes and Hall, 1984) but not the Virginian opossum (Hartman, 1916, 1919). As the difference between the two populations is not always visible in live or fixed whole specimens, and is usually undetectable in sections (Hill, 1910; Selwood and Woolley, 1991), two populations may also be present in early unilaminar blastocysts of other species in which they have not been reported. With further increase in cell number and blastocyst expansion all cells are uniform in appearance (Table III).

Lyne and Hollis (1976) have done the most comprehensive electron microscope study of the unilaminar blastocyst, in bandicoots. In unilaminar blastocysts with cell numbers between 300 and 400 and a total vesicle diameter of between 400 and 500 μm, two populations of cells are visible. Some of these blastocysts appeared to have collapsed either prior to or during processing, so these results should be viewed with caution. The ultrastructure of the cells, however, was assessed as normal. No ultrastructural differences were described for the two cell types, the differences being mainly of size and shape, one being more rounded and larger than the other. In other blastocysts with between 340 and 1500 cells no differences were detected between cells of the blastocyst. The cells were extremely flattened, with microvilli mainly located on the zona-facing cell surface. The cells contained a large nucleus with prominent nucleolus, vacuoles, irregularly shaped mitochondria, and mainly agranular endoplasmic reticulum mostly in a perinuclear location. Junctional complexes were found between cells, which had the appearance of typical, simple, squamous epithelial cells. In the feather-tail glider, blastocyst cells in blastocysts 0.4–2.0 mm in diameter with up to 2000 cells have a similar ultrastructure (Ward and Renfree, 1988). With further blastocyst expansion, the cells become increasingly flattened, agranular endoplasmic reticulum diminishes while small amounts of granular endoplasmic reticulum appear, mitochondria develop a more regular profile than in early blastocysts, and cells and cell fragments diminish in number or are no longer found in the blastocoel (Lyne and Hollis, 1976; Ward and Renfree, 1988).

Even though blastomeres flatten *on* the zona pellucida, the apical side of the blastocyst epithelium is the side apposed to the zona, not the side facing the blastocoel. In this the blastocyst epithelium is similar to the trophoblast epithelium of eutherian mammals as described by Enders and Schlafke (1965). The blastocyst cells also share a number of common characteristics with eutherian trophoblast cells. The apical surface is richly endowed with microvilli and possesses numerous micropinocytotic vesicles. Junctional complexes lie at the apical ends of the intracellular space. Granular endoplasmic reticulum is rare in early blastocyst cells. Marsupial blastocyst epithelial cells possess lipid droplets but also contain granules and degradation bodies similar to those found in eutherian trophoblast cells. Lipid droplets are released into the blastocoel in bandicoots (Lyne and Hollis, 1976). Apart from surface characteristics and certain cytoskeletal characteristics, the inner cell mass (ICM) cells of eutherian species are ultrastructurally similar to those of the trophoblast (Enders and Schlafke, 1965), so perhaps it is not surprising that all cells are similar in ultrastructure in the marsupial blastocyst.

Although the unilaminar blastocyst may sometimes persist as a unilaminar structure for only one cell doubling or less after the blastocyst is completed, as in the Virginian opossum, it is widespread amongst marsupial families (Table III). It has been found in every family examined so far. It is in the unilaminar blastocyst stage that embryonic diapause occurs in marsupials (for reviews, see Tyndale-Biscoe and Renfree, 1987; Tyndale-Biscoe and Hinds, 1989).

3. Blastocyst Expansion and Growth

Expansion of the embryo has two major phases. The first phase is a preliminary expansion of the space inside the zona, which compresses the zona and mucoid layers but does not greatly affect the total dimensions. It begins before the unilaminar blastocyst is completed and thus is not dependent on an intact epithelium for transport of fluid (Fig. 8). Selwood and Sathananthan (1988) have suggested that the extracellular matrix that accumulates prior to the beginning of expansion may be implicated by taking up water after it is released into the cleavage cavity. Preliminary expansion usually starts three or four cleavage divisions before definitive expansion occurs (Fig. 8 and Table IV). Definitive expansion is when the overall dimensions of the embryo are increased. In bandicoots definitive expansion occurs within the same division as preliminary expansion (Fig. 8 and Table IV). In the opossum, which has a very thick mucin envelope, preliminary expansion does not occur until the end of the unilaminar blastocyst stage (Fig. 8). Definitive expansion (increase in outside dimensions) occurs at the early unilaminar stage in brown antechinus but not until the advanced bilaminar stage in the Virginian opossum (Table IV and Fig. 8) (see review by Selwood, 1989a).

Growth in marsupial embryos was estimated on the basis of cell volumes by Selwood (1989a). Growth of cell volume was shown to be independent of preliminary and early definitive expansion and was delayed until the time of primary endoderm formation (Table IV). During the last half of the preimplantation period, growth and expansion occur together (Fig. 8). The rate of development and the relative lengths of each stage have been reviewed extensively in recent years (Tyndale-Biscoe and Renfree, 1987; Selwood, 1989a) and will not be dealt with here. Briefly, 60–80% of the gestation period is occupied by the preimplantation stages. Implantation usually occurs at an early somite, flat embryonic

Table IV Diameters of Embryos at Major Developmental Stages[a]

Animal	Early cleavage	Unilaminar complete	Primary endoderm appears	Bilaminar complete	Trilaminar blastocyst	Ref.
Bennett's wallaby	—	0.21 × 0.25	0.35	1.40	4.80 (early)	Walker and Rose (1981); Selwood (1986a)
Tasmanian bettong	—	0.18	0.27	0.70 × 0.43	1.40 (early) 5.5 (late)	Kerr (1935, 1936)
Brown antechinus	0.16*	0.23*	1.3–1.5	2.8 × 3.1	3.0 × 3.5 (early) 4.1 × 4.5 (late) (764.7)	Selwood (1980; Selwood and Young (1983)
	(1.4)	(1.0)				
Native cat	0.28*	0.39	4.5–5.0	(204.3) 8.5	6.0–6.75 (early)	Hill (1910)
Stripe-faced dunnart	0.27*	0.31*	1.37	2.2–3.0	3.2 × 3.6 (late)	Selwood and Woolley (1991)
Virginian opossum	0.11*	0.11*	0.11–0.34*	0.75	1.4 (early) 7.5 (late)	Selenka (1887); Hartman (1916, 1919); Krause and Cutts (1985a,b); McCrady (1938)
	(0.6)	(0.5)	(2.7)	(6.5)	(—)	
Brush-tailed possum	0.23*	0.31*	0.74	3.2	4.5 (early) 9.8 (late)	Hughes and Hall (1984)
Greater glider	0.12*	0.10*	<1.5	—	>4.4	Bancroft (1973)
Bandicoots	0.22*	0.23*	0.8–1.1	1.5–1.9	2.0 (early) 4.3 (late)	Lyne and Hollis (1976, 1977); Hollis and Lyne (1977)
	(1.7)	(1.3)	(11.4)	(55.2)	(1419)	

[a] Diameters measured in millimeters. Measurements for cleavage and blastocyst stages do not include the egg envelopes where marked with an asterisk. Growth, expressed as volume of cellular material (in nanoliters), is estimated from data in the given references and from Selwood (1989a), and is included (in parentheses) for the bandicoots, brown antechinus, and Virginian opossum.

shield stage. Of all the preimplantation stages, the unilaminar blastocyst stages is the most variable in length (Selwood, 1989a), partly because of variations in the rate of cell division compared to earlier and later stages, partly because of the timing of appearance of the primary endoderm cells (Fig. 8), and because of embryonic diapause in some species, such as the grey kangaroo, *Macropus giganteus* (Clark and Poole, 1967) and the red kangaroo, *Macropus rufus* (Sharman and Pilton, 1964).

B. Bilaminar Blastocyst

Development of the bilaminar blastocyst is initiated when primary endoderm cells appear in one hemisphere of the blastocyst. Formation of the bilaminar blastocyst has been studied in most detail in the native cat and red-neck wallaby (Hill, 1910), the Virginian opossum (Selenka, 1887; Hartman, 1916, 1919; McCrady, 1938), the Tasmanian bettong, *Bettongia cuniculus* (= *B. gaimardi*) (Kerr, 1935), and in the bandicoot (Hollis and Lyne, 1977; Lyne and Hollis, 1977). All these studies had access to a good series of stages of primary endoderm formation and specimens were mostly examined as sections. Hill (1910) demonstrated the usefulness of examination of whole mount as well as sections for study of marsupial blastocysts. Differences that are not detectable in histological sections are visible in whole mounts (Hill, 1910). Hollis and Lyne (1977) and Lyne and Hollis (1977) incorporated an ultrastructural study in their examination of the bandicoot bilaminar blastocyst.

1. Modes of Bilaminar Blastocyst Formation

Modes of formation of the primary endoderm were reviewed by Selwood (1986a), who established that three types of formation have been described (Fig. 13).

a. Type 1: Proliferation of Primary Endoderm via Endoderm Mother Cells. (See Fig. 13a–d.) This mode of formation was first described by Selenka (1887) in the Virginian opossum and subsequently further analyzed by Hartman (1916, 1919). Both Hartman (1916) and Hill (1910) initially discounted the interpretation by Selenka, but after examining more material Hartman (1919) confirmed that Selenka's account was substantially correct. In this form of endoderm proliferation, certain marginal cells in the unilaminar blastocyst, called the endoderm mother cells, increase enormously in size to 35×45 μm, and acquire basophilic staining. At this stage the blastocyst is not expanded (0.11 mm), has between 50 and 60 uniform cells, and its yolk is partly resorbed, although some yolk is accumulated at the pole where endoderm mother cells appear. Some endoderm mother cells then move into the blastocoel and the blastocyst epithelium reseals behind them. Others divide, sometimes several times, and the

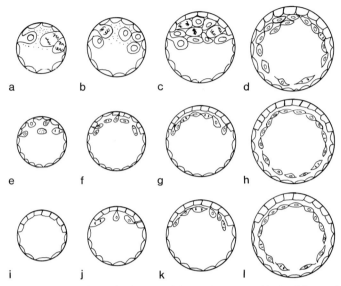

Fig. 13 Primary endoderm formation in marsupials (not drawn to scale). (a–d) In type 1 (proliferation via endoderm mother cells), a blastocyst that is not expanded and is apparently undifferentiated produces greatly enlarged endoderm mother cells (a) that detach and enlarge (b) and accumulate in one hemisphere (c). As they do so, the outermost cells transform into the primitive ectoderm (c). The primary endoderm extends around the blastocoel (d) and the trophectodermal part of the hemisphere expands until the blastocyst is evenly expanded within the investments. (e–h) In type 2 (direct proliferation), a previously uniform population of blastocyst cells distinguished only by the position of the yolk mass in the expanded blastocyst generates primary endoderm cells (e). As primary endoderm cells move into the blastocoel the embryoblast cells above them differentiate as primitive ectoderm (f). As primary endoderm cells line the primitive ectoderm the two layers differentiate from the trophectoderm (g). Eventually with further expansion the blastocyst is complete bilaminar (h). (i–l) In type 3 (proliferation from embryonic area), the blastocyst, which is originally unexpanded with a uniform population of cells, expands and the difference between trophoblast and embryoblast becomes obvious (i). Primary endoderm cells are generated directly from the embryoblast (j) and migrate to line the transformed primitive ectoderm cells (k) and eventually the complete expanded bilaminar blastocyst (l) develops.

innermost cells are detached into the blastocoel, where they lie surrounded by yolk material. The endoderm mother cells continue to grow as they move into and lie in the blastocoel and they develop a characteristic rounded shape. During and after migration into the blastocoel the endoderm mother cells proliferate and decrease in size until they form a cap of primitive endoderm cells, two to three cells deep, lying under the epithelium of one hemisphere, thus defining the embryoblast from the trophoblast.

Within the embryoblast the outer cells lying over the primitive endoderm become more cuboidal and more densely basophilic. These are the cells of the primitive ectoderm or epiblast. The cells of the epiblast and primitive endoderm are more basophilic and contain more yolk granules than cells of the trophoblast, which are extremely flattened and relatively pale. In blastocysts with a diameter of between 0.35 and 0.50 mm, the primitive endoderm cells flatten and begin to migrate out from the embryonic area to form a reticulum under the trophoblast cells. As this process occurs the blastocyst continues to expand eccentrically within the egg envelopes until the embryoblast is near the shell membrane. This indicates an increasing dichotomy between trophoblast and embryoblast functions.

Bennett's wallaby appears to fit into the mode of formation of primary endoderm cells observed in the opossum. The unilaminar blastocyst is small (0.21 mm) with only 62 cells (Table III) and shows uneven expansion in the mucoid layers (Selwood, 1986a). Enlarged endoderm mother cells appear when blastocysts have expanded slightly to 0.26–0.33 mm. Blastocysts, with several hundred cells and 0.35 mm in diameter, are found with several layers of primary endoderm cells accumulated at the embryonic pole (Walker and Rose, 1981).

b. Type 2: Direct Proliferation of Primary Endoderm Cells. (See Fig. 13e–h.) This mode of proliferation is found in blastocysts that have undergone several cell cycles (up to five) (Fig. 8) and have increased in size, sometimes considerably (Table IV). It is not possible to distinguish between embryoblast and trophoblast, either in whole mounts or in sections, at the time that direct proliferation of primary endoderm cells begins in the native cat (Hill, 1910), bandicoot (Hill, 1910; Lyne and Hollis, 1977), brown antechinus (Selwood, 1986a), and the stripe-faced dunnart (Selwood and Woolley, 1991). Once primary endoderm cells appear, the differences between embryoblast and trophoblast become increasingly obvious because of progressive transformation of embryoblast cells at the sites of endoderm cell proliferation. The embryoblast cells become increasingly basophilic and more cuboidal in shape, with more sharply defined margins as they transform into primitive embryonic ectoderm. Once several sites of proliferation are present, the two populations, embryoblast and trophoblast, are again easily distinguished in flat mounts and are separated by a distinct sutural line, as described by Hill (1910). The embryoblast occupies about one-third or less of the blastocyst wall and can be seen in fresh specimens as either an opaque patch (e.g., in the native cat or stripe-faced dunnart) or a transparent window (as in the brown antechinus). In the bandicoot (Hollis and Lyne, 1977) and brown antechinus (Selwood, 1980) the yolk mass disappears prior to the appearance of the primary endoderm. The yolk mass always persists, although reduced in size, until the time of primary endoderm formation in the stripe-faced dunnart (Selwood and Woolley, 1991) and occasionally is also found at this time in the native cat (Hill, 1910). When present the yolk mass is located beneath the embryoblast or in the hemisphere containing the embryoblast; once primary endoderm formation is well established, it disappears.

The ultrastructure of this mode of primary endoderm cell formation has been described by Hollis and Lyne (1977) and Lyne and Hollis (1977) in the bandicoot. Primary endoderm cells are sometimes visible because of staining characteristics or their shape in the embryoblast epithelium, but ultrastructural differences between the two are not marked. Primary endoderm cells become more flattened after they move into the blastocoel. They remain loosely attached to each other and to nearby epiblast cells and eventually form a loose reticulum lying under the thickened epiblast. The epiblast has junctional complexes on the outer surface, but primary endoderm cells, while closely adherent, are not linked to each other or the epiblast by junctional complexes.

Two other species possibly fit into this category. The greater glider has primary endoderm cells, reminiscent of those in bandicoots, in expanded early bilaminar blastocysts, 1.5 mm in diameter (Bancroft, 1973). The brush-tailed possum may also fit into this category because the primary endoderm cells are found in a small area in expanded blastocysts, 0.75 mm in diameter (Hughes and Hall, 1984), and the differentiation of the epiblast extends out from this limited area to occupy about half the blastocyst wall.

c. Type 3: Proliferation of Primary Endoderm from an Already Distinct Embryonic Area. (See Fig. 13i–l.) This type of formation of primary endoderm has been described in the Tasmanian bettong, *Bettongia cuniculus* (= *B. gaimardi*) (Kerr, 1935). In this mode, primary endoderm formation occurs when there is very little expansion of the blastocyst and cell numbers are low. The first sign of endoderm formation is found when blastocysts expand from a 0.18-mm blastocyst with a uniform population of cells, to 0.27 mm. At this stage blastocysts have about 80 cells and the embryoblast has developed as an area of thickened cells covering about one-fourth of the surface of the blastocyst. The primary endoderm cells then appear as more rounded cells that move into the blastocoel and lie on the inner surface of the embryoblast cells. The inward migration and establishment of a loosely fenestrated layer is similar to that described for the native cat.

Early and late bilaminar blastocysts have been described in a number of other species. Either because insufficient detail or insufficient stages are described, it is not possible to put these animals into any category. These animals include the long-nosed potoroo, *Potorous tridactylus* (Kerr, 1935), the quokka (Sharman, 1955b), the common opossum (Hill, 1918), the rock wallaby, *Petrogale penicillata*, the parma wallaby, *Macropus parma* (Selwood, 1986a), and the tammar wallaby, *M. eugenii* (Renfree and Tyndale-Biscoe, 1973). Some of these, such as the rock and parma wallabies, have small blastocysts in early expansion, with a multilayered embryoblast that is reminiscent of type 1 (Fig. 13c), but without stages showing endoderm mother cells it is impossible to categorize them as this type confidently.

5. Development of Pattern in Marsupial Embryos 215

2. Why Endoderm Formation Should be Considered a Renewal of Polarity

Appearance of the primary endoderm marks a development of blastocyst polarity. The two populations that develop during cleavage become indistinguishable as the blastocyst expands; two cell populations again appear as the embryoblast and the trophoblast differentiate. The embryoblast generates primary endoderm cells and thus bifurcates to develop as the primary endoderm and the primitive ectoderm or epiblast. This can be seen as a renewal of polarity rather than initiation of a new polarity for a number of reasons.

1. In some species that do not expand greatly, the two populations become visible prior to endoderm formation (type 3). In other species, such as the Virginian opossum (type 1), that also have minimal expansion the information is conflicting. Hartman (1916, 1919) sometimes reported a prior thickening of embryoblast and sometimes not, and McCrady (1938) concluded that endoderm mother cells were generated without prior differentiation of the embryoblast. Even in greatly expanded blastocysts where no distinction can be made, a differentiation of two separate populations is initiated once primary endoderm cells start to leave the embryoblast. When blastocysts can be flat mounted, as in the native cat (Hill, 1910) or stripe-faced dunnart (Selwood and Woolley, 1991), the two populations are separated by a distinct sutural line, even before the embryoblast is markedly differentiated.

2. The initial polarity in cleavage is related to the distribution of yolk in many species. In some animals, for example, the stripe-faced dunnart (Selwood and Woolley, 1991) and the native cat (Hill, 1910), the renewed polarity has an embryonic (associated with eliminated yolk) and extraembryonic (not associated with eliminated yolk) axis. Even in the Virginian opossum the zygote has some evidence of polarity related to eliminated yolk. McCrady (1938) claimed that the end producing the larger, more advanced blastomere had most of the yolk vesicles (60%). At times of endoderm mother cell formation, yolky material is mostly accumulated under the embryonic pole (Hartman, 1919). This could reflect an initial polarity or be due to uneven rates of yolk utilization at the two poles because of two different cell types.

3. When primary endoderm cells appear, they always appear in one hemisphere or part of a hemisphere. They have never been reported as developing from all parts of the blastocyst epithelium.

3. Origin of Primary Endoderm Cells

Whatever the mode of formation, primary endoderm cells leave the surface epithelium or embryoblast and migrate into the blastocoel (Fig. 13). Mitoses can occur before, during, and after migration. All studies imply that the primary endoderm cells located within the epiblast are a separate population of cells and in some cases they are detectable before migration. Their numbers are increased during this migratory phase and after entry into the blastocoel. The cells remaining in the embryoblast differentiate to form primitive ectoderm cells or epiblast. Differentiation of primitive ectoderm cells takes the form of the cells becoming thicker and more cuboidal, but electron microscope differences are not detectable, at least initially. Primary endoderm cells leave the embryoblast from a number of different sites scattered over the embryoblast area and in no apparent pattern in those species showing type 2 formation of primary endoderm, such as the native cat (Hill, 1910), bandicoot (Hollis and Lyne, 1977), and the stripe-faced dunnart (Selwood and Woolley, 1991). In the Virginian opossum (type 1 species) they are primarily generated from the marginal areas of the embryoblast (Hartman, 1919). This lends support to the idea that generation of primary endoderm cells within the embryoblast area is determined by a positional effect. In the species showing the type 1 formation, the cell population lying over the yolk mass is the first to maximize cell contacts and forms a central disk of cells in the expanding epithelium. Within the disk certain cells are the first to locate their neighbors. The chance of this happening depends on relative cell size, available space on the zona, and order of division (Fig. 12). The figures shown by Hill (1918) for the common opossum and Hartman (1916, 1919) and McCrady (1938) for the Virginian opossum show that the first cells to meet their neighbors tend to be located more marginally on the tier forming the presumptive embryoblast and most primary endoderm cells are generated at the margin of the embryoblast. It is impossible without appropriate experimental evidence to exclude the possibility that a determinant distributed randomly in type 2 embryos and to marginal cells in type 1 embryos specifies the fate of primary endoderm cells, but it seems a less likely hypothesis.

4. "Embryonic Area" or "Medullary Plate?"

During primary endoderm formation the blastocyst differentiates into two areas, an embryonic area (embryoblast) and a trophectoderm (trophoblast). Hill (1910) in the native cat, suggested and Kerr (1935) in the Tasmanian bettong, and Lyne and Hollis (1977) in bandicoots, established that the cells of the embryonic area represent a true primitive ectoderm or epiblast that gives rise to all embryonic lineages. McCrady (1938) in the Virginia opossum determined that the embryonic area gives rise to the embryonic lineages of neuroectoderm, endoderm, and mesoderm, but not to embryonic ectoderm and amnion. McCrady thought the evidence was slight (a few slides), but convincing, and thus called the embryonic area a "medullary plate." In support of this, McCrady found that the embryonic ectoderm could not be distinguished from the extraembryonic ectoderm (McCrady, 1938), a view supported by Krause and Cutts (1985b), but not Hartman (1919), who stated that embryonic ectoderm was distinct from trophectoderm.

From the evidence to date, it seems that the specification of primitive ectodermal components is more restricted, in that it does not include embryonic ectoderm, in the Virginian opossum than in the native cat, Tasmanian bettong, and bandicoot. Despite the conflict between the opinions of Hill, Kerr, Lyne and Hollis, Hartman, and McCrady, the term "medullary plate" has been enthusiastically adopted by many authors (Bancroft, 1973; Hughes and Hall, 1984; Tyndale-Biscoe and Renfree, 1987).

In addition, even if McCrady is correct in the opossum in the assessment of embryonic lineages, it would not be the medullary plate but primitive ectoderm that generates neuroectoderm, embryonic endoderm, and mesoderm. The primitive ectoderm then transforms into a medullary plate after the formation of the mesoderm and embryonic endoderm. It is more appropriate at present to use the less specific term embryoblast or embryonic area for marsupials other than the Virginian opossum unless evidence is provided to support doing otherwise (see Section IV,C).

5. Complete Bilaminar Blastocyst

The primary endoderm cells, once formed, continue to multiply and spread along the inner surface of the epiblast cells and finally the trophectoderm cells. The cells initially form a fenestrated epithelium (Hill, 1910; Hartman, 1919; Selwood, 1980; Selwood and Woolley, 1991) but this is progressively thickened until it forms a complete epithelium of extremely flattened cells. At this stage, there is no obvious difference between the primary endoderm cells lining the epiblast or lining the trophectoderm, so that the embryo consists of three cell types: primitive ectoderm, primary (primitive) endoderm, and trophectoderm. Formation of the bilaminar blastocyst occurs in blastocysts of varying size and cell number (Table IV). The blastocyst wall is completely lined with cells to form a completely double structure at different stages of blastocyst expansion dependent on the species and individual embryo. Completion usually occurs just prior to the appearance of the mesoderm in blastocysts that are less than 1.0 mm in diameter in the Virginian opossum (Hartman, 1919; McCrady, 1938), the bettong (Selwood, 1986a), and Bennett's wallaby (Walker and Rose, 1981; Selwood, 1986a,b) or in very expanded blastocysts up to 8.5 mm in diameter in the native cat (Hill, 1910). In the native cat, the mesoderm appears before the blastocyst wall is completely bilaminar (Table IV).

C. Trilaminar Blastocyst

1. Formation

As with the appearance of the primary endoderm, the onset of mesoderm formation occurs in blastocysts of varying sizes, from 1.4 mm in the Tasmanian bettong and the Virginian opossum to 6.0 mm in the native cat (Table IV). In addition to this variation between species there is also some variation within species and within a batch of embryos. Despite these variations, certain common themes emerge if expansion and growth, calculated as cell volume, are examined (Selwood, 1989a). In the three species where it is possible to estimate growth by estimation of cell volume, it remains static or decreases during cleavage, but increases greatly at primary endoderm and mesoderm formation (Fig. 8). Changes in expansion are less extreme but are also more marked at these times of renewed embryonic differentiation. Investigations of possible hormonal mechanisms for initiating new phases in embryonic differentiation, especially development of the primary endoderm, show changes in hormonal profiles at each new phase (for a review, see Tyndale-Biscoe and Hinds, 1989). Comparisons of development of embryos *in vitro* with that *in vivo* (for a review, see Selwood, 1989a) suggest that development from unilaminar to bilaminar stages requires some unknown stimulus or conditions available *in vivo* but not provided by currently used media *in vitro*. Development from bilaminar to trilaminar stages can proceed *in vitro*, however, implying that external stimuli for each of these two steps in differentiation (primary endoderm and mesoderm formation) are different.

2. Mesoderm Formation

While mesoderm formation and development of the primitive streak have been described in a number of species (Table IV), a more detailed analysis, allowing allocation of cells to different lineages, has been done only for the Tasmanian bettong (Selenka, 1892; Kerr, 1936), the bandicoot (Lyne and Hollis, 1977), and the Virginian opossum (Selenka, 1887; McCrady, 1983; Krause and Cutts, 1985a,b).

The onset of mesoderm formation is indicated when cells begin to increase in density because of proliferation by mitosis of primitive ectoderm in a crescent-shaped area (Hartman, 1919; McCrady, 1938; Lyne and Hollis, 1977) or circular patch (Kerr, 1936) at one side of the embryoblast. These proliferation sites mark the posterior end of the embryos. At this stage, Selenka (1887) in the Virginian opossum, Kerr (1935) in the bettong, and Lyne and Hollis (1977) in the bandicoot found isolated mesenchymal cells leaving the embryoblast layer. Possibly these represent primordial germ cells. At slightly later stages, cell densities begin to increase in the midline, and the embryoblast starts to develop a characteristic "pear shape." An increasing number of cells leave the embryoblast along the midline and form the primitive streak. Two wings of mesoderm move laterally from the more posterior part of the streak to form the mesodermal crescent (McCrady, 1938) and the streak is terminated anteriorly by Hensen's node. In bandicoots at this stage, junctional complexes (tight junctions and desmosomes) are found between ectodermal cells at the apical surface (next to the shell) and between all parts of the endoderm at the surface facing the blastocoel (Lyne and

Hollis, 1977). With the appearance of the primitive streak a basal lamina develops beneath the cells of the embryonic ectoderm. In more advanced primitive streak stages of the opossum with four to five somites, mesodermal cells are linked by junctions similar to the nexus (communicating) type and by the occasional desmosome-like junctions (Krause and Cutts, 1985a).

In the bettong Kerr (1936) located a thickened region of primitive endoderm at the edge of the primitive ectoderm, the "annular zone," that generates mesodermal cells during more advanced stages of streak formation. Kerr assumed that it generated mesoderm because of the proximity of mesodermal cells to it, its mitotic activity, and because of cells apparently leaving it (or entering it?). In bandicoots, Lyne and Hollis (1977) also found an annular zone, but no signs that it contributed to the mesoderm. At an advanced stage in bandicoot primitive streak development, streak cells contributed to embryonic endoderm (L. Selwood, unpublished observation, 1991) and the mesoderm had spread out beyond the epiblast in the posterior region to form extraembryonic mesoderm.

3. Cell Lineages

In advanced trilaminar blastocysts, different cell types can be recognized on the basis of cell shape, staining characteristic, density of ribosomes, and size and shape of lipid inclusions. These cell types are (1) embryonic ectoderm, (2) extraembryonic ectoderm or trophoblast divided into proximal and distal regions, (3) embryonic endoderm, (4) annular zone endoderm, (5) extraembryonic endoderm, (6) embryonic mesoderm, and (7) extraembryonic mesoderm. The likely lineage for bandicoots on the basis of descriptions by Lyne and Hollis (1976, 1977) and Hollis and Lyne (1977) is shown in Fig. 14. As in other marsupials (see Sharman, 1961), the extraembryonic mesoderm extends to the equator of the blastocyst and defines the limits of the trilaminar or vascular yolk sac (Fig. 15). The remainder of the blastocyst wall is the bilaminar yolk sac. It is not vascularized, as blood vessels do not spread further than the sinus terminalis at the limit of the trilaminar yolk sac. The cell lineage as shown for the bandicoot (Fig. 14a) is similar to what is described in the bettong except for the role of the annular zone of embryonic endoderm. Kerr (1936) felt that it also contributed to the mesoderm. Apart from this difference, the description of bandicoot and bettong lineages is very similar and would develop as a structure like that shown in Fig. 15a.

In contrast, the lineages described by McCrady (1938) in the opossum are quite different because the initial bifurcation of cell lineages is different (Fig. 14b) and thus develops as a different structure (Fig. 15b). According to McCrady, the cleaving embryo separates into embryoblast and trophoectoblast. The embryoblast gives rise to the medullary plate (neuroectoderm) and embryonic mesoderm and endoderm. The embryonic ectoderm is derived from a bifurcation of trophoectoblast into embryonic ectoderm and trophectoderm at a much later

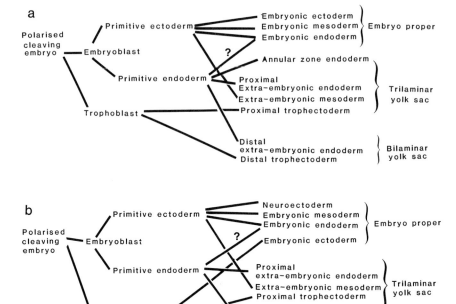

Fig. 14 Proposed cell lineages in the bettong and bandicoots (a) and the Virginian opossum (b) based on descriptions of embryonic development given by Lyne and Hollis (1976, 1977) and Hollis and Lyne (1977) for bandicoots, Kerr (1935, 1936) for the Tasmanian bettong, and Hartman (1928) for the opossum (see text). Note that there is a major different between the two forms in the first step in determination.

stage of development. McCrady based this decision on a few slides of a specimen undergoing late gastrulation, in which the parietal mesoderm extended beyond the limits of the region of thickened blastocyst epithelium called the "embryonic area" or equivalent term by Selenka (1892) and Hartman (1919). Because of the position of the parietal mesoderm relative to the thickened region and because no differences between embryonic ectoderm and trophectoderm could be distinguished at this stage, McCrady called the thickened region the "medullary plate." This interpretation should be treated with some caution because the rapid expansion of the blastocyst at this stage would make major differences difficult to detect; Hartman (1919) did note some differences between embryonic ectoderm and trophectoderm. For these reasons and because of the major differences between some marsupial lineages and what is known for eutherian mammals (see, Rossant, 1986; Beddington, 1986, for review), marsupial lineages should be examined in more detail with appropriate labeling techniques, especially in the Virginian opossum.

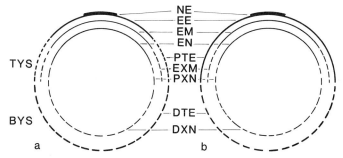

Fig. 15 A diagrammatic reconstruction of the differentiated cell lineages in the trilaminar blastocysts of bandicoots and the Tasmanian bettong (a) and the Virginia opossum (b) based on the references given in Fig. 14. NE, Neuroectoderm; EE, embryonic ectoderm; EM, embryonic mesoderm; EN, embryonic endoderm; PTE, proximal trophectoderm; DTE, distal trophectoderm; EXM, extraembryonic mesoderm; PXN, proximal extraembryonic endoderm; DXN, distal extraembryonic endoderm; TYS, trilaminar yolk sac; BYS, bilaminar yolk sac. Note that in the opossum (b) the embryonic ectoderm in indistinguishable from the proximal trophectoderm, whereas in the bandicoot and bettong the two are distinct.

V. Discussion and Conclusions

A. Polarity and the Investments

From early stages in oogenesis, the marsupial oocyte shows a polarized structure reflected in the position of the nucleus and distribution of cytoplasmic organelles. This polarized state is enhanced or modified during activation following fertilization. In some species, in which cytoplasmic organelles are distributed in a radial pattern, the position of the nucleus and oocyte shape are the only indication of polarity. In other species with a more polarized pattern, the nucleus and yolky storage products lie along a nonyolky–yolky axis. This axis appears to mark the future embryonic–extraembryonic axis that will eventually be transformed into the dorsoventral axis. Polarization of the zygote is transformed into a polarized state in blastomeres. This is very marked in some species, more subtle in others, but detectable in all.

The marsupial investments provide the framework within which the marsupial blastocyst is constructed by blastomere–zona adhesion followed by blastomere–blastomere adhesion. The blastomeres form an epithelium because they are enclosed within a limited framework. If the surface for blastomere–substrate adhesion is relatively unlimited (such as the base of a culture dish) blastomeres disperse and do not form a continuous epithelium (Selwood, 1989b). Because of the particular cleavage pattern, blastomeres populate a particular part of the zona first. In many embryos with polarized yolk elimination, most yolk is accumulated in this part of the blastocoel. Blastomeres are held in this more yolky region by the gellike matrix filling the blastocoel. The zona itself may have a polarized character because of deposition of extracellular matrix material by blastomeres during early cleavage. In some embryos, with a radial pattern of yolk elimination, yolk distribution is slightly polarized (McCrady, 1938) but the order of division probably determines which tier of cells populates the zona first. Initially, blastomeres and their descendants populate the part of the zona nearest to them. If a blastomere is killed, the space occupied by it and its descendants is free of cells (Selwood, 1986b) at least during cleavage. The investments also probably enclose a specific microenvironment necessary for embryonic development. While the envelopes of unfertilized eggs show free movement of quite large molecules (Hughes and Shorey, 1973), this has yet to be investigated in a stage-specific fashion in embryos. The preliminary expansion of many embryos during cleavage before the blastocyst is complete implies that from early stages in cleavage the microenvironment of the blastocoel is separate and distinct. This, and the role of the investments and yolk in nutrition of the embryo, have yet to be investigated.

B. Cleavage Results in Two Cell Populations

There is no doubt that two populations of cells are generated during cleavage and early blastocyst formation in marsupials. These cell populations are characterized by differences in size and shape and position relative to the yolk mass as well as differences in histology in embryos with polarized yolk elimination. In embryos with radial yolk elimination, the populations are distinguished by size and rate of division. During blastocyst formation and expansion, the two populations in the more polarized embryos exhibit differences in position with respect to the yolk mass, cell cycle time, and eventually ability to generate primary endoderm. In embryos with a radial form of yolk elimination the two populations during blastocyst formation and preliminary expansion differ in the time of utilization of yolk; the eccentric location of the embryo within the investments implies that the two populations utilize the mucoid layer at different times, and only one cell population has the ability to generate primary endoderm cells. Polarization of blastomeres and directional orientation of blastomeric flattening lead to specific division planes at right angles to the long axis of the blastomere and occupation of one part of the zona first. This ensures that any differences between blastomeres will be reinforced by the position of a blastomere in the developing epithelium. Blastomeres that occupy a more central location as opposed to the edge of the developing epithelium have more extensive contact with neighbors and have more immediate neighbors than blastocysts at the edge of the epithelial sheet (Figs. 9 and 10).

In the Virginian and common opossums, a distinct cleavage pattern is less apparent and conflict is found in descriptions of cleavage (Section III,C), implying more random cell–cell associations leading to blastocyst formation. One could predict from this, if close association between cells generates an embryonic area, that polyembryonic blastocysts would be found in the opossums. It is interesting to note that a polyembryonic blastocyst has been described in the Virginian opossum (Patterson and Hartman, 1917) and such blastocysts are said to be of frequent occurrence in the common opossum (Bluntschli, 1913). More information is needed on the formation of these blastocysts, to confirm whether or not they form from development of more than one embryoblast as suggested by Patterson and Hartman, (1917).

Johnston and Robinson (1987) showed that late-replicating X chromosomes are present in unilaminar blastocysts with about 1000 cells and in both embryonic and extraembryonic cells of bilaminar and trilaminar blastocytes of the brown antechinus. They suggested that X inactivation is not correlated with cell differentiation in antechinus as it appears to be in the mouse. To confirm this, it would be desirable to examine early cleavage stages, particularly before and after formation of the 2 cell populations (i.e., at the 16-cell stage) and also to distinguish between primitive ectoderm and endoderm in bilaminar blastocysts and between major germ layers in the trilaminar blastocysts. As getting sufficient embryonic material is very difficult in the brown antechinus, it might be better to repeat this study using embryos from the stripe-faced dunnart, in which embryos are more easily obtained and two cell populations are distinct for several generations into blastocyst formation. As in the mouse, high methylation levels are associated with embryonic DNA and lower levels with extraembryonic DNA both in the vascular and avascular yolk sac (Stevens et al., 1988). Again, it would be interesting to follow this in earlier stages to detect whether differences can be shown between the two populations of cells during late cleavage.

The importance of maternal determinants in marsupial embryonic development and establishment of two cell populations have yet to be clarified. There is some evidence suggesting that maternal determinants play a role at least in early cleavage. Some of the events in early cleavage, such as polarization of yolky cytoplasm, elimination of yolk and extracellular matrix material, and (more rarely) cell flattening on the zona, are seen in unfertilized eggs (L. Selwood, unpublished observations, 1991). They are, however, much less organized and more retarded than in fertilized eggs. The polarization and elimination of yolk is less efficient and only some cell fragments flatten on the zona. The stylized pattern of cleavage and the polarized distribution of organelles encountered in marsupials is traditionally associated with uneven distribution of maternal determinants (Jeffery et al., 1986) but could just as easily arise because of the problems created by the relatively large amounts of yolk. If determinants do play a major role in early embryonic development in marsupials, they would have to be distributed in such a way that they are not eliminated during deutoplasmolysis; so that "the baby is not thrown out with the bath water." This is easier to envisage in embryos showing polarized yolk elimination. If determinants are present and important in later stages, then one would expect them to be associated with cytoplasmic skeletal elements. Also, one would expect the cells of the upper tier to be qualitatively different from cells of the lower tier.

C. Capacity to Regulate

The capacity to regulate for loss or addition of blastomeres has been relatively unexplored. In cleavage, loss of a fraction of the blastomeres results in fractional embryos developing (Selwood, 1986b). If one of four blastomeres is killed, three-fourths of the zona is lined with cells, and so on. This may occur because the space in the blastocyst wall was too large to be occupied by the remaining blastomeres during the time available. To confirm this apparent lack of capacity to regulate for loss of blastomere this experiment needs to be repeated using transfer to surrogates to follow further development of these fractional blastocysts. Such fractional blastocysts are occasionally encountered in the Virginian opossum (Hartman, 1919) and the native cat (Hill, 1910), so transfer may not allow for completion of the blastocyst. Marsupials do, however, demonstrate some capacity to regulate for loss of blastomeres. Entire blastomeres and fragments of blastomeres are lost into the blastocoel during early stages of blastocyst formation and expansion (Hill, 1910; Hartman, 1916, 1919; McCrady, 1938; Lyne and Hollis, 1976, 1977; Selwood and Young, 1983; Ward and Renfree, 1988). Some of this cell death may be due to handling techniques, as loss of cells into the interior is a characteristic of developmental failure of blastocyst *in vitro* (L. Selwood, unpublished observations, 1991).

D. Pattern and Order of Division in Specification of Cell Fate

The most likely mechanisms to generate positional signals for the specification of cell fate at the first step in determination in marsupial embryos are cell–cell relationships and the order of division. Both positional signals (reviewed by Johnson and Maro, 1986) and the order of division (Kelly *et al.*, 1978; Spindle, 1982; Surani and Barton, 1984; Garbutt *et al.*, 1987) have been implicated in specification of fate in reconstructed embryos of the mouse. Sutherland *et al.* (1990), working with intact mouse embryos, demonstrated that oriented divisions during the fourth cleavage, influenced slightly by the order of division, affect the specification of cell fate. In marsupials, division planes are more precisely oriented, especially in the dasyurids, until advanced stages of blastocyst formation (Selwood and Smith, 1990). Division planes in early divisions to the fourth division appear to result in allocation of cells to embryonic or extraembryonic lineages. Orders of division and the particular planes of division

5. Development of Pattern in Marsupial Embryos   225

generated by that order is also a possible mechanism for specification of later steps in determination in the embryoblast lineage. Allocation of cells to the two lineages may be established over several divisions but in general the eight cells of the upper tier appear to give rise to embryoblast and the eight cells of the lower tier to trophoblast by the latitudinal divisions at the eight-cell stage. The most convincing evidence of this is in the information on cell cycle time based on cell counts at specified times in the stripe-faced dunnart (Selwood and Woolley, 1991). Initially the 8 cells in each lineage have similar cell cycle times (although they divide in a particular order) and each produced 16, then 32 distinct cells separated by a distinct boundary. The trophoblast cells continue to divide on an 8-hr cell cycle for about 1 day but the embryoblast switches to a 24-hr cell cycle time and the blastocyst expands unevenly. Prior to appearance of the primary endoderm, the embryoblast cells switch to a cell cycle time of about 8 hr and the trophoblast cells slow down and cell numbers become approximately equal. The cell counts of Hartman suggest that something similar may be happening in the opossum. These alterations in cell cycle time in the unilaminar blastocyst may be part of the mechanism that initiates and terminates embryonic diapause in marsupials.

Once cells have been segregated into two populations, one (the embryoblast) crowded into about one-third of the blastocyst wall, the other (the trophoblast) occupying the remaining two-thirds, the order of division provides a possible mechanism to generate positional information. As the order of division in the stripe-faced dunnart (except for the first and last cells to divide) appears to be random (Fig. 12), closer association between cells would be randomly scattered over the embryoblast with some bias toward cells that are descendants of the first cell to divide at the four-cell stage. For instance, during the formation of the embryoblast epithelium adjacent to the yolk mass, during the fifth division in the embryo shown on the left-hand side of Fig. 12, the daughter cells of the second and seventh cell to divide would contact their neighbors and each other before the daughter cells of the thirteenth or sixteenth cell to divide would contact their neighbors and each other. The order of division thus may establish patches of cells in the epithelium in which these cells have contacted more neighbors earlier. As the order of division appears to vary from embryo to embryo (Fig. 12), the position of these patches would vary from embryo to embryo. If primary endoderm cells are generated by this chance association between cells, one would expect them to appear in random arrangement and with some differences in time of appearance over the embryoblast surface. This is exactly what happens (Hill, 1910; Selwood and Smith, 1990). In opossums, in contrast, the cleavage program produces the embryoblast cells as a tightly packed ring of cells (Hartman, 1916; Hill, 1918; McCrady, 1938) so that the cells that are first to have the most contacts lie at the margins of the embryoblast, and it is marginal cells that mainly generate primary endoderm cells (Hartman, 1916, 1919).

E. Axis Formation

Nothing is known about how the posterior–anterior axis is established in marsupial blastocysts, which lie free in the uterus until long after the longitudinal axis is established (see Selwood, 1989a, for review). Once implanted, marsupials are found with the anterior or posterior end of the embryo aligned with the anterior or posterior end of the animal (Sharman, 1961). The sites of primary endoderm cell proliferation in turn could create positional signals to determine mesoderm formation. It is possible that the pattern produced by the order of cell division by creating two foci of the descendants of first and last cells to divide at each division with an axis across the embryo between them is establishing the posterior–anterior axis. In bandicoots, the first mesoderm cells to leave the epiblast are scattered throughout the epiblast, with an increase in incidence at one point, the site of the posterior end of the future primitive streak (Hollis and Lyne, 1977). This pattern is consistent with positional signals derived from division order. The order of division would provide a very convenient mechanism for establishing the posterior–anterior axis. It has flexibility, it is established over a number of cell cycles, and the differences in cell cycle time increase with each cell division (Selwood and Smith, 1990). It implies information exchange between cells on division order and establishment of a gradient or hierarchy. It fits with the common theme in vertebrates, whereby the first sign of the primitive streak is an increase in cell numbers at one place on the epiblast. This is as true in marsupials (Kerr, 1936; McCrady, 1938; Lyne and Hollis, 1977) as it is in the chick (Eyal-Giladi, 1984). Order of division, with the fastest dividing population of cells (to generate the posterior end) and the slowest dividing population of cells (the anterior end), opposite the fastest dividing cells, thus provides a possible mechanism for axis formation.

F. Marsupials in Context

There are obviously a number of alternative hypothesis for mechanisms leading to the specification of cell fate in marsupials. For instance, uneven distribution of maternal determinants could lead to 2 cell populations at the 16-cell stage and then to primary endoderm formation from 1 of these. Alternatively the foci created by the order of division could generate two populations of cells, but this seems unlikely as these populations would overlap with the two populations formed during cleavage. At the present level of knowledge, positional signals generated by the cleavage pattern and then order of division provide a more attractive hypothesis for specification of cell fate in the blastocyst epithelium. This does not exclude a role for uneven distribution of maternal determinants, especially at the fourth division. It is also a hypothesis that can be applied to other mammals without an inner cell mass. If one assumes that positional signals can operate at a two-dimensional level in an epithelium, then any mechanism that ensures that some cells meet their neighbors first and thus have greater cell–cell contact can generate cell diversity. Examination of other mammals with a unilaminar blastocyst (Wimsatt, 1975) supports this. In the tenrec, *Hemicentetes*, at the 16-cell stage the unilaminar blastocyst has a crowded population of cells at 1 pole and this population produces the inner cell mass cells (Bluntschli, 1938; Goetz, 1939). Cleavage in the tenrec produces a more crowded population of near neighbors at one pole, and these give rise to embryonic cell lineages.

What would happen if all blastomeres are without any obvious polarity and all have equal opportunity to populate the blastocyst wall and develop cell–cell associations with their neighbors? One would predict the new lineages could be generated from any part of the blastocyst epithelium because all blastomeres have an equal opportunity. This is what happens in *Elephantulus*, the elephant shrew (Van der Horst, 1942). The blastocyst is formed by four, apparently identical, nonpolar cells and embryonic cells are generated from all parts of the blastocyst epithelium. Specification of cell fate in monotremes could operate in a fashion similar to marsupials. A blastoderm is formed and central cells become distinct from peripheral cells and generate the hypoblast. These two steps could also be driven by positional signals derived from the relative proximity of neighboring cells.

This article has been an attempt to explain how determination of new cell lineages can occur in the marsupial embryo. The mechanism proposed is really an extension of the positional hypothesis proposed for other mammals but adapted for a two dimensional structure, the marsupial blastocyst epithelium. An attempt has been made here to show how examination of marsupial embryonic development can suggest possible mechanisms for specification of cell fate in mammalian embryos without an inner cell mass and for establishment of patterns such as axis formation. It should be possible to study the predictions of this hypothesis by experimental manipulation of cell relationships in the developing blastocyst epithelium or labeling of particular cells for lineage analysis. Further analysis of marsupial embryonic development requires pioneering studies to extend culture into the expanded unilaminar stage and to develop manipulation procedures appropriate to the marsupial embryo, enclosed as it is in a series of envelopes, especially the outer shell. The unique advantages of the marsupial system make these pioneering studies very worthwhile.

Acknowledgments

This work has been supported over many years by the Australian Research Council and their support is gratefully acknowledged. I also want to take this opportunity to thank the many marsupial biologists who have supported and encouraged my work on marsupial developmental biology.

References

Alcorn, G. T., and Robinson, E. S. (1983). Germ cell development in female pouch young of the tammar wallaby (*Macropus eugenii*). *J. Reprod. Fertil.* **67**, 319–325.

Baggott, L. M., and Moore, H. D. M. (1990). Early embryonic development of the grey short-tailed opossum. *Monodelphis domestica*, in vivo and *in vitro*. *J. Zool.* **222**, 623–639.

Baggott, L. M., Davis-Butler, S., and Moore, H. D. M. (1987). Characterisation of oestrus and timed collection of oocytes in the grey short-tailed opossum, *Monodelphis domestica*. *J. Reprod. Fertil.* **79**, 105-114.

Balinsky, B. J. (1981). "An Introduction to Embryology." Saunders, Philadelphia, Pennsylvania.

Bancroft, J. B. (1973). Embryology of *Schoinobates volans* (Kerr) (Marsupialia:Petauridae). *Aust. J. Zool.* **21**, 33–52.

Beddington, R. (1986). Analysis of tissue fate and prospective potency in the egg cylinder. In "Experimental Approaches to Mammalian Embryonic Development" (J. Rossant and R. A. Pedersen, eds.), pp. 121–147. Cambridge Univ. Press, Cambridge.

Bluntschi, H. (1913). Demonstration of embryos of *Didelphys marsupialis*. *Anat. Anz.* **44** (Suppl.), 196–204.

Bluntschli, H. (1938). Le developpement primaire et l'implantation chez un centetine (*Hemicentetes*). *C.R. Assoc. Anat.* **44**, 39–46.

Boyd, J. D., and Hamilton, W. J. (1952). Cleavage, early development and implantation of the egg. In "Marshall's Physiology of Reproduction" (A. S. Parkes, ed.), Vol. 2, pp. 1–126, Longman, Green, and Co., London.

Breed, W. G., and Leigh, C. M. (1988). Morphological observations on sperm–egg interactions during *in vivo* fertilization in the dasyurid marsupial *Sminthopsis crassicaudata*. *Gamete Res.* **19**, 131–149.

Breed, W. G., and Leigh, C. M. (1990). Morphological changes in the oocyte and its surrounding vestments during *in vivo* fertilization in the dasyurid marsupial *Sminthopsis crassicaudata*. *J. Morphol.* **204**, 177–196.

Breed, W. G., Leigh, C. M., and Bennett, J. H. (1989). Sperm morphology and storage in the female reproductive tract of the fat-tailed dunnart, *Sminthopsis crassicaudata* (Marsupialia:Dasyuridae). *Gamete Res.* **23**, 61–75.

Caldwell, M. A. (1887). The embryology of Monotremata and Marsupialia—Part 1. *Philos. Trans. R. Soc. London B* **178**, 463–486.

Clark, M. J. (1966). The blastocyst of the red kangaroo, *Megaleia rufa* (Desm.) during diapause. *Aust. J. Zool.* **14**, 19–25.

Clark, M. J. (1967). Pregnancy in the lactating pigmy possum, *Cercartetus concinnus*. *Aust. J. Zool.* **15**, 673–683.

Clark, M. J., and Poole, W. E. (1967). The reproductive system and embryonic diapause in the female grey kangaroo, *Macropus giganteus*. *Aust. J. Zool.* **15**, 441–459.

Denker, H. W. (1976). Formation of the blastocyst. Determination of trophoblast and embryonic knot. *Curr. Top. Pathol.* **62**, 59–79.

Denker, H. W., and Tyndale-Biscoe, C. H. (1986). Embryo implantation and proteinase activities in a marsupial (*Macropus eugenii*). *Cell Tissue Res.* **246**, 279–291.

Dietl, J. (ed.) (1989). "The Mammalian Egg Coat." Springer-Verlag, Berlin.

Ducibella, T., and Anderson, E. (1975). Cell shape and membrane changes in the eight-cell mouse embryo: Pre-requisites for morphogenesis of the blastocyst. *Dev. Biol.* **47**, 45–58.

Dumont, J. N., and Brummett, A. R. (1985). Egg envelopes in vertebrates. In "Developmental Biology—A Comprehensive Synthesis. Volume 1 Oogenesis" (L. W. Browder, ed.), pp. 235–288, Plenum, New York.

Enders, A. C., and Enders, R. K. (1969). The placenta of the four-eyed opossum *Philander opossum*. *Anat. Rec.* **165**, 431–450.

5. Development of Pattern in Marsupial Embryos

Enders, A. C., and Schlafke, S. J. (1965). The fine structure of the blastocyst: Some comparative studies. *In* "Pre-Implantation Stages of Pregnancy" (G. E. W. Wolstenholme and M. O'Connor, eds.), pp. 29–59, Ciba Foundation, Churchill, London.

Eyal-Giladi, H. (1984). The gradual establishment of cell commitments during the early stages of chick development. *Cell Differ.* **14**, 245–255.

Fadem, B. H., and Rayve, R. (1985). Characteristics of the oestrus cycle and influence of social factors in grey short-tailed opossums (*Monodelphis domestica*). *J. Reprod. Fertil.* **73**, 337–342.

Fraser, E. A. (1919). The development of the urinogenital system in the marsupialia, with special reference to *Trichosurus vulpecula*. Part 2. *J. Anat.* **53**, 97–129.

Garbutt, C. L., Johnson, M. H., and George, M. (1987). When and how does cell division order influence cell allocation to the inner cell mass of the mouse blastocyst? *Development (Cambridge, UK)* **100**, 325–332.

Gardner, R. L. (1983). Cell lineage and cell commitment in the early mammalian embryo. *In* "The Biological Basis of Reproductive and Developmental Medicine" (J. B. Warshaw, ed.), pp. 31–42. Arnold, London.

Godfrey, G. K. (1969). Reproduction in a laboratory colony of the marsupial mouse *Sminthopsis larapinta* (Marsupialia:Dasyuridae). *Aust. J. Zool.* **17**, 637–654.

Goetz, R. H. (1939). On the early development of the Tenrecoidea (*Hemicentetes semispinosus*). *Bio-Morphosis* **1**, 67–79.

Griffiths, M. (1978). "The Biology of the Monotremes." Academic Press, New York.

Guraya, S. S. (1968). Histochemical study of developing ovarian oocyte of the American opossum. *Acta Embryol. Morphol. Exp.* **10**, 181–191.

Hartman, C. G. (1916). Studies in the development of the opossum *Didelphys virginiana*. I. History of the early cleavage. II. Formation of the blastocyst. *J. Morphol.* **27**, 1–83.

Hartman, C. G. (1919). Studies in the development of the opossum *Didelphys virginiana*. III. Description of new material on maturation, cleavage and endoderm formation. IV. The bilaminar blastocyst. *J. Morphol.* **32**, 1–139.

Hartman, C. G. (1920). The freemartin and its reciprocal: Opossum, man, dog. *Science* **52**, 469–471.

Hartman, C. G. (1928). The breeding season of the opossum *Didelphis virginiana* and the rate of intra-uterine and post-natal development. *J. Morphol.* **46**, 143–215.

Hill, J. P. (1910). The early development of the marsupialia, with special reference to the native cat (*Dasyurus viverrinus*). Contributions to the embryology of the marsupialia. IV. *Q. J. Microsc. Sci.* **56**, 1–134.

Hill, J. P. (1918). Some observations on the early development of *Didelphys aurita*. (Contributions to the embryology of the marsupialia, V). *Q. J. Microsc. Sci.* **63**, 91–139.

Hollis, D. E., and Lyne, A. G. (1977). Endoderm formation in the blastocysts of the marsupial *Isoodon macrourus* and *Perameles nasuta*. *Aust. J. Zool.* **25**, 207–223.

Hughes, R. L. (1977). Egg membranes and ovarian function during pregnancy in monotremes and marsupials. *In* "Reproduction and Evolution" (J. H. Calaby and C. H. Tyndale-Biscoe, eds.), pp. 281–291. Australian Academy of Science, Canberra.

Hughes, R. L. (1982). Reproduction in the Tasmanian devil *Sarcophilus harrisii* (Dasyuridae, Marsupialia). *In* "Carnivorous Marsupials" (M. Archer, ed.), Vol. 1, pp. 49–63. Royal Zoological Society of New South Wales, Sydney.

Hughes, R. L., and Hall, L. S. (1984). Embryonic development in the common brushtail possum *Trichosurus vulpecula*. *In* "Possums and Gliders" (A. P. Smith and I. D. Hume, eds.), pp. 197–212. Australian Mammal Society, Sydney.

Hughes, R. L., and Shorey, C. D. (1973). Observations on the permeability properties of the egg membranes of the marsupial, *Trichosurus vulpecula*. *J. Reprod. Fertil.* **32**, 25–32.

Hughes, R. L., Thomson, J. A., and Owen, W. H. (1965). Reproduction in natural populations of the Australian ringtail possum, *Pseudochierus perigrinus* (Marsupialia: Phalangeridae) in Victoria. *Aust. J. Zool.* **13**, 383–406.

Jeffery, W. R., Bates, W. R., Beach, R. L., and Tomlinson, C. R. (1986). Is maternal m-RNA a determinant of tissue-specific proteins in ascidian embryos? *In* "Determinative Mechanisms in Early Development" (C. C. Wylie, ed.), pp. 1–14. The Company of Biologists, Cambridge.

Johnson, M. H. (1981). The molecular and cellular basis of preimplantation mouse development. *Biol. Rev.* **56**, 463–498.

Johnson, M. H., and Maro, B. (1986). Time and space in the mouse early embryo: A cell biological approach to cell diversification. *In* "Experimental Approaches to Mammalian Embryonic Development" (J. Rossant and R. A. Pedersen, eds.), pp. 121–147. Cambridge Univ. Press, Cambridge.

Johnson, M. H., Chisholm, J. C., Fleming, T. P., and Houliston, E. (1986). A role for cytoplasmic determinants in the development of the mouse early embryo. *J. Embryol. Exp. Morphol.* **97** (Suppl.), 97–121.

Johnston, P. G., and Robinson, E. S. (1987). X Chromosome inactivation in female embryos of a marsupial mouse (*Antechinus stuartii*). *Chromosoma* **95**, 419–423.

Kelly, S. J., Mulnard, J. G., and Graham, C. F. (1978). Cell division and cell allocation in early mouse development. *J. Embryol. Exp. Morphol.* **48**, 37–51.

Kerr, T. (1935). Notes on the development of the germ layer in diprotodont marsupials. *Q. J. Microsc. Sci.* **77**, 305–315.

Kerr, T. (1936). On the Primitive streak and associated structures in the marsupial *Bettongia cuniculus*. *Q. J. Microsc. Sci.* **78**, 687–715.

Krause, W. J., and Cutts, J. H. (1983). Ultrastructural observations on the shell membrane of the North American opossum (*Didelphis virginiana*). *Anat. Rec.* **207**, 335–338.

Krause, W. J., and Cutts, J. H. (1985a). Morphological observations on the mesodermal cells of the 8 day opossum embryo. *Anat. Anz.* **158**, 273–278.

Krause, W. J., and Cutts, J. H. (1985b). Placentation in the opossum, *Didelphis virginiana*. *Acta Anat.* **123**, 156–171.

Lintern-Moore, S., Moore, G. P. M., Tyndale-Biscoe, C. H., and Poole, W. E. (1976). The growth of oocyte and follicle in the ovaries of monotremes and marsupials. *Anat. Rec.* **185**, 325–332.

Lyne, A. G., and Hollis, D. E. (1976). Early embryology of the marsupials *Isoodon macrourus* and *Perameles nasuta*. *Aust. J. Zool.* **24**, 361–382.

Lyne, A. G., and Hollis, D. E. (1977). The early development of marsupials with special reference to bandicoots. *In* "Reproduction and Evolution" (J. H. Calaby and C. H. Tyndale-Biscoe, eds.), pp. 293–302. Australian Academy of Science, Canberra.

Lyne, A. G., and Hollis, P. E. (1983). Observations on Graafian follicles and their oocytes during lactation and after the removal of pouch young in the marsupials *Isoodon macrourus* and *Perameles nasuta*. *Am. J. Anat.* **166**, 41–61.

McCrady, E., Jr. (1938). The embryology of the opossum. *Am. Anat. Mem.* **16**, 1–233.

McCrady, E., Jr. (1944). The evolution and significance of the germ layers. *J. Tenn. Acad. Sci.* **19**, 240–251.

McLaren, A. (1982). The embryo. *In* "Reproduction in Mammals: 2. Embryonic and Fetal Development" (C. R. Austin and R. V. Short, eds.), pp. 1–25, Cambridge Univ. Press, Cambridge.

Martinez-Esteve, P. (1942). Observations on the histology of the opossum ovary. *Contrib. Embryol.* **30**, 19–26.

Minot, C. S. (1911). Note on the blastodermic vesicle of the opossum. *Anat. Rec.* **5**, 295–300.

Mintz, B. (1965). Experimental genetic mosaicism in the mouse. *In* "Preimplantation Stages of Pregnancy" (G. E. W. Wolstenholme and M. O'Connor, eds.), pp. 194–207. Churchill, London.

Morgan, C. F. (1943). The normal development of the ovary of the opossum from birth to maturity and its reactions to sex hormones. *J. Morphol.* **72**, 27–85.
Panyaniti, W., Carpenter, S. M., and Tyndale-Biscoe, C. H. (1985). Effects of hypophysectomy on folliculogenesis in the tammar *Macropus eugenii* (Marsupialia: Macropodidae). *Aust. J. Zool.* **33**, 303–11.
Patterson, J. T., and Hartman, C. G. (1917). A polyembryonic blastocyst in the opossum. *Anat. Rec.* **13**, 87–95.
Pedersen, R. A. (1986). Potency, lineage, and allocation in preimplantation mouse embryos. *In* "Experimental Approaches to Mammalian Embryonic Development" (J. Rossant and R. A. Pederson, eds.), pp. 3–33. Cambridge Univ. Press, Cambridge.
Pedersen, R. A., and Spindle, A. I. (1980). Role of the blastocoel microenvironment in early mouse embryo differentiation. *Nature (London)* **284**, 550–552.
Pilton, P. E., and Sharman, G. B. (1962). Reproduction in the marsupial *Trichosurus vulpecula*. *J. Endocrinol.* **25**, 119–36.
Renfree, M. B., and Tyndale-Biscoe, C. H. (1973). Intrauterine development after diapause in the marsupial *Macropus eugenii*. *Dev. Biol.* **32**, 28–40.
Renfree, M. B., and Tyndale-Biscoe, C. H. (1978). Manipulation of marsupial embryos and pouch young. *In* "Methods in Mammalian Reproduction" (J. C. Daniel, Jr. ed.), pp. 308–331. Academic Press, New York.
Robinson, E. S., Johnston, P. G., and Sharman, G. B. (1977). X chromosome activity in germ cells of female kangaroos. *In* "Reproduction and Evolution" (J. H. Calaby and C. H. Tyndale-Biscoe, ed.), pp. 89–94. Australian Academy of Science, Canberra.
Rodger, J. C., and Bedford, J. M. (1982a). Induction of oestrus, recovery of gametes, and the timing of fertilization events in the opossum, *Didelphis virginiana*. *J. Reprod. Fertil.* **64**, 159–169.
Rodger, J. C., and Bedford, J. M. (1982b). Separation of sperm pairs and sperm–egg interaction in the opossum, *Didelphis virginiana*. *J. Reprod. Fertil.* **64**, 171–179.
Rodger, J. C., and Mate, K. E. (1988). A PMSG/GnRH method for the superovulation of the monovulatory brush-tailed possum (*Trichosurus vulpecula*). *J. Reprod. Fertil.* **83**, 885–891.
Rossant, J. (1986). Development of extra-embryonic cell lineages in the mouse embryo. *In* "Experimental Approaches to Mammalian Embryonic Development" (J. Rossant and R. A. Pederson, eds.), pp. 97–120. Cambridge Univ. Press, Cambridge.
Sandes, F. P. (1903). The corpus luteum of *Dasyurus viverrinus* with observations on the growth and atrophy of the Graafian follicle. *Proc. Linn. Soc. N.S.W.* **2**, 364–405.
Selenka, E. (1887). "Studien uber Entwickelungs-geschichte der Thiere. Part 4. Das Opossum (*Didelphys virginiana*)." C. W. Kreidel, Wiesbaden.
Selenka, E. (1892). "Studien uber Entwickelungsgeschichte der Thiere. 5. Beuteltuchs und Kanguruhratte." C. W. Kreidels, Wiesbaden.
Selwood, L. (1980). A timetable of embryonic development of the dasyurid marsupial *Antechinus stuartii* (Macleay). *Aust. J. Zool.* **28**, 649–668.
Selwood, L. (1982). A review of maturation and fertilization in marsupials with special reference to the dasyurid: *Antechinus stuartii*. *In* "Carnivorous Marsupials" (M. Archer, ed.), pp. 65–76. Roy. Zool. Soc. N.S.W., Sydney.
Selwood, L. (1986a). The marsupial blastocyst—a study of the blastocysts in the Hill Collection. *Aust. J. Zool.* **34**, 177–187.
Selwood, L. (1986b). Cleavage *in vitro* following destruction of some blastomeres in the marsupial *Antechinus stuartii* (Macleay). *J. Embryol. Exp. Morphol.* **92**, 71–84.
Selwood, L. (1987). Embryonic development in culture of two dasyurid marsupials, *Sminthopsis crassicaudata* (Gould) and *Sminthopsis macroura* (Spencer) during cleavage and blastocyst formation. *Gamete Res.* **16**, 355–370.
Selwood, L. (1989a). Marsupial pre-implantation embryos *in vivo* and *in vitro*. *In* "Development

of Pre-Implantation Embryos and Their Environment". (K. Yoshinaga and T. Mori, eds.), pp. 225–236. Alan R. Liss., New York.

Selwood, L. (1989b). Development *in vitro* of investment-free marsupial embryos during cleavage and early blastocyst formation. *Gamete Res.* **23,** 399–413.

Selwood, L. and Sathananthan, A. H. (1988). Ultrastructure of early cleavage and yolk extrusion in the marsupial *Antechinus stuartii*. *J. Morphol.* **195,** 327–344.

Selwood, L., and Smith, D. (1990). Time-lapse analysis and normal stages of development of cleavage and blastocyst formation in the marsupials the brown antechinus and the stripe-faced dunnart. *Mol. Reprod. Dev.* **26,** 53–62.

Selwood, L., and Woolley, P. (1991). A timetable of embryonic development and ovarian and uterine changes during pregnancy, in the stripe-faced dunnart, *Sminthopsis macroura* (Marsupialia: Dasyuridae). *J. Reprod. Fertil.* **91,** 213–227.

Selwood, L., and Young, G. J. (1983). Cleavage *in vivo* and in culture in the dasyurid marsupial *Antechinus stuartii* (Macleay). *J. Morphol.* **176,** 43–60.

Sharman, G. B. (1955a). Studies on marsupial reproduction. II. The oestrous cycle of *Setonix brachyurus*. *Aust. J. Zool.* **3,** 44–55.

Sharman, G. B. (1955b). Studies on marsupial reproduction. III. Normal and delayed pregnancy in *Setonix brachyurus*. *Aust. J. Zool.* **3,** 56–70.

Sharman, G. B. (1961). The embryonic membranes and placentation in five genera of diprotodont marsupials. *Proc. Zool. Soc. London* **137,** 197–220.

Sharman, G. B. (1963). Delayed implantation in marsupials. *In* "Delayed Implantation" (A. C. Enders, ed.), pp. 3–14. Chicago Univ. Press, Chicago.

Sharman, G. B., and Pilton, P. E. (1964). The life history and reproduction of the red kangaroo (*Megaleia rufa*). *Proc. Zool. Soc. London* **142,** 29–48.

Spindle, A. (1982). Cell allocation in pre-implantation mouse chimaeras. *J. Exp. Zool.* **219,** 361–367.

Spurgeon, C. H., and Brooks, R. J. (1916). The implantation and early segmentation of the ovum of *Didelphis virginiana*. *Anat. Rec.* **10,** 385–395.

Stevens, M. E., Maidens, P. M., Robinson, E. S., VandeBerg, J. L., Pedersen, R. A., and Monk, M. (1988). DNA methylation in the developing marsupial embryo. *Development (Cambridge, UK)* **103,** 719–724.

Surani, M. A. H., and Barton, S. C. (1984). Spatial distribution of blastomeres is dependent on cell division order and interactions in mouse morulae. *Dev. Biol.* **102,** 335–343.

Sutherland, A. E., Speed, T. P., and Calarco, P. G. (1990). Inner cell allocation in the mouse morula: The role of oriented division during fourth cleavage. *Dev. Biol.* **137,** 13–25.

Talbot, P., and Dicarlantonio, G. (1984). Ultrastructure of opossum oocyte investing coats and their sensitivity to trypsin and hyaluronidase. *Dev. Biol.* **103,** 159–167.

Tarkowski, A. K. (1961). Mouse chimaeras developed from fused eggs. *Nature (London)* **190,** 857–860.

Tarkowski, A. K., and Wroblewska, J. (1967). Development of blastomeres of mouse eggs isolated at the 4- and 8-cell stage. *J. Embryol. Exp. Morphol.* **18,** 155–180.

Tyndale-Biscoe, C. H., and Hinds, L. (1989). The hormonal milieu during early development in marsupials. *In* "Development of Pre-Implantation Embryos and Their Environment" (K. Yoshinaga and T. Mori, eds.), pp. 237–246, Alan R. Liss, New York.

Tyndale-Biscoe, C. H., and Renfree, M. B. (1987). "Reproductive Physiology of Marsupials." Cambridge Univ. Press, Cambridge.

Ullman, S. L. (1978). Observations on the primordial oocyte of the bandicoot *Isoodon macrourus* (Peramelidae, Marsupialia). *J. Anat.* **128,** 619–631.

Ullman, S. L. (1981). Observations on the primordial germ cells of bandicoots (Peramelidae, Marsupialia). *J. Anat.* **132,** 581–595.

Ullman, S. L. (1984). Early differentiation of the testis in the native cat, *Dasyurus viverrinus* (Marsupialia). *J. Anat.* **138,** 675–688.
Van der Horst, C. J. (1942). Early stages in the embryonic development of *Elephantulus*. *S. Afr. J. Med. Sci. Biol.* **7** (Suppl.), 55–65.
Walker, M. T., and Rose, R. (1981). Prenatal development after diapause in the marsupial *Marcropus rufogriseus*. *Aust. J. Zool.* **29,** 167–187.
Ward, S. J., and Renfree, M. B. (1988). Reproduction in females of the feather tail glider *Acrobates pygmaeus* (Shaw) (Marsupialia). *J. Zool.* **216,** 225–239.
Waring, H., Sharman, G. B., Lovat, D., and Kahan, M. (1955). Studies on marsupial reproduction. 1. General features and techniques. *Aust. J. Zool.* **3,** 34–43.
Wimsatt, W. A. (1975). Some comparative aspects of implantation. *Biol. Reprod.* **12,** 1–40.

6
Experimental Chimeras: Current Concepts and Controversies in Normal Development and Pathogenesis

Y. K. Ng and P. M. Iannaccone
Department of Pathology and
Markey Program in Developmental Biology
Northwestern University
Chicago, Illinois 60611

I. Introduction
   A. Definitions
   B. Markers Used to Elucidate Mosaic Patterns
   C. Methods
II. Normal Development
   A. Analysis of Mosaic Pattern
   B. Organogenesis
   C. Computer Simulations of Mosaic Growth
III. Pathogenesis
   A. Neoplasia
   B. Germ Cell and Embryo-Derived Tumors
   C. Immunological Tolerance
   D. Genetic Diseases
IV. Conclusion
   References

I. Introduction

A. Definitions

Experimental chimeras are multizygotic mosaic animals produced by amalgamation of tissues or cells from distinguishable genetic backgrounds (Fig. 1). There are two types of experimental chimeras: primary and secondary. Primary chimeras are usually made by combining tissues or cells from different embryos at a very early stage of development. For example, aggregation chimeras are made by amalgamating two or more embryos at an early four- or eight-cell stage; microinjection chimeras are made by microinjecting tissues or cells from a distinguishable strain into the blastocyst cavity of an embryo. The combined embryo is carried to term by a pseudopregnant surrogate mother (McLaren, 1976; Prather *et al.*, 1989). Secondary chimeras are produced or arise from the combination of fetal tissues after the period of organogenesis has begun. Secondary chimeras may contain composite tissues derived from two or more adult individuals. Such animals can be produced by transfer of cells either between mother and fetus or from one fetus to another or by grafting distinguishable tissues (McLaren, 1976; Austin and Short, 1972). Monozygotic mosaic animals may be obtained from the spontaneous inactivation of one X chromosome in female conceptuses. If the paternal and maternal X chromosomes produce distinguishable proteins, the mosaicism can be identified in females. The introduction of exogenous DNA into the pronuclei of fertilized ova can result in monozygotic mosaic animals if the integration of the DNA is delayed until after the first cleavage. Such mosaics can be identified by examining the distribution of the transgene product, but their usefulness depends on uniform expression of the transgene within the lineages that contain the transgene (Fig. 1). The previous distinction between mosaic animals and chimeric animals seems unjustified because both types of animals comprise distinguishable cell lineages and are therefore mosaic. It is more useful to consider all such animals as mosaic and then distinguish the various types of animals on the basis of the method of their production (Iannaccone, 1987).

Chimeric animals, particularly rats, mice (McLaren, 1976), birds (LeDouarin and McLaren, 1984), frogs (Maeno *et al.*, 1985), nematodes, and insects (Santamaria, 1983; Simcox and Sang, 1983), have been used in the genetic analysis of development in several ways. Lineage analysis has been successful for tracing the fate of marked clones of primordial cells. Chimeras have been used to establish the type and number of founder cells for different organs. The developmental potential of different primordial cells has been established with chimeras. Finally, chimeras are useful for determining the effect of particular mutations (see review by Wilkins 1986; Rossant, 1987).

Fig. 1 Diagrammatic representation of the production and analysis of chimeric animals. Distinguishable strains are mated to produce homozygous embryos. These embryos are amalgamated and returned to the reproductive tract of a surrogate mother. If the two strains have different coat colors, then the chimeric offspring will have striped coats. Mosaic animals are either multizygotic or monozygotic. Multizygotic mosaics, called chimeras, may be produced by pushing early embryos together or by microsurgically inserting inner-cell mass cells from one blastocyst into the blastocoele of another. Monozygotic mosaics include spontaneous X chromosome-linked mosaics. Because one X chromosome is inactivated in early development, female mammals are mosaics. In one population of cells the maternal X chromosome is active; in the other population of cells the paternal X chromosome is active. Microinjection of exogenous DNA will create monozygotic mosaic animals if the injected DNA integrates into the host genome after the injected fertilized egg divides. The cells that carry the foreign DNA (gene) may be visualized with probes that react with either the foreign gene or its product. (Drawing by Tom Herzberg.)

Fig. 1

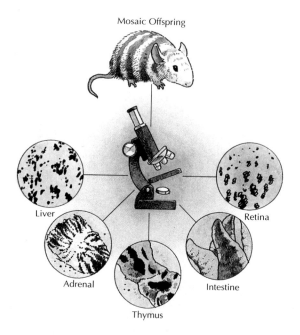

Fig. 2

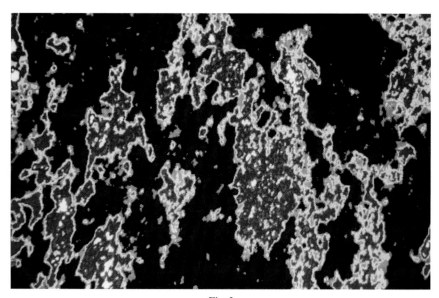

Fig. 3

6. Experimental Chimeras

B. Markers Used to Elucidate Mosaic Patterns

The usefulness of chimerism depends on the availability of suitable cell marker systems to differentiate cells from the different combined lineages. The marker should be inherited stably during cell proliferation and should be detectable at the cellular level in all types of cells at all stages of development (McLaren, 1976; West, 1978). This allows the migration of specific cells and the contribution of different cell lineages to organs in the chimera to be established and in some instances quantified.

Many biochemical markers, for example, autosomal dimorphic glucosephosphate isomerase-1 (GPI-1, EC 5.3.1.9), X chromosome-linked dimorphic phosphoglycerate kinase-1 (PGK-1, EC 2.7.2.3), and autosomal dimorphic isocitrate dehydrogenase-1 (ID-1, EC 1.1.1.42) and carboxyesterase (EC 3.1.1.1) have been utilized in the study of tissues isolated from chimeric animals (Nielsen and Chapman, 1977; Mintz and Baker, 1967; Muhlbacher et al., 1983; Grim et al., 1984; Iannaccone et al., 1978; Iannaccone, 1987; Beddington and Robertson, 1989). In situ histochemical localization of malic enzyme (EC 1.1.1.40) (Gardner, 1984), β-glucuronidase (EC 3.2.1.31) (Hayashi et al., 1964; Condamine et al., 1971; Mullen and Herrup, 1979), and carbonate dehydratase (EC 4.2.1.1) (Hansson, 1967) have been reported. However, not every cytochemically distinguishable enzyme is suitable as a cell lineage marker, for there may be histological or metabolic differences among cells that affect enzyme expression. For example, β-glucuronidase has been found to be transferred from cell to cell (Herrup and Mullen, 1979), so it is not an ideal marker for chimeras between strains differing in β glucuronidase activity. Immunohistochemical staining based on the differential agglutination of specific cell surface carbohydrates or the utilization of monoclonal antibodies against specific major histocompatibility complex (MHC) surface antigens have proved useful in chimeric mice and rats (Ponder et al., 1983, 1986; Iannaccone and Weinberg, 1987; Iannaccone et al., 1987a; Weinberg et al., 1985). Other in situ cell markers, like the T6 chromosomal marker (Ford et al., 1975; Iannaccone et al., 1985) and Mus satellite DNA (Rossant et al., 1983) have also been used. These markers can

Fig. 2 Histological examination of mosaic tissue is possible when the mosaic animals are constructed so that the cell lineages are distinguishable in situ. Results from the retina, small intestine, thymus, adrenal gland, and liver are summarized. (Drawing by Tom Herzberg.)

Fig. 3. False-color representation of density in an autoradiogram from chimeric rat liver. The dark areas represent patches of cells of one lineage, while areas of green, yellow, red, and white present patches of cells of the other lineage. For analysis, the image can be segmented into a binary white and black representation. Autoradiographs display a geometrically complex, "geographic" pattern with fractal features. (Image and image analysis software by L. Berkwits.)

elucidate the migration and allocation of primordial cells as well as the pattern of subsequent clonal expansion in various organs.

Exogenous markers like horseradish peroxidase and rhodamine-conjugated dextran have been injected into single inner cell mass (ICM) cells of preimplantation mouse embryos in an attempt to establish an ICM contribution to trophectoderm and endoderm in the mouse embryo (Winkel and Pedersen, 1988). Tritiated thymidine-labeled cells have been introduced into the primitive streak of early mouse embryos to trace primitive streak-derived cells in the 8-day mouse embryo. The posterior primitive streak cells contributed to extraembryonic mesoderm (Copp et al., 1986), the cells in the middle of the streak contributed to lateral mesoderm, and the anterior portion of the streak gave rise to paraxial mesoderm, gut, and notochord (Tam and Beddington, 1987).

A transgenic mouse that constitutively expresses an exogenous marker gene, lac Z, regulated by rat β-actin promoter, has been used to make chimeras with strains not expressing lac Z in order to visualize the ICM clones during early organogenesis using β-galactosidase histochemical staining (Beddington et al., 1989). Mouse embryo-derived stem cells (ES) containing an exogenous gene construct of lac Z fused to a mouse hsp68 gene promoter has been used to make chimeras in order to study the control of gene expression during development (Gossler et al., 1989). The hsp68 promoter is not sufficient to express lac Z and thus the expression of lac Z is dependent on cis-acting regulatory element(s) close to the exogenous gene integration site in the ES cell. By examining the pattern of lac Z expression during embryogenesis, the expression of the endogenous host gene can be elucidated. Therefore, the endogenous genes expressed in different developmental stages could be identified by cloning the genes next to the site of integration of the actively expressing lac Z exogenous reporter gene.

C. Methods

Methods for producing mouse aggregation chimeras, mouse chimeras by microinjection, and spontaneous X-linked mosaics have been covered extensively in other reviews and books (LeDouarin and McLaren, 1984; McLaren, 1976). Production of transgenics by microinjection of exogenous DNA is well established and treated in detail in a number of published sources. In this article we will briefly present procedures used for the production of aggregation chimeras of the rat because this methodology has not been extensively covered by existing publications. Our laboratory utilized the basic approach adopted from the mouse procedures by Yamamura and Markert (1981). We have taken advantage of the ability to induce pseudopregnancy in the rat by mechanical stimulation of the cervix in the recipient, a strategy that is successful in our hands.

The parental animals are mated without hormone stimulation to yield homozygous embryos of distinguishable types. Pregnancy is established by discovery of sperm in the vaginal smear. The eight-cell stage embryos are removed by flushing the uterus of pregnant rats on the third day following mating. The precompaction eight-cell embryos are flushed into T6' medium (Van Winkle *et al.*, 1990; Ng and Iannaccone, 1992) and the zonae pellucidae are removed by brief (10 sec) exposure to acidified salt solution [T6' without bovine serum albumin (BSA) or serum]. The eight-cell embryos, devoid of zonae pellucidae, are placed in microdrops of T6' containing phytohemagglutinin and are physically apposed. As described by Yamamura and Markert (1981), the eight-cell rat embryo is essentially disk shaped with six blastomeres in a plane and one blastomere centrally located on either surface of this plane. The apposition of two eight-cell stage embryos required that they be stacked one on top of the other in order to maximize the surface area in contact between the embryos. Exposure of rat embryos to T6' for periods longer than 5 hr at the eight-cell stage of development results in failure to obtain live pups when the cultured embryos are transferred to surrogate mothers. Therefore, the amalgamated embryos are then cultured overnight in Markert's modification of Whittingham's medium (Yamamura and Markert, 1981), buffered to pH 7.0 with CO_2. Osmolality is adjusted by addition of water so that the medium has a final osmolality of 300–310 mOsm/kg H_2O. The probability of obtaining live births from rat embryos cultured overnight is critically dependent on the correct osmolality of the medium.

In the middle of the afternoon following overnight culture, the amalgamated embryos are transferred to the uterus of surrogate mothers. The procedure is performed as in the mouse, but the surrogate mothers are not mated. Holtzman strain females are mechanically stimulated with a glass rod attached to a vibrator on the day following mating of the donor parental animals. Forty-eight hours following the first stimulation, cervical smears are obtained and the presence of leukocytes rather than squamous cells in the smear indicates those animals that are pseudopregnant. This shift in cell type from squamous to leukocyte is a discrete event that occurs on day 1 of pseudopregnancy in Holtzman females (DeFeo, 1962, 1966; Yochim, 1984). Thus, ovulation in these animals occurred 24 hr following ovulation of the donor, parental animals. The surrogate mothers are therefore at day 3 of pseudopregnancy at the time the amalgamated chimeric embryos are transferred to their uteri. In a large series from this laboratory these transfer and culture procedures resulted in a 96% pregnancy rate among surrogate mothers and a 34% live birth rate of all embryo pairs transferred; 43% of the pups born were chimeric.

II. Normal Development

A. Analysis of Mosaic Pattern

1. Biochemical Analysis

Early experiments in mammals were performed using biochemical analysis of lineage. By homogenizing tissue, various isozymes marking the two cell lineages could be identified following electrophoresis. Dimeric dimorphic proteins with allelic variations would result in two isozyme bands in chimeras, three isozyme bands in the heterozygous F_1 animals, or one band in the homozygous parental animals. Although this approach could be taken after partial purification of tissue types, the disruption of tissue relationships required to isolate enzymes for electrophoresis presented a major limitation. Nevertheless, important conclusions were reached using this approach. The combination of early embryonic tissues to form experimental chimeras established a tentative fate map of the mouse embryo, established the totipotency of early mouse blastomeres, established the time of commitment of trophectoderm and inner cell mass, and established the origin of extraembryonic ectoderm. These observations pointed to fundamental differences between mammals, amphibians, and insects in embryonic strategies for establishing the body plan. Cytoplasmic localization of determinants was shown not to be an important feature of early development in mammals and pointed to topological mechanisms in early commitment events. As a consequence of these results, the maternal role in establishing the body plan of the early embryo appears to be less important in mammals than in invertebrates, such as insects, in which maternally directed positional gradients direct differential morphogenesis. In early *Drosophila* embryos nuclear migration through zones of differentiation, specifying various transcription factors, can establish committed primordia at the cellularized blastoderm stage. Such positional information begins a hierarchy of gene expression that results in segmental differentiation in the larval stages. In nematodes, oriented cell divisions and consequent compartmentalization of cytoplasmic determinants result in differentiated body parts in larval stages. Similar compartmentalizing strategies are used in amphibians, so it is not exclusively an invertebrate strategy. However, mammalian embryos use neither strategy. Rather, topological events that define early cell collectives (Edelman, 1988) and activation of signal transduction pathways are associated with compaction in the early mouse embryo (Winkel *et al.*, 1990). Commitment to embryonic lineage is a consequence of compaction as those cells on the inside of the embryos are fated to form the inner cell mass (Tarkowski and Wroblewska, 1967; Hillman *et al.*, 1972).

2. Patches

Aggregates of cells of like marker lineage are known as *patches* and considerable effort has gone into their analysis over the years (West, 1975, 1976b). The most common approach to determining the size of patches in tissue on the basis of biochemical markers was to establish the binomial distribution of proportion of isozyme activity. In this approach large numbers of uniform, small samples of an isolated tissue were electrophoresed to reveal the proportion of isozyme activity in each sample. If the analysis is performed in a manner that allows enzyme activity to be proportional to cell number then one gains an estimate of the proportion of the two marker cell types in each sample. If the sample size is small

6. Experimental Chimeras   241

relative to the patch size, then there will be great variation from sample to sample because the sample will be either mostly from one type of patch or the other. If the sample size is large relative to the patch size, then there will be little variation from sample to sample, because each of the samples includes many patches (Hutchison, 1973; Iannaccone, 1980). This approach is limited by several factors. It assumes a normal distribution of patch size, uniform placement of geometrically regular patches, and does not allow consideration of the proportion of cell types in the entire tissue. The first two assumptions are not justified, and the patch size *and* the distribution of patch size are very sensitive to the overall proportion of cell type. Tissue proportion of cell types must be a consideration in indirect methods of determining the patch size (Iannaccone *et al.*, 1987b, 1988, 1989). The binomial approach can be used in a comparative way if the samples can be obtained from an isolated tissue. As discussed previously (Ng *et al.*, 1990) this approach has been used to establish a similarity in epidermal patch size between X chromosome-linked mosaic animals and chimeric animals. The same conclusion, that patch sizes in these two types of mosaic animals are similar, was based on observations in pigmented retinal epithelium (Nesbitt, 1974), another tissue in which a pattern could be observed in a "pure" state without the inclusion of other cell types.

The importance of *in situ* methodology in mosaic pattern analysis has grown in recent years. The use of such markers has been previously reviewed (West, 1984; Iannaccone *et al.*, 1988). The obvious advantage of *in situ* analysis is that it allows direct examination of pattern within tissues of visceral organs (Figs. 2 and 3, see pg. 237). These patterns are the remnants of the primary embryologic processes, such as cell movement, cell division, and cell death (Edelman, 1988), that formed them. Thus, a mosaic pattern provides an independent marker of these primary processes. The principal disadvantage is that mosaic pattern analysis must be retrospective: one must in most cases observe the fully formed pattern and infer what forces might have produced it. In some cases this is straightforward, but in other cases it is not.

The correct interpretation of mosaic patterns requires an understanding of the dynamic and nonlinear nature of its main constituent, patches. Patches are formed in a dynamic, oscillatory manner. Cell division will place daughter cells of primordia in adjacent positions. Cell division among neighboring cells, some from different marker lineages, will cause placement of these cells within a patch of the other marker lineage. This causes fragmentation of the patch. Further division can cause cells previously split away from a patch to rejoin it. Thus, the observation of a patch at any given point in time is the result of clonal expansion, fragmentation, and reaggregation.

The concept that finely variegated patch patterns imply extensive cell movement is the basis of numerous studies in which patch size is related either to the number of founder cells or to the time of immobilization of cells. Extensive computer simulation of the process of patch formation indicates that the pattern may be independent of both primordial cell number and cell movement (Iannaccone et al., 1989). On the other hand, patch pattern is dependent on proportion of cell types in the mosaic and on cell division rules.

The notion that patch size is related to time after fixation of patch identity originates in mosaic pattern analysis of cuticular structures in *Drosophila*. The observations of T. H. Morgan in the first quarter of the twentieth century led to Morgan's description of gynandromorphic fruit flies (Morgan, 1919). Chromosomal instability in the early precellularized blastoderm stage in female (XX) flies causes some nuclei to become phenotypically male (XO). Progeny of these cells develop into cuticular structures of the adult fly, which are morphologically distinguishable from their female counterparts. The areas of contiguous male structures can be mapped and these contiguous areas in gynandromorphic *Drosophila* are called "clones." More recently the process has been experimentally induced by mitotic recombination, a procedure that, within some limits, allows variation in the time at which mosaicism is induced. In *Drosophila* it is clear that clones remain contiguous and their size is related to the time of induction of mosaicism. Early induction results in large clones. The clones that are observed in adult cuticular structures of the fruit fly are actually the progeny of imaginal disks, the primordial structures that survive metamorphosis to develop into adult cuticular structures. It is astonishing that clones maintain their structural integrity and even compartment boundaries throughout this process.

Imaginal disks are laid down in early larval stages in *Drosophila* and it is reasonable to suppose that marked nuclei and their progeny remain contiguous through cellularization and gastrulation until the imaginal disk is produced. The imaginal disk is an epidermal structure like a deflated balloon. The epidermal nature of the disk and the cuticular structures they produce tempt a two-dimensional analysis of mosaic pattern, but this is not really justified. The formation of adult structures in which mosaicism is induced is the result of eversion and "inflation" of the disk structure. The expanding nature of the imaginal disk may counteract division from unmarked neighboring cells, thus preventing fragmentation of clones and keeping them more contiguous than patches in mammalian tissues.

Extensive iteration of cell division and daughter cell placement can result in highly complex mosaic patterns. These iterations produce statistically self-similar objects (those in which a portion of the object is similar but not identical to the entire object). Such statistically self-similar objects are often fractal (objects that display increasing complexity with higher magnification and hence whose topologic parameters are scale dependent). We have recently proven that liver patches in chimeric rats are fractal (Iannaccone, 1990) (Fig. 4). Fractal objects have detail nested within detail, so that as they are viewed closer, more and more complexity becomes apparent (Mandelbrot, 1977). Lower scale observation of fractal objects does not predict complexity at higher scales of observation. When observing the perimeter of an object such as a box, which is not fractal, the length of the side of a box at one scale of observation is completely predictable at

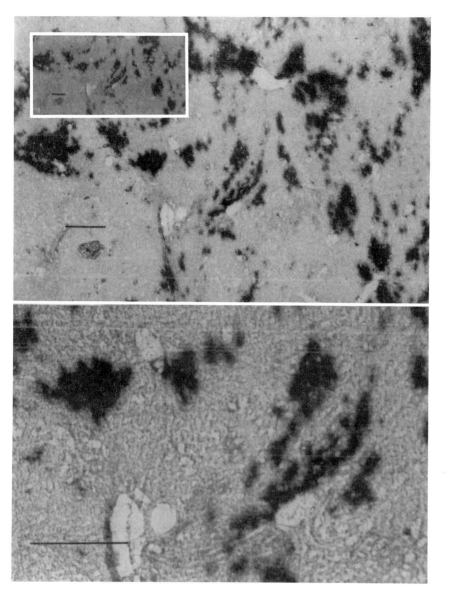

Fig. 4 Unstained autoradiograms of chimeric rat liver. Patches of PVG-$RT1^a$ are the black areas. The area and perimeter length of each of the patches were determined at the three magnifications displayed with computer-assisted digital analysis. The slope of the line relating the log of the area and perimeter length to the log of the reciprocal of magnification estimates the fractal dimension. If the absolute value of the fractal dimension of the patches exceeds the topologic dimension (one for length, two for area) then the length or area has increased with increasing magnification (scale of observation) more than predicted at a lower scale of observation and the patches are fractal. That is the result obtained in the liver of chimeric rats. Bars = 0.25 mm. (From Iannaccone, 1990, reprinted with permission.)

another scale of observation. Thus, if one corrects for the scale of observation the perimeter of a box will be the same at all scales of observation. The perimeter of a box is said to be scale independent, a characteristic of a nonfractal object. A fractal object, on the other hand, will display scale-dependent characteristics with respect to parameters such as perimeter length. That is, the closer one looks, the more perimeter length one finds even if correction for the scale of observation is made. This unsettling property is most intuitively understood for coastlines, for which the length when determined, for example, from a satellite cannot predict the much greater length and detail found when the same coastline is viewed from an airplane. Neither predicts the much greater length or detail observed by walking along the same coastline. Consequently, morphometric parameters such as perimeter length, area, or volume of a fractal object cannot be known absolutely: they are dependent on the scale of observation (Feder, 1988; Iannaccone, 1990).

Because self-similar processes frequently result in fractal objects, the hypothesis that self-similar processes are responsible for the production of mosaic patterns predicts that patches would be fractal. The mathematical definition of a fractal requires that for any parameter, measurement of the topological dimension of the parameter is less than the fractal dimension of the object. The topologic dimension for a line is 1, for a plane is 2, and for a shape is 3. The fractal dimension has been proven to be identical to the capacity dimension and to the Hausdorf–Besicovitch dimension (Farmer et al., 1983). The value of the fractal dimension can be estimated using the equation

$$D_F = \log N/\log 1/m; \quad \text{because } D_C \equiv D_H = \lim_{\epsilon \to 0} [\log N(\epsilon)/(1/\epsilon)]$$

where D_F is the fractal dimension; N is the uncorrected length of a curve, area of a surface, or volume of a shape determined at various scales of observation (m); D_C is the capacity dimension; D_H is the Hausdorf–Besicovitch dimension; ϵ is the size of circles covering a surface; and $N(\epsilon)$ is the number of these circles required to cover the surface.

Thus, if the parameter of interest (perimeter length, area, or volume) is established at many scales of magnification, then the slope of the log of the parameter vs the log of (magnification)$^{-1}$ estimates the fractal dimension. If this procedure is applied to a Euclidean object, for example, a box, the slope of such a plot will equal the topologic dimension, because at each scale of observation the box has a predictable area. The analysis of patch distribution and geometry should extend well beyond simply determining the size of these objects. They contain a rich informational content and present a challenge to experimentalists studying mosaic patterns to come up with meaningful ways of applying principles of image analysis to the pattern.

Recently, small intestinal crypts were determined to be clonal in mice. The crypt structure is a primordial cell population responsible for the formation of

absorptive structures, the villi, which greatly increase the surface area of the epithelium. *In situ* analysis of intact villi in mosaic mice revealed a harlequin pattern itself was subjected to Greig–Smith analysis of distribution (Greig-Smith, 1952). (This procedure is widely used in the ecological sciences to establish whether or not plant progeny in a field are scattered randomly or clustered in some areas.) The crypt analysis revealed a striking scale dependence of distribution of the marked crypts. This observation was taken to mean that the placement of primordia and progeny were nonrandom (Schmidt *et al.*, 1985a,b). An alternative interpretation is that the original marked clones were placed randomly but that the progeny cells and the patches they formed were created from simple, iterating, probabilistic cell division programs that resulted in clustering in the vicinity of the original marked primordia. A pattern created in this way would likely be fractal. This interpretation suggests that random and deterministic mechanisms generating mosaic patterns are not necessarily mutually exclusive, because fractals are complex objects that can be generated deterministically but without apparent order.
complex objects that can be generated deterministically but without apparent order.

Organ formation and the generation of mosaic patterns proceed in three-dimensional space in time, a four-dimensional problem. To establish that the patches generated are fractal in three dimensions (a reasonable *a priori* assumption) it is necessary to perform extensive three-dimensional reconstructions and to determine the fractal dimensions of these objects based on volume measurements. Several laboratories have achieved such reconstructions, most notably with computer assisted procedures. As yet no information concerning the fractal dimension is available. In simulated growth, mosaic patches are fractal in three dimensions (Iannaccone *et al.*, 1992).

B. Organogenesis

Mosaic pattern analyses in aggregation chimeras have been utilized to understand histogenesis of various organs and their maintenance (for a review, see Iannaccone *et al.*, 1987a, 1988). Mosaic pattern in retinal epithelial cells in chimeric mice suggested early random mixing of primordial cells and subsequent clonal growth of committed retinal epithelial cells giving rise to small patches near the optic nerve and larger patches in the periphery, where cell division is more rapid (West, 1976a; Bodenstein and Sidman, 1987a,b). Mosaic patch patterns in chimeras suggested that the random assortment of primordial cells and probabilistic daughter cell placement during growth result in discrete geometric patches in liver (West, 1976b; Iannaccone *et al.*, 1987b). Several studies (Wilson *et al.*, 1985; Ponder *et al.*, 1985; Schmidt *et al.*, 1988) suggest that a single progenitor cell forms an individual small intestinal crypt, while the formation of an intestinal villus requires the cooperative growth of epithelial cells from two adjacent crypts. The histogenesis of the rat adrenal cortex appears to result from the inward centripetal cell division of adrenal cortical cells, based on data from *in situ* mosaic pattern analysis in chimeric rats (Iannaccone and Weinberg, 1987).

Even in unbalanced chimeras, all or most organs show some contribution of the minor marker cell lineage, so it is generally held that progeny of the marker cell lineages are well mixed by cell movement until at least the gastrula stage of development in mouse and rat. However, once the organ primordia are allocated, cell movement does not have to be involved to explain the observed mosaic pattern, at least in the epidermis of the mouse (Ng *et al.*, 1990) or the liver of the mouse and rat (Iannaccone *et al.*, 1987b). The use of the mosaic analysis to study pattern formation in various organ systems is considered below.

1. Muscle

The glucosephosphate isomerase-1 isozyme pattern has been studied in small samples of isolated mosaic muscle fibers in aggregation chimeric mice made from embryos homozygous for distinguishable GPI isozymes. The homogeneous distribution of heterodimeric and monomeric isozymes along the skeletal muscle fiber suggested that there may be a wide distribution of *Gpi-1* mRNA and its protein product along syncytial myofibrils, which allows the mixing of the monomeric enzyme to form heterodimers (Frair and Peterson, 1983). The spatial distribution of the two distinct populations of myonuclei does not seem to control the biochemical function locally, but there is nucleocytoplasmic cooperation along the myofibril that transports the *Gpi* message or proteins randomly along the myofibrils. This finding supports the cell fusion model of skeletal muscle fiber formation, which states that syncytial muscle fibers are formed by fusion of adjacent myofibrils instead of by multiple karyokinesis without cytokinesis within a muscle fiber (Mintz and Baker, 1967).

2. Retina

Studies of mosaic pattern in organ development in secondary chimeras have also been done. Pigmented embryonic eye rudiments were grafted into albino hosts in *Xenopus* and the fates of the different cell types were traced. When half of the chimeric eyeball was removed, the remaining anlagen regenerated into a distinct section of graft-coherent pigmented tissues and the ventral half anlagen healed symmetrically and behaved as a mirror image as compared to the anterior half (Conway and Hunt, 1987). These findings supported the idea of clonal propagation of cells with minimal cell mixing of stem cells residing in the annular germinal zone between the iris and retina. This idea of organ formation without extensive cell mixing and migration is also proposed for liver development in chimeric rats and has been demonstrated by the use of computer simulations

(Iannaccone et al., 1987b; Bodenstein and Sidman, 1987a,b; Iannaccone et al., 1988; 1989).

3. Kidney

Interspecific grafting of mouse embryonic kidney onto quail chorioallantoic membrane has been performed to visualize the contribution of various cell types during the differentiation of mammalian kidney. Antibodies recognizing the extracellular matrix and the endothelium of quail were used (Sariola et al., 1984). Chimeric glomeruli consisted of mouse podocytes and quail endothelial cells. Quail chorioallantoic vessels penetrated the mouse kidney explant in which the growth of the capillaries followed the branching of ureteric buds and nephron formation. Renal capillaries grew into a fibronectin-rich matrix and then attached to and stopped at the epithelial basement membrane of the kidney tubules. Thus, the distribution of extracellular matrix seemed to affect the pattern of renal angiogenesis. These results are compatible with an induction model of kidney development (Vainio et al., 1989).

4. Epidermis and Dermis

Mosaic pattern analysis of dermal and epidermal tissues was used to study coat spotting patterns and the interrelationship between the development of epidermis and dermis. The pigmentation pattern of black-and-tan (a^t/a) ↔ black (a/a) chimeras has been examined. The a^t-allele produced melanocytes with yellow pigment or black pigment, dependent on their ventral or dorsal positioning, respectively. There was extensive lateral migration and mixing of the a/a and a^t/a cells in the ventral dermis during development. However, a^t/a cells have demonstrated the capability of recognizing their location in the dermis regardless of the genotype of adjacent cells (Petters and Markert, 1979). This observation emphasized the importance of positional information on pigmentation.

The development of mosaic patterns in dermis and epidermis in mice is interrelated. The mesodermal cells in the dermis were well mixed but the ectodermal cells affected the differentiating melanocytes in some way, which resulted in their invasion of hair follicles and the formation of discrete pigmented patches. This finding is based on the study of the distribution of melanocytes in the dermis of Sl/Sl^d ↔ $+/+$ mouse aggregation chimeras. Sl^d, the Steel-Dickie gene, is a spontaneously mutated gene located on chromosome 10 in the mouse. This gene has both coat color and hematological effects resulting in a deficiency in melanocytes and mast cells in the skin of the mutant mice. Sl/Sl^d ↔ $+/+$ aggregation chimeras had melanocytes in the chimeric dermis. The dermis of Sl/Sl^d ↔ $+/+$ chimeras has homogeneously mixed Sl/Sl^d cells and $+/+$ cells but the epidermis shows discrete patches. The mast cells and melanocytes are distributed evenly in the chimeric dermis. The melanocytes migrated into the hair follicles composed of $+/+$ cells but not into hair follicles composed of Sl/Sl^d epidermal cells. The results suggested that the mast cell precursors differentiate in the dermal layer, whereas the melanocyte precursors must invade their follicles to complete differentiation (Nakayama *et al.*, 1988b). Patches in epidermis and dermis were independent. Mosaic patterns in isolated epidermis of the mouse were observed directly in chimeras (Schmidt *et al.*, 1987). The observations suggested a discrete ventral boundary with midline discontinuity. The results were further interpreted as demonstrating large patches; however, higher resolution observations demonstrated many smaller patches and no quantitation was made available. Moreover, it is not possible to interpret patch size data meaningfully without knowing the proportion of the two cell types present (Iannaccone *et al.*, 1987b).

5. Nervous System

The feasibility of using chimeras to study nervous system development has been discussed elsewhere (see review by Mullen and Herrup, 1979). The interaction of mutant and normal cells during morphogenesis can be examined using mutant ↔ normal chimeras. The abnormal development in the mutant can be analyzed to determine whether the defects are simply due to abnormality at a cellular level or because of some disturbance of regulation when mutant cells are incorporated during development. Cerebellar abnormalities are widely studied because of their importance in motor coordination and their ease of detection. The cerebellum is composed of few cell types, that is, Purkinje cells, Golgi cells, and granule cells, which can be easily characterized anatomically. The staggerer (*sg*) mutation, on mouse chromosome 9, results in degeneration of granule cells and arrested development of Purkinje cells and is a cerebellar mutant. These animals display a lurching gait and have small cerebella with few if any granule cells. The study of aggregation chimeric mice made between *staggerer* (*sg*) and wild-type revealed that there was a linear relationship between the number of surviving cerebellar granule cells and the functional Purkinje cells in the chimeric cerebella (Herrup and Sunter, 1987). Purkinje cells are the primary postsynaptic targets of cerebellar granule cells. The *staggerer* ↔ wild-type chimeric cerebella suggested that target-related granule cell death is important in maintaining a numerical balance in the cerebella of the mice. The number of surviving granule cells matched that of functional wild-type Purkinje cells present in the chimeric cerebellum.

The study of *lurcher* ↔ wild-type chimeric mice supported the notion of target-related cell death in the regulation of numerical balance of cells in the cerebella. The *lurcher* (*Lc*) mutation on mouse chromosome 6 results in early death of Purkinje cells in the mutants. These Purkinje cell-lethal mutant mice show a characteristic gait, with swaying hind quarters and frequent falling to one

6. Experimental Chimeras 249

side. The number of surviving granule cells was higher in *lurcher* ↔ wild-type chimeric cerebella than in *staggerer* ↔ wild-type chimeras. This increase in granule cell survival in the *lurcher* ↔ wild-type chimeras suggested the presence of some trophic factor produced by the wild-type Purkinje cells that delayed the degeneration of *lurcher* Purkinje cells (Vogel *et al.*, 1989).

The necessity of cell–cell interaction for normal cerebellar cell migration during morphogenesis was shown using mouse chimeras developed from normal and *reeler* (*rl*) strains (Terashima *et al.*, 1986). *reeler* mice have abnormally shaped and distributed cerebellar cells and abnormal motor behavior. The $+/+$ ↔ *rl/rl* chimeras have mosaic cerebella; however, there are normal neuronal and glial subpopulations. The chimeras have normal motor behavior, although the mutant cerebellar cells have abnormal shapes. The results suggest that development of a functional neuronal network requires appropriate cell–cell interactions that appear to be provided by the normal cells.

The utilization of isozyme-specific-anti-GPI-1B antisera in immunocytochemical staining of the cerebella of BALB/c bg^J ↔ C57BL/6J chimeric mice showed chimerism in the Purkinje cell population. Purkinje cells existed as contiguous patches throughout the development process (Oster-Granite and Gearhart, 1981). The concept of cerebellar development with small clones of Purkinje cells was supported by the use of the Purkinje cell-lethal mutant, *lurcher* ($+/Lc$), as a cell lineage marker in *lurcher* ↔ wild-type chimeras. The surviving Purkinje cells were always intermingled with the mutant ones and the sizes of the clones of Purkinje cells were intrinsically strain specific (Herrup and Sunter, 1986).

shiverer ↔ normal chimeric mice have been used to trace lineage in the formation of Schwann cells in the peripheral nervous system (Mikoshiba *et al.*, 1984). The *shiverer* (*shi*) mutant lacks myelin basic protein. Chimeras having relatively equal proportions of normal and *shiverer* genotypes behaved normally. Chimeras having predominantly the shiverer genotype showed typical *shiverer* behavior of tremor and convulsions. Immunohistochemistry showed a linear array of patches of normal and mutant myelin sheaths with intermingling Schwann cells in the chimeras. Schwann cells existed as contiguous groups or cell collectives like the Purkinje cells described above. The patch size varied according to the degree of chimerism; however, the significance of this specific arrangement of Schwann cells is still uncertain.

Chick ↔ quail chimeras have been used to study the migratory pattern and the fate of cells during development of the neural crest. The neural crest cells of the chick in chick ↔ quail chimeras contributed to mesenchymal derivatives like bone, cartilage, dermis in face and neck, and glands in the tongue and pharynx (LeLievre and LeDouarin, 1975; LeDouarin, 1980, 1982). Grafting a neural tube of a quail embryo into a chick embryo resulted in a normal functioning chimeric nervous system in viable chimeric birds (Kinutani and LeDouarin, 1985). This observation suggested that fixation of the fate of cells in the neural tube occurs early in development. When the neural tube cells of a quail are grafted in appropriate embryonic position in a chick embryo, they develop normally to the mature nervous system in the chick embryo (Tanaka and Landmesser, 1986).

6. Sexual Differentiation

Theoretically, half of aggregation chimeras with have XX ↔ XY sex chromosome composition. These chimeras can be identified easily by chromosome analysis (Iannaccone et al., 1985) or by analyzing electrophoretically distinguishable dimorphic phosphoglycerate kinase-1 (PGK-1) variants of this X chromosome-linked enzyme (Nakayama et al., 1988a). XX ↔ XY chimeric mice may have ovotestes with varying proportions of ovarian and testicular tissues (Milet et al., 1972). Histological studies of the development of ovotestes on days 12.5–14.5 of gestation showed that testicular tissue was located in the central region and the ovarian tissue in the caudal region of the embryo. This is similar to the condition in spontaneous hermaphrodites, but in chimeras there is subsequent regression of the ovarian portion of the ovotestes on day 14 of gestation, resulting in testes in most adult intersex chimeras. The regression of the ovarian tissue may be stimulated by müllerian inhibiting substances or anti-müllerian hormones (Bradbury, 1987).

The analyses of the distribution of XX and XY cells in chimeric mice with enzyme markers and *in situ* DNA markers established that Leydig cells and vascular connective tissue of the tunica albugina of the testis were XX, whereas Sertoli cells were XY (Burgoyne et al., 1988). This suggests that Y chromosome genes are important for the differentiation of Sertoli cells, which in turn support spermatogenesis. Genes have been described (e.g., *ZFY* on the 1A portion of the human Y chromosome and *Sry* in mice) (Page et al., 1987; Mardon and Page, 1988; Koopman et al., 1991) that may be involved in this important aspect of the determination of maleness. *Sry* appears a likely candidate for male determination based on the report of two transgenic XX mice with *Sry* expression that are phenotypically male. However, this observation will have to be independently confirmed, particularly because the karyotypes of these two "sex-reversed" mice were not reported.

The effect the of sterility-inducing Sl/Sl^d mutation on the process of spermatogenesis in mice was studied by making aggregation chimeras with normal mouse embryos. The chimeras had patches of normal (differentiated) and abnormal (nondifferentiated) segments in a single seminiferous tubule and no spermatogonia were detected in the patches of mutant Sertoli cells. Thus, the absence of spermatogonia may be attributed to the defective microenvironment of Sertoli cells. A way to study the proportion of functional germ cells from two cell lineages in male chimeric mice is to allow them to mate freely with nonchimeric females and then analyze the variation in progeny frequencies (Iannaccone et al.,

1985); however, variation can occur that is unrelated to the proportion of the two marker cell lineages (Buehr and McLaren, 1984).

7. Hematopoietic Stem Cell Regulation

Chimeric mice have been used to study the developmental potential and dynamic behavior of hematopoietic stem cells. The analyses of electrophoretically distinct variants of isocitrate dehydrogenase, malate dehydrogenase, and H-2K, H-2B antigens of the red blood cells (RBCs) in C57BL/6 ↔ C3H mice showed the coexistence of two different hematopoietic stem cell populations and the absence of any cell hybridization in these populations. However, there was a chimeric shift of erythropoiesis toward C57BL/6 over C3H cells in C3H ↔ C57BL/6 chimeras. Some studies of peripheral RBC and white blood cell (WBC) populations in chimeric mice showed stable chimerism (Warner *et al.*, 1977), while some indicated a chimeric shift to either parental composition (Stephens and Warner, 1980). This phenomenon of instability may be due to different selective pressures on the blood cells that originate from mesoderm as compared to ectodermal-derived tissues, which showed no variation over time (Stephens and Warner, 1980). Shifts in hematopoietic stem cells demonstrated in the peripheral blood cell populations may be the result of strain-dependent differential rates of mitotic activity, particularly because shifts in the populations of the two marker lineages of erythrocytes, platelets, and lymphocytes have been reported to be less in chimeric mice between congenic rather than between noncongenic strains (Behringer *et al.*, 1984). It is not established whether the chimeric shifts are due to genetic differences in the chimeras or if it is an intrinsic factor in the hematopoietic system.

It is generally held that cells of the hematopoietic system are arranged in a hierarchy such that a small number of stem cells are allocated as progenitor blood cells. These cells then differentiate into various types of blood cells (Chan and Winton, 1989). Bone marrow transplants have been employed to introduce specially marked stem cells of bg^J/bg^J strain into wild-type mice to establish the stem cell regulatory volume. Patt and Maloney (1978) studied the marked peripheral neutrophils as a log function of the amount of bone marrow transplanted because the degree of chimerism in hematopoietic stem cells is indicated by peripheral blood chimerism (Francescutti *et al.*, 1985). There was a linear log time response between the dose of marrow transplanted and the replacement by marked neutrophils. These results suggested a restricted stem cell migration resulting in a latent period of at least 3 weeks before replacement peripheral neutrophils were detected.

The *in vivo* dynamics of the hematopoietic stem cell activity are more appropriately revealed by the study of newly synthesized, short-lived circulating reticulocytes rather than the mature RBCs, which exert a damping effect on the overall erythroid population because of their long half-life. The observation of the variance of the proportion of the population of reticulocytes in C57BL/6 ↔ CBA/HT6J chimeric mice having electrophoretically different hemoglobin types, d and s, suggested at least 33 to 118 new stem cell clones were formed during a 3-day period (Harrison et al., 1987). This number is too large to satisfy the clonal succession and selection hypothesis of serial activation and inactivation of stem cells, which was based on studies of bone marrow grafting (Lemischka et al., 1986) and female X-linked mosaics (Burton et al., 1982).

Studies on the population kinetics of peripheral WBCs in congenic chimeric rats may help to reconcile some of the controversial findings discussed above. The peripheral WBCs are short lived and their population is a better representation of the population dynamics of the hematopoietic stem cells, the stem cell population being recruited for blood cell formation. In vivo experiments using chimeras are easier to interpret than in vitro experiments because all of the regulatory elements (for example, macrophages, endothelial cells, and T cells) producing various colony-stimulating factors are present in the chimeras (Chan and Winton, 1989). In our studies, congenic chimeric rats show oscillatory changes in proportions of marked RBCs and WBCs (lymphocytes, neutrophils, and monocytes) on a cycle of several weeks (Ahmad et al., 1992). This observation is incompatible with the notion that stem cell fluctuation is occurring as an artifact of cell selection following the combination of disparate strains.

8. Thymus

The thymus arises from the division and differentiation of thymic subcapsular lymphoblasts, which give rise to small, deep, cortical thymic lymphocytes, medullary lymphocyte (Fathman et al., 1975), or an independent self-renewing population in the medulla that contributed to the precursors of the peripheral T cell pool (Shortman et al., 1987). The study of the development of the thymus by using irradiated mice reconstituted with congenic bone marrow cells revealed the presence of two independent thymus lineages: cortex and medulla, and medulla alone (Ezine et al., 1984). A similar study by Matsumoto and Fujiwara (1987) showed that the thymus medullary cells were mainly reconstituted with donor type cells, and only a few donor cells were found in the cortex. In situ analysis of the thymus of aggregation chimeras may resolve some of the controversies over the histogenesis and maintenance of the thymus without physically disturbing the developed organ. For example, chimeric rat thymus shows very large contiguous patches in the cortex, but not the medulla, irrespective of the proportion of the marker lineages. Such evidence of fixed stem cell proliferation is found both in very young and in older chimeras. Extensive three-dimensional reconstruction of cortical patches reinforces the notion that these patches result from stem cell proliferation (M. Khokha, I. Ahmad, and P.M. Iannaccone, 1992, unpublished observation).

It has been reported that the presence of thymic cells is necessary for maturation of B cells in response to either T cell-dependent or T cell-independent protein antigens (Siskind et al., 1978). As B cells mature, they express specific surface markers, for example, Ia antigens and C3 receptors (Mosier et al., 1978). However, Vainio and Toivanen (1983) showed that the genotype of the B cells plays an important role in controlling their maturation and MHC-restricted capacity to react with T cells, and that the thymic environment has no collaborative effect on B cell maturation. Major histocompatibility products are important for T cell tolerance because MHC-congenic skin grafts were always accepted. Major histocompatibility class I/II antigens are involved in antigen recognition by T cells. Cytotoxic T cell responses required the antigens to be presented on the antigen-presenting cells in association with cell surface MHC molecules, for example, the *Ir* gene product (Moller, 1987). Nakamura (1989) reported the possible involvement of the H-2 MHC gene in T cell recognition. The locus between the S and D regions of the *H-2* gene codes for neuraminidase isozymes that may affect cell surface glycoprotein formation, which acts as a recognition site on the target tissue for the histocompatibility reaction after tissue transplantation. Functional donor and recipient immune competence were found to be very important for the development of tolerance in semiallogenic limb grafts. The allogenic donor strain stem cell may play an important role in tolerance induction (Hewitt et al., 1986). Chimeras made by fetal liver and thymus transplants showed that HLA mismatch still enabled full cooperation of the transplanted T cells and the recipient B cells to restore immunological competence. HLA-matched or HLA-mismatched thymus grafts to nude mice restored the T cell functions in the recipients. Thus, thymus grafting may be a feasible treatment for patients with T cell or thymic function deficiencies (Roncarolo et al., 1986; Furukawa et al., 1988). However, these data from mice seem at odds with the concepts of MHC-restricted antigen recognition by T cells and do not agree with the findings of allogenic B cell development in cooperation with T cell response in birds (Vainio and Toivanen, 1983) or of bone marrow grafting in canine species (Deeg et al., 1987).

C. Computer Simulations of Mosaic Growth

When two marker lineages assort themselves during development, cell division patterns and cell movement result in a mosaic pattern typical of a given tissue, as described above. The detailed analysis of these patterns has followed mathematic rules developed for the analysis of patterns and tilings (Grunbaum and Shepard, 1987). Early simulations of mosaic patterns were performed on two-dimensional random patterns (Whitten, 1978). Attempts at three-dimensional analysis developed close-packing models in which cells were arrayed in patches with regular Euclidean shapes, most often hexagons. More recently it has been possible to study the dynamics of patch formation through the generation of mosaic patterns by creating computer programs that simulate cell division patterns (Iannaccone et al., 1989, 1992; Bodenstein, 1986). These models offer the opportunity to determine what cell division and/or cell placement trends might result in the actual mosaic patterns observed.

Iterated functions (repetitive application of functions to a data set where the output of one round of calculation provides the input for the next) have been used to reproduce images that had been reduced to self-affine functions (i.e., transformation functions that take a finite number of points of a subset of an object and remap them so that following a large number of iterations the entire object is represented) by degenerating the image of interest (Barnsely et al., 1986). This approach, however, could not provide clues about the generation of mosaic patterns within an organ because the organ has grown from a small anlagen, not from decomposition of a large structure. Our two-dimensional analysis, however, suggested that an iterative model of the generation of mosaic pattern beginning with a small anlagen in three dimensions would yield realistic patterns. We undertook such studies because mosaic patterns represent an independent marker of cell division programs and might provide clues to the type of process important in the generation of parenchymal mass in organogenesis.

Patches in the liver of both chimeric mice and chimeric rats suggest that probabilistic decisions were important in their formation. We produced a program of iterating functions that began with a 10-cell anlagen in which the position of cells of a given type was established randomly, but the proportion of the 2-cell types could be selected. The cells all divided in a randomly chosen order and the daughter cell of each division was placed in a randomly chosen adjacent position. This rule set was repeated for 10 generations, where 1 generation was defined as division of all cells in the 2-dimensional field. The patterns obtained were highly variegated even though cell movement was not part of the simulation. Quantitative analysis demonstrated that the size and number of patches were both dependent on the proportion of the two cell types that comprised the mosaic. This pattern could be altered to produce cordlike structures by biasing the probability of daughter cell position toward a center position within the growing organ. Thus, the observed pattern could be altered with minor adjustment of the simulation rule set.

Three-dimensional computer simulations were written in the same way (Fig. 5). These programs demonstrated that the generation of mosaic patterns can be nonlinear. The three-dimensional simulation produced fractal aggregates and demonstrated that simple iterated division rules can produce complex patterns. Results of these simulations suggest that the topological information necessary for correct positioning of cells in time during development could be efficiently expressed as iterating functions. It was further determined that the fractal dimension of patches in three dimensions is not sensitive to the mosaic proportion, but implies the presence of a chaotic attractor (i.e., an overlapping solution set of a nonlinear dynamic process). By this we mean each patch may have a characteristic size and geometric complexity to which it returns with statistical certainty

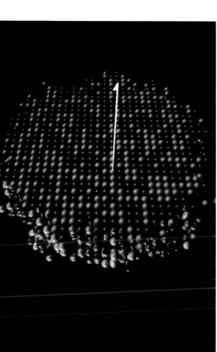 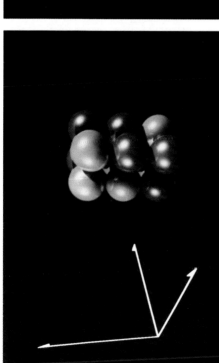

Fig. 5 In computer simulations the generation of mosaic pattern can be modeled in three dimensions. *Left*: The two cell lineages are displayed as green or red spheres. *Right*: The tenth generation from one such simulation. There is rapid evolution of complexity from simple rules. A probabilistic iterating model divided all cells in the matrix in a random order. Positions for the daughter cells were adjacent to the dividing cells and chosen at random. The parenchymal mass after 10 generations was cut in half to simulate the 2-dimensional pattern obtained by sectioning mosaic tissue. (Image and display software by G. Lecsinsky and T. DeFanti; simulation software by J. Lindsay.)

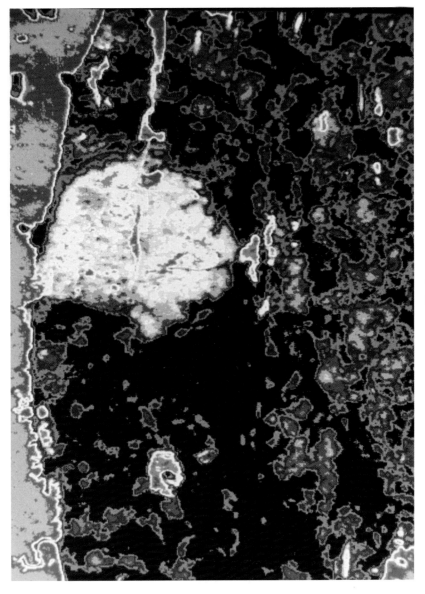

Fig. 6

as a tissue grows. As the patch expands clonally and contracts by fragmentation it will most frequently appear with a characteristic size and shape that is peculiar to a given tissue. The process is nonlinear in that output conditions from each new input are not predictable on the basis of solutions to previous iterations. The solution set that represents these characteristic sizes and shapes of the oscillations (a dynamic process) is the attractor and may define the observed patterns. Because the process is dynamic and nonlinear the attractor is chaotic.

These simulations of the generation of mosaic patterns are useful, but also have some serious limitations. In several respects the growth mechanisms are unrealistic. There is no cell death, the adjacent positions are rastered, there is no provision for cell adhesion, and there is no provision for damping the cell bumping effect at a distance. All of these features are being added to simulations currently in preparation.

Many of the predictions of such computer simulations can be tested by examining the growth of actual mosaic tissue in time. Our laboratory approached this problem by examining uniform growth in a well-characterized mosaic tissue, liver. The simulation established relationships between generation number and number or area of patches as well as fractal dimensions of patches. These relationships could be tested *in vivo* by examining a growing mosaic tissue at various times during growth so long as no focal centers of growth existed. These conditions were met by inducing compensatory growth in the liver of chimeric rats with partial hepatectomy. It was determined that the size of patches both in terms of perimeter length and area increased until the first mitotic wave had passed, while the fractal dimension remained unchanged throughout the regenerative process. These results precisely match the predictions of the simulation, in which iterating self-similar division rules without focal centers of growth or cell movement were responsible for increase in parenchymal mass (Ng and Iannaccone, 1991).

III. Pathogenesis

A. Neoplasia

The cellular origins of tumors have been studied in mosaic tissues by many investigators over the past two decades. The general conclusion of these studies, that tumors arise by clonal expansion of rare, genetically altered cells, was hard

Fig. 6 False-color representation of density from an autoradiogram of chimeric rat liver. White, blue, green, and brown represent grain densities indicating PVG lineage; black represents grain densities indicating PVG-$RT1^a$ lineages. The liver was from a rat treated with a liver carcinogen. Several large areas (white or black) are clearly abnormal. These areas line up precisely with premalignant lesions caused by the carcinogens. The lesions are clonal, and their geometry seems less fractal than the normal patches, suggesting that the normal parenchyma is chaotically derived while the premalignant tissue is not. (From Weinberg and Iannaccone, 1988, reprinted with permission.)

won, proved to have many nuances, and spawned deep controversy that continues to the present. Tumorigenesis is a multistep process and if any of the steps are rare events then tumors appear to be clonal growths. There have been extensive reviews describing the theories of clonal expansion of genetically altered cell populations in neoplasia (Nowell, 1986; Iannaccone et al., 1987c). It is not our intention to rehearse those arguments here, but rather to provide a brief overview of some aspects of tumor progression studied with mosaic individuals.

Early experiments with X chromosome-linked mosaic women indicated that when there were two isoforms of glucose-6-phosphate dehydrogenase (G6PD) produced by separate cell populations, only one of these was present in tumors. Leiomyomas (mesenchymal tumors of the uterine wall) were analyzed extensively. These early studies suffered the disadvantage of low sensitivity of the gel techniques employed, but the general conclusion is that these tumors are derived from single cells. Recently, other smooth muscle tumors (those in atherosclerotic plaques) from experimental chimeras with distinguishable cell types were shown to be clonal (Penn et al., 1986).

Most hepatomas induced in experimental chimeras made between strains that varied in β-glucuronidase expression were found to be entirely of one cell type. However, two tumors were discovered with heterogeneous marker cell populations. Chimeras were made between susceptible and nonsusceptible strains of mice, and the majority of tumors obtained comprised cells of the susceptible strain. The observation that cells from the nonsusceptible strain participated in the formation of tumors remains unexplained. Chimeras are so well suited to the examination of cell-autonomous behavior that it is surprising that more studies of tumor susceptibility have not employed them. Condamine et al. (1971) suggested that hepatomas were composed of diverse clones of transformed cells. This conclusion seems incorrect given the results of another in situ analysis of hepatic tumors in mice by Howell et al. (1985), who used genetic differences in ornithine transcarbamylase expression and demonstrated that all of a large number of hepatomas in mouse chimeras were clonal.

The hypothesis that hepatic tumors are clonal in origin was further reinforced by an analysis of preneoplastic lesions in chimeric rat liver. The chimeric rats were marked histologically with variants of major histocompatibility antigens. These animals, subjected to partial hepatectomy and treated with diethylnitrosamine, developed lesions with extreme phenotypic variation with respect to the expression of γ-glutamyl transpeptidase, glucose-6-phosphatase, ATPase, and glycogen retention. Despite heterogeneous expression of these enzymes (generally taken to be markers of preneoplasia in the rat liver), 96% of over 1100 such lesions were clonal (Fig. 6). Those that could not be definitively shown to be composed of one of the two cell types that made up the chimera were quite small. The conclusion was that lesions early in tumor progression are also clonal and that perhaps the earliest steps in hepatocarcinogenesis are clonal. The observation that such phenotypic heterogeneity is derived from a clonal population suggests that the earliest steps in hepatocarcinogenesis led to genetic instability (Rabes et al.. 1982; Iannaccone et al.. 1987c; Weinberg et al.. 1987; Weinberg and Iannaccone, 1988).

Other premalignant lesions have been analyzed as well. *In situ* carcinomas were revealed as clonal loci in a mosaic cellular environment (Ponder et al., 1986) in experiments with chimeric mice designed to study colonic epithelial dysplasia (Ponder and Wilkinson, 1986). Papillomas induced in chimeric mice were shown to be clonal in origin by using glucosephosphate isomerase as a marker of mosaicism (Iannaccone et al., 1978). Similar findings were obtained utilizing X-linked mosaic animals (Deamant and Iannaccone, 1987; Iannaccone et al., 1987c) and interspecific chimeras (Deamant et al., 1986). Studies of methylcholanthrene and benzpyrene-induced skin papillomas showed that there were mixed tumors, indicating they had arisen from large numbers of cells (Reddy and Fialkow, 1983); but these experiments were performed under conditions that resulted in a mixture of neoplastic and nonneoplastic tissues. Papillomas generally consist of an epithelial neoplastic portion on the surface and a fibroconnective tissue core. The core tissue is the result of a host response to the overgrowth of epithelial tissue and will therefore have the same mosaic proportion as other normal tissues in the animal. By separating these components it was possible to demonstrate that the epithelial neoplastic portion of these tumors was clonal in origin, and was composed of a single population, either PGK-1A or PGK-1B, while the host-derived fibroconnective core of the tumor was composed of both populations (Deamant and Iannaccone, 1987).

More recently Winton et al. (1989) described studies with an *in situ* marker of mosaicism and examined chemically induced skin tumors in chimeric mice. The authors reported numerous mixed tumors implying a nonclonal origin, but examination of the published histologic material clearly shows that the lesions studied were not papillomas but rather interepithelial hyperplasias, very small cellular growths, sometimes known as skin tags. No evidence from well-formed papillomas was presented. Indeed, it has been previously established that hyperplasias are nonclonal. The best-studied system is parathyroid hyperplasia: it has been shown to be nonclonal in chimeras, whereas parathyroid adenomas (capable of autonomous growth) are clonal (Arnold and Kim, 1989; Arnold et al., 1988). Numerous studies utilizing a variety of markers in numerous organ systems have thus established that autonomous growth in neoplasia represents the clonal expansion of rare, genetically altered cells (see Iannaccone et al., 1987c, for review).

B. Germ Cell and Embryo-Derived Tumors

Teratocarcinoma mouse cells expressing GPI-1A isozyme have been injected into normal blastocysts expressing the GPI-1B isozyme. The results proved that teratocarcinoma cells could participate in normal fetal development (Mintz and Cronmiller, 1978; Papaioannou et al.. 1978; Aizawa et al., 1986). Cell fusion hybrids of mouse teratocarcinoma (thymidine kinase deficient) and hepatoma [hypoxanthine–guanine phosphoribosyltransferase (HGPRT) deficient] cells have also been injected into C57BL/6J normal blastocysts. The hybrid cells showed limited integration in liver development and there was differential gene expression from the introduced xenogenic chromosome (Illmensee and Croce, 1979; Duboule et al., 1982). When mouse embryonal carcinoma cells (EC), the undifferentiated stem cells of teratocarcinoma, were injected into the normal blastocyst, they seem to be regulated such that they do not form tumors (Pierce et al., 1989). The signals from normal ICM and trophectoderm cells play a role, as yet undefined, in the regulation of the development of EC cells. Martin et al, (1984) studied the possibility of reprogramming hybrid thymocytes made between normal mouse thymocytes and HGPRT-deficient teratocarcinoma mouse cells to participate in chimerization when they were injected onto normal ICM. It was found that these mammalian somatic cells are not pluripotent, as are some amphibian cells (DiBerardino, 1988; Gurdon, 1974).

Spontaneous parthenogenetic ova may implant but subsequent development does not occur. If they are grafted to extrauterine sites they will survive as teratomas. The production of four- to eight-cell parthenogenetic embryo ↔ normal embryo mouse chimeras has shown that even though the parthenogenetic embryos cannot survive *in utero*, cells from the parthenogenote are capable of participating in normal development and producing functional germ cells when associated with normal cells (Stevens, 1978; Kaufman et al., 1984). Thus, the normal genetic constitution of neighboring cells may play a key role in rescuing parthenogenetic cells during preimplantation and early postimplantation. However, there is a strong selective pressure against parthenogenetic cells later in chimeric fetal development, resulting in unbalanced chimeras. The adult tissues have a majority of cells from the normally fertilized embryonic lineage (Nagy et al., 1987). There is systemic elimination of parthenogenetic cells in the primitive endoderm lineage and some organs (e.g., liver and muscle). Parthenogenetic cells are more consistently detected in brain, heart, kidney, and spleen in chimeras (Fundele et al., 1989); therefore, the paternal genome appears to be important in later fetal cellular differentiation in certain tissues, but not others. The paternal genome also seems to be especially important for extraembryonic development because the parthenogenetic cells in the yolk sac diminish very rapidly and are excluded from the trophoblast. It thus appears the paternal and maternal genes are working in a complementary way to enable normal embryonic development (McGrath and Solter, 1984).

Mouse embryo-derived stem cells (ES) have been shown to be pluripotent (Evans and Kaufman, 1981; Bradley et al., 1984; Evans et al., 1985; Robertson and Bradley, 1986; Nagy et al., 1990). When ES cells are injected into mouse blastocysts, they are incorporated into the germ line of resulting chimeras. This can be exploited to achieve germ line transmission of an exogenous gene that has been inserted into the ES cells. Targeted genetic manipulation of ES cells is possible and procedures have been previously reviewed (Kuehn et al., 1987; Bernstein et al., 1986). The ES cells carrying the target gene are then injected into the blastocyst of another embryo to produce chimeras. Progeny of the manipulated ES cells participate in the production of functional gametes so that the offspring from the chimeras can be used to breed the genetic manipulation to homozygosity. This procedure is most useful in the genetic analysis of loss of function (McMahon and Bradley, 1990).

C. Immunological Tolerance

Aggregation chimeric mice provide a good model for studies of immunological tolerance because mosaic lymphoid tissues are produced "naturally" during the course of embryonic development of the chimeras. There is a good correlation between the proportion of marker lineages in peripheral blood and spleen in chimeras. Chimeras can also demonstrate similar proportions of marker lineages in the thymus and peripheral blood or spleen (Warner et al., 1977). Therefore, it is reasonable to use the peripheral blood population as an indicator of the degree of mosaicism in the spleen and possibly the thymus in a chimera. Partial lymphoid chimerism, including spleen, lymph nodes, Peyer's patches, but not bone marrow or thymus, has been induced in nonirradiated B lymphocyte-deficient CBA/N mice by injecting T6 chromosome-marked normal spleen cells (Volf et al., 1978). Bone marrow transplant has been confirmed to produce hematopoietic chimerism in the stem cell populations, peripheral blood, and immunoglobulins in mice (Wade et al., 1987) as well as in humans (Yam et al., 1987).

Immunological tolerance has been most extensively examined in bone marrow chimeras, especially irradiation chimeras. Bone marrow chimeras can be produced by the antibody-facilitated (AF) method. $P_1 \leftrightarrow (P_1 \times P_2)F_1$ AF chimeras are made by simultaneously injecting P_1 bone marrow cells and anti-P_2 MHC monoclonal antibody into the normal adult $(P_1 \times P_2)F_1$ recipients. This method does not involve irradiation or immunosuppressive drug administration in the host animals. The production of AF bone marrow chimeric mice with the donor cells in the peripheral blood, marrow, and hematopoietic stem cell compartment for over 3 months has also been successful (McCarthy et al., 1985). It has been shown that the donor cells participated in B and T cell formation in the chimeras, but there is no cooperation in the immune response in the aggregation chimeric mice studied (Gorcyca et al., 1982). Thus, the exact mechanism of immunological tolerance in chimeras is still unclear. Immunocompetence and tolerance have been studied in allogenic (Urso and Genozian, 1977; Ildstad et al., 1985; Groves and Singer, 1983) and syngenic irradiation bone marrow chimeras. It has been shown that there were quantitative deficiencies in either T or B cells but these did not relate to the severity of graft-vs-host disease (GVHD), which occurred usually after bone marrow transplantation in irradiation chimeras. Moreover, Yasumizu *et al.* (1985) reported a defect in the macrophages in irradiation chimeric mice reconstituted with allogenic bone marrow. Graft-vs-host disease is commonly studied in skin grafts (Charley *et al.*, 1983), and may be due to recipient T cells responding to donor MHC antigens or their secretion of lymphokines, like interleukin 2, to recruit more donor cells as well as host cells to express cytotoxic effects (Clancy, 1989). However, it is difficult to assess the possible type of natural effector cells involved in either acute or chronic GVHD resulting from allogenic tissue transplantation, because there are problems in distinguishing host cells responsible for natural resistance to graft cells from the donor (graft) cells. This problem may be resolved by studying bone marrow chimeras with specific distinguishable cell markers.

The importance of environment on the inductive differentiation of T cells was studied in *BB* ↔ *WF* chimeric rats. *BB* rats are well studied as a model of type 1 insulin-dependent diabetes mellitus, believed to be due to an autoimmune response to pancreatic β cells. In lethally irradiated *BB* rats reconstructed with T cell-dependent *Wistar–Furth* (*WF*) bone marrow, the incidence of diabetes was reduced or prevented. Such animals had normal lymphocyte count and function in the peripheral blood in contrast to those reconstituted with *BB* bone marrow cells. However, these *BB* ↔ *WF* chimeras still have an abnormal T cell subset distribution similar to the diabetes-prone *BB* rats. When the *WF* rats were irradiated and reconstituted with *BB* bone marrow, they did not develop diabetes and the diabetes-prone *BB* T cells differentiated normally in the normal recipient (Scott *et al.*, 1987). These studies thus demonstrated that while *BB* bone marrow progeny are associated with diabetes (which can be reversed with normal bone marrow progeny) they are not sufficient to cause it.

D. Genetic Diseases

Genetic imbalance, for example, aneuploidy, adversely affects normal development. This imbalance results in early spontaneous abortion of affected fetuses or altered phenotypes in live births. Aneuploidy, the duplication or deletion of chromosomes, produces gametes with chromosomal additions or deletions. Altered gene dosage in the animal results in an abnormal phenotype or perinatal lethality (Gearhart *et al.*, 1987; Suzuki *et al.*, 1986). The length of survival of an aneuploid fetus appears to be inversely related to the size of the affected chromosome. For example, 30% of human trisomy 21 (Ts21) fetuses survive postnatally, whereas human Ts1 fetuses do not survive postimplantation. In mice, any changes resulting in a 2% increase in total chromosome material in the zygote is usually lethal. The smallest autosome in mouse is chromosome 19 and Ts19 mice can survive up to 3 weeks postnatally. However, other aneuploid mouse embryos usually do not survive the gestation period (Epstein, 1986).

6. Experimental Chimeras

Aneuploidies of mouse with homology to human syndromes have been used as animal models to study the mechanisms of developmental pathogenesis and to elucidate the effect of gene dosage on development. Trisomic or monosomic mice are produced by mating animals with appropriate combinations of Robertsonian translocation chromosomes. These mouse embryos usually die but they can be rescued by making chimeras with normal animals, which allows the embryos to survive longer. As a result, it is possible to establish the functional consequences of the presence of aneuploid cells on development (Gropp et al., 1983; Gearhart et al., 1986a). To produce aggregation chimeras between normal and specific aneuploid embryos, aneuploid four-cell embryos were obtained from mice with the appropriate double-heterozygous Robertsonian chromosome. One blastomere of the aneuploid embryo was karyotyped to establish the aneuploidy at the time of aggregation. Mouse Ts16 has been widely used as an animal model of human Ts21 (Down syndrome) because some genes on human chromosome 21 and the 12q22 fragment, which is associated with Down syndrome, map to mouse chromosome 16 (for reviews see Reeves et al., 1986; Epstein, 1986; Oster-Granite, 1986) Ts16 ↔ 2n (diploid) mouse aggregation chimeras with a higher proportion of Ts16 cells in hematopoietic tissues died even though they had a lower proportion of Ts16 cells in other tissues (Cox et al., 1984; Epstein, 1986). Also, Ts16 hematopoietic stem cells derived from live Ts16 fetuses could not restore the hematopoietic system of lethally irradiated normal adult hosts (Gearhart et al., 1986b). It has been found that Ts16 mouse fetuses have a delay in lymphocyte maturation (Berger and Epstein, 1989). This evidence suggested a proliferative disadvantage of Ts16 cells during development that results in hematologic abnormalities and immune deficiency in these trisomic mammals. Further analysis of the neurological development of Ts16 mice provided evidence of morphological abnormalities in the hippocampus with reduced cell density and delayed formation of anlagen of the dentate gyrus (Oster-Granite and Hatzidi-Mitrion, 1985).

It has been proposed that trisomic chromosome 15, especially the distal portion of chromosome 15, and the amplification of the c-*myc* gene, which is located on mouse chromosome 15, are essential in conferring a proliferative advantage to T cells, which results in murine T cell leukemia (Wirschubsky et al., 1984; Silva et al., 1988). Other experiments showed that most of the malignant lymphoid cells induced chemically or that occurred spontaneously in mice were diploid instead of Ts15 (Boggs et al., 1983; Goodenow et al., 1986). Normally Ts15 mouse embryos do not survive to birth, making the assessment of the effect of Ts15 cells in adult animals impossible. The proliferative efficiency of Ts15 cells was assessed in viable Ts15 ↔ 2n aggregation chimeric mice. Ts15 cells were found in all tissues in the chimeras but fewer were found in the thymus and spleen. Thus, Ts15 cells do not exhibit proliferative advantage over diploid cells in fetal development as suggested by T cell leukemia and thymic lymphomas (Epstein et al., 1984). Ts15 may not be necessary for cell proliferation in lymphoma even though the trisomy is associated with the malignancy. A similar conclusion was reached using radiation chimeras made by transplanting Ts19 murine fetal hematopoietic liver cells to lethally irradiated normal adult mice. The engrafted Ts19 cells completely reconstituted the hematopoietic system in the recipient, although Ts19 cells did not show additional chromosomal abnormalities when leukemia was induced in the chimeras (Herbst et al., 1985).

Monosomy in mammals is lethal at periimplantation stages (Epstein, 1986). Any deleterious recessive gene on the single remaining chromosome becomes hemizygous and may be directly expressed phenotypically. If monosomies were viable they would be very useful in locating recessive genes on specific chromosomes. When *Drosophila* monosomic lines are crossed with the homozygote for the gene of interest and the progenies of each cross are examined for the expression of the recessive phenotype, the cross in which the phenotype appears identifies its chromosomal location (Suzuki et al., 1986).

The phenotypic expression of monosomy 19 (Ms19) lethality in mice is an embryo that contains fewer cells than normal at the early morula stage. The Ms19 cells divide slower than the normal cells. The production of Ms19 ↔ $2n$ aggregation mice chimeras that survive past implantation (day 9) indicated the importance of gene dosage balance on some as yet unidentified gene products necessary for early fetal development (Epstein and Travis, 1979; Magnuson et al., 1982).

Aggregation chimeras made between normal embryos and embryos with genetic mutations will sometimes rescue the mutant embryo and result in a normal phenotype. This can serve as a model system to understand the functional consequences of a mutation in specific organ development at a histological, cellular, and molecular level. The congenital cataractous mouse mutant, which has a human counterpart, was studied using normal ↔ mutant aggregation chimeras. The chimeras demonstrated a significant proportion of cataract mutant genotype in the lens, retina, blood and coat color in a phenotypically normal chimera. The normal cells may thus exert a positive regulatory effect on the mutant cells during organogenesis and result in normal crystalline synthesis and osmotic balance in the mutant cells in the lens (Muggleton-Harris et al., 1987).

Aggregation chimeras between *pcd*/*pcd* mutant ↔ normal mouse embryos were utilized to study the site of *pcd* gene action. The autosomal recessive *pcd* gene causes degeneration of Purkinje cells. Histochemical staining of β-glucuronidase, which is a cell marker for mutant *pcd* cells of the cerebellar cortex, revealed only unstained Purkinje cells in the chimera. The surviving Purkinje cells were β-glucuronidase negative (wild-type cells), supporting the hypothesis that there is a direct intrinsic effect of *pcd* on Purkinje cells (Mullen, 1977).

Chimeras made between retinal dystrophic (*rdy*) ↔ normal rats were used as a model for studying the target cells for the *rdy* gene as well as its gene action. The homologous gene causes retinitis pigmentosa and blindness in humans. The chimeric rats had mosaic eyes with a neural retina that contained both normal and degenerated photoreceptors. Degenerated photoreceptors were present only opposite nonpigmented, mutant pigment epithelium. This indicated that the *rdy* gene acts in the pigmented epithelial cells rather than in the photoreceptor cells (Mullen and La Vail, 1976; Sanyal *et al.*, 1988).

A similar approach has been utilized to study the target tissue for the gene causing audiogenic seizures in mice. Chimeras made between DBA/L (susceptible strain) and C57BL/6 (resistant strain) demonstrated a spectrum of susceptibility phenotypes in balanced chimeras, whereas the seizure phenotype was the same as that of the strain most represented in the coat color in unbalanced chimeras. This suggested a relatively small target tissue for the gene influencing the development of audiogenic seizure susceptibility (Dewey and Maxson, 1982).

Muscular dysgenesis in mice is caused by the lethal recessive mutation of the *mdg* gene. Pathogenesis of the mutation involves abnormal neuromuscular system development resulting in death due to gross failure of skeletal muscle development. $mdg/mdg \leftrightarrow +/+$ mouse chimeras provide evidence of an abnormal pattern of innervation resulting in dysgenic diaphragms, skeletal muscle, and extramuscular components (Rieger *et al.*, 1984; Peterson *et al.*, 1979). These findings suggest the contribution of positive regulatory inductive effects exerted by the normal nervous system on the development of the muscular system.

Friend virus-induced leukemia in mice (Friend, 1957) is used as a model system to study viral gene action and the pathogenesis of leukemia. A number of strains of mice sensitive and resistant to this virus have been identified. The resistant and sensitive alleles of the *Fv-2* host gene have been shown to modulate the responsiveness of the animal to infection (Steeves *et al.*, 1975). In chimeric mice consisting of C57BL/6 (resistant strain) and DBA (susceptible strain), the chimeras having 15% or less of the susceptible strain of hematopoietic cells are resistant to the virus infection. The few susceptible marrow cells are apparently protected from the virus by the presence of resistant marrow cells. The protective effect of $Fv-2^{rr}$ seems to operate through a diffusible factor (Yoosook *et al.*, 1980). However, if the chimeras were pretreated with anti-thymocyte antisera before infection, there was a shift in the mosaic RBC population toward the susceptible strain. Thus, the protective effect of the resistant strain is mediated through the immune system. This enabled the selection of the *Fv-2* gene *in vivo* (Steeves *et al.*, 1975; Dewey and Eldridge, 1982; Behringer and Dewey, 1985, 1989; Behringer *et al.*, 1978).

IV. Conclusion

Chimeras having distinguishable cell lineages are useful in the study of cell fate, cell–cell interaction in embryonic development, organ maintenance, and pathogenesis of diseases. Action of specific developmental mutations has been examined by rescuing mutations in chimeras. The resulting animals can establish whether the genetic effects of these mutations are cell dependent or cell autonomous. Mosaic patterns are both conserved and regulated in visceral organs so that mosaic pattern analysis has provided information useful in the generation of hypotheses of cell division patterns during organ formation. Mosaic pattern analysis in chimeras provides valuable cues to tissue organization. The development of new molecular markers useful in prospective cell lineage analysis will provide new approaches to understanding cell commitment and fate in mammalian development.

Acknowledgment

Supported in part by U.S. Public Health Service Grants HD 28992 and ES03498 from the U.S. Department of Health and Human Services and by the Markey Charitable Trust.

References

Ahmad, I., Ng, Y. K., and Iannaccone, P. M. (1992). "Chimeric Drift in Hematogenous Cells in Chimeric Rats Produced between Congenic Strains." In preparation.

Aizawa, S., Suda, Y., and Ikawa, Y. (1986). Chimeric potency of a mutator strain of mouse teratocarcinoma cells. *Jpn. J. Cancer Res.* **77**, 327–329.

Arnold, A., and Kim, H. G. (1989). Clonal loss of one chromosome 11 in a parathyroid adenoma. *J. Clin. Endocrinol. Metab.* **69**, 496–499.

Arnold, A., Staunton, C. E., Kim, H. G., Gaz, R. D., and Kronenberg, H. M. (1988). Monoclonality and abnormal parathyroid hormone genes in parathyroid adenomas. *N. Engl. J. Med.* **318**, 658–662.

Austin, C. R., and Short, R. V. (1972). Reproduction in mammals. *In* "Book 2. Embryonic and Fetal Development" Cambridge Univ. Press, London.

Barnsley, M. F., Ervin, V., Hardin, D., and Lancaster, J. (1986). Solution of an inverse problem for fractals and other sets. *Proc. Natl. Acad. Sci. U.S.A.* **83**, 1975–1977.

Beddington, R. S. P., and Robertson, E. J. (1989). An assessment of the developmental potential of embryonic stem cells in the midgestation mouse embryo. *Development (Cambridge, UK)* **105**, 733–737.

Beddington, R. S. P., Morgernstern, J., Land, H., and Hogan, A. (1989). An *in situ* transgenic enzyme marker for the midgestation mouse embryo and the visualization of inner cell mass clones during early organogenesis. *Development (Cambridge, UK)* **106**, 37–46.

Behringer, R. R., and Dewey, M. J. (1985). Cellular site and mode of *FV-2* gene action. *Cell (Cambridge, Mass.)* **40**, 441–447.

Behringer, R. R., and Dewey, M. J. (1989). Cellular site and mode of *FV-2* gene action. II. Conditional protection of *FV-2ss* cells by admixture with *FV-2rr* cells. *Exp. Hematol.* **17**, 330–334.

Behringer, R. R., Eldridge, P. W., and Dewey, M. J. (1984). Stable genotype composition of blood cells in allophenic mice derived from congenic C57BL/6 strains. *Dev. Biol.* **101**, 251–256.

Behringer, R. R., LoCascio, N. J., and Dewey, M. J. (1987). Erythroid cell fusion in the early phase of Friend virus leukemogenesis. *J. Natl. Cancer Inst.* **79**, 601–603.

6. Experimental Chimeras 265

Berger, C. N., and Epstein, C. J. (1989). Delayed thymocyte maturation in the trisomy 16 mouse fetus. *J. Immunol.* **143**, 389–396.

Bernstein, A., Dick, J. E., Huszar, D., Robson, I., Rossant, J., Magli, C., Estrov, Z., Freedman, M., and Phillips, R. A. (1986). Genetic engineering of mouse and human stem cells. *Cold Spring Harb. Symp. Quant. Biol.* **51**, 1083–1091.

Bodenstein, L. (1986). A dynamic simulation model of tissue growth and cell patterning. *Cell Differ.* **19**, 19–33.

Bodenstein, L., and Sidman, R. L. (1987a). Growth and development of the mouse retinal epithelium I. Cell and tissue morphometrics and topography of mitotic activity. *Dev. Biol.* **121**, 192–204.

Bodenstein, L., and Sidman, R. L. (1987b). Growth and development of the mouse retinal pigment epithelium II. Cell patterning in experimental chimaeras and mosaics. *Dev. Biol.* **121**, 205–219.

Boggs, S. S., Patrene, K. D., Downer, W. R., Schwartz, G. N., and Saxe, D. F. (1983). Trisomy of chromosome 6.15 is not necessary for proliferation of AKR (Rb6.15)1Ald lymphoma cells. *Cancer Genet. Cytogenet.* **16**, 151–166.

Bradbury, M. W. (1987). Testes of XX ↔ XY chimeric mice develop from fetal ovotestes. *Dev. Genet.* **8**, 207–218.

Bradley, A., Evans, M., Kaufman, M. H., and Robertson, E. (1984). Formation of germ-line chimaeras from embryo-derived teratocarcinoma cell lines. *Nature (London)* **309**, 255–256.

Buehr, M., and McLaren, A. (1984). Interlitter variation in progeny of chimaeric male mice. *J. Reprod. Fertil.* **72**, 213–221.

Burgoyne, P. S., Buehr, M., Koopman, P., Rossant, J., and McLaren, A. (1988). Cell autonomous action of the testis-determining gene: Sertoli cells are exclusively XY in XX ↔ XY chimaeric mouse testes. *Development (Cambridge, UK)* **102**, 443–450.

Burton, D. I., Ansell, J. D., Gray, R. A., and Micklem, H. S. (1982). A stem cell for stem cells in murine haematopoesis. *Nature (London)* **298**, 562–563.

Chan, W. C., and Winton, E. F. (1989). The natural immune system and hematocytopenias. *In* "Functions of the Natural Immune System" (C. W. Reynolds and R. H. Wiltront, eds.), pp. 381–410. Plenum, New York.

Charley, M. R., Bangert, J. L., Hamilton, B. L., Gilliam, J. N., and Sontheimer, R. D. (1983). Murine graft-versus-host skin disease: A chronologic and quantitative analysis of two histologic patterns. *J. Invest. Dermatol.* **81**, 412–417.

Clancy, J., Jr. (1989). Involvement of natural effector cells in graft versus host disease. *In* "Functions of the Natural Immune System" (C. W. Reynolds and R. H. Wiltront, eds.), pp. 361–379. Plenum, New York.

Condamine, H., Custer, R. P., and Mintz, B. (1971). Pure-strain and genetically mosaic liver tumors histochemically identified with the β-glucuronidase marker in allophenic mice. *Proc. Natl. Acad. Sci. U.S.A.* **68**, 2032–2036.

Conway, K. M., and Hunt, R. K. (1987). Whole eye reconstruction from embryonic half anlagen: Alternations in donor-derived territories in *Xenopus* pigment chimeras *J. Exp. Zool.* **244**, 231–241.

Copp, A. J., Roberts, H. M., and Polani, P. E. (1986). Chimaerism of primordial germ cells in the early postimplantation mouse embryo following microsurgical grafting of posterior primitive streak cells *in vitro*. *J. Embryol. Exp. Morphol.* **95**, 95–115.

Cox, D. R., Smith, S. A., Epstein, L. B., and Epstein, C. J. (1984). Mouse trisomy 16 as an animal model of human trisomy 21 (Down syndrome): Production of viable trisomy 16 ↔ diploid mouse chimeras. *Dev. Biol.* **101**, 416–424.

Deamant, F. D., and Iannaccone, P. M. (1987). Clonal origin of chemically induced papillomas: Separate analysis of epidermal and dermal components. *J. Cell Sci.* **88**, 305–312.

Deamant, F. D., Vijh, M., Rossant, J., and Iannaccone, P. M. (1986). *In situ* identification of host derived infiltrating cells in chemically induced fibrosarcomas of interspecific chimeric mice. *Int. J. Cancer* **37**, 283–286.

Deeg, H. J., Severns, E., Raff, R. F., Sale, G. E., and Storb, R. (1987). Specific tolerance and immunocompetence in haploidentical, but not in completely allogeneic, canine chimeras treated with methotrexate and cyclosporine. *Transplantation* **44**, 621–632.

DeFeo, V. J. (1962). Temporal aspect of uterine sensitivity in the pseudo-pregnant or pregnant rat. *Endocrinology (Baltimore)* **72**, 305–316.

DeFeo, V. J. (1966). Vaginal-cervical vibration: A simple and effective method for the induction of pseudopregnancy in the rat. *Endocrinology (Baltimore)* **79**, 440–442.

Dewey, M. J., and Eldridge, P. W. (1982). Friend viral pathogenesis in C576BL/6 reversible DBA/2 allophenic mice. *Exp. Hematol.* **10**, 723–731.

Dewey, M. J., and Maxson, S. C. (1982). Audiogenic seizure susceptibility of C57BL/6 in equilibrium DBA/2 allophenic mice. *Brain Res.* **246**, 154–156.

DiBerardino, M. A. (1988). Genomic multipotentiality of differentiated somatic cells. In "Regulatory Mechanisms in Developmental Processes" (G. Eguchi, T. S. Okada, L. Saxen, eds.), pp. 129–136, Elsevier, Ireland.

Duboule, D., Croce, C. M., and Illmensee, K. (1982). Tissue preference and differentiation of malignant rat × mouse hybrid cells in chimaeric mouse fetuses. *EMBO J.* **1**, 1585–1603.

Edelman, G. M. (1988). "Topobiology." Basic Books, New York.

Epstein, C. J. (1986). Developmental genetics. *Experientia* **42**, 1117–1128.

Epstein, C. J., and Travis, B. (1979). Preimplantation lethality of monosomy for mouse chromosome 19. *Nature (London)* **280**, 144–145.

Epstein, C. J., Smith, S., and Cox, D. R. (1984). Production and properties of mouse trisomy 15 ↔ diploid chimeras. *Dev. Genet.* **4**, 159–165.

Evans, M. J., and Kaufman, M. M. (1981). Establishment in culture of pluripotent cells from mouse embryos. *Nature (London)* **292**, 154–156.

Evans, M. J., Bradley, A., Kuehn, M. R., and Robertson, E. J. (1985). The ability of EK cells to form chimeras after selection of clones in G418 and some observations on the integration of retroviral vector proviral DNA into EK cells. *Cold Spring Harbor Symp. Quant. Biol.* **50**, 685–689.

Ezine, S., Weissman, I. L., and Rouse, R. V. (1984). Bone marrow cells give rise to distinct cell clones within the thymus. *Nature (London)* **309**, 629–631.

Farmer, J. D., Ott, E., and Yorke, J. A. (1983). The dimension of chaotic attractors. *Physica Part D* **7**, 153–180.

Fathman, C. G., Small, M., Herzenberg, L. A., and Weissman, I. L. (1975). Thymus cell maturation. II. Differentiation of three mature subclasses *in vivo*. *Cell. Immunol.* **15**, 109–128.

Feder, J. (1988). "Fractals." Plenum, New York.

Ford, C. E., Evans, E. P., and Gardner, R. L. (1975). Marker chromosome analysis of two mouse chimaeras. *J. Embryol. Exp. Morphol.* **33**, 447–457.

Frair, P. M., and Peterson, A. C. (1983). The nuclear-cytoplasmic relationship in "mosaic" skeletal muscle fibers from mouse chimaeras. *Exp. Cell Res.* **145**, 167–178.

Francescutti, L. H., Gambel, P., and Wegmann, T. G. (1985). Characterization of hemopoietic stem cell chimerism in antibody-facilitated bone marrow chimeras. *Transplantation* **40**, 7–11.

Friend, C. (1957). Cell free transmission in adult Swiss mice of a disease having the character of a leukemia. *J. Exp. Med.* **105**, 307–318.

Fundele, R., Norris, M. L., Barton, S. C., Reik, W., and Surani, M. A. (1989). Systemic elimination of parthenogenetic cells in mouse chimeras. *Development (Cambridge, UK)* **106**, 29–35.

Furukawa, F., Ikehara, S., Good, R. A., Nakamura, I., Tanaka, H., Imamura, S., and Hamashina, Y. (1988). Immunological status of nude mice engrafted with allogenic and syngeneic thymus. *Thymus* **12**, 11–26.

6. Experimental Chimeras

Gardner, R. L. (1984). An *in situ* marker for clonal analysis of development of the extraembryonic endoderm in the mouse. *J. Embryol. Exp. Morphol.* **80**, 251-288.

Gearhart, J. D., Davisson, M. T., and Oster-Granite, M. L. (1986a). Autosomal aneuploidy in mice: Generation and developmental consequences. *Brain Res. Bull.* **16**, 789-801.

Gearhart, J. D., Singer, H. S., Moran, T. H., Tiemeyer, M., Oster-Granite, M. L., and Coyle, J. T. (1986b). Mouse chimeras composed of trisomy 16 and normal (2N) cells: Preliminary studies. *Brain Res. Bull.* **16**, 815-824.

Gearhart, J. D., Oster-Granite, M. L., Reeves, R. H., and Coyle, J. T. (1987). Developmental consequences of autosomal aneuploidy in mammals. *Dev. Genet.* **8**, 249-265.

Goodenow, M., Kessler, K., Leinwand, L., and Lilly, F. (1986). Absence of trisomy 15 in chemically induced murine T-cell lymphomas. *Cancer Genet. Cytogenet.* **19**, 205-211.

Gorcyca, D. E., McIvor-Nayer, J. L., Maurer, P. H., and Warner, C. M. (1982). The immune response of allophenic mice to the synthetic polymer Glϕ. *J. Immunogenet.* **9**, 83-92.

Gossler, A., Joyner, A. L., Rossant, J., and Skarnes, W. C. (1989). Mouse embryonic stem cells and reporter constructs to detect developmentally regulated genes. *Science* **244**, 463-465.

Greig-Smith, P. (1952). The use of random and contiguous quadrants in the study of the structure of plant communities. *Ann. Bot.* **16**, 293-316.

Grim, H., Christ, B., Jacob, H. J., Kulich, J., and Parizek, J. (1984). Isoelectrically focused carboxyesterase as a biological marker in chimeras. *Experientia* **40**, 1142-1144.

Gropp, A., Winking, H., Herbst, F. W., and Claussen, C. P. (1983). Murine trisomy. Developmental profiles of the embryo, and isolation of trisomic cellular systems. *J. Exp. Zool.* **228**, 253-269.

Groves, E. S., and Singer, A. (1983). Role of the H-2 complex in the induction of T-cell tolerance to self minor histocompatibility antigens. *J. Exp. Med.* **158**, 1483-1497.

Grunbaum, B., and Shephard, G. C. (1987). "Tilings and Patterns." Freeman, New York.

Gurdon, J. B. (1974). "The Control of Gene Expression in Animal Development." Harvard Univ. Press, Cambridge, Massachusetts.

Hansson, H. P. (1967). Histochemical demonstration of carbonic anhydrase activity. *Histochemie* **11**, 112-128.

Harrison, D. E., Lerner, C., Hoppe, P. C., Carlson, G. A., and Alling, D. (1987). Large number of primitive stem cells are active simultaneously in aggregated embryo chimeric mice. *Blood* **69**, 773-777.

Hayashi, M., Nakajima, Y., and Fishman, W. H. (1964). The cytologic demonstration of β-glucuronidase employing naphthol AS-BI glucuronide and hexazonium pararosanilin; a preliminary report. *J. Histochem. Cytochem.* **12**, 293-297.

Herbst, E. W., Gropp, A., and Pluznik, D. H. (1985). Susceptibility of mice reconstituted with trisomy 19 hematopoietic cells to infection with Rauscher leukemia virus. *Cancer Res. Clin. Oncol.* **108**, 107-114.

Herrup, K., and Mullen, R. J. (1979). Intercellular transfer of β-glucuronidase in chimeric mice. *J. Cell Sci.* **40**, 21-31.

Herrup, K., and Sunter, K. (1986). Cell lineage dependent and independent control of Purkinje cell number in the mammalian CNS: Further quantitative studies of *lurcher* chimeric mice. *Dev. Biol.* **117**, 417-427.

Herrup, K., and Sunter, K. (1987). Numerical matching during cerebellar development: Quantitative analysis of granule cell death in *staggerer* mouse chimeras. *J. Neurosci.* **7**, 829-836.

Hewitt, C. W., Black, K. S., Dowdy, S. F., Gonzalez, G. A., Achaner, B. M., Martin, D. C., Furnas, D. W., and Howard, E. B. (1986). Composite tissue (limb) allografts in rats III. Development of donor-host lymphoid chimeras in long term survivors. *Transplantation* **41**, 39-43.

Hillman, N., Sherman, M. I., and Graham, C. (1972). The effect of spatial arrangement on cell determination during mouse development. *J. Embryol. Exp. Morphol.* **28**, 263-278.

Howell, S., Wareham, K. A., and Williams, E. D. (1985). Clonal origin of mouse liver cell tumors. *Am. J. Pathol.* **121**, 426–432.

Hutchison, H. T. (1973). A model for estimating the extent of variegation in mosaic tissues. *J. Theor. Biol.* **38**, 61–79.

Iannaccone, P. M. (1980). Clone and patch size in chimeric mouse skin. *Math. Biosci.* **51**, 117–123.

Iannaccone, P. M. (1987). The study of mammalian organogenesis by mosaic pattern analysis. *Cell Differ.* **21**, 79–91.

Iannaccone, P. M. (1990). Fractal geometry in mosaic organs: A new interpretation of mosaic pattern. *FASEB J.* **4**, 1508–1512.

Iannaccone, P. M., and Weinberg, W. C. (1987). The histogenesis of the rat adrenal cortex: A study based on histologic analysis of mosaic pattern in chimeras. *J. Exp. Zool.* **243**, 217–223.

Iannaccone, P. M., Gardner, R. L., and Harris, H. (1978). The cellular origins of chemically induced tumors. *J. Cell Sci.* **29**, 249–269.

Iannaccone, P. M., Evans, E. P., and Burtenshaw, M. D. (1985). Chromosomal sex and distribution of functional germ cells in a series of chimeric mice. *Exp. Cell Res.* **156**, 471–477.

Iannaccone, P. M., Howard, J. C., Weinberg, W. C., Berkwits, L., and Deamant, F. D. (1987a). Models of organogenesis based on mosaic pattern analysis in the rats. *In* "Banbury Report: Developmental Toxicology: Mechanism and Risk" (J. M. McLaren, R. M. Pratt, and C. L. Markert, eds.), Vol. 26, pp. 73–92. Cold Spring Harbor Laboratory, Cold Spring Harbor, New York.

Iannaccone, P. M., Weinberg, W. C., and Berkwits, L. (1987b). A probabilistic model of mosaicism based on the histological analysis of chimaeric rat liver. *Development (Cambridge, UK)* **99**, 187–196.

Iannaccone, P. M., Weinberg, W. C., and Deamant, F. D. (1987c). On the clonal origin of tumors: A review of experimental models. *Int. J. Cancer* **39**, 778–784.

Iannaccone, P. M., Howard, J. C., and Berkwits, L. (1988). Mosaic pattern and lineage analysis in chimeras. *In* "Regulatory Mechanisms in Developmental Process" (G. Equchi, T. S. Okada, and L. Saxen, eds.), pp. 77–90, Elsevier, Amsterdam.

Iannaccone, P. M., Berkwits, L., Joglar, J., Lindsay, J., and Lunde, A. (1989). Probabilistic division systems modeling the generation of mosaic fields. *J. Theor. Biol.* **141**, 363–377.

Iannaccone, P. M., Lindsay, J., Berkwits, L., Lescinsky, G., DeFanti, T., and Lunde, A. (1992). Fractal geometry in virtual organ growth. *In* "Biomedical Modeling and Simulation" (J. Eisenfeld, M. Witten, and D. S. Levine, eds.). Elsevier, Amsterdam, in press.

Ildstad, S. T., Wren, S. M., Bluestone, J. A., Barbieri, S. A., and Sachs, D. H. (1985). Characterization of mixed allogeneic chimeras. Immunocompetence, *in vitro* reactivity, and genetic specificity of tolerance. *J. Exp. Med.* **162**, 231–244.

Illmensee, K., and Croce, C. M. (1979). Xenogeneic gene expression in chimeric mice derived from rat-mouse hybrid cells. *Proc. Natl. Acad. Sci. U.S.A.* **76**, 879–883.

Kaufman, M. H., Evans, M. J., Robertson, E. J., and Bradley, A. (1984). Influence of injected pluripotential (EK) cells on haploid and diploid parthenogenetic development. *J. Embryol. Exp. Morphol.* **80**, 75–86.

Kinutani, M., and LeDouarin, N. M. (1985). Avian spinal cord chimeras. I. Hatching ability and posthatching survival in homo- and heterospecific chimeras. *Dev. Biol.* **111**, 243–255.

Koopman, P., Gubbay, J., Vivian, N., Goodfellow, P., and Lovell-Badge, R. (1991). Male development of chromosomally female mice transgenic for *Sry*. *Nature (London)* **351**, 117–121.

Kuehn, M. R., Bradley, A., Robertson, E. J., and Evans, M. J. (1987). A potential animal model for Lesch-Nyhan syndrome through introduction of HPRT mutations into mice. *Nature (London)* **326**, 295–298.

6. Experimental Chimeras 269

LeDouarin, N. M. (1980). The ontogeny of the neural crest in avian embryo chimeras. *Nature (London)* **286,** 663-669.
LeDouarin, N. M. (1982). "The Neural Crest." Cambridge Univ. Press, London.
LeDouarin, N. M., and McLaren, A. (eds.) (1984). "Chimeras in Developmental Biology." Academic Press, New York.
LeLievre, C. S., and LeDouarin, N. M. (1975). Mesenchymal derivatives of the neural crest: Analysis of chimaeric quail and chick embryos. *J. Embryol. Exp. Morphol.* **34,** 125-154.
Lemischka, I. R., Raulet, D. H., and Mulligan, R. C. (1986). Developmental potential and dynamic behavior of hematopoietic stem cells. *Cell (Cambridge, Mass.)* **45,** 917-927.
McCarthy, S. A., Griffith, I. J., Gambal, P., Francescutti, L. H., and Wegmann, T. G. (1985). Characterization of host lymphoid cells in antibody-facilitated bone marrow chimeras. *Transplantation* **40,** 12-17.
McGrath, J., and Solter, D. (1984). Completion of mouse embryogenesis requires both the maternal and paternal genomes. *Cell (Cambridge, Mass.)* **37,** 178-183.
McLaren, A. (1976). "Mammalian Chimaeras." Cambridge Univ. Press, Cambridge.
McMahon, A. P., and Bradley, A. (1990). The *Wnt*-1 (*int*-1) proto-oncogene is required for development of a large region of the mouse brain. *Cell (Cambridge, Mass.)* **62,** 1073-1085.
Maeno, M., Tochinai, S., and Katagiri, C. H. (1985). Differential participation of ventral and dorsolateral mesoderms in the hemopoiesis of *Xenopus*, as revealed in diploid-troploid or interspecific chimeras. *Dev. Biol.* **110,** 503-508.
Magnuson, T., Smith, S., and Epstein, C. J. (1982). The development of monosomy 19 mouse embryos. *J. Embryol. Exp. Morphol.* **69,** 223-236.
Mandelbrot, B. B. (1977). "The Fractal Geometry of Nature." Freeman, San Francisco, California.
Mardon, G., and Page, D. C. (1988). The sex-determining region of the mouse Y chromosome encodes a protein with a highly acidic domain and 13 zinc fingers. *Cell (Cambridge, Mass.)* **56,** 765-770.
Martin, G. M., Ogburn, C. E., An, K., and Disteche, C. M. (1984). Altered differentiation, indefinite growth potential, diminished tumorigenicity, and suppressed chimerization potential of hybrids between mouse teratocarcinoma cells and thymocytes. *J. Exp. Pathol.* **1,** 103-133.
Matsumoto, Y., and Fujiwara, M. (1987). Absence of donor-type major histocompatibility complex class I antigen-bearing microglia in the rat central nervous system of radiation bone marrow chimeras. *J. Neuroimmunol.* **17,** 71-82.
Mikoshiba, K., Yokoyama, M., Takamatsu, K., Tsukada, Y., and Nomura, T. (1984). Cell lineage analysis of Schwann cells in the PNS determined by *shiverer*-normal mouse chimeras. *Dev. Biol.* **105,** 221-226.
Milet, R. G., Mukherjee, B. B., and Whitten, W. K. (1972). Cellular distribution and possible mechanism of sex-differentiation in XX-XY chimeric mice. *Can. J. Gent. Cytol.* **4,** 933-941.
Mintz, B., and Baker, W. M. (1967). Normal mammalian muscle differentiation and gene control of isocitrate dehydrogenase synthesis. *Proc. Natl. Acad. Sci. U.S.A.* **58,** 592-598.
Mintz, B., and Cronmiller, C. (1978). Normal blood cells of anemic genotype in teratocarcinoma-derived mosaic mice. *Proc. Natl. Acad. Sci. U.S.A.* **75,** 6247-6251.
Moller, G. (1987). Antigenic requirements for activation of MHC restricted response. *Immunol. Rev.* **98,** 187.
Morgan, T. H. (1919). "The Physical Basis of Heredity." Lippincott, Philadelphia, Pennsylvania.
Mosier, D. E., Zitron, I. M., Mond, J. J., and Paul, W. E. (1978). Requirement for the induction of antibody formation in the newborn mouse. *In* "Developmental Immunobiology" (G. W. Siskind, S. D. Litwin, and M. E. Weksler, eds.), pp. 25-32. Grune & Stratton, New York.
Muggleton-Harris, A. L., Hardy, K., and Higbee, N. (1987). Rescue of developmental lens abnormalities in chimaeras of noncataractous and congenital cataractous mice. *Development (Cambridge, UK)* **99,** 473-480.

Muhlbacher, C., Kuntz, G. W., Haedenkamp, G. A., and Krietsch, W. K. (1983). Comparison of the two purified allozymes (1B and 1A) of X-linked phosphoglycerate kinase in the mouse. *Biochem. Genet.* **21**, 487–496.

Mullen, R. J. (1977). Site of *pcd* gene action and Purkinje cell mosaicism in cerebella of chimaeric mice. *Nature (London)* **270**, 245–247.

Mullen, R. J., and Herrup, K. (1979). Chimeric analysis of mouse cerebellar mutants. *In* "Neurogenetics: Genetic Approaches to the Nervous System" (X. O. Breakefield, ed.), pp. 173–196. Elsevier, New York.

Mullen, R. J., and La Vail, M. M. (1976). Inherited retinal dystrophy: Primary defect in pigment epithelium determined with experimental rat chimeras. *Science* **192**, 799–801.

Nagy, A., Paldi, A., Dezso, L., Varga, L., and Magyar, A. (1987). Prenatal fate of parthenogenetic cells in mouse aggregation chimaeras. *Development (Cambridge, UK)* **101**, 67–71.

Nagy, A., Gocza, E., Diaz, E. M., Prideaux, V. R., Ivanyi, E., Markkule, M., and Rossant, J. (1990). Embryonic stem cells alone are able to support fetal development in the mouse. *Development (Cambridge, UK)* **110**, 815–821.

Nakamura, I. (1989). Involvement of natural effector cells in bone marrow transplantation and hybrid resistance. *In* "Functions of the Natural Immune System" (C. W. Reynolds and R. H. Wiltront, eds.), pp. 321–339. Plenum, New York.

Nakayama, H., Kuroda, H., Fujita, J., and Kitamura, Y. (1988a). Fast and easy detection of mouse sex chimeras using electrophoretic polymorphism of phosphoglycerate kinase-1, and X chromosome-linked enzyme. *Biol. Reprod.* **39**, 923–927.

Nakayama, H., Kuroda, H., Ru, X. M., Fujita, J., and Kitamura, Y. (1988b). Studies of Sl/Sl^d ↔ +/+ mouse aggregation chimeras. I. Different distribution patterns between melanocytes and mast cells in the skin. *Development (Cambridge, UK)* **102**, 107–116.

Nesbitt, M. N. (1974). Chimeras vs. X inactivation mosaics: Significance of differences in pigment distribution. *Dev. Biol.* **38**, 202–207.

Ng, Y. K., and Iannaccone, P. M. (1992). Fractal geometry of mosaic pattern demonstrates liver regeneration is a self-similar process. *Dev. Biol.*, **151**, 419–430.

Ng, Y. K., Ohaki, Y., Deamant, F., and Iannaccone, P. M. (1990). Comparison of patch size in X-chromosome linked mosaic and dizygotic chimeric animals. *Cell Differ. Dev.* **30**, 27–34.

Nielsen, J. T., and Chapman, V. M. (1977). Electrophoretic variation for sex-linked phosphoglycerate kinase (PGK-1) in the mouse. *Genetics* **87**, 319–325.

Nowell, P. C. (1986). Mechanisms of tumor progression. *Cancer Res.* **46**, 2203–2207.

Oster-Granite, M. L. (1986). The neurobiologic consequences of autosomal trisomy in mice and men. *Brain Res. Bull.* **61**, 767–771.

Oster-Granite, M. L., and Gearhart, J. (1981). Cell lineage analysis of cerebellar Purkinje cells in mouse chimeras. *Dev. Biol.* **85**, 199–208.

Oster-Granite, M. L., and Hatzidi-Mitrion, G. (1985). Development of the hippocampal formation in trisomy 16 mice. *Pediatr. Res.* **19**, 328A.

Page, D. C., Mosher, R., Simpson, E. M., Fisher, E. M. C., Mardon, G., Pollack, J., McGillvray, B., de la Chapelle, A., and Brown, L. (1987). The sex-determining region of the human Y chromosome encodes a finger protein. *Cell (Cambridge, Mass.)* **51**, 1091–1104.

Papaioannou, V. E., Gardner, R. L., McBurney, M. W., Babinet, C., and Evans, M. J. (1978). Participation of cultured teratocarcinoma cells in mouse embryogenesis. *J. Embryol. Exp. Morphol.* **44**, 93–104.

Patt, H. M., and Maloney, M. A. (1978). The bg^J/bg^J:W/W^v bone marrow chimera. *Blood Cells* **4**, 27–35.

Penn, A., Garte, S. J., Warren, L., Nesta, D., and Mindich, B. (1986). Transforming gene in human atherosclerotic plaque DNA. *Proc. Natl. Acad. Sci. U.S.A.* **83**, 7951–7955.

Peterson, A. C., Frair, P., Rayburn, H., and Cross, D. (1979). Development and disease in the

6. Experimental Chimeras

neuromuscular system of muscular dystrophic ↔ normal mouse chimaeras. *Soc. Neurosci. Symp.* **4,** 248–273.

Petters, R. M., and Markert, C. L. (1979). Pigmentation pattern of a black-and-tan (a^t) ↔ black chimera. *J. Hered.* **70,** 65–67.

Pierce, G. B., Lewellyn, A. L., and Parchment, R. E. (1989). Mechanism of programmed cell death in the blastocyst. *Proc. Natl. Acad. Sci. U.S.A.* **86,** 3654–3658.

Ponder, B. A., and Wilkinson, M. M. (1986). Direct examination of the clonality of carcinogen-induced colonic epithelial dysplasia in chimeric mice. *J. Natl. Cancer Inst.* **77,** 967–976.

Ponder, B. A. J., Wilkinson, M. M., and Wood, M. (1983). H-2 antigens as markers of cellular genotype in chimeric mice. *J. Embryol. Exp. Morph.* **76,** 83–93.

Ponder, B. A. J., Schmidt, G. H., and Wilkinson, M. M., Wood, M. J., Monk, M., and Reid, A. (1985). Derivation of mouse intestinal crypts from single progenitor cells. *Nature (London)* **313,** 689–691.

Ponder, B. A. J., Schmidt, G. H., and Wilkinson, M. M. (1986). Immunohistochemistry in the analysis of mouse aggregation chimaeras. *Histochem. J.* **18,** 217–227.

Prather, R. S., Hagemann, L. J., and First, N. L. (1989). Preimplantation mammalian aggregation and injection chimeras. *Gamete Res.* **22,** 233–247.

Rabes, H. M., Bucher, T., Hartmann, A., Linke, I., and Dunnwald, M. (1982). Clonal growth of carcinogen-induced enzymes-deficient preneoplastic cell populations in mouse. *Liver Cancer Res.* **42,** 3220–3227.

Reddy, A. L., and Fialkow, P. J. (1983). Papillomas induced by initiation-promotion differ from those induced by carcinogen alone. *Nature (London)* **304,** 69–71.

Reeves, R. H., Gearhart, J. D., and Littlefield, J. W. (1986). Genetic basis for a mouse model of Down syndrome. *Brain Res. Bull.* **16,** 803–814.

Rieger, F., Cross, D., Peterson, A., Pinson-Raymond, M., and Tretjakoff, I. (1984). Disease expression in ±/± ↔ *mdg/mdg* mouse chimeras: Evidence for an extramuscular component in the pathogenesis of both dysgenic abnormal diaphragm innervation and skeletal muscle 16 3 acetylcholinesterase deficiency. *Dev. Biol.* **106,** 296–306.

Robertson, E. J., and Bradley, A. (1986). Production of permanent cell lines from early embryos and their use in studying developmental problems. *In* "Experimental Approaches to Mammalian Embryonic Development" (J. Rossant and R. A. Pedersen, eds.), pp. 475–508. Cambridge Univ. Press, New York.

Roncarolo, M. G., Touraine, J. L., and Banchereau, J. (1986). Cooperation between major histocompatibility complex mismatched mononuclear cells from a human chimera in the production of antigen-specific antibody. *J. Clin. Invest.* **77,** 673–680.

Rossant, J. (1987). Cell lineage analysis in mammalian embryogenesis. *Curr. Top. Dev. Biol.* **23,** 115–146.

Rossant, J., Vijh, M., Siracusa, L. D., and Chapman, V. M. (1983). Identification of embryonic cell lineages in histological sections of *M. musculus* ↔ *M. caroli* chimaeras. *J. Embryol. Exp. Morphol.* **78,** 179–191.

Santamaria, P. (1983). Analysis of haploid mosaics in *Drosophila*. *Dev. Biol.* **96,** 285–295.

Sanyal, S., Hawkins, R. K., and Zeilmaker, G. H. (1988). Development and degeneration of retina in *rds* mutant mice: Analysis of interphotoreceptor matrix staining in chimeric retina. *Curr. Eye Res.* **7,** 1183–1190.

Sariola, H., Peault, B., LeDouarin, N., Buck, C., Dieterlen-Lievre, F., and Saxen, L. (1984). Extracellular matrix and capillary in growth in interspecies chimeric kidneys. *Cell Differ.* **15,** 43–51.

Schmidt, G. H., Garbutt, D. J., Wilkinson, M. M., and Ponder, B. A. J. (1985a). Clonal analysis of intestinal crypt populations in mouse aggregation chimaeras. *J. Embryol. Exp. Morphol.* **85,** 121–130.

Schmidt, G. H., Wilkinson, M. M., and Ponder, B. A. J. (1985b). Detection and characterization of spatial pattern in chimaeric tissue. *J. Embryol. Exp. Morphol.* **88,** 219–230.

Schmidt, G. H., Blount, M. A., and Ponder, B. A. J. (1987). Immunochemical demonstration of the clonal organization of chimaeric mouse epidermis. *Development (Cambridge, UK)* **100**, 535–541.

Schmidt, G. H., Winton, D. J., and Ponder, B. A. J. (1988). Development of the pattern of cell renewal in the crypt-villus unit of chimaeric mouse small intestine. *Development (Cambridge, UK)* **103**, 785–790.

Scott, J., Engelhard, V. H., and Benjamin, D. C. (1987). Bone marrow irradiation chimeras in the *BB* rat: Evidence suggesting two defects leading to diabetes and lymphopoenia. *Diabetologia* **30**, 774–781.

Shortman, K., Wilson, A., Van Ewijk, W., and Scollay, R. (1987). Phenotype and localization of thymocytes expressing the homing receptor-associated antigen MEL-14: Arguments for the view that most mature thymocytes are located in the medulla. *J. Immunol.* **138**, 342–351.

Silva, S., Babonits, M., Wiener, F., and Klein, G. (1988). Further studies on chromosome 15 trisomy in murine T-cell lymphomas: Mapping of the relevant chromosome segment. *Int. J. Cancer* **41**, 738–743.

Simcox, A. A., and Sang, J. H. (1983). When does determination occur in *Drosophila* embryos? *Dev. Biol.* **97**, 212–221.

Siskind, G. W., Goidl, E. A., Kim, Y. T., Sherr, D. H., Sogn, D. D., and Szewczuk, M. R. (1978). A thymus cell influence on the ontogeny of B-lymphocyte function. *In* "Developmental Immunobiology" (G. W. Siskind, D. S. Litwin, and M. E. Weksler, eds.), pp. 201–211 Grune & Stratton, New York.

Steeves, R. A., Blank, K. J., and Lilly, F. (1975). Genetic control of Friend virus-transformed colony-forming cells by the *Fv-2* resistant gene in mice. *Bibl. Haematol.* **43**, 151–153.

Stephens, T. J., and Warner, C. M. (1980). Chimeric drift in allophenic mice. II. Analysis of changes in red blood cell and white blood cell populations in C57BL/10Sn ↔ A mice. *Cell. Immunol.* **56**, 132–141.

Stevens, L. C. (1978). Totipotent cells of parthenogenetic origin in a chimaeric mouse. *Nature (London)* **276**, 266–267.

Suzuki, D. T., Griffiths, A. J. F., Miller, J. H., and Lewontin, R. C. (1986). "An Introduction to Genetic Analysis," 3rd Ed. Freeman, New York.

Tam, P. P. L., and Beddington, R. S. P. (1987). The formation of mesodermal tissues in the mouse embryo during gastrulation and early organogenesis. *Development (Cambridge, UK)* **99**, 109–126.

Tanaka, H., and Landmesser, L. T. (1986). Interspecies selective motorneuron projection patterns in chick-quail chimeras. *J. Neurosci.* **6**, 2880–2888.

Tarkowski, A. K., and Wroblewska, J. (1967). Development of blastomeres of mouse eggs isolated at the 4- and 8-cell stage. *J. Embryol. Exp. Morphol.* **18**, 155–180.

Terashima, T., Inoue, K., Inoue, Y., Yokoyama, M., and Mikoshiba, K. (1986). Observations on the cerebellum of normal ↔ *reeler* mutant chimeras. *J. Comp. Neurol.* **252**, 264–278.

Tran, D., Mensy-Desolle, N., Josso, N. (1977). Anti-mullerian hormone is a functional marker of foetal sertoli cells. *Nature (London)* **303**, 338–340.

Urso, P., and Genozian, N. (1977). Immune competence of splenic lymphocytes following graft-versus-host disease in mouse allogenic radiation chimeras. *J. Immunol.* **118**, 657–661.

Vainio, O., and Toivanen, A. (1983). B cell genotype determines interaction preference with T cells: No effect of maturation environment. *J. Immunol.* **131**, 9–12.

Vainio, S., Lehtonen, E., Jalkanen, M., Bernfield, M., and Saxen, L. (1989). Epithelial-mesenchymal interactions regulate the stage-kidney expression of a cell surface proteoglycan, syndecan, in the developing kidney. *Dev. Biol.* **134**, 382–391.

Van Winkle, L. J., Iannaccone, P. M., Campione, A. L., and Garton, R. L. (1990). Transport of cationic and zwitterionic amino acids in preimplantation rat conceptuses. *Dev. Biol.* **142**, 184–193.

6. Experimental Chimeras

Vogel, W. M., Sunter, K., and Herrup, K. (1989). Numerical matching between granule and Purkinje cells in lurcher chimeric mice: A hypothesis for the trophic rescue of granule cells from target-related cell death. *J. Neurosci.* **9,** 3454–3462.

Volf, D., Sensenbrenner, L. L., Sharkis, S. J., Elfenbein, G. J., and Scher, I. (1978). Induction of partial chimerism in nonirradiated B-lymphocyte-deficient CBA/N mice. *J. Exp. Med.* **147,** 940–945.

Wade, A. C., Luchert, P. H., Tazume, S., Niedbalski, J. L., and Pollard, M. (1987). Characterization of xenogenic mouse-to-rat bone marrow chimeras. I. Examination of hematologic and immunologic function. *Transplantation* **44,** 88–92.

Warner, C. M., McIvor, J. L., and Stephens, T. J. (1977). Chimeric drift in allophenic mice *Transplantation* **24,** 183–193.

Weinberg, W. C., and Iannaccone, P. M. (1988). Clonality of preneoplastic liver lesions: Histological analysis in chimeric rats. *J. Cell Sci.* **89,** 423–431.

Weinberg, W. C., Howard, J. C., and Iannaccone, P. M. (1985). Histological demonstration of mosaicism in a series of chimeric rats produced between congenic strains. *Science* **227,** 524–527.

Weinberg, W. C., Berkwits, L., and Iannaccone, P. M. (1987). The clonal nature of carcinogen-induced altered foci of γ-glutamyl transpeptidase expression in rat liver. *Carcinogenesis* **8,** 565–570.

West, J. D. (1975). A theoretical approach to the relation between patch size and clone size in chimaeric tissue. *J. Theor. Biol.* **50,** 153–160.

West, J. D. (1976a). Distortion of patches of retinal degeneration in chimaeric mice. *J. Embryol. Exp. Morphol.* **36,** 145–149.

West, J. D. (1976b). Patches in the livers of chimaeric mice. *J. Embryol. Exp. Morphol.* **36,** 151–161.

West, J. D. (1978). Analysis of clonal growth using chimaeras and mosaics. *In* "Developments in Mammals" (M. H. Johnson, ed.), Vol. 3, pp. 413–460. North-Holland, Amsterdam.

West, J. D. (1984). Cell markers. *In* "Chimeras in Developmental Biology" (N. LeDouarin and A. McLaren, eds.), pp. 39–67. Academic Press, New York.

Whitten, W. K. (1978). Combinatorial and computer analysis of random mosaics. *In* "Genetic Mosaics and Chimeras in Mammals" (L. B. Russell, ed.), pp. 445–463. Plenum, New York.

Wilkins, A. S. (1986). "Genetic Analysis of Animal Development," pp. 546. Wiley, New York.

Wilson, T. J., Ponder, B. A. J., and Wright, N. A. (1985). Use of mouse chimaeric model to study cell migration patterns in the small intestinal epithelium. *Cell Tissue Kinet.* **18,** 333–344.

Winkel, G. K., and Pedersen, R. A. (1988). Fate of the inner cell mass in mouse embryos as studied by microinjection of lineage tracers. *Dev. Biol.* **127,** 143–156.

Winkel, G. K., Fergusen, J. E., Takeichi, M., and Nuccitelli, R. (1990). Activation of protein kinase C triggers premature compaction in the four cell stage mouse embryo. *Dev. Biol.* **138,** 1–15.

Winton, D. J., Blount, M. A., and Ponder, B. A. (1989). Polyclonal origin of mouse skin papillomas. *Br. J. Cancer* **60,** 59–63.

Wirschubsky, Z., Wiener, F., Spira, J., Sumegi, J., and Klein, G. (1984). Triplication of one chromosome No. 15 with an altered c-*myc* containing *Eco*RI fragment and elimination of the normal homologue in a T-cell lymphoma line of AKR origin (TIKAUT). *Int. J. Cancer* **33,** 477–481.

Yam, P. Y., Petz, L. D., Knowlton, R. G., Wallace, B. R., Stock, A. D., DeLange, G., Brown, V. A., Donis-Keller, H., and Blume, K. G. (1987). Use of DNA restriction fragment length polymorphisms to document marrow engraftment and mixed hematopoietic chimerism following bone marrow transplantation. *Transplantation* **43,** 399–405.

Yamamura, K. I., and Markert, C. L. (1981). The production of chimeric rats and their use in the analysis of the hooded pigmentation pattern. *Dev. Genet.* **2,** 131–146.

Yasumizu, R., Onoe, K., Iwabuchi, K., Ogasawara, M., Fujita, M., Okuyama, H., Good, R. A., and Morikawa, K. (1985). Characteristics of macrophages in irradiation chimeras in mice reconstituted with allogeneic bone marrow cells. *J. Leukocyte Biol.* **38**, 305–315.

Yochim, J. M. (1984). Modulation of uterine sensitivity to decidual induction in the rat by nicotinamide; challenge and extension of a model of progestation differentiation. *Biol. Reprod.* **30**, 637–645.

Yoosook, C., Steeves, R., and Lilly, F. (1980). Fv-2r-mediated resistance of mouse bone-marrow cells to Friend spleen focus-forming virus infection. *Int. J. Cancer* **26**, 101–106.

7
Genetic Analysis of Cell Division in *Drosophila*

Pedro Ripoll, Mar Carmena, and Isabel Molina
Centro de Biología Molecular (CSIC-UAM)
Campus de Cantoblanco
28049 Madrid, Spain

I. Introduction
   A. General
   B. The Genetic Approach
II. Effect of Mitotic Mutations on Embryonic Development
   A. Preblastoderm Embryonic Development
   B. The Maternal Effect
   C. Postblastoderm Embryonic Development
   D. Genes Flies Borrowed from Yeast
III. Mitosis during Postembryonic Development
   A. Imaginal Development
   B. Effect of Mitotic Mutations in Imaginal Tissues
   C. Male Meiosis and Spermatogenesis in Mitotic Mutants
IV. Tubulin and Kinesin Gene Families
V. General Considerations
   References

I. Introduction

A. General

Reviews dealing with diverse aspects of the cell cycle in several quite different organisms (from clams to mammals, passing through slime molds, fungi, and flies) have become familiar in scientific journals during the last few years. Lately, there has been an avalanche of such reviews. The immediate interpretation of such a burst is that an overwhelming body of information about the cell cycle is continuously being released in the form of papers, articles, communications in scientific meetings, and the like. In all, this means only that the general interest in the cell cycle remains alive and healthy.

The cell cycle and its control continues to be one of the central problems of cell biology. Obviously, it is also a key issue in the development of multicellular organisms. In spite of the many years devoted to its study (remembering that the pioneering description of mitosis by Flemming is well over a century old) we are all aware that, even if the advances in the field have been enormous, we are still far from fully understanding the basic mechanisms leading to the equitable segregation of subcellular organelles from a mother cell to its daughter cells.

Although there are obvious differences in the way different cell types divide, the basic features of mitosis are universally conserved in all eukaryotes from unicellular fungi to mammals. All mitoses share two characteristics: complexity and accuracy. The former is but the evolutionary consequence of the need for the latter. Because it was known that many structural components essential for the cell cycle, such as tubulins, had been evolutionarily conserved, the evolutionary conservation of genes controlling the entrance into, or progression through, the cell cycle should not have surprised us. The evolutionary conservation of such genes indicates that by the time unicellular organisms gave rise to multicellular organisms, the need for accurate cell division had already been established.

Cell division functions to partition the genetic material so that daughter cells are genetically identical to mother cells. To this end mechanisms have evolved that pack the DNA of the cell into chromosomes that can condense during prophase (to ease the transport of identical sister chromatids to opposite poles) and decondense during interphase to perform their biochemical functions. The identity of sister chromatids is maintained through the faithful replication of complementary strands of DNA during the S phase. The existence of DNA repair mechanisms ensures their integrity. To avoid errors in segregation of the sister chromatids, they are held together through specific contact regions until the onset of anaphase. Special arrays of microtubules, nucleated by microtubule organizing centers, mediate the separation of sister chromatids in all cells so that the DNA can be equationally partitioned. All chromosomes have specialized structures (centromeres or kinetochores) that play an active role in chromatid migration. Finally, except when a syncytium is formed, sister cells must be individualized after telophase by cytokinesis. Cell division is a highly dynamic process that requires a complex series of enzyme-dependent events to proceed correctly: dyneins, kinases, phosphatases, topoisomerases, kinesin-like motors, and polymerases have all been identified as essential components of the cell cycle.

Researchers have used a wide range of organisms and an extensive range of techniques to approach problems related to the cell cycle. Different model systems have been chosen for their suitability for particular techniques. Some organisms are amenable to mechanical manipulation; others to biochemical analyses or drug treatments. Some are particularly useful for cytology and some for molecular genetics. Finally (and primarily for historical reasons), a few can be subject to classical genetic approaches. There is no such thing as a perfect model system. Models appropiate to one technique are often unapproachable by other technologies. Systems where the genetic dissection is easy, such as the budding yeast, are cytologically poor; systems where cytology or mechanical manipulation is easy, such as unicellular algae, have not been studied genetically in depth;

7. Genetic Analysis of Cell Division in *Drosophila*

systems where cytology, biochemistry, and isolation of conditional mutations can be easily undertaken, such as mammalian cell cultures, are not the appropriate material for complex genetics; systems where biochemistry, genetics, and cytology can be undertaken, such as the fission yeast, are considered by some as inadequate models for cell division in higher eukaryotes. Fortunately, as research progresses and techniques become more sophisticated, we are finding more common features that help to build a general view of cell division and its genetic control that is independent of the specific details of each different model system.

Several reviews have dealt with different aspects of cell division in fruit flies since the first mitotic mutants were found in *Drosophila* (Baker *et al.*, 1976, 1978). In a previous review from our laboratory (Ripoll *et al.*, 1987) we made an effort to include all the literature and information available at that time; a large fraction of the mutants had then only been superficially described and most of the information came from unpublished results (see Gatti *et al.*, 1983a). Since then, the mitotic phenotypes of many mutations have been published (see, e.g., Gatti and Baker, 1989) while a few others have been subject to deeper genetic and cytological analyses (see, e.g., Glover, 1989, and literature cited therein). We still lack molecular data for most mutations because investigators isolating the mutations did not have molecular cloning in mind when inducing them. Now, most laboratories are searching for mitotic mutants using molecularly oriented screens. So far, only a few genes involved in somatic cell divisions and for which mutant alleles are available have been molecularly characterized: *cyclin A* (Lehner and O'Farrell, 1989, 1990a; Whitfield *et al.*, 1989, 1990), *PP1 87B* (Dombrádi *et al.*, 1989; Axton *et al.*, 1990), α4t (see review by Fyrberg and Goldstein, 1990), *string* (Edgar and O'Farrell, 1989), *spaghetti-squash* (Karess *et al.*, 1991), and *polo* (Llamazares *et al.*, 1992). In contrast, we have substantial molecular information about genes involved in cell division (some homologous to cell division genes of unicellular fungi); however, we still lack mutations confirming their mitotic roles in flies: topoisomerase II (Nolan *et al.*, 1986), the 205K MAP located in 100EF (Goldstein *et al.*, 1986), the centrosomal antigen Bx63 (Whitfield *et al.*, 1988), Cyclin B (Whitfield *et al.*, 1989, 1990; Lehner and O'Farrell, 1990a), Cyclin C (Léopold and O'Farrell, 1991), Dm cdc2 (Jiménez *et al.*, 1990; Lehner and O'Farrell, 1990b), Dm cdc2c (Lehner and O'Farrell, 1990b), CDC25Dm2 (Jiménez *et al.*, 1990), PP1 9C, PP1 13C, and PP1 96A (Dombrádi *et al.*, 1989), γ-tubulin (Zheng *et al.*, 1991), and several α- and β-tubulin genes (reviewed by Fyrberg and Goldstein, 1990).

Most of the available descriptions of mutant phenotypes are rather preliminary and still rely on single alleles. Thus we must proceed with caution and await molecular characterization before interpreting the function of the mutated genes. Even so, there are enough mitotic mutants described so far to allow us to group them into well-defined phenotypic classes (Gatti and Baker, 1989). When we do so, with few exceptions new mutations fall into previously established classes. This means that, at least phenotypically, the system might be close to saturation.

It should be noted that, as a whole, the phenotypic classes into which Gatti and Baker grouped their mutations coincide with phenotypes previously described for mitotic mutants in mammalian systems, suggesting that the range of possible mitotic phenotypes in higher eukaryotes is limited. Our intent with the present article is to update the information on *Drosophila* mutations phenotypically involved in the cell cycle and to try to correlate what we now know about mitotic mutations in *Drosophila* with their effect in development. The field is moving so fast that we are sure we will miss some important articles recently published; we apologize for these omissions. Several reviewers have described mutations in genes known to be needed for mitosis in fruit flies and have compiled the pertinent literature (Gatti *et al.*, 1983a; Ripoll *et al.*, 1987; Glover, 1989, 1990, 1991; Glover *et al.*, 1989; Gatti and Baker, 1989; O'Farrell *et al.*, 1989).

B. The Genetic Approach

The need for the genetic dissection of the cell cycle has been emphasized before. There are advantages to the genetic dissection of any genetically controlled process, because (1) it interferes at a specific point, providing a way to understand the function of an individual gene product, so long as pleiotropy is avoided or controlled; (2) it permits the construction of doubly mutant individuals, thus revealing the temporal and hierarchical order of different genes; (3) it allows the identification of genes even if the amount of gene product is too low for conventional biochemical analyses; and (4) it lets the organism tell us what it considers to be important.

Recent developments in molecular genetics, such as the polymerase chain reaction (PCR) or "enhancer trap" techniques, make virtually any organism accessible to molecular genetic manipulation. However, complex genetic studies are still restricted to those organisms with a long history of classical genetic manipulation such as yeast and *Drosophila melanogaster*.

Yeast (*Saccharomyces cerevisiae* and *Schizosaccharomyces pombe*) are the leading organisms in the study of the genetic control of the cell cycle. Although many aspects of the genetic control of cell division have been evolutionarily conserved from unicellular fungi to mammals (Lee and Nurse, 1987; O'Farrell *et al.*, 1989; Nurse, 1990) there are some aspects of the cell cycle that must be peculiar to complex multicellular organisms. For instance, genetic decisions leading to cell differentiation are frequently coupled with cell division and are commonly asymmetric.

Drosophila still remains the most reliable model system for genetic dissection of the cell cycle in multicellular organisms. Not only is there an impressive amount of genetic knowledge about *Drosophila*, but also its development has been described in great detail. Thus it has become the ideal higher organism to marry classical and molecular genetics. On the other hand, even the largest

7. Genetic Analysis of Cell Division in Drosophila

Drosophila diploid cells are too small to make cytogenetics an enjoyable task: despite its reduced number of chromosomes, *Drosophila* would never have been chosen as a system for cytology had there not been the need to complement the existing genetic studies. Nevertheless, (1) the resolving power of cytological techniques has improved dramatically in the last few years, due primarily to the introduction of confocal indirect immunofluorescence microscopy (see, e.g., González *et al.*, 1990; Axton *et al.*, 1990: Whitfield *et al.*, 1990) and (2) immunocytology is becoming increasingly important to understanding the function of genes involved in cell division.

Assuming that mutations in genes essential for the cell cycle were bound to result in cell lethality, investigators looked for conditional mutations that would show a mutant phenotype under restrictive conditions but that would be viable under permissive ones. Drug resistance and, more importantly, temperature sensitivity turned out to be the ideal tools to isolate mutations in cell cycle events, at least in unicellular systems such as yeast and mammalian cell cultures. The seminal work of Hartwell and co-workers opened the way to the genetic dissection of the cell cycle using conditional mutations. It provided the essential tools to order yeast cell cycle genes in a temporal and hierarchical order (for reviews, see Pringle and Hartwell, 1981; Hayles and Nurse, 1986) and it has proved to be, via structural or functional homology, an invaluable source for molecular studies in other organisms.

Unfortunately, conditional lethality cannot be used in *Drosophila* for selective screens of cell division mutants. It is true that conditional mutations (mutagen or temperature sensitive) have served to identify mitotic mutants (Baker *et al.*, 1978; Baker and Smith, 1979; Smith *et al.*, 1985) but their conditional nature is not useful for the understanding of their mitotic roles. This, together with our erroneous conviction that mitotic mutations would lead to early embryonic death, delayed for many years the search for mitotic mutants in fruit flies. One argument favoring this pessimistic view (pointed out by Baker and Hall, 1976) was the near absence of mutations altering chromosome disjunction during the second meiotic division. Because this division was almost identical to a mitotic division, both processes were expected to be under the same genetic control, and the absence of meiotic mutants indicated lethality of mitotic mutations.

II. Effect of Mitotic Mutations on Embryonic Development

The first mutants affecting mitosis in *Drosophila* were found while studying the effect on mitotic recombination of mutations altering the frequency and distribution of exchanges during female meiosis. This analysis revealed that a majority of these genes did not affect somatic crossing over, but they were needed to ensure chromosome integrity during somatic divisions in both sexes, and their gene products were involved in DNA repair pathways (Baker et al., 1976, 1978; Baker and Smith, 1979; Gatti, 1979; Gatti et al., 1980). The existence of these viable mitotic mutants together with the isolation of lethal alleles with effective phases in late larval/early pupal stages (Smith et al., 1985) made obvious that at least some *Drosophila* mutations affecting cell division could be studied.

The presence of abundant wild-type products left in the oocyte by the mother explained the late death of mutants with altered mitoses. Maternally provided products carried the embryo through embryonic divisions and formation of a normal first instar larva. Because most larval tissues grow through polyploidization rather than cell division, the larva does not require many of the genes normally needed for cell division for its survival. Individuals mutant for cell division genes would be expected to die at the late larval/early pupal period, when tissues remaining diploid and mitotically active (those needed for imaginal development) are essential for the completion of metamorphosis. Following this argument a large majority of the mitotic mutants known to date in *Drosophila* have been isolated among late lethals showing abnormal mitotic figures in neuroblasts of the mature larval central nervous system (Ripoll et al., 1985, 1987; Smith et al., 1985; González et al., 1988; Sunkel and Glover, 1988; Karess and Glover, 1989; Axton et al., 1990; Glover, 1989; Gatti and Baker, 1989). As will be discussed below, not all cell division mutants die late in development because some, such as mutants for *string* (Edgar and O'Farrell, 1989) or *cyclin A* (Lehner and O'Farrell, 1989), die earlier, during embryogenesis.

A. Preblastoderm Embryonic Development

The first signs of activity in the mature oocyte normally occur immediately after oviposition when female meiosis, arrested during the first meiotic metaphase, resumes, giving rise to the female pronucleus and the three haploid polar bodies. The completion of meiosis is independent of fertilization because meiosis also takes place in unfertilized eggs. Moreover, because some eggs start development before being laid (a common event in crowded cultures) oviposition does not seem to be a requirement for resumption of female meiosis. What triggers the completion of meiosis in *Drosophila* oocytes is unknown, but it could well be the limiting amount of a homolog of the maturation promoting factor (MPF) of amphibian oocytes (Masui and Markert, 1971) because conserved homologs of the two components of MPF are also found in *Drosophila* (Lehner and O'Farrell, 1989, 1990b; Whitfield et al., 1989; Jiménez et al., 1990). As in amphibian oocytes, there could be a factor in *Drosophila* responsible for the arrest of female meiosis at metaphase I. A possible component of the cytostatic factor (CSF; Masui and Markert, 1971) is a conserved serine/threonine protein kinase encoded by the c-*mos* protooncogene (see Sagata et al., 1989, and literature cited therein). Molecular techniques could address the question of whether or not a c-*mos* homolog exists in fruit flies and, if it does, make it accessible to conventional genetics.

The earliest phenotypic signs of the effect of mutations in development are found when studying eggs laid by females carrying viable heteroallelic combinations of the gene *abnormal spindle* (*asp*). Mutations in *asp* lead to chromosome nondisjunction in the soma and the germ line (Ripoll *et al.*, 1985) thought to result from a specific increase in the stability of the microtubules of the spindle (Casal *et al.*, 1990a; González *et al.*, 1990). A fraction of the eggs laid by mutant *asp* females (ASP embryos) does not have any detectable DNA within them (González *et al.*, 1990). This phenotype is also seen in unfertilized ASP eggs and could be due to premeiotic abnormalities in chromosome segregation. The oocyte is recruited among 16 oogonia derived through 4 mitotic divisions from an original precursor gonial cell (see Mahowald and Kambysellis, 1980). Cytokineses during these oogonial divisions are incomplete, so that all the products of division remain joined by large cytoplasmic bridges. At the end of the fourth division 2 of the 16 cells show 4 intercellular bridges and 1 of these 2 cells will eventually mature into a functional oocyte (Mahowald and Kambysellis, 1980). The existence of ASP eggs devoid of detectable DNA indicates that the determination and maturation of the oocyte does not require the presence of a nucleus. Another early phenotype shown by a fraction of ASP embryos (also seen in both fertilized and unfertilized eggs) consists of the presence of one to four large nuclei with no signs of mitotic activity, probably resulting from segregational mistakes during premeiotic or meiotic divisions, followed by uncoordinated DNA synthesis (González *et al.*, 1990).

Following meiosis the three haploid polar bodies frequently fuse to form a triploid nucleus that remains arrested at metaphase. Under normal conditions no mitotic activity occurs unless fertilization takes place. However, in unfertilized eggs laid by females homozygous for *giant nuclei* (*gnu*; Freeman *et al.*, 1986), *plutonium* (*plu*), or *pan gu* (*png*) (Shamanski and Orr-Weaver, 1991), continuous rounds of DNA synthesis start in both the female pronucleus and the polar bodies; because there are no nuclear divisions this synthesis results in eggs with very large nuclear masses. Because all three mutants show similar phenotypes, in what follows we will restrict our discussion to *gnu*. This phenotype is somewhat similar to the one described above for some ASP unfertilized eggs. As discussed in González *et al.* (1990), it is possible that the mechanisms controlling embryonic DNA synthesis are coupled in some way to the meiotic process and are altered in both ASP and GNU eggs. In any case, the GNU phenotype appears to be specific to this early developmental event while this aspect of the ASP phenotype seems to be an indirect consequence of the involvement of the *asp* gene in chromosome segregation. Something in the cytoplasm of GNU eggs allows the initiation of uncoordinated DNA synthesis in a system where this synthesis is repressed unless fertilization triggers embryonic divisions. The facts that *gnu* homozygous individuals are viable and that the males are fertile seems to indicate that the gene is needed only during this very early step of embryogenesis. Through an elegant combination of molecular tools, classical genetics, and cytology, Freeman and Glover (1987) have demonstrated that in fertilized GNU eggs giant nuclei are also formed by the male pronucleus, independently of whether or not syngamy takes place.

Syngamy is not a prerequisite to start coordinated DNA synthesis followed by nuclear divisions during early steps of embryogenesis. In eggs laid by females homozygous for the mutation *maternal haploid* (*mh*; Zalokar et al., 1975) syngamy does not take place, and only the maternal pronucleus divides, giving rise to a haploid embryo. When these eggs are derived from females homozygous for *mh* and *gnu* both the paternal and maternal pronuclei, as well as the polar bodies, give rise to giant nuclei. It follows that under normal circumstances fertilization, but not syngamy, triggers the onset of nuclear divisions in only one of the nuclei present in the egg cytoplasm: in wild-type eggs DNA synthesis remains repressed in the polar bodies; in *mh* eggs this repression affects both the polar bodies and the male pronucleus. A similar situation might take place in embryos fertilized by *ms(3)K81* (Fuyama, 1984), in which the mutant sperm can provide the signals needed for initiation of embryonic divisions but the male pronucleus does not divide, which again results in a haploid embryo derived solely from the maternal pronucleus. Thus, *gnu* (and, similarly, *plu* and *png*) may play a role in the selective derepression of DNA synthesis in only one of the nuclei present in the egg cytoplasm or, alternatively, may participate in the selective repression of DNA synthesis in all but one of the nuclei.

The signal carried by the sperm that is responsible for the initiation of nuclear divisions remains unclear. It is known that the sperm carries a basal body that can generate the centrioles that for the rest of the mitotic and meiotic divisions will be associated with the centrosomes. Female centrioles degrade during the maturation of the oocyte (Mahowald and Strassheim, 1970). Therefore the paternally inherited basal body could be a candidate for the initiator of nuclear mitotic activity. However, in unfertilized GNU and ASP eggs there are centrosomes with the ability to nucleate asters of microtubules (Freeman and Glover, 1987; González et al., 1990) and no nuclear divisions take place in these eggs even if they are fertilized. Besides, there are some *Drosophila* strains in which parthenogenesis can take place under some conditions, meaning that fertilization as a requirement for nuclear division can be circumvented.

If the signal carried by the sperm is enough to trigger division, GNU fertilized eggs should show signs of cell division in at least one nucleus, but they do not. Possibly, the uncoordinated DNA synthesis caused by *gnu* precedes fertilization and the subsequent sperm-carried signal cannot reverse this uncoordinated phenomenon. Similarly, fertilized ASP eggs that show no detectable DNA also show no signs of cell division. Because these eggs should have a male pronucleus (that, in principle, would be enough to start haploid nuclear divisions) it follows

7. Genetic Analysis of Cell Division in *Drosophila* 283

that the presence of a maternally derived pronucleus (or something associated with it) is essential for the activation of nuclear divisions in fertilized embryos.

After syngamy the wild-type zygote undergoes a series of rapid nuclear divisions, giving rise to a syncytium, until cellularization takes place at interphase of cycle 14. Experiments performed by injecting aphidicolin into syncytial embryos (Raff and Glover, 1988) have shown the existence in *Drosophila* of "oscillators" independent of DNA replication, similar to those described in other embryonic systems (for references see Raff and Glover, 1988). We do not know much about these oscillators except that they are dependent on fertilization, because no sign of oscillation is seen in unfertilized eggs. Therefore, in rapidly dividing systems, some events that seem to be coupled (and at first sight under common genetic control) can be experimentally shown to be independent. GNU embryos show a dramatic example of uncoupling of mitotic events. In these embryos DNA replication proceeds in the absence of karyokinesis, giving rise to massive nuclei. In spite of the lack of nuclear divisions, the centrosome cycle continues to take place and the centrosomes not only divide normally but they also migrate as they would in normal embryos. These centrosomes conserve their ability to nucleate asters of microtubules and, when confronted with chromatin fragments, to form spindle fibers (Freeman *et al.*, 1986). Raff and Glover (1988) saw the same independence of nuclear and centrosomal divisions in aphidicolin-treated wild-type embryos, suggesting that the behavior of centrosomes is independent not only of karyokinesis but also of DNA replication. Some ASP blastoderm embryos show areas where centrosomes, but not nuclei, are present; this again indicates that centrosomes can migrate without the accompanying nuclei (González *et al.*, 1990). As will be shown later, some mutant phenotypes indicate that centrosome duplication and migration are independent of cytokinesis in larval neuroblasts as well (Gatti and Baker, 1989; Karess *et al.*, 1991).

Several viable *Drosophila* mutations alter chromosome behavior during very early embryonic divisions and give rise to mosaic individuals via chromosome loss. Eggs laid by females homozygous for *mitotic loss-inducer* (*mit*; Gelbart, 1973) produce early chromosome loss of both paternally and maternally inherited chromosomes. Eggs fertilized by males homozygous for *paternal-loss* (*pal*; Baker, 1975) produce early chromosome loss of paternally inherited chromosomes. Eggs laid by females homozygous for *nonclaret disjunctional* (*ncd*; Sequeira *et al.*, 1989), the element responsible for the disjunctional phenotype of the doubly mutant *claret nondisjunctional* (*cand*) allele (Lewis and Gencarella, 1952), show early chromosome loss only of maternally inherited chromosomes. Eggs laid by females homozygous for *no distributive disjunction* (*nod*; Carpenter, 1973; Zahng *et al.*, 1990) also show early somatic chromosome loss although with lower frequency. In this case the lost chromosomes not only must be of maternal origin but they also must come from nonexchange tetrads (Zahng and Hawley, 1990). We still do not know the exact mechanisms leading to these early chromosome disjunctional failures but it may be related to the way the first division takes place: syngamy in *Drosophila* actually occurs after the first mitosis. During the first embryonic division both pronuclei divide independently and two parallel spindles are formed, one for each pronucleus, so that the actual syngamy occurs only when the chromatids have reached the spindle poles. Three of the mutations causing chromosome loss (*pal*, *ncd*, and *nod*) also cause abnormal meiotic disjunction, and they may participate in cytoskeletal dynamics (Fyrberg and Goldstein, 1990). In fact, *ncd* and *nod* code for kinesin-like motors (reviewed in Fyrberg and Goldstein, 1990; Carpenter, 1991).

Until the end of division 9, all the nuclei remain in the central part of the embryo. After that, most nuclei start migrating to the periphery to form the blastoderm, while a few remain in the interior. The latter continue synthesizing DNA in the absence of karyokinesis, thus becoming polyploid. Apart from the early formation of a triploid nucleus after fusion of the three haploid polar bodies, this is the first instance during *Drosophila* development when some nuclei become polyploid, an event that will be frequent later. This early nuclear decision requires some sort of unknown signal aimed at halting karyokinesis, but not DNA replication, in specific nuclei at a specific time even within a syncytium and in the absence of transcription.

Although later in development polyploidy will be a common cellular event, we still know little about its genetic control. Gatti and Baker (1989) have described a mutation, *l(3)13m-281*, producing endoreduplication in larval neuroblasts. In mutant cells the DNA has replicated several times (up to 16 sister chromosomes) but the chromatids remain attached to a single centromeric region. As argued by Gatti and Baker, this gene might be involved in polytenization of wild-type cells, an attractive possibility that deserves further investigation.

The nuclei that reach the posterior part of the egg are soon surrounded by cell membranes giving rise to the first embryonic cells. After two rounds of cell division they bud out of the embryo to become the pole cells that eventually originate the adult germ line. Interestingly, Raff and Glover (1988) have shown that pole cells do not require for their formation the presence of nuclei at the posterior pole, but rather the presence of centrosomes or other centrosome-associated cytoskeletal elements. This does not seem to be the case for the rest of the cortex, where cellularization appears to be triggered by the nuclear-to-cytoplasmic ratio (Edgar *et al.*, 1986). In fact, haploid embryos derived from *mh* homozygous females go through an extra round of cell division before cellularization, so that these cellular blastoderms are formed by twice as many cells (but with the same nuclear-to-cytoplasmic ratio) as a regular wild-type blastoderm (Sullivan, 1987).

Several mutations cause mitotic abnormalities during preblastoderm development. We have previously referred to the very early defects resulting from alterations of the maternal contribution of *asp*, *gnu*, *plu*, and *png*. The requirement for *gnu*, *plu*, and *png* wild-type product seems to be restricted to very early events. Later phenotypes shown by GNU embryos, such as the degradation of the giant

7. Genetic Analysis of Cell Division in Drosophila

nuclei and the subsequent formation of occasional spindles, are secondary consequences of the primary effect caused by the mutation (Freeman et al., 1986). α4t is another gene that, like *gnu*, is required only maternally. It codes for a tubulin subunit that is required for spindle formation only during early embryonic divisions. This gene is transcribed and translated exclusively during oogenesis (reviewed by Fyrberg and Goldstein, 1990). The absence of *asp* product, or the presence of abnormal *asp* products during early development, has longer lasting consequences (González et al., 1990). Syncytial nuclear divisions are asynchronous, the spindles are morphologically abnormal (with long wavy arrays of fibers associated in bundles), and mistakes in the distribution of chromatin to opposite poles during mitoses are frequent. This results in abnormal blastoderms in which not only the nuclei have different genetic constitutions but also some areas of the blastoderm are devoid of nuclei. Another mutation showing abnormal preblastoderm nuclear behavior is *polo* (Sunkel and Glover, 1988). Llamazares et al. (1992) have demonstrated that *polo* encodes a cell cycle-related serine–threonine kinase. Viable heteroallelic combinations for *polo* are female sterile. The cause of this sterility is the death of POLO embryos before blastoderm formation, with mostly polyploid nuclei, and abnormal spindle poles lacking distinct centrosomes (Sunkel and Glover, 1988).

Female sterile alleles of other essential mitotic genes have a preblastodermic mitotic phenotype but they have not been described in detail. Such is the case of *mus-101*, involved in condensation of heterochromatin in larval neuroblasts (Gatti et al., 1983b), which shows a mitotic phenotype in embryos originated from *mus-101* homozygous females (Glover et al., 1989). Similarly, embryos laid by females homozygous for the mutation *aurora* have centrosomes that are affected during preblastoderm divisions (Glover et al., 1989). In embryos laid by females homozygous for the mutation *lodestar*, preblastoderm divisions display lagging chromatids during anaphase and chromatid bridges (Glover et al., 1989). Females homozygous for *rough deal* (*rod*; Karess and Glover, 1989), a mutation that causes chromosome nondisjunction and anaphase bridges during somatic anaphases, are sterile and the embryos they produce show signs of preblastodermic mitotic defects (Glover et al., 1989). Finally, mutant females for the tubulin gene α*1t* show semidominant sterility (Matthews and Kaufman, 1987).

These results show that weak alleles of zygotically essential genes involved in mitotic events result in female sterility due to abnormal embryonic development. This sterility probably indicates that the rapid embryonic nuclear divisions that occur in the absence of transcription are more sensitive to reduced levels of products required for mitosis than somatic cells. Somatic cells not only have noticeable G_2 phases (so they can synthesize their own gene products even if these are less effective than wild-type products) but may be able to participate in regenerative processes so that the "healthier" cells can substitute for phenotypically "sicker" cells.

Once the first viable alleles of essential mitotic loci were shown to result in female sterility (such as *mus-101*, *mus-105*, and *mus-109*; Baker and Smith, 1979), it was reasoned that female sterile mutants could be a source of mitotic mutations. Gatti and Baker studied 36 autosomal female sterile mutations and found that 6 of them also affected chromosome behavior in somatic (larval neuroblast) cells (Gatti *et al.*, 1983a). We still do not know if these mutations affect embryonic divisions, but it is likely because some of the mutations that affect embryonic as well as somatic divisions (*rod*, *lodestar*, *aurora*) were first selected because of their female sterile phenotype (Glover *et al.*, 1989).

B. The Maternal Effect

The embryonic phenotypes of eggs laid by females mutant for several cell division genes indicate that these genes are also essential for imaginal development. This suggests that the oocyte is loaded with maternal products that can ensure development at least until the time when most mitotic genes are no longer needed for larval development. There are, however, instances in which this maternal effect can last longer. For instance, individuals mutant for *mei-9* or *mei-41* (mutations involved in DNA repair pathways; Baker *et al.*, 1976, 1978) die when they are exposed to ionizing radiation. When a cross is established in such a way that the hemizygous (tester) males are derived from females carrying two doses of the wild-type gene (used as control), the mortality of the males does not depart from that of the control females during the first larval instar, but it does increase considerably afterward. This observation means that at least until the end of the first larval instar (roughly 48 hr at 25°C) there are enough wild-type products provided by the mother in the egg cytoplasm to repair a substantial number of the chromosome breaks caused by the irradiation. Another example is *dotted* (*dot*; Ripoll *et al.*, 1987), a mutation causing polyploid cells with extreme chromosome condensation in larval neuroblasts. Culture in colchicine of homozygous mutant second instar larval brains results in an elevation of the mitotic index and the appearance of diploid cells, meaning that at this stage some cells are normal and retain their ability to divide. However, the same treatment in third instar larval brains does not result in either a higher number of metaphases or in the appearance of diploid cells, showing that all cells are abnormal near the end of larval development. These results indicate that wild-type maternal products can partially sustain normal cell division at least until the end of the second larval instar (3 days at 25°C).

To determine whether maternal effects are indeed a function of the amount of product synthesized by the mother during oogenesis we have quantified the maternal effect of *asp* (Carmena *et al.*, 1991). Because mutations in *asp* are known to give rise to somatic chromosome nondisjunction we monitored the occurrence of nondisjunction during the development of the imaginal wing disk by coupling a duplication of a cuticular cell marker gene to the Y chromosome of homozygous *asp* individuals. Loss of this marked chromosome gives rise to clones of marked epidermal cells in the adult wing. Because large clones must derive from early nondisjunctional events and small clones from later events, the largest clones will represent the time in development when the presence of maternally provided wild-type products is not enough to completely ensure normal cell divisions. By varying the doses of the wild-type gene in the maternal genome we could address the question of whether or not an increase of maternal doses results in a delay in the occurrence of the first mitotic mistakes. We have found that an increase of one maternal dose of the wild-type allele is accompanied by a delay of three cell divisions in the appearance of the first errors in chromosome segregation. Therefore, more gene product in the oocyte ensures proper chromosome segregation during more cell divisions. We have also found that the maternal products influence the development of the mutant disks even during the third larval instar. Because there is no reason to consider *asp* a peculiar case, the maternal contribution of many other genes involved in cell division may also have effects until very late in development.

C. Postblastoderm Embryonic Development

Because the first nine divisions (requiring 8 to 10 min/cycle) are successions of S and M phases and take place nearly synchronously in the absence of recognizable transcription, they are fully dependent on the supply of products provided by the mother during oogenesis. Zygotic transcription starts during cycle 10, increasing during subsequent cycles (Edgar and Schubiger, 1986). O'Farrell *et al.* (1989) have shown by injecting α-amanitin in *Drosophila* embryos that this early zygotic transcription is dispensable for nuclear divisions prior to cycle 14. The increase in zygotic transcription is accompanied by a progressive lengthening of the duration of the cycle from around 9 min in cycle 10 to 21 min in cycle 13 (Foe and Alberts, 1983). At the onset of gastrulation, during the interphase of cycle 14, several dramatic changes occur. From then on the cell cycle lengthens and a noticeable interphase becomes evident (Foe and Alberts, 1983). Following cellularization the embryo begins gastrulation and the ensuing morphogenetic movements that will generate the first instar larva. During this period some cells, like the amnioserosa, do not divide and become polyploid; most cells in the nervous system divide six more times and differentiate (a small pool remains mitotically active to give rise to the imaginal central nervous system); the remaining embryonic cells go through three rounds of cell division (cycles 14 to 16) before their terminal differentiation. Even if fertilized by normal sperm, ASP embryos that survive until cellularization present severe abnormalities in the size and distribution of nuclei but, despite these abnormalities, most of the morphogenetic movements proceed and abnormal mature embryos can occasionally be found (González *et al.*, 1990). These results suggest that morphogenetic movements are independent of the number of blastoderm cells and their chromosome content.

During cell cycles 14 to 16, well-defined groups of cells (the mitotic domains) enter into mitosis following a very precise temporal and spatial pattern (Foe and Alberts, 1983; Hartenstein and Campos-Ortega, 1985; Foe, 1989; Foe and Odell, 1989). The temporal sequence in which different mitotic domains enter mitosis depends on the duration of the G_2 phase (which varies from 30 min to more than 120 min) in each domain, because during the interphase of cycle 14 all cells start and finish DNA duplication synchronously. Although the divisions following cellularization are necessary for the survival of the embryo, they are not needed either to complete morphogenetic movements or for the differentiation of a variety of cell types, as revealed by the phenotypes of embryos in which these divisions are abnormal (ASP embryos; González et al., 1990) or they do not take place: null mutations in the gene *string* (*stg*; Edgar and O'Farrell, 1989) block embryonic divisions from cycle 14 on. The mutant mature embryos have a reduced number of cells, but they form all possible cell types. Mutations in the gene coding for cyclin A block cell divisions partially during cycle 15 and entirely during cycle 16. In spite of this, nearly normal embryos are formed (Lehner and O'Farrell, 1989). Edgar and O'Farrell (1990) have shown, by using a *stg* construct under the control of a heat shock promoter, that altering the spatiotemporal pattern of postblastoderm divisions does not have any effect on the formation of first instar larvae. Mutations in the genes coding for cyclins B and C, or some homologs of the yeast genes involved in control of division (*cdc2*, *wee1*, *suc1*, etc.) will probably have similar effects; therefore, embryonic lethals might be a productive source of new cell division-related mutations.

D. Genes Flies Borrowed from Yeast

The marriage between biochemical approaches in different systems (clams, amphibians, sea urchins), and genetic manipulation in yeast has given birth to the discovery of a highly conserved system for initiation and completion of mitosis in all cells (Nurse, 1990). The central core of this trigger mechanism involves several genes in yeast (for simplicity we will use the nomenclature used for the fission yeast, *Schizosaccharomyces pombe*). The key element is a protein kinase ($p34^{cdc2}$) encoded by the gene *cdc2*. The amount of *cdc2* gene product does not vary during cell division but its kinase activity is positively regulated by other gene products that appear and disappear during specific stages of the cell cycle. These are cyclin ($p56^{cdc13}$, encoded by the gene *cdc13*) and the product of the gene *cdc25* ($p80^{cdc25}$). The system is also regulated by the products coded by the genes *wee1*, *nim1*, and *suc1*. The combination of $p34^{cdc2}$ and cyclin, regulated by *cdc25* and other genes, results in an active complex (mitotic or meiotic promoting factor, MPF) (Masui and Markert, 1971) that is able to trigger the start of mitosis as well as the transition from G_2 to M phase and the accompanying cellular events (breakdown of the nuclear membrane, chromatin condensation, and spindle formation) leading to cell division. The degradation of cyclin results in the inactivation of MPF activity necessary for the initiation of anaphase (see the review by Nurse, 1990). The involvement of MPF in cell division in *Drosophila* has been ascertained only for the G_2–M transition (see reviews by O'Farrell *et al.*, 1989; Glover, 1991).

The *Drosophila* gene homologous to *cdc2* has been molecularly isolated in two ways. Jiménez *et al.* (1990) used the elegant approach followed by Lee and Nurse (1987) to identify the *cdc2* homolog in humans. They searched for the *Drosophila* cDNA that was able to rescue the phenotype of *cdc2* mutant yeast cells and isolated a gene localized by *in situ* hybridization to region 31E of the polytene salivary chromosomes. (Note that with this system function, not structural homology, is the selective requirement of the screen.) Lehner and O'Farrell (1990b) have cloned the same gene using structural homology (through PCR amplification) as their selective criterion. These authors later showed the functional ability of the cloned *Dm cdc2* homolog to rescue the phenotype of mutant yeast cells. Using structural homology Lehner and O'Farrell (1990b) have identified a cognate localized in 92F (*Dm cdc2c*) that, although it shows the same pattern of transcription during development as *Dm cdc2*, is unable to complement *cdc2* mutations in yeast. No mutations are available for either the *Dm cdc2* or *Dm cdc2c* genes in *Drosophila*. On the other hand, functional tests have served to identify two *Drosophila* genes complementing *cdc25* yeast mutant cells (Jiménez *et al.*, 1990). One of these genes corresponds to the *string* gene studied by Edgar and O'Farrell (1989), localized to region 99A in the salivary chromosomes. In spite of its apparently identical function in yeast, the other gene (*CDC25Dm2*) is molecularly different (with respect to restriction map and size) and it is localized in another chromosomal region. Although both *cdc25* functional homologs can rescue the mutant phenotype in yeast, *CDC25Dm2* cannot substitute for *string* in *Drosophila*, indicating that, although essentially conserved, the control of cell division in fruit flies must be much more complicated than in yeast.

A similar situation is found when studying the *Drosophila* homologs of *cdc13* in yeast. Two genes for cyclin have been identified in fruit flies based on structural homology with *cdc13* (Lehner and O'Farrell, 1989, 1990a; Whitfield *et al.*, 1989). Depending on their conserved homology with cyclins in other organisms they have been classified as cyclin A and cyclin B. *Drosophila* cyclin A is structurally more similar to cyclin A of other organisms than it is to *Drosophila* cyclin B, and vice versa (Whitfield *et al.*, 1989, 1990; Lehner and O'Farrell, 1989, 1990a). In spite of the structural conservation, cloned *Drosophila* cyclin genes are unable to complement *cdc13* mutations in yeast (Jiménez *et al.*, 1990). It seems that while *Dm cdc2* gene products can interact with yeast cyclin to form a functional complex, *Drosophila* cyclins cannot form a functional MPF with the yeast *cdc2* gene product. It follows that neither structural homology nor interspecific complementation are definitive criteria, and that both must be used if we want to isolate fly genes similar to genes from other organisms.

The proof that *Drosophila* cyclins A and B are functionally different comes from the observation that mutations in *cyclin A* are not rescued by the presence of *cyclin B* gene products (Lehner and O'Farrell, 1990a). Besides, many aspects of the behavior of both *cyclin A* and *B* products (mRNA as well as protein) indicate that in *Drosophila* the functions of cyclins A and B are significantly different. In both cases maternal mRNA is found distributed over the entire embryo up to nuclear division 13 (Whitfield *et al.*, 1989, 1990; Lehner and O'Farrell, 1990a), although *cyclin B* mRNA is preferentially accumulated at the posterior pole during oogenesis and in the germ line during later development (Whitfield *et al.*, 1989). The preferential accumulation of *cyclin B* mRNA in the pole plasm, with still unknown function, is reflected in its association with polar granules, which may protect it from degradation. In contrast, the association of cyclin B with the nuclei migrating to the cortex is dependent on microtubules (Raff *et al.*, 1990); this cortical *cyclin B* mRNA is degraded during cycle 14.

The amount of *cyclin A* and *B* mRNAs does not change during preblastoderm development, so their role during these early nuclear divisions is unclear. It has been postulated that a nondetectable cyclic variation in the amount or activity of cyclin proteins might be taking place (Whitfield *et al.*, 1989; Lehner and O'Farrell, 1990a). Translation of mRNA from the two cyclins is quite different. Cyclin A protein is barely detectable during the first hour of development (Lehner and O'Farrell, 1989; Whitfield *et al.*, 1990), later increasing and maintaining high levels until 10 hr, then being degraded afterward. In unfertilized eggs cyclin B, but not cyclin A, proteins are detectable; in fertilized eggs cyclin B proteins are maintained at high levels during the first 10 hr (Whitfield *et al.*, 1990). After cellularization (and equally in embryonic and imaginal cells) both cyclins undergo quantitative variation but their pattern of accumulation–degradation is different. Both cyclins accumulate during interphase and peak during prophase. Cyclin B protein abruptly disappears at the metaphase–anaphase transition (Whitfield *et al.*, 1990). It is not clear whether cyclin A disappears before (Whitfield *et al.*, 1990) or during (Lehner and O'Farrell, 1990a) metaphase. Lehner and O'Farrell (1990a) have concluded that cyclins are not responsible for the entrance into mitosis during cycle 14 because cyclins A and B accumulate at equal levels in all blastoderm cells, even cells that, like the amnioserosa, will never enter mitosis.

Leopold and O'Farrell (1991) have cloned a new *Drosophila* cyclin (cyclin C) functionally and structurally homologous to the G_1 cyclins of budding yeast. Cyclin C may regulate the interphase function of *Dm cdc2* in a similar way as the three G_1 cyclins regulate *cdc2* in yeast cells.

The gene responsible for the temporal entrance into mitosis of specific mitotic domains (groups of embryonic cells that divide synchronously) has been shown to be *string* (*stg*), which is structurally and functionally homologous to the gene *cdc25* of fission yeast (Edgar and O'Farrell, 1989, 1990; O'Farrell *et al.*, 1989).

7. Genetic Analysis of Cell Division in *Drosophila*

As happens with many other mitotic loci, the egg cytoplasm has abundant maternal *stg* transcripts. We still do not know for sure the exact role played by these maternal transcripts during preblastoderm divisions. The case of *stg* is one of the few where the maternal transcripts are not gradually diluted and substituted by zygotically synthesized products; rather, they abruptly disappear during cycle 13 so that cycle 14 is fully dependent on the zygotic transcription of *stg*, which in turn may depend on combinations of segmentation and homeotic genes (O'Farrell *et al.*, 1989). Accumulation of *stg* products strictly precedes, with the same temporal and spatial patterns, the entrance into mitosis of the different mitotic domains during cycle 14 (Edgar and O'Farrell, 1989). Moreover, the ectopic expression of *stg* in all the cells of the embryo (using *stg* constructs under the control of a ubiquitous heat shock promoter) causes the disappearance of mitotic domains so that all the cells enter mitosis 14 simultaneously (Edgar and O'Farrell, 1990).

III. Mitosis during Postembryonic Development

A. Imaginal Development

Two types of cells form the first instar larva that hatches from the egg once the embryo has matured. Most of the larval cells will go through rounds of duplication of their genetic material in the absence of cytokinesis. A classical example is the cells of the salivary glands that will carry the well-known and experimentally invaluable giant chromosomes. The other type are the cells that remain diploid and mitotically active, the majority of which will give rise, after metamorphosis, to the adult tissues. Among these imaginal cells we can broadly distinguish several kinds of mitotically active cells.

The abdominal histoblasts appear as nests of epidermal cells intermingled with larval cells. They do not divide during larval development, proliferating exponentially after puparium formation at a rather fast mitotic rate (2–4 hr/division) to give rise to the abdominal tergites and sternites of the imago.

The imaginal disk cells will give rise to most of the rest of the adult cuticle. They grow exponentially during larval development and stop dividing shortly after pupariation. Although as a whole these imaginal disk cells divide in a rather uniform fashion (an average of 8–10 hr/cycle in all disks) different imaginal disks present differences in the way in which cell divisions proceed. For instance, after puparium formation the eye disk is transversed by what has been called a differentiation furrow: the population of cells ahead of this furrow divides exponentially while cells behind this furrow will never divide again, but they differentiate and become involved in the complex morphogenetic events leading to the adult eye (see Tomlinson, 1988). Therefore, at a given time of the eye development a well-defined band of cells irreversibly leaves mitosis synchronously. Contrarily, the cells in the imaginal wing disk go through two waves of mitoses (one transverse, the other longitudinal) late in development (for a review and bibliography, see García-Bellido and Ripoll, 1978). What drives cells to leave mitosis in synchrony and what signal specifies mitotic waves remain unsolved questions. The genital disk provides an interesting case of control of entry into mitosis. The male and female genitalia derive from two independent primordia, both present in every mature embryo irrespective of its sex. Whether or not cells in these primordia divide depends in a cell-autonomous way on the sex of the cells. If a cell in the male genital primordium is male, it will divide and differentiate male genital structures; if the same cell is female it will not divide or differentiate. The opposite process takes place in the female primordium (Nöthiger et al., 1977). Therefore some signal associated with gender can in some instances either trigger or repress cell division. How this control is established is not known but it poses very interesting questions.

Every imaginal disk carries adepithelial cells, the precursors of the imaginal muscles. These cells are associated with the epidermal cells, which will give rise to the adult cuticle and the sensory organs. As happened earlier with embryonic muscle precursors, these cells will give rise to syncytial muscle cells, and this process requires another unknown signal to uncouple karyokinesis from cytokinesis. The distribution of pericentrosomal material might play a role in this transition from mono- to polynucleated cells, as has been described by Tassin et al. (1985) studying the distribution of a centrosome-associated antigen in human myoblasts. In these cells the centrosome-associated staining changes from the usual discrete compact spots to a more dispersed perinuclear distribution. They interpreted the change in antigen distribution as a reorganization of microtubule organizing centers to direct the formation of longitudinal microtubules in the myotube. This reorganization might also result in the loss of bipolarity by the myoblast; because, in turn, the establishment of bipolarity is thought to be a requisite for cytokinesis (Mazia, 1961), the dispersion of centrosomal material can result in inhibition of cytokinesis. Casal et al. (1990b) found dispersion of centrosomal material during meiosis in *Drosophila*. We may interpret this dispersion in a similar way, because spermiogenesis might also require the presence of multiple microtubule organizing centers.

The cells that eventually give rise to the germ line behave in a specific way. They are derived from the pole cells, which are set apart very early in development. Gametogenesis in both sexes starts from a small pool of stem cells. Each stem cell will give rise, following cell division, to a cell inheriting the status of stem cell and to a cell destined to be the precursor of the gametes (Mahowald and Kambysellis, 1980; Lindsley and Tokuyasu, 1980). In both sexes each gonial cell goes through 4 rounds of cell division, giving rise to 16 cells joined by large intercellular bridges. From then on there is a clear difference between both sexes. In the female, 15 of the 16 cells will become polyploid and supply the sixteenth cell with large amounts of the products they synthesize. This remaining cell will enter into meiosis, accumulate the products synthesized by the surrounding cells and develop and mature, giving rise to the oocyte. In males, all 16 cells will go through both meiotic divisions and originate bundles of 64 spermatozoa.

Most of the cytological studies performed in *Drosophila* have traditionally used the late third instar larval central nervous system (CNS)—in reality a mixture of the larval and the presumptive imaginal nervous systems—because their cells are large for *Drosophila* standards, mitoses are frequent, and they can be studied with relative ease. Meiosis has been studied preferentially in males; recent developments in immunocytological technique (Casal *et al.*, 1990a,b) are serving to complement classical cytological techniques (Cooper, 1950) for studies of *Drosophila* meiosis.

B. Effect of Mitotic Mutations in Imaginal Tissues

With the exception of *gnu*, *plu*, *png*, $\alpha t4$, *cyclin A*, and *string*, the phenotypes of most of the mitotic mutants known to date in fruit flies have been described in the neuroblasts of the larval CNS. There is a wide range of possible mutant phenotypes in the roughly 70 different mutations isolated so far. We can find examples among the comprehensive collection of mutations described by Gatti and Baker (1989). Following the classification suggested by these authors, we can group the alterations produced by mutations in genes involved in mitosis essentially into a reduced number of mitotic phenotypes: low or high mitotic indices, endoreduplication, abnormal chromosome condensation, chromosome breaks, abnormal anaphases, and aneuploidy (for simplicity we will define as "aneuploid" all cells with more than the diploid number of chromosomes, including true polyploid cells). To the classification of Gatti and Baker we can add only the presence of circular mitotic figures, a phenotype that they either never found or did not consider abnormal because it is common in wild-type mitoses in other organisms. We will refer here only to the general characteristics of mitotic mutants. A detailed list of most of the mitotic genes described to date, as well as pertinent literature, can be found in the review by Glover (1989).

Contrary to other systems, where the mitotic index is measured as percentage of cells in mitosis, in *Drosophila* the mitotic index is usually measured as mitotic figures (or as metaphases) per microscopic field, or as number of mitotic figures per larval brain. Gatti and Baker (1989) have interpreted a low mitotic index as resulting from an arrest of cells in interphase, or from constraints on the entrance of cells into mitosis. High mitotic indices are frequent among mitotic mutants, and they may reflect overall unspecific increases in mitotic activity or longer duration or arrest of specific stages of mitosis. Endoreduplication is an interesting phenomenon that might be related to polytenization but it has been described only in the mutation *l(3)13m-281* (Gatti and Baker, 1989); it may result from failures in the progression through G_2 phase.

Defects in chromosome condensation might have diverse origins. We assume that overcondensation derives from lengthening of (or arrest at) pro- or metaphase, on the grounds that the degree of condensation is proportional to the time the cells spend in these phases (Mazia, 1961). Mutants with overcondensed chromosomes are good candidates for defects in spindle function. With few exceptions such as *metaphase arrest* (*mar*; Ripoll et al., 1987), and strong allelic combinations of *asp* and *l(3)13m-230* (Gatti and Baker, 1989), they also show varying degrees of aneuploidy. In many mutants the chromosomes are irregularly condensed, swollen, or fuzzy. An interesting case is that of *mus-101*, a gene needed for the condensation of heterochromatic, but not of euchromatic, regions (Gatti et al., 1983b).

Mutants showing chromosome breaks are frequent and in some especially interesting instances these breaks show regional specificity (*mus-105*, *mus-101*, *fs(3)820*, *l(1)d deg-3*, *l(1)d deg-10*, and *breakage during anaphase* (*bra*; Ripoll et al., 1987). Different events can give rise to these breaks, for example, failures in the repair of spontaneous lesions in DNA, chromatid bridges and breakage during anaphase, and secondary consequence of failures in chromatin condensation.

Abnormal anaphases are frequently associated with mutants showing overcondensed chromosomes and aneuploidy. The most common types are colchicine-like anaphases (mitotic figures in which chromatids are separated but their centromeres are not properly oriented toward the poles) and anaphases with broad poles. Other abnormal anaphases show chromatid bridges (*bra*, *lodestar*, *fs(3)-2755*, and *l(1)TW-6^{CS}*), or failures in the attachment of chromatids to the spindle (*rod*). Circular mitotic figures, such as those found in brains mutant for *merry-go-round* (*mgr*, González et al., 1988), *polo* (Sunkel and Glover, 1988), or *aurora* (Glover et al., 1989), could be considered monopolar anaphases if (as discussed in González et al., 1988) anaphase is defined as the phase when chromatids (chromosomes in the case of these mutants) are synchronously transported toward the poles.

One of the most frequent cytological phenotypes associated with mitotic mutations is aneuploidy. Low levels of aneuploidy result from precocious separation of sister chromatids [*l(1)zw10*], multipolar spindles (*polo*), failures in chromosome attachment to the spindle fibers (*rod*), spindle defects (*asp*), and so on. More elevated degrees of ploidy can come from accumulation (through successive cell divisions of these same defects), from monopolar mitoses, or from cycles of DNA synthesis in the absence of cytokinesis. Mutant *asp* brains have an elevated mitotic index, due to cells arresting at, or spending a long time at, metaphase (Ripoll et al., 1985). Actually, when careful measurements are made, the number of anaphases per microscopic field or larval CNS happens to be normal in *asp* mutants. This means that even if many cells have been delayed at metaphase, the overall mitotic activity of the brains is normal (I. Molina, C. González, and P. Ripoll, unpublished). Many of the *asp* anaphases are grossly aneuploid, and eventually will give rise to other, more aneuploid, cells. The largest cells found in *asp* brains have 60–80 chromosomes.

Similar chromosome numbers are observed in larval neuroblasts mutant for *mgr* (González *et al.*, 1988). The mutant phenotype may be due to failures in the separation of centrosomes; extensive aneuploidy may derive from successive monopolar divisions. In these mutant brains monopolar divisions are associated with the appearance of circular mitotic figures. A circular mitotic figure is defined as one in which the chromosomes are oriented so that their centromeres form a circle and their chromatids are pointing outward, suggesting that they are being attracted toward a single pole in the middle of the circle. The possibility that there is an actual attraction force was tested by treating *mgr* mutant brains with taxol, a microtubule-stabilizing drug. In these treated brains the centromeres in circular mitotic figures remained as a circle, but the chromatids ceased pointing outward. The requirement of functional microtubules for the formation of circular mitotic figures was tested in two ways: (1) constructing *mgr asp* doubly mutant combinations, and (2) treating *mgr* mutant brains with colchicine, a microtubule-depolymerizing drug. In both cases the circular mitotic figures disappeared. González *et al.* (1990) also found monopolar mitoses in *asp* neuroblasts, but in this case no circular figures are formed, and the condensation and orientation of the chromosomes as well as the morphology of the spindle suggest that the spindle is not functional.

The most dramatic cases of mutation-induced polyploidy are shown by mutants for *l(1)d deg 11* and *l(3)7m-62* (Gatti and Baker, 1989) and *spaghetti-squash* (*sqh*; Karess *et al.*, 1991). In all cases enormous cells with as many as 500–1000 chromosomes can be found, but neither the mitotic index nor the condensation of chromosomes are altered. These giant cells most probably arise from failures in cytokinesis. We have found (and unfortunately lost) another mutation in the second chromosome with an identical phenotype, which suggests that cytokinesis must be (directly or indirectly) controlled by several genes. Karess *et al.* (1991) have shown that *sqh* codes for the regulatory light chain of nonmuscle myosin, known to be needed for cytokinesis in other organisms. In these cells the duplication of both chromosomes and centrosomes seems to have proceeded normally but cytokinesis has not. Anaphases in these giant cells show many spindle poles, indicating that the multiple centrosomes are functionally normal. Therefore, centrosome duplication and migration can be uncoupled from cytokinesis in somatic cells. Even if polyploid, *asp* cells never show more than two centrosomes (González *et al.*, 1990). This may indicate that centrosome duplication in somatic cells is dependent on the completion, or at least initiation, of anaphase.

Because the maternal products can have an effect on cell division rather late in development, and the mutant alleles studied might be leaky, it is uncommon to find mutant larval brains with a uniform mitotic phenotype. Usually, there is a variety of cellular phenotypes associated with a given mutation. In many cases it is easier to interpret mutants with these variable phenotypes than mutants with a unique defect. For instance, all the mitotic figures found in *dot* or *podgy* (*pod*; Ripoll et al., 1987) present polyploid cells with extreme chromosome condensation so that each chromosome is reduced to a spot; the only conclusion to be reached from this uniform terminal phenotype is that mitosis is affected, but it neither provides clues as to how it is affected nor suggests experiments to interfere with it (besides the induction of weaker alleles). On the other hand, very complex phenotypes do not always provide useful information. For instance, arguing that dephosphorylation had to play an important role in cell division, Axton et al. (1990) have studied mutations in the gene coding for the serine–threonine phosphatase 1 located in region 87B of the salivary gland chromosomes (*PP1 87B*, one of the four PP1 isoenzymes identified in *Drosophila*) (Dombrádi et al., 1989). Mutant *PP1 87B* neuroblasts show a variety of phenotypic defects such as fuzzy chromosome condensation, overcondensation, different degrees of aneuploidy, chromatid separation, anaphases with overcondensed chromatids, multipolar spindles, and so on. It is very difficult to interpret such a complex phenotype in a simple way, as unfortunately happens with many other mitotic mutants. Even knowing what type of protein *PP1 87B* encodes, it was quite difficult for Axton et al. (1990) to correlate the mutant phenotype with the presumed function of the protein. We at least know now what phenotype to expect from mutations in other phosphatase 1 genes.

Mutations in genes needed for mitoses often lead to larval/pupal lethality. A common consequence of mutations in these mitotic genes is that mutant larvae do not have imaginal disks or have small or degenerate imaginal disks. For example, 19 out of the 20 mutants studied by Gatti and Baker (1989) that had missing or degenerate imaginal disks turned out to be defective for functions needed for mitosis. These mutants usually lack gonads as well, which makes the simultaneous study of their mitosis and meiosis impossible. In most cases it is essential to study different alleles of the same gene to reach accurate interpretation of its function. We have tried to study a variety of alleles for the *asp* gene, including strong as well as weak allelic combinations. The strongest combinations are by themselves rather uninformative: imaginal disks and gonads are missing, and the only mitotic figures to be found in the CNS are diploid metaphases with condensed chromosomes. Furthermore, the addition of colchicine does not affect the phenotype, suggesting a terminal arrest at metaphase. At the other end of the spectrum, very weak alleles are nearly wild type, presenting only a slightly elevated mitotic index and separated chromatids with normal condensation. All the valuable information we have about this locus comes from the study of different alleles and heteroallelic combinations with intermediate phenotypes. The molecular characterization of the gene is currently in progress (Glover et al., 1989) and it will provide tools to better understand the function of *abnormal spindle* in cell division.

Some mitotic loci have mutant alleles that permit the recovery of at least some mutant adults. In most cases surviving individuals show cuticular abnormalities in eyes, wings, and abdomens indicative of cell death during development. In some cases these abnormalities are not restricted to cuticular structures but also affect muscles and the nervous system, leading to behavioral defects (Casal et al., 1990a).

Females homozygous for mutations in some loci needed to ensure chromosome integrity (that are viable unless treated with ionizing radiation) are defective in meiotic recombination (Baker et al., 1978). These mutations offer an interesting example of how some loci can have different roles in somatic cells (the repair of random DNA damage in both sexes) and in the germ line (recombination between homologs at specific places during a specific time in a specific tissue). As we have seen above, in the case of many mutations the adult females are sterile due to abnormal divisions during embryonic development. Male fertility as well as meiotic chromosome disjunction (and, as a consequence, fecundity) are also affected by some cell cycle mutations.

C. Male Meiosis and Spermatogenesis in Mitotic Mutants

In the male germ line mitoses cease after the gonial cells, surrounded by 2 cyst cells that isolate them from the rest of the testis, have undergone 4 rounds of cell division to give rise to a cyst of 16 interconnected primary spermatocytes (see Lindsley and Tokuyasu, 1980). After a period of high metabolic activity, during which these cells grow in volume, the mature primary spermatocytes enter the meiotic process. The genetic trigger that governs this transition from mitosis to meiosis in the germ line of *Drosophila* remains unknown. Male meiosis can be cytologically analyzed with relative ease in *Drosophila* (Cooper, 1950); because some lethal mitotic mutants have normally developed larval testes, mutant meiosis, and at times some early stages of spermatogenesis, can be studied. In instances where some fertile male adults can be recovered even meiotic chromosome segregation can be analyzed by studying their progeny when crossed to chromosomally appropriate tester females. The second (equational) meiotic division is in essence a regular mitosis, so it is not surprising that loci needed for correct somatic divisions are also used during this meiotic division. Although the first meiotic division is somewhat different (homologous chromosomes pair during prophase and chromosomes, not chromatids, disjoin at anaphase) it shares common features with mitosis, and some genes needed for somatic cell divisions are also essential during reductional meiotic divisions. The mutation *polo* (Sunkel and Glover, 1988) alters meiotic chromosome disjunction of both the sex chromosomes and the autosomes. Mutant males also show multipolar spindles during meiosis, which may be the cause of the nondisjunction observed. Sunkel and Glover studied the segregation of the sex chromosomes during both meiotic divisions. They recovered 12 XY gametes (product of nondisjunction during the first division) and 71 XX gametes (product of nondisjunction during the second division). The asymmetry of recovery of XY vs XX gametes suggests that most of the nondisjunctional events took place during the second meiotic division.

We have studied in depth the meiotic behavior of chromosomes during meiosis in *asp* males (Ripoll *et al.*, 1985; González *et al.*, 1989a; I. Molina, C. González, and P. Ripoll, unpublished observations). Although chromosome pairing during the first meiotic division is normal, alterations in chromosome segregation are already seen during this division due to abnormal disjunction of homologs in the presence of defective spindles (González *et al.*, 1990; Casal *et al.*, 1990a). As expected for a mitotic mutation, the second meiotic division is also affected in *asp* males. However, contrary to the process in *polo* males, in *asp* males there is a difference favoring the recovery of nondisjunctional gametes produced during the first meiotic division. For instance, we studied the segregation of the second chromosomes and we recovered 376 diplo-2 gametes due to nondisjunction during the first meiotic division, and 220 diplo-2 gametes due to nondisjunction during the second division. By crossing *asp* males to genotypically different tester females we have found that these males can produce any conceivable kind of gamete. Many aneuploid gametes derive from failures in chromosome disjunction during one or both meiotic divisions. Other gametes come from regular disjunction in cells that entered meiosis already being aneuploid. Finally, we have found that diploid gametes preferentially come from failures in cytokinesis after the second meiotic division, by studying the offspring produced by triploid females recovered from *asp* fathers (I. Molina, C. González, and P. Ripoll, unpublished observations).

Meiosis has been cytologically studied in mutants for two other loci that produce sterile surviving males (*rod*) or for which no viable allelic combinations are available (*mgr*). Karess and Glover (1989) found in *rod* males defects of presumably premeiotic origin as well as defects during both meiotic divisions, and alterations during the second meiotic division seem to be more frequent than alterations during the first meiotic division. These alterations in chromosome or chromatid distribution to the meiotic poles are consistent with the most common phenotype (failures in the attachment of kinetochores to the spindle fibers) found in somatic cells. Similarly, the cytology of meiosis in *mgr* males also confirms the somatic mutant phenotype (González *et al.*, 1988). The first meiotic division seems to proceed normally until anaphase; at this point, instead of normal anaphases one finds circles of condensed chromatin that have been interpreted as circular meiotic figures equivalent to the circular figures found in mutant mitoses. As a consequence, the second division metaphase is always diploid instead of haploid. The resulting spermatid nuclei have the size expected for a chromosome content equivalent to two succesive failures in both meiotic divisions; namely, four times that of a normal gamete (González *et al.*, 1988).

Alterations in chromosome and/or chromatid disjunction during meiosis result in abnormalities during spermiogenesis. The easiest stage in which to recognize

7. Genetic Analysis of Cell Division in *Drosophila*

meiotic defects is the so-called "onion stage" (Lindsley and Tokuyasu, 1980). During this period all the early spermatids within a cyst show (under phase-contrast optics) a single nucleus associated with a single mitochondrial derivative. All the nuclei within a cyst are of equal size, slightly smaller than the also uniformly sized mitochondrial derivatives. By studying males that produce, with known frequencies, gametes of quantitatively different genetic constitutions, González et al. (1989a) showed that the size of the nuclei in onion stage spermatids is proportional to their chromosome content. Therefore, the variations in size of the onion stage spermatid nuclei can be used to roughly estimate meiotic nondisjunction in lethal or male sterile mutations. This method has been used for mutations in *polo*, *asp*, *rod*, and *mgr*. The first three cases, in which the phenotypes includes meiotic nondisjunction of individual chromosomes, produce a wide range of nuclear sizes. In the case of *mgr*, where nondisjunction involves the whole chromosome complement due to monopolar meioses (each postmeiotic cyst is formed by only 16 spermatids), we find uniformly sized nuclei equivalent to the chromosome content of a diploid somatic cell. The number of nuclei per spermatid is also a reflection of abnormal meioses, and multinucleated spermatids may be the origin, at least partially, of meiotic drive (Hardy, 1975; González et al., 1989b; Casal et al., 1990a). For instance, *asp* males contain an average of 1.7 nuclei per early spermatid (Casal et al., 1990b), and they show a high incidence of meiotic drive (Ripoll et al., 1985). Sunkel and Glover (1988) also found multinucleated spermatids in *polo* males; this kind of abnormal spermatid does not occur in *rod* (Karess and Glover, 1989) or *mgr* (González et al., 1988) mutants. Variations in the size of the mitochondrial derivatives, also dependent on the spindle for their distribution, are observed in *asp* and *polo*, but not in *rod* spermatids. This is consistent with the interpretation that, while *asp* and *polo* are needed for correct spindle functioning, the defects found in *rod* cells are attributable to the chromosomes. The mitochondrial derivatives are never properly formed in *mgr*, showing a phenotype similar to the one found when cells go through meiosis in the presence of colchicine (Kemphues et al., 1982). In *mgr* mutants the mitochondrial derivatives disintegrate soon after meiosis and only rarely can elongating cysts be found. Spermiogenesis proceeds more or less normally in many cysts in *asp*, *polo*, and *rod* males, in which motile sperm can be found. This suggests that the primary defects caused by these mutations are restricted to meiosis, and that the abnormalities seen during spermiogenesis are secondary consequences of meiotic defects. Casal et al. (1990a) confirmed this interpretation by studying spermiogenesis ultrastructurally in *asp*. During elongation the most common mutant phenotypes were differences in the size and number of the mitochondrial derivatives, and elongating spermatids with more than one axoneme; both phenotypes could be traced back to defects either in the segregation of mitochondria or in cytokinesis during meiosis. During later stages, the most abnormal spermatids failed to individualize, a common consequence of meiotic alterations (Hardy, 1975). Casal et al. (1990a) concluded that the defects cause by *asp* were spindle specific because both the cytoplasmic microtubules and the microtubules forming the axonemes were normal, and functional sperm was eventually produced.

IV. Tubulin and Kinesin Gene Families

Fyrberg and Goldstein (1990) have reviewed tubulins in *Drosophila* and have provided updated references. We will try to give here a superficial summary. There is no doubt that tubulins are essential for mitosis; however, because they are also essential for many other microtubule-dependent processes, mutations in tubulins are expected to result in cell death rather than in classical mitotic phenotypes. The gene located in 23CD coding for *Drosophila* γ-tubulin, homologous to genes in *Aspergillus nidulans* and humans, has been cloned by Zheng *et al.* (1991); this tubulin is a centrosome-associated protein that may participate in microtubule assembly during cell division. All four genes coding for α-tubulin and four genes for β-tubulin subunits in *Drosophila* have been molecularly identified, and their transcription patterns described. Mutations in two α-tubulins are available, α*1t* and α*4t*. The former gene product is abundant, ubiquitous, and, as expected, the gene is essential for cell viability. The latter gene is stage specific; it is transcribed only during oogenesis and, being an essential component of the spindle during preblastoderm divisions, mutations in this tubulin gene cause female sterility and a maternally dependent embryonic phenotype.

Mutations in two β-tubulin genes have been described: *B3t* and *B2t*. Mutations in *B3t* affect cell viability; the gene is thought to be important for the formation of transient arrays of microtubules. *B2t* is expressed only in the testis; mutations in this gene result in male sterility and do not have any effect in oogenesis or in somatic tissues. Fuller and co-workers, and Kemphues and co-workers, have studied this tubulin in depth: it is essential for the formation of the meiotic spindle, as well as for the many other microtubule-dependent processes taking place during spermiogenesis. A very interesting feature of mutations in *B2t* is that they have provided a powerful tool to isolate mutations in other genes (such as microtubule-associated proteins) that, through their association with β_2-tubulin, are also essential for microtubule function (see Fuller *et al.*, 1989). Not surprisingly, one of the genes identified because of its interaction with *B2t* is the gene coding for α_1-tubulin (Matthews and Kaufman, 1987; Hays *et al.*, 1989). Other nontubulin genes identified as second-site noncomplementors of *B2t* are *haywire* (*hay*; Regan and Fuller, 1988, 1990) and *whirligig* (*wrl*; Green *et al.*, 1990). The *wrl* gene product is essential for the organization of the flagellar axoneme, and its function is hence restricted to spermatogenesis. *hay* is involved in several events taking place during spermatogenesis, including formation of the meiotic spindle. Because null *hay* mutations lead to zygotic lethality, the gene must also be needed for essential microtubule-dependent processes in somatic cells. The discovery of the type of trans interactions shown by these mutations constitutes a most promising approach to identifying gene products in other systems where protein–protein interactions are crucial (Fuller et al., 1989).

Kinesins have long been known as molecules that are able to elicit movement of a variety of particles and vesicles and that are likely to be involved in the movement of chromosomes during mitosis and meiosis. Endow and Hatsumi (1991) have recently discovered that the *Drosophila* genome harbors a large kinesin gene family. They used PCR amplification to recover DNAs homologous to the ATP- and microtubule-binding domains of kinesin. By "*in situ*" hybridization of the amplified DNAs to the salivary chromosomes, they identified around 30 different sites presumably coding for kinesin-like proteins. All the genes that had previously been identified as kinesin-like motors, through mutation and molecular characterization, were recovered in the search by Endow and Hatsumi. These genes have been reviewed by Fyrberg and Goldstein (1990) and Carpenter (1991). *Drosophila* D-kinesin was first recognized as a ubiquitously expressed protein; mutations in the gene result in lethality (Saxton et al., 1988; Saxton and Raff, 1989). Investigators have identified two kinesin-like motors in *Drosophila*, defined because of their structural homology with the heavy chain of kinesin; both have functions related to chromosome segregation during female meiosis: *nonclaret disjunctional* (*ncd*) and *no distributive disjunction* (*nod*). Mutations in both genes have in common that they are female specific, that they affect distributive disjunction of nonexchange tetrads during meiosis, and that they result in somatic chromosome loss during very early stages of development. However, their molecular properties are different: *ncd* is a "minus" directed motor with the motor domain in the carboxy-terminal end of the protein (McDonald and Goldstein, 1990; Walker et al., 1990; McDonald et al., 1990); *nod* appears to be a "plus" directed motor molecule in its N-terminal domain (Zhang et al., 1990). Carpenter (1991) has produced a model to explain the role of kinesin-like molecules with different polarities during distributive disjunction of noncrossover chromosomes in females. The model proposes the existence of three types of fibers (and therefore the forces generated by motor molecules) acting on the chromosomes during meiosis I: (1) kinetochore microtubules pulling the chromosomes toward the poles and using minus-directed motors; (2) astral microtubules exerting forces (the "polar wind") pushing chromosomes away from the poles (this force must be generated by plus-directed motors); and (3) interchromosomal microtubules generating forces (requiring both minus- and plus-directed motors) aimed to neutralize kinetochore-to-pole forces at metaphase. The first two forces should be common to meiosis and mitosis; the third one should be specific to the first meiotic division in females (or at least dispensable in other types of division) so that its absence through mutation would result in a female-specific meiotic phenotype. Although this model is plausible, there is still no evidence for the involvement of kinesin-like motors in chromosome movement during somatic cell divisions. Interestingly, even though null *nod* mutations do not have any detectable effect on somatic cells, a previously characterized mitotic mutation [*l(1)TW-6^{CS}*] has turned out to be an antimorphic allele of *nod* (Zhang and Hawley, 1990). Hardy *et al.* (1981), Goldstein *et al.* (1982), and Porter *et al.* (1987) have described cytoplasmic and sperm-specific forms of dynein, another motor molecule, in fruit flies. We still do not know if they play any specific role during cell division.

V. General Considerations

With this article we have tried to give a general impression of the genetics of cell division in *Drosophila* and the effect of different alterations in cell division on *Drosophila* development. These subjects are still a complicated matter that will eventually become clearer as molecular studies of mitotic genes advance and, more importantly, when protein biochemistry of their gene products is started. *Drosophila* developmental genetics accommodate many old-fashioned techniques that have not been, but should be, routinely used to complement our knowledge of mitotic genes and their function. We can use duplications (or deficiencies) to increase (or decrease) maternal effects; we can use mitotic recombination to study the effect of mitotic mutations in imaginal developing systems; germ line mosaics will tell us about the embryonic phenotype of lethal cell cycle mutants; and doubly mutant individuals will tell us about relationships between mitotic genes.

Only multidisciplinary approaches are going to let us solve the many problems still waiting to be unraveled concerning cell division in higher organisms. Genetics alone cannot shed enough light into cell division to make us understand the basics of this complex process. On the other hand, any other approach is bound to be incomplete without the aid of classical genetics, in the same way that *Drosophila* molecular genetics would have been incomplete without the extensive knowledge available about protein biochemistry in other organisms. For still some time fungi and fruit flies will remain the leading organisms in the study of the genetic control of cell division, as long as investigators continue to profit from what both old and new techniques can offer.

That mammals share many cell division genes with unicellular eukaryotes came as a surprise not long ago. We are now familiar with the systematic screen of fungi-related genes in higher eukaryotes. For many reasons *Drosophila* has occupied a privileged position as a model system for genetics in our reduced historical time scale. It also happens to occupy a privileged position in the evolutionary time scale. We are confident that many of the future relevant findings on the genetic control of cell division will come from laboratories working with fruit flies, and that these findings will be extrapolated with ease to other eukaryotes, including humans.

7. Genetic Analysis of Cell Division in *Drosophila*

Acknowledgments

We are indebted to E. Sánchez-Herrero for comments on the manuscript and help with references, and to the editor of this book for infinite patience. The work from our laboratory was financed by grants from the Dirección General de Investigación Científica y Técnica and an institutional grant to the CBM by Fundación Ramón Areces. I.M. and M.C. were supported by fellowships from Plan Nacional de Formación de Personal Investigador (CAICYT).

References

Axton, J. M., Dombrádi, V., Cohen, P. T. W., and Glover, D. M. (1990). One of the protein phosphatase 1 isoenzymes in *Drosophila* is essential for mitosis. *Cell (Cambridge, Mass.)* **63**, 33–46.

Baker, B S. (1975). Paternal-loss (*pal*): A meiotic mutant in *Drosophila melanogaster* causing loss of paternal chromosomes. *Genetics* **80**, 267–296.

Baker, D. S., and Hall, J. C. (1976). Meiotic mutants: Genetic control of meiotic recombination and chromosome segregation. *In* "The Genetics and Biology of *Drosophila*" (M. Ashburner and E. Novitsky, eds.), Vol. 1a, pp. 352–429. Academic Press, London.

Baker, B. S., and Smith, D. A. (1979). The effects of mutagen-sensitive mutants of *Drosophila melanogaster* in nonmutagenized cells. *Genetics* **92**, 833–847.

Baker, B. S., Boyd, J. B., Carpenter, A. T. C., Green, M. M., Nguyen, T. D., Ripoll, P., and Smith, P. D. (1976). Genetic controls of meiotic recombination and somatic DNA metabolism in *Drosophila*. *Proc. Natl. Acad. Sci. U.S.A.* **73**, 4140–4144.

Baker, B. S., Carpenter, A. T. C., and Ripoll, P. (1978). The utilization during mitotic cell division of loci controlling meiotic recombination and disjunction in *Drosophila melanogaster*. *Genetics* **90**, 531–578.

Carmena, M., González, C., Casal, J., and Ripoll, P. (1991). Dosage dependence of maternal contribution to somatic cell division in *Drosophila melanogaster*. *Development (Cambridge, UK)* **113**, 1357–1364.

Carpenter, A. T. C. (1973). A meiotic mutant defective in distributive disjunction in *Drosophila melanogaster*. *Genetics* **73**, 393–428.

Carpenter, A. T. C. (1991). Distributive segregation: Motors in the polar wind? *Cell (Cambridge, Mass.)* **64**, 885–890.

Casal, J., González, C., and Ripoll, P. (1990a). Spindles and centrosomes during male meiosis in *Drosophila melanogaster*. *Eur. J. Cell Biol.* **51**, 38–44.

Casal, J., González, C., Wandosell, F., Avila, J., and Ripoll, P. (1990b). Abnormal meiotic spindles cause a cascade of defects during spermatogenesis in *asp* males of *Drosophila*. *Development (Cambridge, UK)* **108**, 251–260.

Cooper, K. W. (1950). Normal spermiogenesis in *Drosophila*. *In* "Biology of *Drosophila*" (M. Demerec, ed.), pp. 1–61. Wiley, New York.

Dombrádi, V., Axton, J. M., Glover, D. M., and Cohen, P. T. W. (1989). Cloning and chromosomal localisation of *Drosophila* cDNA encoding the catalytic subunit of protein phosphatase 1α. *Eur. J. Biochem.* **183**, 603–610.

Edgar, B. A., and O'Farrell, P. (1989). Genetic control of cell division patterns in the *Drosophila* embryo. *Cell (Cambridge, Mass.)* **57**, 177–187.

Edgar, B. A., and O'Farrell, P. H. (1990). The three postblastoderm cell cycles of *Drosophila* embryogenesis are regulated in G2 by *string*. *Cell (Cambridge, Mass.)* **62**, 469–480.

Edgar, B. A., and Schubiger, G. (1986). Parameters controlling transcriptional activation during early *Drosophila* development. *Cell (Cambridge, Mass.)* **44**, 871–877.
Edgar, B. A., Kiehle, C. P., and Schubiger, G. (1986). Cell cycle control of the nucleocytoplasmic ratio in early *Drosophila* development. *Cell (Cambridge, Mass.)* **44**, 365–372.
Endow, S. A., and Hatsumi, M. (1991). A multimember kinesin gene family in *Drosophila*. *Proc. Natl. Acad. Sci. U.S.A.* **88**, 4424–4427.
Foe, V. E. (1989). Mitotic domains reveal early commitment of cells in *Drosophila* embryos. *Development (Cambridge, UK)* **107**, 1–22.
Foe, V. E., and Alberts, B. M. (1983). Studies of nuclear and cytoplasmic behavior during the five mitotic cycles that precede gastrulation in *Drosophila* embryogenesis. *J. Cell Sci.* **61**, 31–70.
Foe, V. E., and Odell, G. M. (1989). Mitotic domains partition fly embryos, reflecting early cell biological consequences of determination in progress. *Am. Zool.* **29**, 617–652.
Freeman, M., and Glover, D. M. (1987). The *gnu* mutation of *Drosophila* causes inappropriate DNA synthesis in unfertilized and fertilized eggs. *Genes Dev.* **1**, 924–930.
Freeman, M., Nüsslein-Volhard, C., and Glover, D. M. (1986). The dissociation of nuclear and centrosomal division in *gnu*, a mutation causing giant nuclei in *Drosophila*. *Cell (Cambridge, Mass.)* **46**, 457–468.
Fuller, M. T., Regan, C. L., Green, L. L., Robertson, B., Deuring, B., and Hays, T. S. (1989). Interacting genes identify interacting proteins involved in microtubule function in *Drosophila*. *Cell Motil. Cytoskeleton.* **14**, 128–135.
Fuyama, Y. (1984). Gynogenesis in *Drosophila melanogaster*. *Jpn. J. Genet.* **59**, 91–96.
Fyrberg, E. A., and Goldstein, L. S. B. (1990). The *Drosophila* cytoskeleton. *Annu. Rev. Cell Biol.* **6**, 559–596.
García-Bellido, A., and Ripoll, P. (1978). Cell lineage and differentiation in *Drosophila*. In "Results and Problems in Cell Differentiation" (W. J. Gehring, ed.), Vol. 9, pp. 120–156. Springer-Verlag, Berlin.
Gatti, M. (1979). Genetic control of chromosome breakage and rejoining in *Drosophila melanogaster*: Spontaneous chromosome aberrations in X-linked mutants defective in DNA metabolism. *Proc. Natl. Acad. Sci. U.S.A.* **76**, 1377–1381.
Gatti, M., and Baker, B. S. (1989). Genes controlling essential cell-cycle functions in *Drosophila melanogaster*. *Genes Dev.* **3**, 438–452.
Gatti, M., Pimpinelli, S., and Baker, B. S. (1980). Relationships among chromatid interchanges, sister chromatid exchanges, and meiotic recombination in *Drosophila melanogaster*. *Proc. Natl. Acad. Sci. U.S.A.* **77**, 1575–1579.
Gatti, M., Pimpinelli, S., Bove, C., Baker, B. S., Smith, D. A., Carpenter, A. T. C., and Ripoll, P. (1983a). Genetic control of mitotic cell division in *Drosophila melanogaster*. In "Genetics: New Frontiers," Vol. 3, pp. 193–204. Proc. XV Int. Congress of Genetics, New Delhi.
Gatti, M., Smith, D. A., and Baker, B. S. (1983b). A gene controlling condensation of heterochromatin in *Drosophila melanogaster*. *Science* **221**, 83–85.
Gelbart, W. M. (1973). A new mutant controlling mitotic chromosome disjunction in *Drosophila melanogaster*. *Genetics* **76**, 51–63.
Glover, D. M. (1989). Mitosis in *Drosophila*. *J. Cell Sci.* **92**, 137–146.
Glover, D. M. (1990). Abbreviated and regulated cell cycles in *Drosophila*. *Curr. Opin. Cell Biol.* **2**, 258–261.
Glover, D. M. (1991). Mitosis in the *Drosophila* embryo—in and out of control. *Trends Genet.* **7**, 125–132.
Glover, D. M., Alphey, L., Axton, J. M., Cheshire, A., Dalby, B., Freeman, M., Girdham, C., Gonzalez, C., Karess, R. E., Leibowitz, M. H., LLamazares, S., Maldonado-Codina, M. G., Raff, J. W., Saunders, R., Sunkel, C. E., and Whitfield, W. G. F. (1989). Mitosis in *Drosophila* development. *J. Cell Sci.* **12**(Suppl.), 277–291.

Goldstein, L. S. B., Hardy, R. W., and Lindsley, D. L. (1982). Structural genes on the Y chromosome of *Drosophila melanogaster*. *Proc. Natl. Acad. Sci. U.S.A.* **79**, 7405–7409.

Goldstein, L. S. B., Laymon, R. A., and McIntosh, J. R. (1986). A microtubule-associated protein in *Drosophila melanogaster*. Identification characterization, and isolation of coding sequences. *J. Cell Biol.* **102**, 2076–2078.

González, C., Casal, J., and Ripoll, P. (1988). Functional monopolar spindles caused by mutation in *mgr*, a cell division gene of *Drosophila melanogaster*. *J. Cell Sci.* **69**, 39–47.

González, C., Casal, J., and Ripoll, P. (1989a). Relationship between chromosome content and nuclear diameter in early spermatids of *Drosophila melanogaster*. *Genet. Res.* **54**, 205–212.

González, C., Molina, I., Casal, J., and Ripoll, P. (1989b). Gross genetic dissection and interactions of the chromosomal region 95E;96F of *Drosophila melanogaster*. *Genetics* **123**, 371–377.

González, C., Saunders, R. D. C., Casal, J., Molina, I., Carmena, M., Ripoll, P., and Glover, D. M. (1990). Mutations at the asp locus of *Drosophila* lead to multiple free centrosomes in syncytial embryos but restrict centrosome duplication in larval neuroblasts. *J. Cell Sci.* **96**, 605–616.

Green, L. L., Wolf, N., McDonald, K. L., and Fuller, M. T. (1990). Two types of genetic interaction implicate the *whirligig* gene of *Drosophila melanogaster* in microtubule organization in the flagellar axoneme. *Genetics* **126**, 961–973.

Hardy, R. W. (1975). The influence of chromosome content on the size and shape of sperm heads in *Drosophila melanogaster* and the demonstration of chromosome loss during spermiogenesis. *Genetics* **79**, 231–264.

Hardy, R. W., Tokuyasu, K. T., and Lindsley, D. L. (1981). Analysis of spermatogenesis in *Drosophila melanogaster* bearing deletions for Y-chromosome fertility genes. *Chromosoma* **83**, 593–617.

Hartenstein, V., and Campos-Ortega, J. A. (1985). Fate-mapping in wild-type *Drosophila melanogaster*. I. The spatio-temporal pattern of embryonic cell divisions. *Wilhelm Roux's Arch. Dev. Biol.* **194**, 181–193.

Hayles, J., and Nurse, P. (1986). Cell cycle regulation in yeast. *J. Cell Sci.* **4**(Suppl.), 155–170.

Hays, T. S., Deuring, R., Robertson, B., Prout, M., and Fuller, M. T. (1989). Interacting proteins identified by genetic interactions: A missense mutation in α-tubulin fails to complement alleles of the testis-specific β-tubulin gene of *Drosophila melanogaster*. *Mol. Cell. Biol.* **9**, 875–884.

Jiménez, J., Alphey, L., Nurse, P., and Glover, D. M. (1990). Complementation of fission yeast $cdc2^{ts}$ and $cdc25^{ts}$ mutants identifies two cell cycle genes from *Drosophila*: A *cdc2* homologue and *string*. *EMBO J.* **9**, 3565–3571.

Karess, R. E., and Glover, D. M. (1989). *rough deal*: A gene required for proper mitotic segregation in *Drosophila*. *J. Cell Biol.* **109**, 2951–2961.

Karess, R. E., Chang, X., Edwards, K. A., Kulkarni, S., Aguilera, I., and Kiehart, D. P. (1991). The regulatory light chain of nonmuscle myosin is encoded by *spaghetti-squash*, a gene required for cytokinesis in *Drosophila*. *Cell (Cambridge, Mass.)* **65**, 1177–1189.

Kemphues, K. J., Kaufman, T. C., Raff, R. A., and Raff, E. C. (1982). The testis-specific β-tubulin subunit in *Drosophila melanogaster* has multiple functions in spermatogenesis. *Cell (Cambridge, Mass.)* **31**, 655–670.

Lee, M. G., and Nurse, P. (1987). Complementation used to clone a human homologue of the fission yeast cell cycle control gene *cdc2*. *Nature (London)* **327**, 31–35.

Lehner, C. F., and O'Farrell, P. H. (1989). Expression and function of *Drosophila* cyclin A during embryonic cell cycle progression. *Cell (Cambridge, Mass.)* **56**, 957–968.

Lehner, C. F., and O'Farrell, P. H. (1990a). The roles of *Drosophila* cyclins A and B in mitotic control. *Cell (Cambridge, Mass.)* **61**, 535–547.

Lehner, C. F., and O'Farrell, P. H. (1990b). *Drosophila cdc2* homologs: A functional homolog is coexpressed with a cognate variant. *EMBO J.* **9**, 3573–3581.

Léopold, P., and O'Farrell, P. H. (1991). An evolutionarily conserved cyclin homolog from *Drosophila* rescues yeast deficient in G1 cyclins. *Cell (Cambridge, Mass.)* **66,** 1207–1216.

Lewis, E. B., and Gencarella, W. (1952). Claret and nondisjunction in *Drosophila melanogaster*. *Genetics* **37,** 600–601.

Lindsley, D. L., and Tokuyasu, K. (1980). Spermatogenesis. *In* "The Genetics and Biology of *Drosophila*" (M. Ashburner and T. R. F. Wright, eds.), Vol. 2b, pp. 225–294. Academic Press, New York.

Llamazares, S., Moreira, A., Tavares, A., Girdham, C., Spruce, B. A., González, C., Karess, R. E., Glover, D. M., and Glover, D. M. (1991). *polo* encodes a protein kinase homolog required for mitosis in *Drosophila*. *Genes Dev.* **5,** 2153–2165.

McDonald, H. B., and Goldstein, L. S. B. (1990). Identification and characterization of a gene encoding a kinesin-like protein in *Drosophila*. *Cell (Cambridge, Mass.)* **61,** 991–1000.

McDonald, H. B., Stewart, R. J., and Goldstein, L. S. B. (1990). The kinesin-like *ncd* protein of *Drosophila* is a minus end-directed microtubule motor. *Cell (Cambridge, Mass.)* **63,** 1159–1165.

Mahowald, A. P., and Kambysellis, M. P. (1980). Oogenesis. *In* "The Genetics and Biology of *Drosophila*" (M. Ashburner and T. R. F. Wright, eds.), Vol. 2, pp. 141–209. Academic Press, New York.

Mahowald, A. P., and Strassheim, J. M. (1970). Intercellular migration of centrioles in the germarium of *Drosophila melanogaster*. An electron microscopy study. *J. Cell Biol.* **45,** 306–320.

Masui, Y., and Markert, C. L. (1971). Cytoplasmic control of nuclear behaviour during meiotic maturation of frog oocytes. *J. Exp. Zool.* **177,** 129–146.

Matthews, K. A., and Kaufman, T. C. (1987). Developmental consequences of mutations in the 84B α-tubulin gene of *Drosophila melanogaster*. *Dev. Biol.* **119,** 100–114.

Mazia, D. (1961). Mitosis and the physiology of cell division. *In* "The Cell" (J. Brachet and A. E. Mirsky, eds.), pp. 77–412. Academic Press, New York.

Nolan, J. M., Lee, M. P., Wyckoff, E., and Tao, S. H. (1986). Isolation and characterization of the gene encoding *Drosophila* DNA topoisomerase II. *Proc. Natl. Acad. Sci. U.S.A.* **83,** 3664–3668.

Nöthiger, R., Dübendorfer, A., and Epper, F. (1977). Gynandromorphs reveal two separate primordia for male and female genitalia in *Drosophila melanogaster*. *Wilhem Roux's Arch. Dev. Biol.* **181,** 367–373.

Nurse, P. (1990). Universal control mechanism regulating onset of M-phase. *Nature (London)* **344,** 503–508.

O'Farrell, P. H., Edgar, B. A., Lakich, D., and Lehner, C. H. (1989). Directing cell division during development. *Science* **246,** 635–640.

Porter, M. E., Hays, T. S., Grissom, P. M., Fuller, M. T., and McIntosh, J. R. (1987). Characterization of a high molecular weight, ATP sensitive, microtubule-associated polypeptide from *Drosophila* embryos. *J. Cell Biol.* **105,** 121a.

Pringle, J., and Hartwell, L. (1981). The *Saccharomyces cerevisiae* cell cycle. *In* "The Molecular Biology of the Yeast *Saccharomyces*" (S. Strathern, E. Jones and J. Broach, eds.), pp. 97–142. Cold Spring Harbor Laboratory, Cold Spring Harbor, New York.

Raff, J. W., and Glover, D. M., (1988). Nuclear and cytoplasmic cycles continue in *Drosophila* embryos in which DNA synthesis is inhibited with aphidicolin. *J. Cell Biol.* **107,** 2009–2019.

Raff, J. W., Whitfield, W. G. F., and Glover, D. M. (1990). Two distinct mechanisms localise cyclin B transcripts in syncytial *Drosophila* embryos. *Development (Cambridge, UK)* **110,** 1249–1261.

Regan, C. L., and Fuller, M. T. (1988). Interacting alleles that affect microtubule function: The *nc2* allele of the *haywire* locus fails to complement mutations in the testis-specific β-tubulin gene of *Drosophila*. *Genes Dev.* **2,** 82–92.

Regan, C. L., and Fuller, M. T. (1990). Interacting genes that affect microtubule function in *Drosophila melanogaster*: Two classes of mutation revert the failure to complement between *hay^{nc2}* and mutants in tubulin genes. *Genetics* **125,** 77–90.
Ripoll, P., Pimpinelli, S., Valdivia, M. M., and Avila, J. (1985). A cell division mutant with a functionally abnormal spindle. *Cell (Cambridge, Mass.)* **41,** 907–912.
Ripoll, P., Casal, J., and González, C. (1987). Towards the genetic dissection of mitosis in *Drosophila*. *BioEssays* **7,** 204–210.
Sagata, N., Watanabe, N., Vande Woude, G. F., and Ikawa, Y. (1989). The c-*mos* protooncogene product is a cytostatic factor responsible for meiotic arrest in vertebrate eggs. *Nature (London)* **342,** 512–518.
Saxton, W. M., and Raff, E. C. (1989). *Weltchementz* displayed by kinesin mutants: The kinesin heavy chain is essential in *Drosophila*. *J. Cell Biol.* **109,** 281a.
Saxton, W. M., Porter, M. E., Cohn, S. A., Scholey, J. A., Raff, E. C., and McIntosh, J. R. (1988). *Drosophila* kinesin: Characterization of microtubule motility and ATPase. *Proc. Natl. Acad. Sci. U.S.A.* **85,** 1109–1113.
Sequeira, W., Nelson, C. R., and Szauter, P. (1989). Genetic analysis of the *claret* locus of *Drosophila melanogaster*. *Genetics* **123,** 511–524.
Shamanski, F. L., and Orr-Weaver, T. L. (1991). The *Drosophila plutonium* and *pan gu* genes regulate entry into S phase at fertilization. *Cell (Cambridge, Mass.)* **66,** 1289–1300.
Smith, D. A., Baker, B. S., and Gatti, M. (1985). Mutations in genes encoding essential mitotic functions in *Drosophila melanogaster*. *Genetics* **110,** 647–670.
Sullivan, W. (1987). Independence of *fushi tarazu* expression with respect to cellular density in *Drosophila* embryos. *Nature (London)* **327,** 164–167.
Sunkel, C. E., and Glover, D. M. (1988). *polo*, a mitotic mutant of *Drosophila* displaying abnormal spindle poles. *J. Cell Sci.* **89,** 25–38.
Tassin, M., Maro, B., and Bornens, M. (1985). Fate of microtubule-organizing centers during myogenesis *in vitro*. *J. Cell Biol.* **100,** 35–46.
Tomlinson, A. (1988). Cellular interactions in the developing *Drosophila* eye. *Development (Cambridge, UK)* **104,** 183–193.
Walker, R. A., Salmon, E. D., and Endow, S. A. (1990). The *Drosophila claret* segregation protein is a minus-end directed motor molecule. *Nature (London)* **347,** 780–782.
Whitfield, W. G. F., Millar, S. E. Saumweber, H., Frasch, M., and Glover, D. M. (1988). Cloning of a gene encoding an antigen associated with the centrosome in *Drosophila*. *J. Cell Sci.* **89,** 467–480.
Whitfield, W. G. F., González, C., Sánchez-Herrero, E., and Glover, D. M. (1989). Transcripts of one of two *Drosophila* cyclin genes become localized in pole cells during embryogenesis. *Nature (London)* **338,** 337–340.
Whitfield, W. G. F., González, C., Maldonado-Codina, G., and Glover, D. M. (1990). The A- and B-type cyclins of *Drosophila* are accumulated and destroyed in temporally distinct events that define separable phases of the G2-M transition. *EMBO J.* **9,** 2563–2572.
Zalokar, M. C., Audit, C., and Erk, J. (1975). Developmental defects of female-sterile mutations of *Drosophila melanogaster*. *Dev. Biol.* **47,** 419–432.
Zhang, P., and Hawley, R. S. (1990). The genetic analysis of distributive segregation in *Drosophila melanogaster*. II. Further genetic analysis of the *nod* locus. *Genetics* **125,** 115–127.
Zhang, P., Knowles, B. A., Goldstein, L. S. B., and Hawley, R. S. (1990). A kinesin-like protein required for distributive chromosome segregation in *Drosophila*. *Cell (Cambridge, Mass.)* **62,** 1053–1062.
Zheng, Y., Jung, M. K., and Oakley, B. R. (1991). γ-Tubulin is present in *Drosophila melanogaster* and *Homo sapiens* and is associated with the centrosome. *Cell (Cambridge, Mass.)* **65,** 817–823.

8
Retinoic Acid Receptors: Transcription Factors Modulating Gene Regulation, Development, and Differentiation

Elwood Linney
Department of Microbiology
Duke University Medical Center
Durham, North Carolina 27710

I. Introduction
II. Modular Structure of the Steroid Receptor-like Retinoic Acid and Retinoid X Receptors
III. Molecular Specificity of the Retinoic Acid Receptors and Retinoid X Receptors
   A. Studies Describing the Presence of Specific Retinoic Acid Receptor or Retinoid X Receptor mRNAs (or Proteins)
   B. Potential Dimerization Specificities
   C. DNA Sequence Binding Specificity of Retinoic Acid Receptors
   D. Specificity Differences Based on Preferences of Specific Retinoic Acid Receptors or Retinoid X Receptors for Specific Retinoids
   E. The Availability of Different Retinoids to Different Tissue and at Different Developmental Times
IV. Biological Roles of Retinoic Acid
   A. Retinoic Acid and the Developing Chick Wing Bud
   B. The Use of Cell Culture Models to Study Retinoic Acid-Induced Differentiation
   C. Retinoic Acid as a Teratogen
   D. Recent Retinoic Acid-Related Studies in Medicine
V. Relationships between the Retinoic Acid Receptors and Other Transcription Factors and DNA Binding Proteins
VI. Current and Future Approaches for Dissecting Developmental Specificity and Function of the Retinoic Acid Receptors and Retinoid X Receptors
References

I. Introduction

In 1987, cDNA molecules were described that represented the genetic information coding for a new subclass of steroid receptor-like transcription factors, the retinoic acid receptors or RARs (Petkovich *et al.*, 1987; Giguere *et al.*, 1987). For scientists who were studying the myriad effects of vitamin A and retinoic acid on biological systems, this gene was an attractive candidate for producing at least one mediator molecule of events influenced by retinoic acid. In the short time since those initial reports, additional receptors have been identified and many discoveries have been made regarding RARs. The RARs remain a very attractive set of candidate genes that may play pivotal roles in differentiation, development, and gene regulation. This article will attempt to bring together this relatively new field with that of the biological effects of retinoids and to suggest roles for these receptors and strategies for experimentally examining RAR function and specificity.

Because at this time most of the available information is descriptive (as opposed to functional) the article will first cover molecular information regarding retinoic acid receptors [both RARs and the newly identified retinoid X receptors (RXRs); Mangelsdorf *et al.*, 1990; Hamada *et al.*, 1989; Henrich *et al.*, 1990; Oro *et al.*, 1990] before suggestions linking the two are presented. The emphasis of this article will be on RAR and RXR work with mouse and human genes. The article will not attempt to cover the cellular retinoic acid-binding proteins (CRABP I and II) because of the very different molecular structure of these binding proteins and the absence of any information concerning their roles in transcriptional regulation.

II. Modular Structure of the Steroid Receptor-like Retinoic Acid and Retinoid X Receptors

The RARs are members of a superfamily of ligand-inducible transcription factors that includes the steroid receptors, the thyroid hormone receptors, and the vitamin D_3 receptor (for review, see Evans, 1988; Green and Chambon, 1988; Beato, 1989; O'Malley, 1990; Packer, 1990a,b; De Luca, 1991; Mangelsdorf and Evans, 1992). As interacting molecules, the receptors have regions for binding to specific DNA sequences, for binding to retinoic acid, for dimerization of receptor molecules, and very possibly for interacting with other transcription factors. The receptor proteins have been described as containing different domains; the most common nomenclature is that of Krust *et al.* (1986), in which the receptors are divided into domains A, B, C, D, E, and F based originally on homologies between the chicken estrogen receptor, the human estrogen and glucocortocoid receptors, and the v-*erbA* oncogene. The C domain contains two zinc fingers that are responsible for the DNA binding activity of the receptors. The ligand-binding function is contained within the E domain. Because this division into domains was originally based on amino acid homologies, it does not necessarily correlate with the intron–exon genomic structure of the members of the steroid receptor-like gene family.

After the initial discovery of RAR-α (Petkovich *et al.*, 1987; Giguere *et al.*, 1987) it was shown that a previously published report of a steroid receptor-like gene (Dejean *et al.*, 1986; de The *et al.*, 1987) described a second class of RAR, RAR-β (Brand *et al.*, 1988). This gene was originally reported as a hepatitis B viral integration site (Dejean *et al.*, 1986) and was also independently isolated as RAR-ε (Benbrook *et al.*, 1988). A third class of RARs (RAR-γ) was described

8. Retinoic Acid Receptors

for mouse (Zelent et al., 1989; Giguere et al., 1990) and human (Krust et al., 1989; Ishikawa et al., 1990). A comparison of the amino acid homologies of the original members of the three human (hRAR) and mouse (mRAR) receptor classes is illustrated in Fig. 1.

Between the receptor classes there are strong homologies in the B, C, D, and E domains. There is considerable variance in domains A and F, suggesting that specificity differences between RARs may be conferred by these regions. The fact that within a class of RARs there is considerable homology between human and mouse across the receptor suggests, from an evolutionary standpoint, the importance of each region of the receptors. The noticeable difference in the F domain of the mouse and human RAR-γ is due to an apparent frameshift mutation in the coding region of the F domain of hRAR-γ.

Additional complexity was discovered for hRAR-γ (Krust et al., 1989) and for mRAR-γ (Kastner et al., 1990; Giguere et al., 1990). Seven isoforms of mRAR-γ have been described: γ-1 (or -A), γ-2 (or -B), and γ-3 through γ-7. These isoforms vary in their sequences 5′ to the B domain sequences. The B, C, D, E, F, and non-coding 3′ sequences are identical. The different combinations appear to be due to alternative splicing and possibly two promoters (Kastner et al., 1990; Giguere et al., 1990). The different isoforms can have different and specific patterns of expression.

This form of genomic complexity for the 5′ end of the RAR-γ gene has recently also been described for the RAR-β and RAR-α genes. Leroy et al. (1991a) have reported seven isoforms of mRAR-γ generated presumably by alternative splicing and two different promoters, one of which could itself be retinoic acid inducible. In a companion study from the Chambon laboratory, Zelent et al. (1991) have reported the isolation of three different mRAR-β cDNA forms: mRAR-β-1, mRAR-β-2, and mRAR-β-3. The mRAR-β-2 is the isoform

COMPARISON	A	B	C	D	E	F
H-α/M-α	98%	100%	98%	98%	99%	90%
H-β/M-β	94%	100%	100%	98%	99%	92%
H-γ/M-γ	98%	100%	100%	100%	100%	58%
H-α/H-β	<15%	79%	97%	74%	90%	20%
H-γ/H-α	25%	75%	97%	72%	84%	20%
H-γ/H-β	<15%	86%	94%	62%	90%	<15%

Fig. 1 Amino acid homologies between human and mouse retinoic acid receptors. The retinoic acid receptor (RAR) has been described in terms of domains A through F as originally used for the estrogen receptor by Krust et al. (1986). The percentage homologies are those originally published by Krust et al. (1989). These values are for the first isoform of each class to be described.

previously described as mRAR-β by Zelent et al. (1989). This RAR-β mRNA has been shown to be retinoic acid inducible (de The et al., 1989; Zelent et al., 1989; Hu and Gudas, 1990) and the inducibility is conferred via a retinoic acid response element (RARE) (de The et al., 1990a; Sucov et al., 1990; Hoffman et al., 1990). In the Zelent et al. (1991) study it has been shown that mRAR-β-1 and mRAR-β-3 are generated from a region upstream of the mRAR-β-2 promoter and that they also are retinoic acid inducible.

That the three different classes of RARs are indeed different genes is illustrated from the homologies described in Fig. 1, by the description of the different isoforms, and by the mapping of the genes to three different chromosomes. The human RAR-α gene has been mapped to chromosome 17 (Mattei et al., 1988a), the human RAR-β gene to chromosome 3 (Mattei et al., 1988b), and the human RAR-γ gene to chromosome 12 (Ishikawa et al., 1990).

While the domain designation of the steroid receptor-like proteins is based on amino acid homologies (Krust et al., 1986) and not on exon–intron structure, the published information currently available indicates that there are intron–exon borders between the B domain and the various isoform sequences 5' to the B domain for all three RAR classes. However, for RAR-α, it has been reported that the next intron is between the two zinc finger regions of the DNA binding or C domain (Ponglikitmongkol et al., 1988). Further information concerning the genomic organization of RARs remains to be published, but if there are additional parallels with the steroid receptor family, it might be expected that the remaining parts of the coding region will be separated by several intronic regions.

Three points can be made with regard to the amino acid comparisons between the human and mouse RARs: (1) the strong homologies between the human and mouse RAR forms for any specific class suggest, from an evolutionary standpoint, that each region of the receptor must provide important function; (2) from receptor to receptor there is considerable homology in domain C, which contains two zinc finger motifs shown to be responsible for DNA binding (Giguere et al., 1987; Petkovich et al., 1987); and (3) the major differences between RAR classes appear to be in the A domains and the F domains.

In addition to the three major classes of RARs and their respective isoforms, Mangelsdorf et al. (1990) have reported another class of retinoid receptors, RXRs. The RXRs are also steroid receptor-like in structure, but the amino acid homology between hRXR-α and the RARs is not strong (61% amino acid homology in the DNA-binding domain when compared with hRAR-α, 27% amino acid homology in the E and F domains when compared with hRAR-α). Mangelsdorf et al. (1990) showed that the RXR-α did respond to retinoic acid and to a receptor gene construct containing a thyroid receptor element (TRE) to which the RARs had been shown to respond (Umesono et al., 1988). However, RXR-α and RAR-α responded to other retinoids in different ways.

Independent of Mangelsdorf et al. (1990), Hamada et al. (1989) reported a cDNA, H2-RIIBP, which had been isolated from a mouse liver cDNA expression

8. Retinoic Acid Receptors

library through its binding to a radioactive probe to the RII region of the *H2* promoter. When the manuscript was written, the authors did not know the identity of the ligand for H2-RIIBP. Mangelsdorf *et al.* (1990) suggested that this gene was the mouse RXR-β because of its homology to the hRXR-β gene. This is indeed the case. Mangelsdorf and Evans (1992) report on three mouse RXR genes: RXR-α, -β, and -γ. This laboratory has independently isolated mRXR-α and mRXR-γ (C. Hoopes, Q. Liu, and E. Linney, unpublished observations, 1991) and the amino acid homologies of the three mRXRs are compared with hRXR-α in Fig. 2. In addition, we have mapped mRXR-α, mRXR-β, and mRXR-γ to mouse chromosomes 2, 17, and 1, respectively (Hoopes *et al.*, 1992), illustrating that they are clearly separate genes.

Two recent reports of *Drosophila* cDNAs with strong homologies to hRXR-α (XR2C; Oro *et al.*, 1990) and H2-RIIBP (2C; Henrich *et al.*, 1990) suggest that RXRs might be present in other species. These RXRs presumably represent the same gene. Oro *et al.* (1990) reported the XR2C gene as the ultraspiracle (*usp*) locus in *Drosophila*, a locus required both maternally and zygotically for pattern formation. A ligand has yet to be identified for these *Drosophila* receptors.

Recently, Thaller and Eichele (1990) have isolated and identified a new retinoid from the chick wing bud, 3,4-didehydroretinoic acid. Additionally, in a companion manuscript from the same laboratory, Wagner *et al.* (1990) identified polarizing activity in the floor plate of the neural tube similar to that which has already been identified for the chick wing bud (Tickle *et al.*, 1975, 1982; Sum-

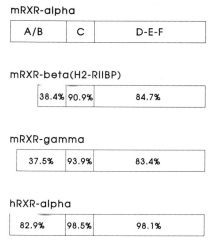

Fig. 2 Amino acid similarities (using the program of Smith and Waterman, 1981) between mouse RXR-α, mouse RXR-β (H2-RIIBP), mouse RXR-γ, and human RXR-α, respectively. The hRXR-α sequence was from Mangelsdorf *et al.* (1990) and the H2-RIIBP sequence was from Hamada *et al.* (1989). The mouse RXR-α sequence is from C. Hoopes and E. Linney (unpublished observations, 1991) and the mouse RXR-γ sequence is from Q. Liu and E. Linney (unpublished observations, 1991).

merbell, 1983; Thaller and Eichele, 1987) and showed that this region can synthesize retinoic acid and 3,4-didehydroretinol, the precursor to 3,4-didehydroretinoic acid. Heyman *et al.* (1992) and Levin *et al.* (1992) have recently shown that 9-*cis*-retinoic acid binds and activates the RXR-α receptor at a much lower concentration that all-*trans*-retinoic acid. Therefore, the actual metabolism of retinoids may play a significant indirect role in the regulation of the RAR- and RXR-mediated activity.

In summary, there are three major classes of RARs, each with different isoforms that result in variances at the amino terminus. There is also a parallel set of retinoid receptors, RXRs, of which three different members have been identified. These receptors have a modular structure with regions for interaction with other molecules.

III. Molecular Specificity of Retinoic Acid Receptors and Retinoid X Receptors

The question of specificity arises when one considers the large number of individual receptor molecules derived from th RARs and RXRs. Are these all independent ligand-inducible transcription factors that interact independently with different promoters, or is there some overlap in specificity or interaction between different receptor molecules to create a unique specificity?

We can discuss the specificity of the receptors at several levels: (1) the different receptors are expressed in different tissues and at different developmental times; (2) the receptors, like steroid receptors, probably dimerize when binding to their specific target sequence, and there may be some specificity to this dimerization that would allow for a hierarchical affinity for different dimer states, similar to the Fos/Jun family; (3) the RARs and the RXRs appear to interact with specific repeat sequences, but the biological response to this interaction is different and dependent on the arrangement of the DNA sequences; (4) the specific RARs and/or RXRs may have specificity preferences for different retinoids; (5) retinoic acid gradients, or retinoids preferred by a specific RAR or RXR, could impose specificity on this system; and (6) part or all of the individual specificity differences between receptors may indicate the presence or absence of interacting cellular factors. These issues will be discussed below.

A. Studies Describing the Presence of Specific Retinoic Acid Receptor or Retinoid X Receptor mRNAs (or Proteins)

The presence of RARs in different cells at different times has been examined through (1) direct analysis of extracted RNA from different tissues using either Northern gel analysis or RNase protection techniques, (2) *in situ* analysis of

8. Retinoic Acid Receptors

RNA, and (3) (to a much lesser extent) analysis of proteins present in cells. While these studies are quite valuable in directing investigators to the source of different RARs, we should consider several aspects of these approaches so that the data are not overinterpreted: (1) in some of the Northern analyses, rehybridization of the Northern blots with a standard probe has not been done, allowing for the possibility that absence of a transcript might also mean that the particular RNA preparation is degraded; (2) the presence of an mRNA for a protein does not always mean that the protein is present in the tissue; (3) most of the published *in situ* studies were performed or published before the various isoforms of the RARs were described, with the result that extensive *in situ* data concerning expression differences between different isoforms of the same RAR are not available at this time; (4) three isoforms of RAR-β and at least one isoform of RAR-α are retinoic acid inducible, so that metabolism of retinoids *in vivo* might modulate the presence of these transcripts over time. While there is growing information about retinoid metabolism (for review, see Blomhof *et al.*, 1990), little is know about the localization and metabolism of retinoic acid or retinoids in the developing embryo.

Before different isoforms of the RARs were clearly defined, Northern gel analysis of RAR expression suggested that RAR-α was expressed in most tissues (Giguere *et al.*, 1987; Zelent *et al.*, 1989). While RAR-β or -ε had a more limited expression profile (Brand *et al.*, 1988; Benbrook *et al.*, 1988; Zelent *et al.*, 1989), RAR-γ (Zelent *et al.*, 1989; Krust *et al.*, 1989) showed distinct expression in the skin of newborn and adult mice. In F9 EC cells, RAR-α and RAR-γ mRNAs are present in undifferentiated and retinoic acid-induced cells (Zelent *et al.*, 1989; Hu and Gudas, 1990) while RAR-β mRNA is induced by retinoic acid in F9 EC cells (Zelent *et al.*, 1989; Hu and Gudas, 1990). This retinoic acid induction of RAR-β was originally described by de The *et al.* (1989) in human hepatoma cells.

While RXR-α is expressed relatively strongly in kidney, liver, lung, and muscle (Mangelsdorf *et al.*, 1990), H2-RIIBP (or mouse RXR-β) is expressed relatively strongly in brain and thymus and weakly in spleen and liver (Hamada *et al.*, 1989). Strong expression of RXR-γ is reported in skeletal muscle and heart (Mangelsdorf and Evans, 1992).

With the discovery of isoforms of RAR-α (Leroy *et al.*, 1991a), RAR-β (Zelent *et al.*, 1991), and RAR-γ (Krust *et al.*, 1989; Kastner *et al.*, 1990; Giguere *et al.*, 1990), the specific expression of different RARs increases in complexity.

The expression of mRNAs for RAR-α, RAR-β, RAR-γ and the cellular retinoic acid binding protein (CRABP) has been examined in the developing limbs of the mouse (Dolle *et al.*, 1989). The mRAR-α and mRAR-γ are present and uniformly distributed in the day-10 embryonic limb bud. At later stages mRAR-γ transcripts become specific to cartilage tissue and differentiating skin while mRAR-β transcripts are restricted mainly to the interdigital mesenchyme. These differences in expression suggest different roles for the RARs in the developing limb bud. Because the probes used were full-length cDNAs, the results did not distinguish between the different isoforms of the three classes of RARs.

Ruberte et al. (1990) have examined the spatial and temporal distribution of RAR-γ transcripts from days 6.5 to 15.5 of mouse embryonic development. Transcripts of RAR-γ appear as early as day 8 in the presomitic posterior region. From days 9.5 and 11.5, the transcripts are uniformly distributed in the mesenchyme of the frontonasal region pharyngeal arches, limb buds, and sclerotomes. As development proceeds, transcripts are detected in all cartilages and differentiating squamous keratinizing epithelia along with developing teeth and whisker follicles. The patterns of expression in various tissues, some of which have been shown to be sensitive to retinoic acid as a teratogen, suggest important roles for the RAR-γ at these developmental stages. The probe used for RAR-γ included the whole coding region of RAR-γ-1 (or -A), so one could not distinguish between different expression patterns for different RAR-γ isoforms.

Recently, Dolle et al. (1990) and Ruberte et al. (1991) have reported on the expression of RAR-α, -β, -γ, and the cellular retinoic acid and retinol-binding proteins during mouse organogenesis and early morphogenesis. The probes used for the RARs in these studies would not distinguish between the different isoforms of the RARs.

Expression of the receptors as protein remains to be documented in detail. Anti-peptide antisera to hRAR-α and hRAR-β have been generated by Gaub et al. (1989), but because anti-peptide antisera are usually not effective for immunohistological analysis, the antisera have been primarily used in Western analysis of protein from mammalian cells in culture. An alternative approach has recently been published by Mendelsohn et al. (1991): they have produced transgenic mice that have the RAR-β-2 promoter driving the expression of the reporter gene, bacterial β-galactosidase. Accepting certain assumptions (such as the investigators having all of necessary regulatory sequences in their promoter fragment) the expression pattern of β-galactosidase should represent the expression patterns of the RAR-γ mRNAs that are driven from this promoter.

In summary, there are distinct differences in expression of different RARs in developing mouse embryos and in adult tissues. However, a detailed *in situ* analysis of the RXRs and the individual isoforms of each of the RARs remains to be reported.

B. Potential Dimerization Specificities

Forman and Samuels (1990a,b) have reviewed the dimerization signals for the steroid receptor family of receptors. Other investigators have also considered the dimerization signals in general reviews of these receptors (Beato, 1989; Green and Chambon, 1988; O'Malley, 1990; Nunez, 1989; Ham and Parker, 1989;

8. Retinoic Acid Receptors 317

Mangelsdorf and Evans, 1992). By extrapolation from the data on dimerization of steroid receptors (Kumar and Chambon, 1988; Tsai *et al.*, 1988; Guiochon-Mantel *et al.*, 1989; Fawell *et al.*, 1990) we may conclude that there is probably a major region for dimerization in the ligand binding domain of the mRNAs. Two independent studies involving deletion mutants of the thyroid receptors (c-ErbA-α and c-ErbA-β receptors) provide evidence for this conclusion. Forman *et al.* (1988) showed that c-ErbA-α mediated thyroid hormone-independent gene expression in certain cell types. Using deletion mutants of c-ErbA-α, they determined that a region in the hormone binding domain that contains eight potential α-helical heptads was important for activity. Overexpression of this region (without a DNA binding domain) inhibited ligand-inducible activation of endogenous thyroid receptor and exogenously added RAR-α. Forman *et al.* (1989) also suggested that the formation of heterodimers between the deleted c-ErbA-α receptor fragment and hRAR-α caused the inhibition of the hRAR-α activity. Glass *et al.* (1989) provided additional evidence for a dimerization domain within the ligand-binding domain; RAR-α and c-ErbA-β were translated *in vitro* and shown to cooperatively interact in the binding of RARs to thyroid response elements (TREs). Through deletion analysis of the β-thyroid receptor and the RAR-α, they were also able to identify a region in the ligand-binding domain responsible for this cooperative interaction. The regions defined by these two studies overlap with the dimerization region identified for the estrogen receptor (Fawell *et al.*, 1990).

The mapping of the dimerization domains of the receptors has taken on increased significance with the finding that the RXRs can heterodimerize with RARs, the vitamin D receptor, and thyroid hormone receptors (Yu *et al.*, 1991; Leid *et al.*, 1992; Kliewer *et al.*, 1992; Zhang *et al.*, 1992; Marks *et al.*, 1992). This *in vitro* work suggests that there are multiple possibilities of receptor interaction, at least among the RXRs and other receptors. Understanding the hierarchical interrelationships between these receptors and the DNA sequences with which they interact could aid in understanding how different genes are regulated by these receptors. The observation also complicates the interpretation of transfection experiments with receptor expression vectors, because the signal obtained from a receptor construct may be due not only to the expression of the receptor gene transfected into the cell, but also due to something in the cell heterodimerizing with this added receptor.

Dimerization sites could be present in the DNA binding domains of the receptors, based on studies with the estrogen receptor (Kumar and Chambon, 1988; Fawell *et al.*, 1990). Recent nuclear magnetic resonance (NMR) and/or X-ray analysis of the DNA binding domain of the glucocortocoid receptor (Hard *et al.*, 1990; Luisi *et al.*, 1991) and the estrogen receptor (Schwabe *et al.*, 1990) suggests potential contact sites for dimerization in the DNA binding domains of these receptors. This is also supported by deletion studies with the glucocorticoid and estrogen receptors. Deleting the ligand binding domains of these receptors results in weak constitutive transcriptional activity when they are transfected into cells in culture (Godowski et al., 1987; Kumar et al., 1987; Hollenberg et al., 1987).

C. DNA Sequence Binding Specificity of Retinoic Acid Receptors

The original studies describing the hRAR-α receptor (Petkovich et al., 1987; Giguere et al., 1987) showed that the cDNAs code for a retinoic acid-inducible transcription factor by using the technique of domain-swapping; that is, the DNA binding domain of either the estrogen receptor or the glucocortocoid receptor was used to replace the putative DNA binding domain of the RAR. Cotransfection experiments with reporter plasmids coupled to DNA elements—to which either the estrogen receptor or the glucocortocoid receptor bound—resulted in retinoic acid induction from these chimeric receptors.

Umesono et al. (1988) showed that RAR-α could trans-activate via the DNA binding element of the TRE. There are various natural and artificial TREs. Umesono et al. (1988) used (1) a natural TRE found in the growth hormone promoter (5'-CAGGGACGTGACCGCA-3'; Glass et al., 1987), and (2) an artificial palindromic TRE (TREp: 5'-TCAGGTCATGACCTGA-3'). In both cases they could detect a retinoic acid induction of chloramphenicol acetyltransferase (CAT) activity from their reporter gene. This discovery allowed for more detailed study of transcriptional activation and retinoic acid sensitivity of the different RARs. All three of the RARs have been shown to trans-activate through TREs, as might be expected from their homologous DNA binding domains (see Fig. 1).

The RAR-β has also been shown to be retinoic acid inducible (de The et al., 1989; Zelent et al., 1989; Hu and Gudas, 1990) and when one of the RAR-β promoter regions was examined, an RARE was identified as conferring retinoic acid inducibility to reporter genes (de The et al., 1990a; Sucov et al., 1990; Hoffman et al., 1990). Rather than a palindromic sequence, this RARE is a direct repeat sequence just 5' to a "TATA" element: 5'-<u>GTTCAC</u> CGAAA <u>GTTCAC</u> TCGCA TATA-3'. Transfection assays with the RARE coupled to a promoter and a reporter gene have shown that RAR-α, -β, and -γ all respond to this sequence (de The et al., 1990a; Sucov et al., 1990; Hoffman et al., 1990) although RAR-γ is a poor activator of this RARE (Hoffman et al., 1990). While the thyroid hormone receptor does not activate reporter genes coupled to this RARE (Sucov et al., 1990; Hoffman et al., 1990) the thyroid hormone receptor has been shown to bind to it (Hoffman et al., 1990).

Several additional RAREs have been functionally identified from genes that are retinoic acid inducible (see Table I). Some of these sequences have been examined both for DNA binding and transcriptional activation. In most cases, the sequences appear to have direct repeats rather than palindromic arrangements of binding sites. In Table I, the arrows represent 5'-TCA-3' sequences because

Table I Regulatory Sequences for Retinoic Acid Receptors and Retinoid X Receptors[a]

Receptor	Regulatory sequence	Ref.
TREp	TCAGGTCATGACCTGA →→ AGTCCAGTACTGGACT ←←	Umescono et al. (1988)
RAR-β-2	AGGGTTCACCGAAAGTTCACTCG →→ TCCCAAGTGGCTTTCAAGTGAGC	de The et al. (1990) Sucov et al. (1990) Hoffmann et al. (1990)
RAR-α-2	GCGAGTTCAGCAAGAGTTCAGCC →→ CGCTCAAGTCGTTCTCAAGTCGG	Leroy et al. (1991b)
ADH3	AAAACTGAACTCTGAATGACCCCTGTGG TTTTGACTTGAGACTTACTGGGGACACC ←←←	Duester et al. (1991)
B1LAM	CCAGACAGGTTGACCCTTTTTCTAAGGGCTTAACCTAGCTCACCTC → GGTCTGTCCAACTGGGAAAAAGATTCCCGAATTGGATCGAGTGGAG ←	Vasios et al. (1989) Vasios et al. (1991)
CRBPI	TTTAGTAGGTCAAAAGGTCAGACAC →→ AAATCATCCAGTTTTCCAGTCTGTG	Smith et al. (1991)
PEPCK	CCCTTCTCATGACCTTTGGCCGTGGGAGTGACACCT → GGGAAGAGTACTGGAAACCGGCACCCTCACTGTGGA ←←	Lucas et al. (1991)
Factor H	CAATCCAGCAGGTCACTGACAGGGC → GTTAGGTCGTCCAGTGACTGTCCCG ←	Munoz-Canoves et al. (1990)
APOA1	ACTGAACCCTTGACCCCTGCCCT TGACTTGGGAACTGGGGACGGGA ←←	Rottman et al. (1991)
CRBPII	GCTGTCACAGGTCACAGGTCACAGGTCACAGTTCA →→→→ CGACAGTGTCCAGTGTCCAGTGTCCAGTGTACAAGT	Mangelsdorf et al. (1991)
DR+3	TCAGGTCACTGTCAGGTCA →→→→ AGTCCAGTGACAGTCCAGT	Naar et al. (1991)
DR-5	AGGTCACCAGGAGGTCA → TCCAGTGGTCCTCCAGT	Umesono et al. (1991)

[a]Regulatory sequences for RARs and RXRs identified through transcriptional activation experiments. Arrows represent the presence of a 5'-TCA-3' sequence, because this has been identified as part of the half-site recognition sequence in RAREs and RXREs. 5'-GGTCA-3' and 5'-AGGTCA-3' have been commonly described as half-site recognition sequences.

these have been functionally associated with direct repeats in the RAR-β RARE (as 5′-GTTCAC-3′ by Sucov et al., 1990) in laminin B_1 sequences by Vasios et al., (1991) (as 5′-TGACC-3′ or the complementary strand 5′-GGTCA-3′), in the cellular retinol binding protein I (CRBPI) by Smith et al. (1991) (as 5′-GGTCA-3′), and in artificially constructed RAREs [as 5′-AGGTCA-3′ in DR+3 of Naar et al. (1991) and DR-5 of Umesono et al. (1991)].

Regions of the DNA binding C domain have been mutagenized in the steroid receptors to examine what amino acids are critical for the DNA binding specificity of the different steroid receptors. Some general suggestions of elements critical to binding specificity have emerged from this work. The two zinc fingers of the C domain appear to have separate functions. The DNA sequences to which the steroid receptors bind are usually palindromic in nature or have different spacing of the DNA sequence between the two elements of the palindrome. The different binding specificities can be divided into specificity for one-half of the palindrome plus specificity for the size of the DNA sequence spacing between the palindromic elements. Discrimination of specific DNA binding sites appears to be determined by three amino acids at the base of the first zinc finger (Mader et al., 1989; Danielsen et al., 1989; Umesono and Evans, 1989; Schena et al., 1989). This has been termed the P box by Umesono and Evans (1989). They also described the D box, a region of five amino acids in the second zinc finger that appears to discriminate the spacing between the two elements of a DNA binding site. When the RARs, RXRs, vitamin D_3 receptor, and estrogen receptor are compared in these P and D domains one can seen strong homologies in these specific regions (see Table II). Given the above differences in the the P and D boxes, the various members of the steroid/thyroid/retinoic acid/vitamin D_3 receptors have been classified into families based on the sequences in the zinc fingers, and on the sequences to which they have been shown to functionally respond (Forman and Samuels, 1990a). While this type of organization is useful for future experiments, the molecular details of how the same receptor (RAR) can activate through either a palindromic sequence such as TREp or a direct repeat sequence such as that found in the RAR-β-2 promoter RARE underline how little is known molecularly about these receptor–DNA interactions.

The similarities in the P box of the RARs, vitamin D receptor, and thyroid hormone receptor compared with their dissimilar D boxes (see Table II) suggest that half-site sequence binding of these receptors may be similar while the arrangement of the half-sites may be different for the different receptors. Naar et al. (1991) and Umesono et al. (1991) recently examined this through a series of constructions using 5′-AGGTCA-3′ as a half-site and constructing direct repeat sequences with this half-site but varying in the spacer sequences between the half-site. When these sequences were coupled to simple promoters and reporter genes and cotransfected with expression vectors for the different receptors, a pattern of specificity developed: RARs has a transcriptional activation preference

8. Retinoic Acid Receptors

Table II C Domain P and D Boxes: Amino Acid Comparisons[a]

	5' Zinc finger P box	3' Zinc finger D box
RAR-α	EGckG	HRDKN
RAR-β	EGckG	HRDKN
RAR-γ	EGckG	HRDKN
mRXR-α	EGckG	RDNKD
mRXR-β (H2-RIIBP)	EGckG	RDNKD
mRXR-γ	EGckG	RDNKD
Vitamin D_3 receptor	EGckG	PFNGD
c-ErbA-α	EGckG	KYDSC
c-ErbA-β	EGckG	KYEGK
Estrogen receptor	EGckA	PATNQ
Glucocorticoid receptor	GSckV	AGRND

[a]Classification of P and D box regions of various receptors. Capital letters indicate those amino acids shown by mutationl analysis to be critical to DNA binding specificity of the glucocorticoid and estrogen receptors. The other receptor sequences are placed here for comparison. Sequence information for mRXR-α and mRXR-γ (C. Hoopes, Q. Liu, and E. Linney, unpublished observations, 1991) H2-RIIBP (Hamada *et al.*, 1989) and RAR-γ (Krust *et al.*, 1989; Zelent *et al.*, 1989) has been added to the information organized and referenced in Umesono and Evans (1989) describing P and D boxes.

for an artificial sequence having 5 bp separating the AGGTCA repeats [this was named DR+3 by Naar *et al.*, (1991) and DR-5 by Umesono *et al.* (1991)]. These sequences are shown in Table I and compared with other RAREs and RXREs.

The above observations were named the 3-4-5 rule by Umesono *et al.* (1991) for the base pair separation of the AGGTCA direct repeats that the vitamin D receptor, the thyroid hormone receptor, and the RARs preferred, respectively. Recent studies have identified promoter sequences through which RXR-α transcriptionally activates. Rottman *et al.* (1991) have identified a DNA sequence in the apolipoprotein A_1 promoter to which RXR-α binds. This sequence is shown in Table I. When coupled to a thymidine kinase promoter and a reporter gene, there was a distinct transcriptional activation preference for RXR-α versus RAR-α. Mangelsdorf *et al.* (1991) have identified a repeated AGGTCA sequence in the cellular retinol binding protein II (CRBPII) promoter. This sequence functions as a retinoid X receptor response element (RXRE) showing a transcription activation preference for RXR-α over RAR-α even though both RXR-α and RAR-α bind to the sequence. Cotransfection of RAR-α inhibits the transcriptional activation effected by RXR-α through this RXRE sequence (Mangelsdorf *et al.*, 1991). In keeping with the 3-4-5 rule, the CRBPII RXRE element has 1 bp separating the AGGTCA repeat; therefore, RXRs may join the formulation with a name change: the 1-3-4-5 rule.

To summarize, there are DNA sequences (TREs) to which both thyroid hormone receptors, RARs and RXR-α, have been shown to functionally respond. There is a growing number of DNA sequences to which RARs functionally respond (RAREs) and to which RXRs respond (RXREs) (see Table I). However, the basic molecular biology of the interaction of RARs, RXRs, or heterodimers of the six different receptor gene products with specific half-site sequences or more complicated arrays of DNA elements has yet to be described. It is apparent with the discovery that RXRs can heterodimerize with several different receptors (Yu *et al.*, 1991; Leid *et al.*, 1992; Kliewer *et al.*, 1992; Zhang *et al.*, 1992; Marks *et al.*, 1992) that we know very little about the molecular picture at the DNA binding site. It is unclear what distinguishes simple DNA binding from functional DNA binding and whether there will be other transcription factor heterodimer partners with RXRs. Certainly, it can be anticipated that new DNA response elements will be identified or known response elements will be more carefully defined to take into account the possibility of heterodimer interactions.

D. Specificity Differences Based on Preferences of Specific Retinoic Acid Receptors or Retinoid X Receptors for Specific Retinoids

Various natural retinoids have been identified and large numbers of synthetic retinoids have been produced. Some synthetic analogs have been compared with retinoic acid for their ability to trans-activate the three RARs (Astrom *et al.*, 1990). As the individual receptors are purified as protein, it should be possible to determine specific binding constants and/or transcriptional activation preferences of different RARs and RXRs for different retinoids. Specific retinoids might then be used to identify the major receptor class affecting a teratogenic or developmental event. Two very different areas on retinoid studies suggest that this might be possible. In one system, the developing chick limb bud, a zone of polarizing activity (ZPA) plays a major role in the distinct digitation of the limb. When such a zone is transplanted to the opposite side of the bud, it produces a mirror image in the digitation of the limb. This effect can be mimicked by creating an opposing gradient of retinoic acid in the limb bud. This suggests either that retinoic acid is a morphogen in this system, or that it stimulates a morphogen. The second possibility is most likely, based on recent studies by Noji *et al.* (1991) and Wanek *et al.* (1991). In both these studies data were presented that supported the hypothesis that retinoic acid induces a morphogen. Recently, Thaller and Eichele (1990) have shown a new retinoic acid form to be present in the chick lamb bud: didehydroretinoic acid. A companion study showed that the floor plate of the neural tube of the chick has ZPA activity in the wing bud system and is capable of producing this new retinoid (Wagner et al., 1990).

The other system is described by Mangelsdorf et al. (1990), identifying the new class of retinoid receptors, RXRs. This name was given to this new class because, even though the RXR-α responded to retinoic acid in cotransfection assays with reporter genes, different responses were obtained when different retinoids were compared with RXR-α and RAR-α. Recently, Heyman et al. (1992) and Levin et al. (1992) have identified 9-cis-retinoic acid as a high-affinity ligand for RXR-α. This ligand binds to RXR-α and in transient transfection experiments is a much more effective inducer of RXR-α compared with all-trans-retinoic acid. The studies suggested that RARs can also bind to and be activated by 9-cis-retinoic acid. Therefore, the distinguishing characteristics between these two ligands and RARs and RXR-α is that considerably higher concentrations of all-trans-retinoic acid are required for RXR activation.

E. The Availability of Different Retinoids to Different Tissue and at Different Developmental Times

Because retinoids are natural substances with specific functions in normal metabolism and development, the extremely pleiomorphic effects of additional retinoic acid administered to embryos suggest that under normal conditions there are distinct and controlled ways of providing retinoids to specific cells (see Blomhoff et al., 1990, for a review). In the studies describing 9-cis-retinoic acid as a high-affinity ligand of RXR α it was shown that this ligand could be isolated from tissue culture cells [COS cells (Levin et al., 1992); CV-1 and *Drosophila* Schneider cells (Heyman et al., 1992)] and from mouse kidneys and liver (Heyman et al., 1992). Therefore, this ligand can be isolated from cells and its metabolism may be very important to the function of RXRs. During embryonic development in the mouse it is difficult to detect the presence of specific retinoids or retinoid gradients because of the limited amount of material. It may be easier to determine this indirectly through the detection of activated RARs or RXRs. The last section of this article will discuss this.

At this early point in our understanding of RARs and RXRs, many conceptual possibilities for the specificity of a specific receptor remain. Some of these are mundane, for example, the differential expression of different receptors in different tissue. Other might be tied directly to the multidomain structure of the receptors: individual domains might confer either unique DNA binding preferences, ligand binding preferences, dimerization preferences, or sites for interaction with other unique transcription factors. Alternatively, the specificity may be combinatorial: for example, a receptor may have a preference for a specific retinoid but this retinoid may be present only at specific developmental times and in specific tissues. These multiple possibilities will require examination at various levels.

IV. Biological Roles of Retinoic Acid

Retinoic acid and other vitamin A derivatives have several functions in the adult and in the developing embryo. These functions could be mediated through a wide variety of different proteins that can bind to retinoids (see Blomhoff *et al.*, 1990, for a review of retinoid transport and storage). This section will cover four areas of research: (1) retinoic acid and the developing chick wing bud; (2) the use of cell culture models to study retinoic acid-induced differentiation; (3) retinoic acid as a teratogen; and (4) recent retinoic acid-related studies in medicine. In all of these areas there is a reasonable probability that RARs and/or RXRs are the molecules mediating retinoic acid-induced function.

A. Retinoic Acid and the Developing Chick Wing Bud

Retinoic acid plays a distinct role as a developmental morphogen or an inducer of a morphogen in the developing chick wing bud (for reviews see Brockes, 1989, 1990; Summerbell and Maden, 1990; Hoffman, 1990; Tabin, 1991). Grafting experiments (Zwilling, 1956; Saunders and Gasseling, 1968) had shown that the apical ectodermal ridge plays a role in the subsequent growth of the wing bud. A small region at the posterior margin of the wing bud, when grafted to an anterior position, was sufficient to cause mirror image reduplication of the anterior–posterior axis. This region was subsequently named the "zone of polarizing activity," or ZPA. It has been suggested that the ZPA releases a morphogen that diffuses across the limb bud to form a gradient (Tickle *et al.*, 1975) that specifies positional information. When retinoic acid is added to the anterior margin of the chick wing bud, it mimics the effect of grafting a ZPA (Tickle *et al.*, 1982; Summerbell, 1983). This provided circumstantial evidence that there is a natural retinoic acid gradient that either is itself a morphogen or the inducer of a morphogen. Recently, two studies have presented data that suggest that the role for retinoic acid in this system is as an inducer of a morphogen rather than being a morphogen itself: Wanek *et al.* (1991) showed that treatment of anterior cells with retinoic acid converts them into ZPA cells. Noji *et al.* (1991) used the retinoic acid inducibility of the RAR-β gene to suggest that ZPA cells do not induce this gene in the chick limb bud. Retinoic acid (Thaller and Eichele, 1987) and 3,4-didehydroretinoic acid (Thaller and Eichele, 1990) have been isolated from chick wing buds. Therefore, retinoids appear to play a distinct direct or indirect morphogenetic role in this system. Retinoic acid also has an effect on regenerating limbs of amphibia (Maden, 1982), on axis formation in amphibia

(Durston et al., 1989; Sive et al., 1990), and on the expression of genes (Cho and De Robertis, 1990). The effects on regenerating limbs of urodele amphibia have recently been reviewed (Brockes, 1989; Summerbell and Maden, 1990; Tabin, 1991). It will be of great interest to determine whether RARs and/or RXRs are mediators of these retinoic acid effects in the wing bud and in the regenerating limbs.

B. The Use of Cell Culture Models to Study Retinoic Acid-Induced Differentiation

Cell culture lines have been the source of considerable information regarding the actions of retinoic acid. Since Strickland and Mahdavi (1978) discovered that the F9 embryonal carcinoma (EC) line would differentiate in the presence of retinoic acid, EC cell lines have been used to study the differential regulation of various genes. In most cases the study of retinoic acid-induced differentiation of EC cell lines has not focused on the molecular reasons for the retinoic acid induction. Instead, the systems have provided convenient sources of material for studying the differential regulation of genes during the progression from EC stem cell to differentiated cell.

These studies support two general statements: (1) the expression of many genes changes during the differentiation induced by retinoic acid; (2) the timing of these changes can be detected either early (within 6 hr) or much later (2 to 3 days). For details of gene expression changes during retinoic acid-induced EC cell differentiation see the recent review by Jetten (1990). The differences in gene expression can be identified within the context of a single cell line. However, many EC cell lines have been isolated, each with its own potential to differentiate. Some of the lines were derived and maintained on nondividing fibroblast feeder layers (Martin and Evans, 1975) while others were isolated to be feeder independent. The EC cells received considerable attention when it was shown that they could contribute to embryonic development when introduced into normal embryos (Brinster, 1974; Mintz and Illmensee, 1975; Papaioannou et al., 1975). More recent studies have focused instead on embryonic stem (ES) cells derived directly from embryos (Martin, 1981; Evans and Kaufman, 1981).

One intriguing aspect of the feeder-independent lines is that the different EC cell lines can produce different cells after retinoic acid treatment. For example, clones of F9 cells can differentiate into parietal endoderm (Strickland and Mahdavi, 1978) or visceral endoderm (Hogan et al., 1981), depending on the experimental protocol. A clone of P19 EC cells from different cell types depending on the retinoic acid concentration (Edwards and McBurney, 1983). This same clone can also differentiate into muscle cells on treatment with dimethyl sulfoxide (Edwards et al., 1983). Various EC cell lines may respond differently to retinoic acid because they represent different stem cell states. However, an alternative possibility is that the process of isolating and selecting clones of EC cells might place unknown pressures on the cells, resulting in different lines with characteristics of EC cells, but with mutations in genes (possibly even RARs or RXRs) that restrict their potential to differentiate. Retinoic acid differentiation studies on PCC4aza1 EC cells required the isolation of a clone that could withstand retinoic acid concentrations of 10^{-6} M. This was a stepwise selection of increasing concentrations of retinoic acid (Jetten et al., 1979). It is unclear whether this selection produced a mutant in the retinoic acid regulatory network or a mutant deficient in retinoic acid transfer.

The EC cell system provides an opportunity for testing whether RARs or RXRs are involved in the retinoic acid-induced differentiation process. The strategy for using these cells in this manner is illustrated in Fig. 3. Retinoic acid induction of F9 EC cells results in differentiated cells with limited division potential. If the RARs are responsible for this differentiation, and if they could be inhibited, the result would be F9 EC cells that are retinoic acid resistant (RA^r) and clonable in retinoic acid. We have chosen to use dominant-negative (Herskowitz, 1987) approaches to inhibit RAR function. The separate, functional domains of the receptors provide excellent targets for this approach. Overexpression of receptor domains for DNA binding, ligand binding, and dimer formation could inhibit endogenous receptor function. A initial report using this approach for hRAR-α-derived vectors has been published (Espeseth et al., 1989). We selected RA^r F9 EC clones, and the RA^r correlated with overexpression of part of this receptor. This study made an interesting observation: some retinoic acid-inducible genes were inhibited in the RA^r clones (tissue plasminogen activator, endo-B cytokeratin, and laminin B_1), while other retinoic acid-inducible genes were not (Hox-1.3, α-1 type IV collagen). This approach is currently being used to examine the ability of other dominant-negative constructs, derived from RARs and RXRs, to inhibit retinoic acid-induced differentiation, and/or to affect the retinoic acid regulation of other genes.

The above study derived its use of retinoic acid selection from earlier studies that selected retinoic acid-resistant clones from EC cell lines after mutagenesis of the cells. Because the differentiated cells have a limited division potential under cell culture conditions, one can select for the continued growth of EC cells in the presence of retinoic acid (Schindler et al., 1981; Sherman et al., 1981; McCue et al., 1983; Jones-Villeneuve et al., 1983; Wang and Gudas, 1984). Examination of these lines might provide clues as to what is affected when the cells become resistant to retinoic acid. Analysis of the cells has not yielded specific answers. Pratt et al. (1990) have reported that a P19 EC cell retinoic acid-resistant clone, RAC65, has a mutant RAR-α gene that has lost its terminal 70 amino acids. Investigators isolated this gene from the RAC65 line and transferred it into normal P19 cells. They were not successful at conferring retinoic acid resistance

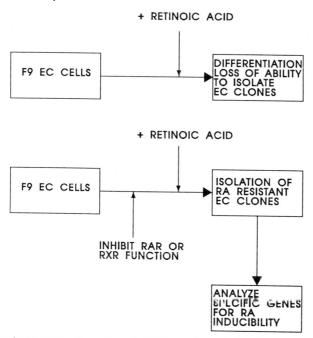

Fig. 3 The scheme used by Espeseth *et al.* (1989) to select retinoic acid-resistant F9 embryonal carcinoma (EC) clones. F9 EC cells can be retinoic acid induced to parietal endodermal cells. These cells eventually cease to grow. If RARs are involved in mediating this retinoic acid-induced differentiation, then interfering with their function should inhibit differentiation, resulting in EC cells that continue to grow. This growth selection then allows one the opportunity to examine which specific retinoic acid-inducible genes are inhibited when differentiation is inhibited. Receptor function was inhibited through overexpression of a *LacZ* fusion protein containing only part of hRAR-α. Clones were then analyzed for the retinoic acid inducibility of several genes. This general scheme is being used with fragments from other receptors.

to the P19 cells with this mutant gene. Therefore, the retinoic acid-resistant phenotype of the RAC65 line is either unrelated to the mutant RAR gene they isolated, or an additional and as yet unidentified change is necessary to produce the retinoic acid-resistant phenotype.

Retinoic acid affects many other cell lines (Lotan, 1980; Haussler *et al.*, 1983; Nakayama *et al.*, 1990). The HL-60 cell line will differentiate to granulocytes in the presence of retinoic acid (Breitman *et al.*, 1980). This line also expresses RAR-α and RAR-β (Hashimoto *et al.*, 1989), and a retinoic acid-resistant subclone of this this line has been isolated (Collins *et al.*, 1990). Retroviral vector transfer of an expressing RAR-α gene to this resistant clone (Collins *et al.*, 1990) overcame this retinoic acid resistance, suggesting that the RAR-α plays a normal role in the differentiation of the HL60 line.

C. Retinoic Acid as a Teratogen

Teratogenicity studies with rodents have illustrated some of the more dramatic effects of retinoic acid. Kochhar (1967) originally showed that retinoic acid was a relatively strong teratogen in rats. In golden hamsters, Shenefelt (1972) reported relationships between the teratogenic effect and the dose and stage of treatment of pregnant females. Near-term fetuses had over 70 types of malformations in many parts of the embryos.

The mouse has been a source of more detailed analysis of retinoic acid teratogenicity. Retinoic acid induces malformations in the craniofacial area (Webster et al., 1986; Sulik et al., 1987, 1988) and limbs (Kochhar, 1973; Kochhar and Agnish, 1977; Kwasigroch and Koshhar, 1980) when administered to pregnant mice and when added to the medium used for whole-embryo culture (Goulding and Pratt, 1986). Administration of retinoic acid at earlier times to pregnant mice produced a high incidence of spina bifida in near-term fetuses (Alles and Sulik, 1990). Sulik and co-workers have suggested that retinoic acid teratogenicity could be due to induced excessive cell death in regions of programmed cell death (Sulik et al., 1987, 1988; Sulik and Dehart, 1988; Alles and Sulik, 1989, 1990). We do not yet have a molecular explanation for how this might occur.

Improper usage of an acne medicine (isotretinoin, 13-*cis*-retinoic acid) by pregnant women has produced evidence that retinoic acid is also a teratogen in humans. Lammer et al. (1985) observed various human malformations in the craniofacial, cardiac, thymic, and central nervous systems.

The basis for the diverse effects of retinoic acid should develop as one begins to understand the various interactions between RARs, RXRs, and other genes involved in developmental regulation. One potential example is represented by research on the *Splotch* mutation in mice. This mutation, in homozygous form, results in neural tube defects (Auerbach, 1954). Retinoic acid administered at specific developmental times can affect the frequency of neural tube defects in the embryos (Kapron-Bras and Trasler, 1988). Recently, Epstein et al. (1991) have identified mutations in the *Pax-3* gene as being for the *Splotch* mutation. Because the *Pax-3* gene has already been shown to be retinoic acid inducible (Goulding et al., 1991) in cell culture, the aberrant expression of *Pax-3* may play a role in some of the teratogenic effects of retinoic acid.

Recent transgenic mouse studies point out potential regions where activated RARs exist during embryonic development (Rossant et al., 1991; Balkan et al., 1992a). In these studies, the RARE element from the RAR-β-2 promoter was coupled to a promoter 5' to the bacterial β-galactosidase gene. Transgenic embryos derived from transgenic lines were examined for β-galactosidase expression in the absence and presence of retinoic acid administration to the pregnant females. Distinct β-galactosidase activity patterns were observed in the closed neural tube at day 8.5, and the patterns became more complex during later

8. Retinoic Acid Receptors   329

development. Presumably, the β-galactosidase activity represents the response of activated RARs to the transgene. The transgenes were shown to be retinoic acid inducible both in cell culture and in embryos in both of the studies. Neural tube expression and expression in the eye and the face support the proposition that at least some of the teratogenic targets of retinoic acid are the result of activated RARs. In the Rossant *et al.* (1991) study there was a dramatic change in the β-galactosidase expression patterns when retinoic acid was administered to the females carrying the transgenic embryos. These changes point out the breadth of potential gene regulation changes that might be associated with the administration of retinoic acid to pregnant females.

D. Recent Retinoic Acid-Related Studies in Medicine

Retinoids and retinoic acid have found many uses in medicine. Besides the above-mentioned use of isotretinoin for acne, several recent applications are intriguing in the light of recent discoveries about RARs and RXRs. A few of these will be outlined below.

Larson *et al* (1984) linked acute promyelocytic leukemia (APL) to a 15;17 translocation. There is evidence that complete remission of APL patients can be obtained through retinoic acid treatment (Huang *et al.*, 1988; Castaigne *et al.*, 1990; Miller *et al.*, 1990; Warrell *et al.*, 1991), presumably by inducing differentiation (Koeffler, 1983). Several teams have cloned the translocation breakpoint region (Borrow *et al.*, 1990; de The *et al.*, 1990b; Lemons *et al.*, 1990; Kakizuka *et al* , 1991) and have located the translocation in an intron of the RAR-α gene (Borrow *et al.*, 1990; de The *et al.*, 1990b; Kakizuka *et al.*, 1991). A chimeric transcript has been identified in promyelocytic cells that contains, at its 5' end, sequences from a previously unidentified gene from chromosome 15 [first named *myl* by de The *et al.* (1990b), now named *PML* (de The *et al.*, 1991; Kakizuka *et al.*, 1991)] 5' and in reading frame with the B, C, D, E, and F domains of RAR-α. This *PML* gene produces multiple spliced mRNAs and the *PML*–RAR-α chimeric transcripts described by de The *et al.* (1991) and Kakizuka *et al.* (1991) contain different-sized *PML* components.

It was originally suggested by de The *et al.* (1990b) that this chimeric gene product might be creating a dominant-negative effect on the normal RAR-α homolog. The *PML* sequences contain potential domains for DNA binding, transcriptional activation, and dimerization. Therefore, the *PML*–RAR chimeric gene product could be acting on either an RAR-α regulatory system, and/or a *PML* regulatory system. In either case, because normal homologs of RAR-α and *PML* should exist in the cells with the t(15;17) translocation, the mechanism would probably involve dominant-negative inhibition of the normal homologue. Because retinoic acid can cause remission, the dominant-negative effect would presumably be inhibited or eliminated in the presence of retinoic acid. A simple model would predict a conformational change in the chimeric protein when retinoic acid associates with the RAR ligand binding domain, resulting in the inactivation of the chimeric protein.

Common variable immunodeficiency (CVI) is an acquired syndrome associated with the inability to produce an antibody response (Hermans *et al.*, 1976). When Sheer *et al.* (1988) produced hybridomas from CVI patients and treated them with retinoic acid, they observed an increase of total Ig mRNA and μ heavy chain mRNA processing shifted to the secreted form, with concomitant increases in secreted antibodies (Sheer *et al.*, 1988). We do not yet know the specific molecular basis for this result, but it suggests yet another role for retinoic acid.

Retinoids have been used for several years as anti-cancer agents, with particular focus on the skin. However, increasingly the retinoids are finding use in the treatment of other cancers (for a review, see Lippman *et al.*, 1987). This is clearly an area that should benefit from a molecular understanding of the function and specificity of RARs and RXRs. It is also an area where we have data that quantitate the therapeutic index of various retinoids. With six different retinoic acid and retinoid X receptor genes having been cloned and identified, a molecular understanding of the roles of the receptors in various normal and abnormal processes should follow. This should be aided by the identification of 9-*cis*-retinoic acid as a high-affinity ligand of RXR-α (and presumably the two other RXR receptors). This ligand and derivatives might be useful in separating RAR-mediated events from RXR-mediated events.

V. Relationships between Retinoic Acid Receptors and Other Transcription Factors and DNA Binding Proteins

The variety of developmental, clinical, and teratological effects of retinoic acid suggests that the mediators of retinoic acid-induced events are interacting at many levels. Some, if not most, of these effects will be manifested through transcriptional regulation. Given that the RARs and RXRs are ligand-inducible transcription factors, one must examine the possible ways in which these receptors can directly or indirectly control genes. Developmentally one can distinguish between primary effects—direct activation or repression of transcription of specific genes or direct protein–protein interactions with other transcription factors—and the secondary effects of regulating the transcription of other transcription factors. Candidate gene families that RARs or RXRs might directly or indirectly affect are the homeobox genes, at least some of the members of the *OCT* gene family, and the *fos/jun* family of genes associated with AP-1 activity.

This section will present evidence that links retinoic acid to various effects of gene regulation: (1) the retinoic acid regulation of *Hox, OCT*, and *AP-1* gene families, and (2) the possible interaction of retinoic acid receptors at promoter regions for other transcription factors, or directly at a protein–protein level with other receptors or transcription factors (heterodimerism between the receptors and other receptor families).

One major family of genes that has been implicated in developmental regulation is the homeobox (*Hox*) family. Researchers have isolated many mammalian homeobox genes based on homology with their counterparts in *Drosophila* and have published a considerable amount of descriptive information on their developmental expression. The homeobox genes appear in clusters on different mammalian chromosomes. When Graham *et al.* (1989) examined expression in the central and peripheral nervous system with individual members of a particular *Hox* locus, the specific region of expression of a particular homeobox gene paralleled its genomic position in the cluster. That is, the 5'-most members of the *Hox-2* cluster are expressed posteriorly and each successive member of the cluster is expressed more anteriorly. Duboule and Dolle (1989) found a similar relationship for members of the *Hox-5* cluster. Wilkinson *et al.* (1989) also found this type of progressive expression of adjacent members of the *Hox-2* locus in the developing mouse hindbrain.

An intriguing study with a human teratocarcinoma cell line that differentiates in response to retinoic acid has shown that there is a sequential activation of members of the *Hox-2* family with increasing time after the addition of retinoic acid to the cultures. In addition, there is a difference in the sensitivity of the different members of the *Hox-2* locus to specific concentrations of retinoic acid. Genes located at the 3' end of the cluster are induced at peak levels by 10^{-8} M retinoic acid whereas genes at the 5' end of the cluster require a retinoic acid concentration of 10^{-6}–10^{-5}. At a fixed retinoic acid concentration, the *Hox-2* genes are sequentially activated in the embryonal carcinoma cells in the 3' to 5' direction (Simeone *et al.*, 1990). These cell culture observations, in combination with the *in situ* analysis presented above, argue for retinoic acid gradients in the mouse embryo similar to those identified in the chick.

Transgenic experiments with ectopic expression of the *Hox-1.1* gene provide additional circumstantial evidence of a possible relationship between retinoic acid and the *Hox* genes (Balling *et al.*, 1989; Kessel *et al.*, 1990). When the *Hox-1.1* gene is coupled to the chicken β-actin promoter and injected into mouse eggs, mice are produced with craniofacial abnormalities (Balling *et al.*, 1989). This suggests that inappropriate expression of this *Hox-1.1* gene can cause abnormalities, at least some of which resemble malformations identified after administration of retinoic acid to pregnant female mice. Moreover, mice expressing the *Hox-1.1* gene showed variations of cervical vertebrae (Kessel *et al.*, 1990).

No one has yet proven that RARs or RXRs transcriptionally regulate the *Hox* genes. Whereas published data have shown that steady state RNA levels of several homeobox genes increase on the addition of retinoic acid to various EC cell lines, there are presently no data documenting RAREs in homeobox promoters. *Hox-1.3* steady state mRNA increases with retinoic acid induction (Murphy *et al.*, 1988; Fibi *et al.*, 1988) and we have shown through nuclear run-on experiments that there is a dramatic *Hox-1.3* transcription rate change when retinoic acid is administered to F9 cells (Murphy *et al.*, 1988). Furthermore, in the previously described dominant-negative RAR-α study (Espeseth *et al.*, 1989) that produced RA^r F9 cells, retinoic acid continued to regulate the *Hox-1.3* gene. Dominant-negative vectors derived from RXRs are now being used to examine whether the regulation of *Hox-1.3* is under the control of this set of receptors (E. Linney, unpublished observations, 1992).

Another family of transcription factors that has members expressed at different points in development is the OCT family of proteins, in which there is a large, conserved POU domain (Herr *et al.*, 1988) responsible for DNA binding specificity. There are at least two *OCT* genes that are expressed in embryonal carcinoma cells and then repressed on retinoic acid induction of the cells. Unfortunately the different laboratories working with these genes have each used different names. *NFA-3* (Lenardo *et al.*, 1989) is the same gene as that called *OCT-3* by Okamoto *et al.* (1990) and Rosner *et al.* (1990). However, this same gene is named *OCT-4* by Scholer *et al.* (1989) and there is an isoform of the same gene called *OCT-5* (Scholer *et al.*, 1989, 1990). Suzuki *et al.* (1990) have isolated another gene, *OCT-6*, which is also expressed in EC stem cells. These are potentially important genes that retinoic acid may regulate in a negative way. We do not yet know whether the promoters of all of these genes might have RAR- or RXR-repressible signals. Recently, one of the laboratories studying the *OCT-3* gene (Okazawa *et al.*, 1991) isolated and analyzed the *OCT-3* promoter. This promoter has a cis element that binds two unidentified cellular factors. While this sequence has been shown to be a cell type-specific RA-repressible enhancer, it does not appear to be directly regulated by RARs (Okazawa *et al.*, 1991). Therefore, its regulation could very possibly be a secondary effect of activated RARs with retinoic acid.

The *Hox* genes and some of the *OCT* genes may thus play roles in elaborating the response of cells to retinoic acid indirectly. That is, RAR activation or repression of specific genes may then result through these changes in the initiation of a cascade of developmental regulation functioning to secondarily regulate the *Hox* or *OCT* genes.

Other genes of potential DNA binding or regulatory nature that are differentially expressed in retinoic acid-treated teratocarcinoma cells are the *MK1* gene (Kadomatsu *et al.*, 1988), a zinc finger-containing *REX-1* gene (Hosler *et al.*, 1989), and the transcription factor *AP-2* (Luscher *et al.*, 1989). A yet to be identified gene that may be very important from a regulatory standpoint is the E1A-like activity that Imperiale *et al.* (1984) identified in undifferentiated F9 EC cells. They showed that this E1A-like activity is present in undifferentiated, but not retinoic acid-induced, F9 EC cells. Because adenovirus E1A proteins dissociated heteromeric complexes involving the E2F transcription factor (Bagchi *et al.*, 1990), the retinoic acid down regulation of this gene may be extremely important to the differentiation process of the cells.

8. Retinoic Acid Receptors

Retinoic acid regulation is also relevant to analysis of the AP-1 transcription factor elements, that is, Fos, Jun, and related transcription factors. Muller and Wagner (1984) originally showed that expression of c-*fos* in F9 cells caused the cells to be morphologically altered and to have limited growth potential. This suggested that artificial expression of c-*fos* caused the cells to differentiate. Since that early observation, Fos and Jun and their related proteins have been shown to form the AP-1 transcription complex that binds to AP-1 sites. c-*fos* is not expressed in undifferentiated F9 (Yang-Yen et al., 1990a) and P19 (de Groot et al., 1990a) EC cells while retinoic acid induces c-*jun* in both of these cell lines. de Groot et al. (1990b) reported that ectopic expression of c-*jun* leads to differentiation of P19 cells. Thus, two members of the AP-1 transcription family have been implicated in the differentiation process of EC cells: c-*fos* and c-*jun*. How RARs or RXRs might play a role in this process remains to be determined.

The relationships between AP-1 factors and RARs or RXRs may turn out to be complex, and may occur at various levels. Schule et al. (1990a) recently reported that the RAR-α and the vitamin D_3 receptors can both recognize a common response element in the human osteocalcin gene promoter. Delineation of this site showed that its core sequence was an AP-1 site. Using promoter constructs driving a reporter gene, the investigators showed by transfection in osteosarcoma cells that they could get both retinoic acid and vitamin D_3 induction of the promoter. Expression of Fos and Jun in the cells inhibited this induction. Therefore, in at least some cases, and possibly in a cell-specific way, RARE sequences may overlap with AP-1 sites. Nicholson et al. (1990) reported that an AP-1 site mediated the retinoic acid-induced negative regulation of the rat stromelysin gene. Schule et al. (1991) provided *in vitro* evidence that RARs might inhibit AP-1-mediated events through inhibition of c-Jun binding to AP-1 sites. Desbois et al. (1991) showed that both RAR-α and c-ErbA-α can repress AP-1-mediated transcriptional activation by decreasing the activity of the AP-1 transcription factor. v-ErbA fails to repress this activity; therefore the maintenance of AP-1 activity may be one mode by which this oncogene functions and this could be due to protein–protein interactions and not through direct transcriptional control.

The above studies are intriguing because the dominant-negative hRAR-α study with F9 EC cells (Espeseth et al., 1989) showed that the retinoic acid induction of cytokeratin endo-B (K-18 for the human gene) was inhibited in the RAR^r clones. Oshima et al. (1990) have recently shown that this gene has an AP-1 site in its first intron and that Fos and Jun can activate it. Therefore, there are three potential mechanisms for the inhibition of retinoic acid induction of the cytokeratin endo-B gene in the dominant-negative clones: (1) direct binding of the dominant-negative protein to the intronic AP-1 site with inhibition of AP-1 activity, (2) inhibition of c-*jun* induction in the clones, or (3) interference with AP-1 activity through protein–protein interactions with c-Jun.

The glucocorticoid receptor interacts with AP-1 transcription factors and AP-1 sites (Jonat et al., 1990; Yang-Yen et al., 1990b; Schule et al., 1990b; Diamond *et al.*, 1990). There is a functional antagonism between the glucocorticoid receptor and AP-1 transcription activity, possibly owing to protein–protein interactions between the factors and receptors or to changes at the region of DNA binding. It would not be surprising to find similar types of interactions with RARs. The Schule *et al.* (1991) study suggests that this might be happening with the c-Jun product and RAR-α. Their *in vitro* data does not support this happening with RXR-α.

Two early reports have suggested that the RAR-α can dimerize with thyroid hormone receptors (Forman *et al.*, 1989; Glass *et al.*, 1989). The dimerization with a thyroid hormone receptor affected both the binding of the thyroid hormone receptor to specific DNA elements and the transcriptional activation function of the receptor (Glass *et al.*, 1989). Forman *et al.* (1989) were able to inhibit RAR-α-mediated retinoic acid induction with truncated regions of the thyroid hormone receptor. Both of these reports support the possibility of a functional dimerization domain in the ligand binding domain of RAR-α.

Most recently, it has been shown that RXRs will heterodimerize *in vitro* with RARs, the vitamin D receptor, and thyroid receptors (Yu *et al.*, 1991; Leid *et al.*, 1992; Kliewer *et al.*, 1992; Zhang *et al.*, 1992; Marks *et al.*, 1992). While these studies were discussed previously, they do point out that RXRs may be potential coregulators of various other transcription factors.

To summarize, there is evidence that retinoic acid receptors can affect various developmentally important gene families. Because these families have their own internal, autoregulatory capabilities, triggering the activation or repression of these gene sets may be a major role of activated RARs or RXRs. Additionally, there is growing evidence for dual usage of DNA sequences by various transcription factors and for protein–protein interactions between different transcription factors. Determining which of these combinations play a specific role in a single developmental event will require careful experimental planning in future studies.

VI. Current and Future Approaches for Dissecting Developmental Specificity and Function of Retinoic Acid Receptors and Retinoid X Receptors

Given the large number of biological processes affected by retinoic acid and the variety of potential RARs and RXRs, two major questions in this area of research concern (1) where and when retinoic acid is present in the embryo, and (2) whether there are individual specificities associated with the individual receptors. Section IV of this article described a number of biological events that are directly or indirectly affected by retinoic acid, and Section III outlined several different possible means for obtaining specificity with receptors. This section will suggest approaches for examining relationships between the biological effects of retinoic

8. Retinoic Acid Receptors

acid and the specific function of individual receptors. They will be preceded by questions of relevance to the study of these receptors. These approaches reflect the individual preferences of this author and in many cases are currently being tested.

Where and when is retinoic acid present in the embryo? The direct determination of this would involve detection of specific retinoids at various developmental times and in different tissues. The Eichele laboratory has been pursuing this approach in the chick embryo (Thaller and Eichele, 1987, 1990; Wagner *et al.*, 1990). In the absence of a significant breakthrough in direct measurement techniques, the direct determination of retinoic acid and other related retinoids will have limited usefulness in the mouse embryo because of tissue limitations.

There are indirect approaches for determining the presence of functional ligands for the receptors. As we identify genes that are induced only by retinoic acid, the presence of their RNA at different developmental times and in different tissues can suggest that retinoic acid is available to this tissue. Alternatively, using the defined regulatory elements (RAREs) in promoters of such genes in transgenic experiments can provide valuable indirect information regarding the presence of retinoic acid in animals.

As discussed briefly in Section IV,C, transgenic lines of mice have been produced using the RARE from the retinoic acid-inducible mRAR-β-2 promoter (Rossant *et al.*, 1991; Balkan *et al.*, 1992a). The experimental design, in both cases, involved placing the RARE next to a simple promoter that was coupled to the bacterial β-galactosidase (*LacZ*) gene. A parallel approach has been reported by Mendelsohn *et al.* (1991), in which the whole RAR-β-2 promoter was coupled to *LacZ*. β-Galactosidase activity in the transgenic animals would represent activated RARs. Rossant *et al.* (1991) used the heat shock protein 68 promoter (*hsp68*) while we used the thymidine kinase promoter. The DNA constructs both work extremely well in distinguishing retinoic acid-treated from nontreated transfected cells in culture. The transgenic progeny from these founder mice are being used as "indicator mice" for the presence of activated receptors and, indirectly, for the presence of functional retinoid ligand. Specific *LacZ* staining patterns have been detected in progeny transgenic embryos. The patterns that Rossant *et al.* (1991) report are very similar to ours (Balkan *et al.*, 1992a) at earlier developmental stages; for example, we both detect β-galactosidase activity in the closed neural tube of day-8.5 transgenic embryos. The respective transgenes are inducible when retinoic acid is administered to pregnant females carrying the embryos. While this approach shows promise in identifying the copresence of retinoic acid and RARs, the patterns may, in part, be affected by the choice of promoters used in the transgenic studies. Through this type of analysis it is hoped that the presence of activated receptors can be identified at various developmental times. If regulatory sequences are identified with RAREs specific to individual receptors, this type of analysis could be used to obtain specific information about individual receptors. We are currently producing transgenic lines with RXREs to determine whether indicator transgenic mice may be constructed for activated RXRs.

An entirely different approach toward studying retinoic acid is to examine the enzymes involved in regulating its presence. Duester *et al.* (1991) have functionally identified an RARE in the promoter of the human alcohol dehydrogenase-3 gene (see Table I for the sequence as it compares with other RAREs). The observation that this gene is regulated by retinoic acid brings up the intriguing possibility that this enzyme might be involved in a positive feedback loop for the eventual production of retinoic acid from retinol (Duester *et al.*, 1991). Differential regulation of enzymes involved in retinoid metabolism may be the primary determining step in creating potential gradients of retinoids during development. Therefore, molecular analysis of these enzymes, their regulation, and their distribution could provide a basis for understanding the normal role of retinoids during differentiation and development.

What is the function and specificity of each RAR and RXR? The number of identified RAREs and RXREs has dramatically increased in just the past year (as can be seen in Table I). Cotransfection experiments with these sequences coupled to reporter genes might provide limited information, because part of the specificity of an individual receptor might be the requirement for specific transcription factors that are themselves differentially expressed in different cell types (Glass *et al.*, 1989, 1990). Transient transfection experiments in cell culture might introduce more of the reporter gene construct and/or more of the cotransfected receptor product to "titrate out" the specific interacting transcription factors effectively.

In addition, with the clear demonstration of *in vitro* heterodimerization of RXRs with RARs, the vitamin D receptor, and thyroid hormone receptors (Yu *et al.*, 1991; Leid *et al.*, 1992; Kliewer *et al.*, 1992; Zhang *et al.*, 1992; Marks *et al.*, 1992) there is the distinct possibility that heterodimeric combinations of RXRs may be occurring in transient transfection experiments without the investigator knowing it. Certain heterodimeric combinations may show distinct and unique DNA binding preferences. Therefore, it is expected that two of the more focused directions that investigators will take in this field are (1) the mapping of the dimerization domains of the RXRs and RARs, along with an identification of heterodimer preferences between receptor molecules, and (2) the identification of new DNA response elements that may represent the different heterodimeric combinations.

For studying dimerization domains, the λ repressor system described by Hu *et al.* (1990) may be useful for mapping the regions in an individual receptor. This system allows one to identify dimerization regions of protein in bacteria. An alternative system for studying heterodimeric interface regions of two interactive proteins has been described by Fields and Song (1989). As different receptor–receptor or receptor–transcription factor combinations are identified, it would be useful to be able to select DNA sequences to which these heterodimers bind. One

8. Retinoic Acid Receptors

useful approach toward identifying such sequences would be the SAAB (selected and amplified binding) procedure described by Blackwell and Weintraub (1990). This procedure should allow for the selection, amplification, cloning, and identification of strong heterodimer binding sites.

Because it is now possible to produce sufficient quantities of the receptors or subdomains, the physical examination of binding constants for the receptors to DNA sequences, potential ligands, and to other receptors should also be achievable. It is the appropriate time to consider the development of monoclonal reagents to different forms of the receptor or receptors: to unique regions formed on dimerization of the receptors or direct interaction with RA, and to heterodimerization sites between RARs and RXRs. Monoclonal reagents of this type may be very useful as *in situ* probes for activated or heterodimeric receptors.

The classical approach for testing the function of a gene has been to mutagenize it and examine remaining functions. As a prelude to this type of analysis, the experimenter can use cell culture lines to test various approaches for either restricting the function of a receptor, or for identifying its function in that cell line. A specific RAR or RXR could be used in this approach. This laboratory has directed it efforts toward using dominant-negative repression (Herskowitz, 1987) of receptor function because the RARs and RXRs provide several individual target functions for this approach. It might also be possible to overexpress a DNA binding domain, a ligand binding domain, a dimerization domain, or a region of the receptor required for specific interaction with other transcription factors. Through overexpression, these receptor fragments might then inhibit the function of endogenous, activated receptors. Vectors representing this type of approach are illustrated in Fig. 4. Vector A produces RA^r F9 EC cells (Espeseth *et al*., 1989). We have not yet determined the molecular basis, although two obvious mechanisms are possible: it may be binding to RARE sequences and/or titrating transcription factors that interact with the A/B domain region. Vector B does not produce RA^r clones. This appears to be because the fusion protein remains in the cytoplasm (A. Espeseth, H. Steinfeldt, and E. Linney, unpublished observations, 1991). Vector C is producing RA^r clones (A. Espeseth, B. Kelly, and E. Linney, unpublished observation, 1991) and this is probably due to binding to RARE sequences because we have also shown that the protein will bind to RAREs and to a TRE *in vitro*. Vector D is currently being analyzed for the mRAR-β receptor and the mRAR-γ receptor (D. Cash, M. Colbert, B. Kelley, and E. Linney, unpublished observations, 1991) and is based on a similar construct derived from the thyroid hormone receptor that functions in a dominant-negative manner (Forman *et al.*, 1989). Vector D also produces RA^r F9 EC cells. This could be functioning through either titration of retinoic acid away from endogenous receptor(s) and/or through the production of inactive heterodimers with endogenous receptor(s). Vector E does not yield RA^r F9 EC cells but the fusion protein is localized in the cytoplasm. Vector F has the simian virus 40 (SV40) nuclear translocation signal in reading frame with the fusion protein,

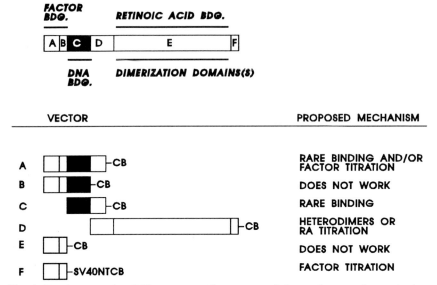

Fig. 4 Dominant-negative RAR constructs. Constructs are being used to examine mechanisms involved in dominant-negative inhibition of RAR function. Vector A has been reported by Espeseth et al. (1989). The -CB represents a collagen-repeat, LacZ cassette that is in reading frame with the RAR coding sequence. This allows for the identification of cells expressing the fusion protein, for subcellular localization, and for quantitation. SV40NT represents the nuclear localization signal of the SV40 T antigen.

which allows for its translocation into the nucleus (A. Espeseth and E. Linney, unpublished findings, 1991). This vector produces RA^r clones.

Analysis of the genes inhibited in the RA^r clones should permit identification of genes regulated by different receptor constructs. Vectors similar to those described in Fig. 4 are being constructed for various RARs and RXRs. Depending on how interchangeable the different subdomains are between receptors, the vectors might have specificity against several receptor classes, or perhaps against a single receptor.

The developmental usefulness of the dominant-negative approach might have greater value in transgenic mice. We are attempting to use these vectors with regulatory sequences that are known to confer expression to retinoic acid-sensitive tissue. A complementary approach, which might be more useful in transgenic analysis than in cell culture, is a dominant-positive approach. When a LacZ fusion protein with the hRAR-α is expressed in F9 cells, the cells differentiate (E. Linney, unpublished observations, 1991). Transient transfection experiments suggest that this fusion protein is a weak, constitutive receptor. When transgenic mouse lines are constructed with this fusion vector driven by a crystallin promoter, they produce the phenotype of microphthalmia (Balkan et *al.*, 1992b), one of the phenotypes detected when normal pregnant mice are treated with retinoic acid.

The ultimate test of the function of a gene requires mutations in the germ line. Quite accidentally, a mutation, *legless*, has been identified in a transgenic line that may have some relevance to retinoic acid regulation and/or metabolism (McNeish *et al.*, 1988, 1990). In homozygous form, skeletal, craniofacial, and visceral malformations occur. The isolation of the gene into which the transgene has integrated might provide clues to retinoic acid regulation.

An alternate approach that we and others are pursuing is to target receptor genes specifically in ES stem cells (Capecchi, 1989). Through homologous recombination targeting vectors, it is now possible to functionally inactivate each gene in ES cells to produce mice defective in each receptor gene. Obviously, with at least six different RAR and RXR genes, this will take time and may result in little if any phenotype when only a single gene is inactivated. It is expected that this technology will be used in the future to create modifications of genes *in situ* through homologous recombination in the germ-line progenitors. The information learned from the dominant-negative approach described above should provide useful data for designing such targeting vectors.

Acknowledgments

The author wishes to thank all the members of his laboratory and particularly Dr. Wayne Balkan and Dr. Makoto Taketo for aid with this review. The review was enhanced by information provided by Drs. Chambon, de Groot, Hamada, Mangelsdorf, Evans, and Pfahl before publication. The research of this laboratory is supported by NIH Grants CA39066 and HD24130. Specific thanks go to Ms. Brenda Hanna for her aid in preparing the reference list.

References

Alles, A. J., and Sulik, K. K. (1989). Retinoic-acid-induced limb-reduction defects: Perturbation of zones of programmed cell death as a pathogenetic mechanism. *Teratology* **40**, 163–171.

Alles, A. J., and Sulik, K. K. (1990). Retinoic acid-induced spina bifida: Evidence for a pathogenetic mechanism. *Development (Cambridge, UK)* **108**, 73–81.

Astrom, A., Pettersson, U., Krust, A., Chambon, P., and Voorhees, J. J. (1990). Retinoic acid and synthetic analogs differentially activate retinoic acid receptor dependent transcription. *Biochem. Biophys. Res. Commun.* **173**, 339–345.

Auerbach, R. (1954). Analysis of the developmental effects of a lethal mutation in the house mouse. *Exp. Zool.* **127**, 305–329.

Bagchi, S., Raychaudhuri, P., and Nevins, J. R. (1990). Adenovirus E1A proteins can dissociate heteromeric complexes involving the E2F transcription factor: A novel mechanism for E1A trans-activation. *Cell (Cambridge, Mass.)* **62**, 659–669.

Balkan, W., Colbert, M., Bock, C., and Linney, E. (1992a). Transgenic indicator mice for studying activated retinoic acid receptors during development. *Proc. Natl. Acad. Sci. U.S.A.* **89**, 3347–3351.

Balkan, W., Klintworth, G. K., Bock, C. B., and Linney, E. (1992b). Transgenic mice expressing a constitutively active retinoic acid receptor in the lens exhibit ocular defects. *Dev. Biol.* **151**, 622–625.

Balling, R., Mutter, G., Gruss, P., and Kessel, M. (1989). Craniofacial abnormalities induced by ectopic expression of the homeobox gene *Hox*-1.1 in transgenic mice. *Cell (Cambridge, Mass.)* **58**, 337–347.

Beato, M. (1989). Gene regulation by steroid hormones. *Cell (Cambridge, Mass.)* **56**, 335–344.

Benbrook, D., Lernhardt, E., and Pfahl, M. (1988). A new retinoic acid receptor identified from a hepatocellular carcinoma. *Nature (London)* **333**, 669–672.

Blackwell, T. K., and Weintraub, H. (1990). Differences and similarities in DNA-binding preferences of MyoD and E2A protein complexes revealed by binding site selection. *Science* **240**, 1103–1110.

Blomhoff, R., Green, M. H., Berg, T., and Norum, K. R. (1990). Transport and storage of vitamin A. *Science* **250**, 399–404.

Borrow, J., Goddard, A. D., Sheer, D., and Solomon, E. (1990). Molecular analysis of acute promyelocytic leukemia breakpoint cluster region on chromosome 17. *Science* **249**, 1577–1580.

Brand, N., Petkovich, M., Krust, A., Chambon, P., de The, H., Marchio, A., Tiollais, P., and Dejean, A. (1988). Identification of a second human retinoic acid receptor. *Nature (London)* **332**, 850–853.

Breitman, T. R., Selonick, S. E., and Collins, S. J. (1980). Induction of differentiation of the human promyelocytic leukemia cell line (HL-60) by retinoic acid. *Proc. Natl. Acad. Sci. U.S.A.* **77**, 2936–2940.

Brinster, R. L. (1974). The effect of cells transferred into the mouse blastocyst on subsequent development. *J. Exp. Med.* **140**, 1049–1056.

Brockes, J. P. (1989). Retinoids, homeobox genes, and limb morphogenesis. *Neuron* **2**, 1285–1294.

Brockes, J. P. (1990). Reading the retinoid signals. *Nature (London)* **345**, 766–768.

Capecchi, M. R. (1989). Altering the genome by homologous recombination. *Science* **244**, 1288–1292.

Castaigne, S., Chomienne, C., Daniel, M. T., Ballerini, P., Berger, R., Fenaux, P., and Degos, L. (1990). All *trans*-retinoic acid as a differentiation therapy for acute promyelocytic leukaemia. I. Clinical results. *Blood* **76**, 1704–1709.

Cho, K. W. Y., and De Robertis, M. D. (1990). Differential activation of *Xenopus* homeo box genes by mesoderm-inducing growth factors and retinoic acid. *Genes Dev.* **4**, 1910–1916.

Collins, S. J., Robertson, K. A., and Mueller, L. (1990). Retinoic acid-induced granulocytic differentiation of HL-60 myeloid leukemia cells is mediated directly through the retinoic acid receptor (RAR-α). *Mol. Cell. Biol.* **10**, 2154–2163.

Danielson, M., Hinck, L., and Ringold, G. M. (1989). Two amino acids within the knuckle of the first zinc finger specify DNA response element activation by the glucocorticoid receptor. *Cell (Cambridge, Mass.)* **57**, 1131–1138.

de Groot, R., Schoorlemmer, J., Van Geneseen, S., and Kruijer, W. (1990a). Differential expression of *jun* and *fos* genes during differentiation of mouse P19 embryonal carcinoma cells. *Nucleic Acids Res.* **18**, 3195–3202.

de Groot, R., Kruyt, F. A. E., Van Der Saag, P. T., and Kruijer, W. (1990b). Ectopic expression of c-*jun* leads to differentiation of P19 embryonal carcinoma cells. *EMBO J.* **9**, 1831–1837.

De Luca, L. M. (1991). Retinoids and their receptors in differentiation, embryogenesis, and neoplasia. *FASEB J.* **5**, 2924–2933.

de The, H., Marchio, A., Tiollais, P., and Dejean, A. (1987). A novel steroid thyroid hormone receptor related gene inappropriately expressed in human hepatocellular carcinoma. *Nature (London)* **330**, 667–670.

8. Retinoic Acid Receptors

de The, H., Marchio, A., Tiollais, P., and Dejean, A. (1989). Differential expression and ligand regulation of the retinoic acid receptor α and β genes. *EMBO J.* **8,** 429-433.

de The, H., Vivanco-Ruiz, M. D. M., Tiollais, P., Stunnenberg, H., and Dejean, A. (1990a). Identification of a retinoic acid responsive element in the retinoic acid receptor β gene. *Nature (London)* **343,** 177-180.

de The, H., Chomienne, C., Lanotte, M., Degos, L., and Dejean, A. (1990b). The t(15;17) translocation of acute promyelocytic leukaemia fuses the retinoic acid receptor α gene to a novel transcribed locus. *Nature (London)* **347,** 558-561.

de The, H., Lavau, C., Marchio, A., Chomienne, C., Degos, L., and Dejean, A. (1991). The PML-RAR-α fusion mRNA generated by the t(15;17) translocation in acute promyelocytic leukemia encodes a functionally altered RAR. *Cell (Cambridge, Mass.)* **66,** 675-684.

Dejean, A., Bougueleret, L., Grzeschik, K., and Tiollais, P. (1986). Hepatitis B virus DNA integration in a sequence homologous to v-*erb*-A and steroid receptor genes in a hepatocellular carcinoma. *Nature (London)* **322,** 70-72.

Desbois, C., Aubert, D., Legrand, C., Pain, B., and Samarut, J. (1991). A novel mechanism of action for v-Erba: abrogation of the inactivation of transcription factor AP-1 by retinoic acid and thyroid hormone receptors. *Cell (Cambridge, Mass.)* **67,** 731-740.

Diamond, M. I., Miner, J. N., Yoshinaga, S. K., and Yamamoto, K. R. (1990). Transcription factor interactions: Selectors of positive or negative regulation from a single DNA element. *Science* **249,** 1266-1272.

Dolle, P., Ruberte, E., Kastner, P., Petkovich, M., Stoner, C. M., Gudas, L. J., and Chambon, P (1989). Differential expression of genes encoding α, β and γ retinoic acid receptors and CRABP in the developing limbs of the mouse. *Nature (London)* **342,** 702-705.

Dolle, P., Ruberte, E., Leroy, P., Morriss-Kay, G., and Chambon, P. (1990). Retinoic acid receptors and cellular retinoid binding proteins. *Development (Cambridge, UK)* **110,** 1133-1151.

Duboule, D., and Dolle, P. (1989). The structural and functional organization of the murine *HOX* gene family resembles that of *Drosophila* homeotic genes. *EMBO J.* **8,** 1497-1505.

Duester, G., Shean, M. L., McBridge, M. S., and Stewart, M. J. (1991). Retinoic acid response element in the human alcohol dehydrogenase gene *ADH3*: Implications for regulation of retinoic acid synthesis. *Mol. Cell Biol.* **11,** 1638-1646.

Durston, A. J., Timmermans, J. P. M., Hage, W. J., Hendriks, H. F. J., Vries, N. J. D., Heideveld, M., and Nieuwkoop, P. D. (1989). Retinoic acid causes an anteroposterior transformation in the developing central nervous system. *Nature (London)* **340,** 140-144.

Edwards, M. K. S., and McBurney, M. W. (1983). Concentration of retinoic acid determines the differentiated cell types formed by a teratocarcinoma cell line. *Dev. Biol.* **98,** 187-191.

Edwards, M. K. S., Harris, J. F., and McBurney, M. W. (1983). Induced muscle differentiation in an embryonal carcinoma cell line. *Mol. Cell Biol.* **3,** 2280-2286.

Epstein, D. J., Vekemans, M., and Gros, P. (1991). *Splotch* (Sp^{2H}), a mutation affecting development of the mouse neural tube, shows a deletion within the paired homeodomain of *Pax-3*. *Cell (Cambridge, Mass.)* **67,** 767-774.

Espeseth, A. S., Murphy, S. P., and Linney, E. (1989). Retinoic acid receptor expression vector inhibits differentiation of F9 embryonal carcinoma cells. *Genes Dev.* **3,** 1647-1656.

Evans, M. J., and Kaufman, M. H. (1981). Establishment in culture of pluripotential cells from mouse embryos. *Nature (London)* **292,** 154-156.

Evans, R. M. (1988). The steroid and thyroid hormone receptor superfamily. *Science* **240,** 889-895.

Fawell, S. E., Lees, J. A., White, R., and Parker, M. G. (1990). Characterization and colocalization of steroid binding and dimerization activities in the mouse estrogen receptor. *Cell (Cambridge, Mass.)* **60,** 953-962.

Fibi, M., Zink, B., Kessel, M., Colberg-Poley, A. M., Labeit, S., Lehrach, H., and Gruss, P.

(1988). Coding sequence and expression of the homeobox gene *Hox* 1.3. *Development (Cambridge, UK)* **102**, 349-359.
Fields, S., and Song, O.-K. (1989). A novel genetic system to detect protein-protein interactions. *Nature (London)* **340**, 245-246.
Forman, B. M., and Samuels, H. H. (1990a). Dimerization among nuclear hormone receptors. *New Biol.* **2**, 587-594.
Forman, B. M., and Samuels, H. H. (1990b). Interactions among a subfamily of nuclear hormone receptors: The regulatory zipper model. *Mol. Endocrinol.* **4**, 1293-1301.
Forman, B. M., Yang, C., Stanley, F., Casanova, J., and Samuels, H. H. (1988). c-*erb* protooncogenes mediate thyroid hormone-dependent and independent regulation of the rat growth and prolactin genes. *Mol. Endocrinol.* **2**, 902-911.
Forman, B. M., Yang, C., Au, M., Casanova, J., Ghysdael, J., and Samuels, H. H. (1989). A domain containing leucine-zipper-like motifs mediate novel *in vivo* interactions between the thyroid hormone and retinoic acid receptors. *Mol. Endocrinol.* **3**, 1610-1626.
Gaub, M. P., Lutz, Y., Ruberte, E., Petkovich, M., Brand, N., and Chambon, P. (1989). Antibodies specific to the retinoic acid human nuclear receptors α and β. *Proc. Natl. Acad. Sci. U.S.A.* **86**, 3089-3093.
Giguere, V., Ong, E. S., Sequi, P., and Evans, R. M. (1987). Identification of a receptor for the morphogen retinoic acid. *Nature (London)* **330**, 624-628.
Giguere, V., Shago, M., Zirngibl, R., Tate, P., Rossant, J., and Varmuza, S. (1990). Identification of a novel isoform of the retinoic acid receptor gamma expressed in the mouse embryo. *Mol. Cell. Biol.* **10**, 2335-2340.
Glass, C. K., Franco, R., Weinberger, C., Albert, V., Evans, R. M., and Rosenfeld, M. G. (1987). A c-*erb*A binding site in the rat growth hormone gene mediates transactivation by thyroid hormone. *Nature (London)* **329**, 738-741.
Glass, C. K., Lipkin, S. M., Devary, O. V., and Rosenfeld, M. G. (1989). Positive and negative regulation of gene transcription by a retinoic acid-thyroid hormone receptor heterodimer. *Cell (Cambridge, Mass.)* **59**, 697-708.
Glass, C. L., Devary, O. V., and Rosenfeld, M. G. (1990). Multiple cell type-specific proteins differentially regulate target sequence recognition by the alpha retinoic acid receptor. *Cell (Cambridge, Mass.)* **63**, 729-738.
Godowski, P. J., Rusconi, S., Miesfeld, R., and Yamamoto, K. R. (1987). Glucocorticoid receptor mutants that are constitutive activators of transcriptional enhancement. *Nature (London)* **325**, 365-368.
Goulding, E. H., and Pratt, R. M. (1986). Isotretinoin teratogenicity in mouse whole embryo culture. *J. Craniofac. Genet. Dev. Biol.* **6**, 99-112.
Goulding, M. D., Chalepakis, G., Deutsch, U., Erselius, J. R., and Gruss, P. (1991). Pax-3, a novel murine DNA binding protein expressed during early neurogenesis. *EMBO J.* **10**, 1135-1147.
Graham, A., Papalopulu, N., and Krumlauf, R. (1989). The murine and *Drosophila* homeobox gene complexes have common features of organization and expression. *Cell (Cambridge, Mass.)* **57**, 367-378.
Green, S., and Chambon, P. (1988). Nuclear receptors enhance our understanding of transcription regulation. *Trends Genet.* **4**, 309-314.
Guiochon-Mantel, A., Loosfelt, H., Lescop, P., Sar, S., Atger, M., Perrot-Applanat, M., and Milgrom, E. (1989). Mechanisms of nuclear localization of the progesterone receptor: Evidence for interaction between monomers. *Cell (Cambridge, Mass.)* **57**, 1147-1154.
Ham, J., and Parker, M. G. (1989). Regulation of gene expression by nuclear hormone receptors. *Curr. Opinion Cell Biol.* **1**, 503-511.
Hamada, K., Gleason, S. L., Levi, B.-Z., Hirschfeld, S., Appella, E., and Ozata, K. (1989). H-2RIIBP, a member of the nuclear hormone receptor superfamily that binds to both the

8. Retinoic Acid Receptors

regulatory element of major histocompatibility class I genes and the estrogen response element. *Proc. Natl. Acad. Sci. U.S.A.* **86,** 8289–8293.

Hard, T., Kellenbach, E., Boelens, R., Maler, B. A., Dahlman, K., Freedman, L. P., Carlstedt-Duke, J., Yamamoto, K. R., Gustafsson, J. A., and Kaptein, R. (1990). Solution structure of the glucocorticoid receptor DNA-binding domain. *Science* **249,** 157–160.

Hashimoto, Y., Petkovich, M., Gaub, M. P., Kagechika, H., Shudo, K., and Chambon, P. (1989). The retinoic acid receptors α and β are expressed in the human promyelocytic leukemia cell line HL-60. *Mol. Endocrinol.* **3,** 1046–1052.

Haussler, M., Sidell, N., Kelly, M., Donaldson, C., Altman, A., and Mangelsdorf, D. (1983). Specific high-affinity binding and biologic action of retinoic acid in human neuroblastoma cells lines. *Proc. Natl. Acad. Sci. U.S.A.* **80,** 5525–5529.

Henrich, V. C., Sliter, T. J., Lubahn, D. B. MacIntyre, A., and Gilbert, L. I. (1990). A steroid/thyroid hormone receptor member in *Drosophila melanogaster* that shares extensive sequence similarity with a mammalian homologue. *Nucleic Acids Res.* **18,** 4143–4148.

Hermans, P. E., Diaz-Buxo, J. A., and Stubo, J. D. (1976). Idiopathic late-onset immunoglobin deficiency. Clinical observations in 50 patients. *Am. J. Med.* **61,** 221–237.

Herr, W., Sturm, R. A., Clerc, R. G., Corcoran, L. M., Baltimore, D., Sharp, P. A., Ingraham, H. A., Rosenfeld, M. G., Finney, M., Ruvkun, G., and Horvitz, H. R. (1988). The POU domain: A large conserved region in the mammalian *pit*-1, *oct*-1, and *oct*-2, and *Caenorhabditis elegans unc*-86 gene products. *Genes Dev.* **2,** 1513–1516.

Herskowitz, I. (1987). Functional inactivation of genes by dominant negative mutations. *Nature (London)* **329,** 219–222.

Heyman, R. A., Mangelsdorf, D. J., Dyck, J. A., Stein, R. B., Eichele, G., Evans, R. B., Eichele, G., Evans, R. M., and Thaller, C. (1992). 9-*cis* retinoic acid is a high affinity ligand for the retinoid X receptor. *Cell (Cambridge, Mass.)* **68,** 397–406.

Hoffman, B., Lehmann, J. M., Zhang, X., Hermann, T., Husmann, M., Graupner, G., and Pfahl, M. (1990). A retinoic acid receptor-specific element controls the retinoic acid receptor-β promoter. *Mol. Endocrinol.* **4,** 1727–1736.

Hoffman, M. (1990). The embryo takes its vitamins. *Science* **250,** 372–373.

Hogan, B. L., Taylor, A., and Adamson, E. (1981). Cell interactions modulate embryonal carcinoma cell differentiation into parietal or visceral endoderm. *Nature (London)* **291,** 235–237.

Hoopes, C., Taketo, M., Ozato, K., Liu, Q., Howard, T., Linney, E., and Seldin, M. F. (1993). Mapping of the mouse *Rxr* loci encoding nuclear retinoid X receptors RXRα, RXRβ and RXRγ. *Genomics* (in press).

Hollenberg, S. M., Giguere, W., Segui, P., and Evans, R. M. (1987). Colocalization of DNA binding and transcriptional activation functions in the human glucocorticoid receptor. *Cell (Cambridge, Mass.)* **49,** 39–46.

Hosler, B. A., LaRosa, G. J., Grippo, J. F., and Gudas, L. J. (1989). Expression of *REX*-1, a gene containing zinc finger motifs, is rapidly reduced by retinoic acid in F9 teratocarcinoma cells. *Mol. Cell. Biol.* **9,** 5623–5629.

Hu, L., and Gudas, L. J. (1990). Cyclic AMP analogs and retinoic acid influence the expression of retinoic acid receptor α, β, and γ mRNAs in F9 teratocarcinoma cells. *Mol. Cell Biol.* **10,** 391–396.

Hu, J. C., O'Shea, E. K., Kim, P. S., and Sauer, R. T. (1990). Sequence requirements for coiled-coils: analysis with γ repressor-GCN4 leucine zipper fusions. *Science* **250,** 1400–1403.

Huang, M. E., Ye, Y. C., Chen, S. R., Chai, J. R., Lu, J. X., Zhoa, L., Gu, L. J., and Wang, Z. Y. (1988). Use of all-*trans* retinoic acid in the treatment of acute promyelocytic leukemia. *Blood* **72,** 567–572.

Imperiale, M. J., Kao, H. T., Feldman, L. T., Nevins, J. R., and Strickland, S. (1984). Common control of the heat shock gene and early adenovirus genes: Evidence for a cellular E1A-like activity. *Mol. Cell. Biol.* **4,** 867–874.

Ishikawa, T., Umesono, K., Mangelsdorf, D. J., Aburatani, H., Stanger, B. Z., Shibasaki, Y., Imawari, M., Evans, R. M., and Takaku, F. (1990). A functional retinoic acid receptor encoded by the gene on human chromosome 12. *Mol. Endocrinol.* **4,** 837–844.

Jetten, A. M. (1990). Regulation of gene expression by retinoic acid-embryonal carcinoma cell differentiation. *In* "Mechanisms of Differentiation" (P. B. Fisher, ed.), Vol. 1, pp. 48–74. CRC Press, Boca Raton, Florida.

Jetten, A. M., Jetten, M. E. R., and Sherman, M. I. (1979). Stimulation of differentiation of several murine embryonal carcinoma cell lines by retinoic acid. *Exp. Cell Res.* **124,** 381–391.

Jonat, C., Rahmsdorf, H. J., Park, K., Cato, A. C. B., Gebel, S., Ponta, H., and Herrlich, P. (1990). Anti tumor promotion and antiinflammation: Down-modulation of AP-1 (Fos/Jun) activity by glucocorticoid hormone. *Cell (Cambridge, Mass.)* **62,** 1189–1204.

Jones-Villeneuve, E. M. V., Rudnicki, M. A., Harris, J. F., and McBurney, M. W. (1983). Retinoic acid-induced neural differentiation of embryonal carcinoma cells. *Mol. Cell. Biol.* **3,** 2271–2279.

Kadomatsu, K., Tomomura, M., and Muramatsu, T. (1988). cDNA cloning and sequencing of a new gene intensely expressed in early differentiation stages of embryonal carcinoma cells and in midgestation period of mouse embryogenesis. *Biochem. Biophys. Res. Commun.* **151,** 1312–1318.

Kakizuka, A., Miller, W. H., Jr., Umesono, K., Warrell, R. P., Jr., Grankel, S. R., Murty, V. V. V. S., Dmitrovsky, E., and Evans, R. M. (1991). Chromosomal translocation t(15;17) in human acute promyelocytic leukemia fuses RARα with a novel putative transcription factor, PML. *Cell (Cambridge, Mass.)* **66,** 663–674.

Kapron-Bras, C. M., and Trasler, D. G. (1988). Interaction between the *splotch* mutation and retinoic acid in mouse neural tube defects *in vitro*. *Teratology* **38,** 165–173.

Kastner, P. H., Krust, A., Mendelsohn, C., Garnier, J. M., Zelent, A., Leroy, P., Staub, A., and Chambon, P. (1990). Murine isoforms of retinoic acid receptor γ with specific patterns of expression. *Proc. Natl. Acad. Sci. U.S.A.* **87,** 2700–2704.

Kessel, M., Balling, R., and Gruss, P. (1990). Variations of cervical vertebrae after expression of a *Hox*-1.1 transgene in mice. *Cell (Cambridge, Mass.)* **61,** 301–308.

Kliewer, S. A., Umesono, K., Mangelsdorf, D. J., and Evans, R. M. (1992). Retinoid X receptor interacts with nuclear receptors in retinoic acid, thyroid hormone and vitamin D_3 signalling. *Nature (London)* **355,** 446–449.

Kochhar, D. M. (1967). Teratogenic activity of retinoic acid. *Acta Pathol. Microbiol. Scand.* **70,** 398–404.

Kochhar, D. M. (1973). Limb development in mouse embryos. I. Analysis of teratogenic effects of retinoic acid. *Teratology* **7,** 289–298.

Kochhar, D. M., and Agnish, N. D. (1977). "Chemical surgery" as an approach to study morphogenetic events in embryonic mouse limb. *Dev. Biol.* **61,** 388–394.

Koeffler, H. P. (1983). Induction of differentiation of human acute myelogenous leukemia cells: Therapeutic implications. *Am. Soc. Hematol.* **62,** 709–721.

Krust, A., Green, S., Argos, P., Kumar, V., Walter, P., Bornert, J., and Chambon, P. (1986). The chicken oestrogen receptor sequence: Homology with v-ErbA and the human oestrogen and glucocorticoid receptors. *EMBO J.* **5,** 891–897.

Krust, A., Kastner, P. H., Petkovich, M., Zelent, A., and Chambon, P. (1989). A third human retinoic acid receptor, hRAR-γ. *Proc. Natl. Acad. Sci. U.S.A.* **86,** 5310–5314.

Kumar, V., and Chambon, P. (1988). The estrogen receptor binds tightly to its responsive element as a ligand-induced homodimer. *Cell (Cambridge, Mass.)* **55,** 145–156.

Kumar, V., Green, S., Stack, G., Berry, M., Jin, J.-R., and Chambon, P. (1987). Functional domains of the human estrogen receptor. *Cell (Cambridge, Mass.)* **51,** 941–951.

Kwasigroch, T. E., and Kochhar, D. M. (1980). Production of congenital limb defects with retinoic acid: Phenomenological evidence of progressive differentiation during limb morphogenesis. *Anat. Embryol.* **161,** 105–113.

8. Retinoic Acid Receptors

Lammer, E. J., Chen, D. T., Hoar, R. M., Agnish, N. D., Benke, P. J., Braun, J. T., Curry, C. J., Fernhoff, P. M., Grix, A. W., Lott, I. T., Richard, J. M., and Sun, S. C. (1985). Retinoic acid embryopathy. *N. Engl. J. Med.* **313**, 837–841.

Larson, R. A., Kondo, K., Vardman, J. W., Butler, A. E., Golomb, H. M., and Rowley, J. D. (1984). Evidence for a 15;17 translocation in every patient with acute promyelocytic leukemia. *Am. J. Med.* **76**, 827–841.

Leid, M., Kastner, P., Lyons, R., Nakshatri, H., Saunders, M., Zacharewski, T., Chen, J.-Y., Staub, A., Garnier, J.-M., Mader, S., and Chambon, P. (1992). Purification, cloning, and RXR identity of the HeLa cell factor with which RAR or TR heterodimerizes to bind target sequences efficiently. *Cell (Cambridge, Mass.)* **68**, 377–395.

Lemons, R. S., Eilender, D., Waldmann, R. A., Rebentisch, M., Frej, A. K., Ledbetter, D. H., Willman, C., McConnell, T., and O'Connell, P. (1990). Cloning and characterization of the t(15;17) translocation breakpoint region in acute promyelocytic leukemia. *Genes Chromosomes Cancer* **2**, 79–87.

Lenardo, M. J., Staudt, L., Robbins, P., Kuang, A., Mulligan, R. C., and Baltimore, D. (1989). Repression of the IgH enhancer in teratocarcinoma cells associated with a novel octamer factor. *Science* **243**, 544–546.

Leroy, P., Krust, A., Zelent, A., Medelsohn, C., Garnier, J.-M., Kastner, P., Dierich, A., and Chambon, P. (1991a). Multiple isoforms of the mouse retinoic acid receptor α are generated by alternative splicing and differential induction by retinoic acid. *EMBO J.* **10**, 59–69.

Leroy, P., Nakshatri, H., and Chambon, P. (1991b). Mouse retinoic acid receptor α_2 isoform is transcribed from a promoter that contains a retinoic acid response element. *Proc. Natl. Acad. Sci. U.S.A.* **88**, 10138–10142.

Levin, A. A., Sturzenbecker, L. J., Kazmer, S., Bosakowski, T., Huselton, C., Allenby, G., Speck, J., Kratzeisen, Cl., Rosenberger, M., Lovey, A., and Grippo, J. F. (1992). 9-*cis* retinoic acid stereoisomer binds and activates the nuclear receptor RXR α. *Nature (London)* **355**, 359–361.

Lippman, S. M., Kessler, J. F., and Meyskens, F. L., Jr. (1987). Retinoids as preventive and therapeutic anticancer agents (Part I). *Cancer Treat. Rep.* **71**, 391–405.

Lotan, R. (1980). Effects of vitamin A and its analogs (retinoids) on normal and neoplastic cells. *Biochim. Biophys. Acta* **605**, 33–91.

Lucas, P. C., O'Brien, R. M., Mitchell, J. A., Davis, C. M., Imai, E., Forman, B. M., Samuels, H. H., and Granner, D. K. (1991). A retinoic acid response element is part of a pleiotropic domain in the phosphoenolpyruvate carboxykinase gene. *Proc. Natl. Acad. Sci. U.S.A.* **88**, 2184–2188.

Luisi, B. F., Xu, W. X., Otwinowski, A., Freedman, F. P., Yamamoto, K. R., and Sigler, P. B. (1991). Crystallographic analysis of the interation of the glucocorticoid receptor with DNA. *Nature (London)* **352**, 497–505.

Luscher, B., Mitchell, P. J., Williams, T., and Tjian, R. (1989). Regulation of transcription factor AP-2 by the morphogen retinoic acid and by second messengers. *Genes Dev.* **3**, 1507–1517.

Maden, M. (1982). Vitamin A and pattern formation in the regenerating limb. *Nature (London)* **295**, 672–675.

Mader, S., Kumar, V., Verneuil, H. D., and Chambon, P. (1989). Three amino acids of the oestrogen receptor are essential to its ability to distinguish an oestrogen from a glucocorticoid responsive element. *Nature (London)* **338**, 271–274.

Mangelsdorf, D. J., and Evans, R. M. (1992). Retinoids as transcription factors. In *"Transcription Regulation"* (K. R. Yamamoto and S. L. McKnight, eds.). Cold Spring Harbor Laboratories, Cold Spring Harbor, New York (in press).

Mangelsdorf, D. J., Ong, E. S., Dyck, J. A., and Evans, R. M. (1990). Nuclear receptor that identifies a novel retinoic acid response pathway. *Nature (London)* **345**, 224–229.

Mangelsdorf, D. J., Umesono, K., Kliewer, S. A., Borgmeyer, U., Ong, E. S., and Evans, R. M. (1991). A direct repeat in the cellular retinol-binding protein type II gene confers differential regulation by RXR and RAR. *Cell (Cambridge, Mass.)* **66,** 555–561.

Marks, S. M., Hallenbeck, P. L., Nagata, T., Segars, J. H., Appella, E., Nikodem, V. M., and Ozato, K. (1992). H-2RII P (RXR β) heterodimerization provides a mechanism for combinatorial diversity in the regulation of retinoic acid and thyroid hormone response genes. *EMBO J.* **11** 1419–1435.

Martin, G. R. (1981). Isolation of a pluripotent cell line from early mouse embryos cultured in medium conditioned by teratocarcinoma stem cells. *Proc. Natl. Acad. Sci. U.S.A.* **78,** 7634–7638.

Martin, G. R., and Evans, M. J. (1975). Differentiation of clonal lines of teratocarcinoma cells: Formation of embryoid bodies *in vitro. Proc. Natl. Acad. Sci. U.S.A.* **72,** 1441–1445.

Mattei, M.-G., Petkovich, M., Mattei, J.-F., Brand, N., and Chambon, P. (1988a). Mapping of the human retinoic acid receptor to the q21 band of chromosome 17. *Hum. Genet.* **80,** 186–188.

Mattei, M.-G., de The, H., Mattei, J.-F., Marchio, A., Tiollais, P., and Dejean, A. (1988b). Assignment of the human hap retinoic acid receptor RAR β gene to the p24 band of chromosome 3. *Hum. Genet.* **80,** 189–190.

McCue, P. A., Matthei, K. I., Taketo, M., and Sherman, M. I. (1983). Differentiation-defective mutants of mouse embryonal carcinoma cells: Response to hexamethylenebisacetamide and retinoic acid. *Dev. Biol.* **96,** 416–426.

McNeish, J. D., Scott, W. J., and Potter, S. S. (1988). *Legless,* a novel mutation found in PHT1-1 transgenic mice. *Science* **241,** 837–839.

McNeish, J. D., Thayer, J., Walling, K., Sulik, K. K., Potter, S. S., and Scott, W. J. (1990). Phenotypic characterization of the transgenic mouse insertional mutation, *legless. J. Exp. Zool.* **253,** 151–162.

Mendelsohn, C., Ruberte, E., LeMeur, M., Morriss-Kay, G., and Chambon, P. (1991). Developmental analysis of the retinoic acid-inducible RAR-$β_2$ promoter in transgenic animals. *Development (Cambridge, UK)* **113,** 723–734.

Miller, W. H., Jr., Warrell, R. P., Jr., Grankel, S., Jakubowski, A., Gabriloce, J. L., Muindi, J., and Dmitrovsky, E. (1990). Novel retinoic acid receptor-α transcripts in acute promyelocytic leukemia responsive to all-*trans* retinoic acid. *Natl. Cancer Inst.* **82,** 1932–1933.

Mintz, B., and Illmensee, K. (1975). Normal genetically mosaic mice produced from malignant teratocarcinoma cells. *Proc. Natl. Acad. Sci. U.S.A.* **72,** 3585–3589.

Muller, R., and Wagner, E. F. (1984). Differentiation of F9 teratocarcinoma stem cells after transfer of c-*fos* protooncogenes. *Nature (London)* **311,** 438–442.

Munoz-Canoves, P., Vik, D. P., and Tack, B. F. (1990). Mapping of a retinoic acid-responsive element in the promoter region of the complement factor H gene. *J. Biol. Chem.* **265,** 20065–20068.

Murphy, S. P., Garbern, J., Odenwald, W. F., Lazzarini, R. A., and Linney, E. (1988). Differential expression of the homeobox gene *Hox*-1.3 in F9 embryonal carcinoma cells. *Proc. Natl. Acad. Sci. U.S.A.* **85,** 5587–5591.

Naar, A. M., Boutin, J.-M., Lipkin, S. M., Yu, C. V., Holloway, J. M., Glass, C. K., and Rosenfeld, M. G. (1991). The orientation and spacing of core DNA-binding motifs dictate selective transcriptional responses to three nuclear receptors. *Cell (Cambridge, Mass.)* **65,** 1267–1279.

Nakayama, Y., Takahashi, K., Noji, S., Muto, K., Nishijima, K., and Taniguchi, S. (1990). Functional modes of retinoic acid in mouse osteoblastic clone MC3T3-E1, proved as a target cell for retinoic acid. *FEBS Lett.* **261,** 93–96.

Nicholson, R. C., Mader, S., Nagpasl, S., Leid, M., Rochette-Egly, C., and Chambon, P. (1990). Negative regulation of the rat stromelysin gene promoter by retinoic acid is mediated by an AP1 binding site. *EMBO J.* **9,** 4443–4454.

8. Retinoic Acid Receptors

Noji, S., Nohno, T., Koyama, E., Muto, K., Ohyama, K., Aoki, Y., Tamura, K., Ohsugi, K., Ide, H., Taniguchi, S., and Saito, T. (1991). Retinoic acid induces polarizing activity but is unlikely to be a morhogen in the chick limb bud. *Nature (London)* **350**, 83–86.

Nunez, E. A. (1989). The Erb-A family receptors for thyroid hormones, steroids, vitamin D and retinoic acid: Characteristics and modulation. *Curr. Opin. Cell Biol.* **1**, 177–185.

O'Malley, B. (1990). The steroid receptor superfamily: More excitement predicted for the future. *Mol. Endocrinol.* **4**, 363–369.

Okamoto, K., Okazawa, H., Okuda, A., Sakai, M., Muramatsu, M., and Hamada, H. (1990). A novel octamer binding transcription factor is differentially expressed in mouse embryonic cells. *Cell (Cambridge, Mass.)* **60**, 461–472.

Okazawa, H., Okamoto, K., Ishino, F., Kaneko, T.-I., Takeda, S., Toyoda, Y., Muramatsu, M., and Hamada, H. (1991). The *oct3* gene, a gene for an embryonic a transcription factor, is controlled by a retinoic acid repressible enhancer. *EMBO J.* **10**, 2997–3005.

Oro, A. E., McKeown, M., and Evans, R. M. (1990). Relationship between the product of the *Drosophila* ultraspiracle locus and the vertebrate retinoid X receptor. *Nature (London)* **347**, 298–301.

Oshima, R. G., Abrams, L., and Kulesh, D. (1990). Activation of an intron enhancer within the keratin 18 gene by expression of c-*fos* and c-*jun* in undifferentiated F9 embryonal carcinoma cells. *Genes Dev.* **4**, 835–848.

Packer, L. (1990a). Retinoids. A. Molecular and metabolic aspects. *In* "Methods in Enzymology," Vol. 189. Academic Press, Orlando, Florida.

Packer, L. (1990b). Retinoids. B. Cell differentiation and clinical applications. *In* "Methods in Enzymology," Vol. 190. Academic Press, Orlando, Florida.

Papioaonnou, V. E., McBurney, M. W., and Gardner, R. L. (1975). Fate of teratocarcinoma cells injected into early mouse embryos. *Nature (London)* **258**, 70–73.

Petkovich, M., Brand, N. J., Krust, A., and Chambon, P. (1987). A human retinoic acid receptor which belongs to the family of nuclear receptors. *Nature (London)* **330**, 444–450.

Ponglikitnongkol, M., Green, S., and Chambon, P. (1988). Genomic organization of the human oestrogen receptor gene. *EMBO J.* **7**, 3385–3388.

Pratt, M. A. C., Kralova, J., and McBurney, M. W. (1990). A dominant negative mutation of the α retinoic acid receptor gene in a retinoic acid-nonresponsive embryonal carcinoma cell. *Mol. Cell. Biol.* **10**, 6445–6453.

Rosner, M. H., Vigano, M. A., Ozato, K., Timmons, P. M., Poirier, F., Rigby, P. W. J., and Staudt, L. M. (1990). A POU-domain transcription factor in early stem cells and germ cells of the mammalian embryo. *Nature (London)* **345**, 686–692.

Rossant, J., Zirngibl, R., Cado, D., Shago, M., and Giguere, V. (1991). Expression of a retinoic acid response element-*hsplacZ* transgene defines specific domains of transcriptional activity during mouse embryogenesis. *Genes Dev.* **5**, 1333–1344.

Rottman, J. N., Widom, R. L., Ginard, B.-N., Mahdavi, V., and Krathanasis, S. K. (1991). A retinoic acid-responsive element in the apolipoprotein AI gene distinguishes between two different retinoic acid response pathways. *Mol. Cell. Biol.* **11**, 3814–3820.

Ruberte, E., Dolle, P., Krust, A., Zelent, A., Morris-Kay, G., and Chambon, P. (1990). Specific spatial and temporal distribution of retinoic acid receptor gamma transcripts during mouse embryogenesis. *Development* **108**, 213–222.

Ruberte, E., Dolle, P., Chambon, P., and Morris-Kay, G. (1991). Retinoic acid receptors and cellular retinoid binding proteins. *Development (Cambridge, UK)* **111**, 45–60.

Saunders, J. W., and Gasseling, M. T. (1968). Ectodermal-mesenchymal interactions in the origin of limb symmetry. *In* "Epithelial-Mesenchymal Interactions" (R. Fleischmayer and R. E. Billingham, eds.). Williams & Wilkins, Baltimore, pp. 78–97.

Schena, M., Freedman, L. P., and Yamamoto, K. R. (1989). Mutations in the glucocorticoid receptor zinc finger region that distinguish interdigitated DNA binding and transcriptional enhancement activities. *Genes Dev.* **3**, 1590–1601.

Schindler, J., Matthaei, K. I., and Sherman, M. I. (1981). Isolation and characterization of mouse mutant embryonal carcinoma cells which fail to differentiate in response to retinoic acid. *Proc. Natl. Acad. Sci. U.S.A.* **78,** 1077–1080.

Scholer, H. R., Hatzopoulos, A. K., Balling, R., Suzuki, N., and Gruss, P. (1989). A family of octamer-specific proteins present during mouse embryogenesis: Evidence for germline-specific expression of an Oct factor. *EMBO J.* **8,** 2543–2550.

Scholer, H. R., Ruppert, S., Suzuki, N., Chowdhury, K., and Gruss, P. (1990). New type of POU domain in germ line-specific protein Oct-4. *Nature (London)* **344,** 435–439.

Schule, R., Umesono, K., Mangelsdorf, D. J., Bolado, J., Pike, J. W., and Evans, R. M. (1990a). Jun-Fos and receptors for vitamins A and D recognize a common response element in the human osteocalcin gene. *Cell (Cambridge, Mass.)* **61,** 497–504.

Schule, R., Rangarajan, P., Kliewer, S., Ransone, L. J., Bolado, J., Yang, N., Verma, I. M., and Evans, R. M. (1990b). Functional antagonism between oncoprotein c-Jun and the glucorcorticoid receptor. *Cell (Cambridge, Mass.)* **62,** 1217–1226.

Schule, R., Rangarajan, P., Yang, N., Kliewer, S., Ransone, L. J., Bolado, J., Verma, I. M., and Evans, R. M. (1991). Retinoic acid is a negative regulator of AP-1-responsive genes. *Proc. Natl. Acad. Sci. U.S.A.* **88,** 6092–6096.

Schwabe, J. W. R., Neuhaus, D., and Rhodes, D. (1990). Solution structure of the DNA-binding domain of the oestrogen receptor. *Nature (London)* **348,** 458–461.

Sheer, E., Adelman, D. C., Saxon, A., Gilly, M., Wall, R., and Sidell, N. (1988). Retinoic acid induces the differentiation of B cell hybridomas from patients with common variable immunodeficiency. *J. Exp. Med.* **168,** 55–71.

Shenefelt, R. E. (1972). Morphogenesis of malformations in hamsters caused by retinoic acid: Relation to dose and stage at treatment. *Teratology* **5,** 104–118.

Sherman, M. I., Matthaei, K. I., and Schindler, J. (1981). Studies on the mechanism of induction of embryonal carcinoma cell differentiation by retinoic acid. *Ann. N.Y. Acad. Sci.* **359,** 192–199.

Simeone, A., Acampora, D., Arcioni, L., Andrews, P. W., Boncinelli, E., and Mavilio, F. (1990). Sequential activation of *HOX2* homeobox genes by retinoic acid in human embryonal carcinoma cells. *Nature (London)* **346,** 763–766.

Sive, H. L., Draper, B. W., Harland, R. M., and Weintraub, H. (1990). Identification of a retinoic acid-sensitive period during primary axis formation in *Xenopus laevis*. *Genes Dev.* **4,** 932–942.

Smith, L. J., and Waterman, M. S. (1981). Comparison of biosequences. *Adv. Appl. Math.* **2,** 484–489.

Smith, W. C., Nakshatri, H., Leroy, F., Rees, J., and Chambon, P. (1991). A retinoic acid response element is present in the mouse cellular retinol binding protein I (mCRBPI) promoter. *EMBO J.* **10,** 2223–2230.

Strickland, S., and Mahdavi, V. (1978). The induction of differentiation in teratocarcinoma stem cells by retinoic acid. *Cell (Cambridge, Mass.)* **15,** 393–403.

Sucov, H. M., Murakami, K. K., and Evans, R. M. (1990). Characterization of an autoregulated response element in the mouse retinoic acid receptor type β gene. *Proc. Natl. Acad. Sci. U.S.A.* **87,** 5392–5396.

Sulik, K. K., and Dehart, D. B. (1988). Retinoic-acid-induced limb malformations resulting from apical ectodermal ridge cell death. *Teratology* **37,** 527–537.

Sulik, K. K., Johnston, M. C., Smiley, S. J., Speight, K. S., and Jarvis, B. E. (1987). Mandibulofacial dysostosis (Treacher Collins syndrome): A new proposal for its pathogenesis. *Am. J. Med. Genet.* **27,** 359–372.

Sulik, K. K., Cook, C. S., and Webster, W. S. (1988). Teratogens and craniofacial malformations: Relationships to cell death. *Development (Cambridge, UK)* **103** (Suppl.) 213–232.

Summerbell, D. (1983). The effect of local application of retinoic acid to the anterior margin of the developing chick limb. *J. Embryol. Exp. Morphol.* **78,** 269–289.

Summerbell, D., and Maden, M. (1990). Retinoic acid, a developmental signalling molecule. *Trends Neurosci.* **13**, 142–147.
Suzuki, N., Rohdewohld, R., Newman, T., Gruss, P., and Scholer, H. R. (1990). Oct-6: A POU transcription factor expressed in embryonal stem cells and in the developing brain. *EMBO J.* **9**, 3723–3732.
Tabin, C. J. (1991). Retinoids, homeoboxes, and growth factors: Toward molecular models for limb development. *Cell (Cambridge, Mass.)* **66**, 199–217.
Thaller, C., and Eichele, G. (1987). Identification and spatial distribution of retinoids in the developing chick limb bud. *Nature (London)* **327**, 625–628.
Thaller, C., and Eichele, G. (1990). Isolation of 3,4-didehydroretinoic acid, a novel morphogenetic signal in the chick wing bud. *Nature (London)* **345**, 815–819.
Tickle, C., Summerbell, D., and Wolpert, L. (1975). Positional signalling and specification of digits in chick limb morphogenesis. *Nature (London)* **254**, 199–202.
Tickle, C., Alberts, B., Wolpert, L., and Lee, J. (1982). Local application of retinoic acid to the limb bond mimics the action of the polarizing region. *Nature (London)* **296**, 564–565.
Tsai, S. Y., Carlstedt-Duke, J., Weigel, N. L., Dahlman, K., Gustafsson, J. A., Tsai, M.-J., and O'Malley, B. W. (1988). Molecular interactions of steroid hormone receptor with its enhancer element: Evidence for receptor dimer formation. *Cell (Cambridge, Mass.)* **55**, 361–369.
Umesono, K., and Evans, R. M. (1989). Determinants of target gene specificity for steroid/thyroid hormone receptors. *Cell (Cambridge, Mass.)* **57**, 1139–1146.
Umesono, K., Giguere, V., Glass, C. K., Rosenfeld, M. G., and Evans, R. M. (1988). Retinoic acid and thyroid hormone induce gene expression through a common responsive element. *Nature (London)* **336**, 262–265.
Umesono, K., Murakami, K. K., Thompson, C. C., and Evans, R. M. (1991). Direct repeats as selective response elements of the thyroid hormone, retinoic acid, and vitamin D_3 receptors. *Cell (Cambridge, Mass.)* **65**, 1255–1266.
Vasios, G. W., Gold, J. D., Petkovich, M., Chambon, P., and Gudas, L. J. (1989). A retinoic acid-responsive element is present in the 5′ flanking region of the laminin B1 gene. *Proc. Natl. Acad. Sci. U.S.A.* **86**, 9099–9103.
Vasios, G. W., Mader, S., Gold, J. D., Leid, M., Luta, Y., Gaub, M-P., Chambon, P., and Gudas, L. (1991). The late retinoic acid induction of laminin B1 gene transcription involves RAR binding to the responsive element *EMBO J.* **10**, 1149–1158.
Wagner, M., Thaller, C., Jessell, T., and Eichele, G. (1990). Polarizing activity and retinoid synthesis in the floor plate of the neural tube. *Nature (London)* **345**, 819–823.
Wanek, N., Gardiner, D. M., Muneoka, K., and Bryant, S. V. (1991). Conversion by retinoic acid of anterior cells into ZPA cells in the chick wing bud. *Nature (London)* **350**, 81–86.
Wang, S., and Gudas, L. J. (1984). Selection and characterization of F9 teratocarcinoma stem cell mutants with altered responses to retinoic acid. *J. Biol. Chem.* **259**, 5899–5906.
Warrell, R. P., Jr., Frankel, S. R., Miller, W. H., Jr., Scheinberg, D. A., Itri, L. M., Hittelman, W. N., Vyas, R., Andreeff, M., Tafuri, A., Jakubowski, A., Gabrilove, J., Gordon, M. S., and Dmitrovsky, E. (1991). Differentiation therapy of acute promyelocytic leukemia with tretinoin (all-*trans* retinoic acid). *N. Engl. J. Med.* **324**, 1385–1393.
Webster, W. S., Johnston, M. C., Lammer, E. J., and Sulik, K. K. (1986). Isotretinoin embryopathy and the cranial neural crest: An *in vivo* and *in vitro* study. *J. Craniofac. Genet. Dev. Biol.* **6**, 211–222.
Wilkinson, D. G., Bhatt, S., Cook, M., Boncinelli, E., and Krumlauf, R. (1989). Segmental expression of *Hox*-2 homeobox-containing genes in the developing mouse hindbrain. *Nature (London)* **341**, 405–409.
Yang-Yen, H., Chiu, R., and Karin, M. (1990a). Elevation of AP1 activity during F9 cell differentiation is due to increased c-*jun* transcription. *New Biol.* **2**, 351–361.
Yang-Yen, H., Chambard, J., Sun, Y., Smeal, T., Schmidt, T. J., Drouin, J., and Karin, M. (1990b). Transcriptional interference between c-Jun and the glucocorticoid receptor: Mutual

inhibition of DNA binding due to direct protein-protein interaction. *Cell (Cambridge, Mass.)* **62,** 1205–1215.

Yu, V. C., Delsert, C., Anderson, B., Holloway, J. M., Devary, O. V., Naar, A. M., Kim, S. Y., Boutin, J.-M., Glass, C. K., and Rosenfeld, M. G. (1991). RXRβ: A coregulator that enhances binding of retinoic acid, thyroid hormone, and vitamin D receptors to their cognate response elements. *Cell (Cambridge, Mass.)* **67,** 1251–1266.

Zelent, A., Krust, A., Petkovich, M., Kastner, P., and Chambon, P. (1989). Cloning of murine α and β retinoic acid receptors and a novel receptor γ predominantly expressed in skin. *Nature (London)* **339,** 714–717.

Zelent, A., Medelsohn, C., Kastner, P., Krust, A., Garnier, J.-M., Ruffenach, F., Leroy, P., and Chambon, P. (1991). Differentially expressed isoforms of the mouse retinoic acid receptor β are generated by usage of two promoters and alternative splicing. *EMBO J.* **10,** 71–81.

Zhang, X-K., Hoffmann, B., Tran, P. B.-V., Graupner, G., and Pfahl, M. (1992). Retinoid X receptor is an auxiliary protein for thyroid hormone and retinoic acid receptors. *Nature (London)* **355,** 441–445.

Zwilling, E. (1956). Interaction between ectoderm and mesoderm in the chick embryo. I. Axis relationships. *J. Exp. Zool.* **132,** 157–172.

9
Transcription Factors and Mammalian Development

Corrinne G. Lobe *
Department of Molecular Cell Biology
Max-Planck-Institute for Biophysical Chemistry
D-3400 Göttingen, Germany

I. Introduction
II. Mouse Embryogenesis Reviewed
III. Regulatory Factors in Early Embryogenesis
  A. Octamer-Binding Proteins
  B. Peptide Growth Factors
IV. Transcription Factors through Midembryogenesis
  A. Positional Information along the Anteroposterior Axis
  B. Retinoic Acid and Its Receptors
  C. Hox Genes
  D. More *Drosophila*-like Genes: *Engrailed*, *Pax*, and *Evx*
V. Conclusions
  References

I. Introduction

The characteristics of a cell are determined by specific regulation of gene expression. In the process of development, cells divide, migrate, and interact with one another to eventually differentiate and generate an entire complex organism from a single cell. This process of embryogenesis sets an intriguing stage on which to study gene regulation in response to outside influences and the consequences of this, both on the fate of the cell and possibly on the fate of its neighboring cells.

In the past few years, there has been a burst in the identification of transcription factors involved in embryo development. The majority of these have been isolated on the basis of homology to developmental genes of *Drosophila melanogaster*. In this article the role of these genes in two major aspects of development will be discussed: one is how the anteroposterior axis of the embryo is determined so that the embryo obtains its correct spatial arrangement, and the second is the process of dorsoventral patterning induced by the notochord.

*Present address: Institute for Molecular Biology and Biotechnology, and Department of Biochemistry, McMaster University, Hamilton, Ontario, Canada L8S 4K1.

II. Mouse Embryogenesis Reviewed

A description of murine embryogenesis is given here because, as a model for studying mammalian development, the mouse (*Mus musculus*) is most commonly used. It is also the organism on which most molecular biology and genetic experiments have been based. However, other systems are often more amenable to study. For instance, *Xenopus laevis* and chicken embryos are more plentiful and develop outside the mother, so they are easier to manipulate and observe. Therefore, where relevant experiments have been done using other vertebrate organisms, cross-reference will be made.

The mouse embryo undergoes three major stages of development: cleavage, gastrulation, and organogenesis (Fig. 1). The oocytes reside in the ovaries and after ovulation they lie in the oviduct where they are fertilized. The fertilized egg, or zygote, then proceeds to divide at a relatively slow rate compared to nonmammalian embryos, taking 20 to 24 hr between divisions in the first two cell cycles and 10 to 12 hr between subsequent divisions (Hogan et al., 1986). By the mid-two-cell stage, transcription of many embryonic genes begins. Cleavage continues for approximately 1 day to the eight-cell stage. Until that time, each cell is equipotent and, because there is still no polarization of the embryo, it remains radially symmetrical. However, at the eight-cell stage compaction occurs, wherein the cells polarize and form a tighter aggregation (the morula) connected with intercellular gap junctions. At the 16-cell stage the embryo becomes further polarized in the sense that outer cells become distinct from inner cells (Johnson and Ziomek, 1981) and are destined to become trophectoderm and the inner cell mass (ICM), respectively. Further cell division results in the formation of a blastocyst that consists of an outer cell layer, the trophectoderm, surrounding an ICM and a hollow, fluid-filled cavity, the blastocyst cavity (Fig. 1B). Up to this point the embryo has not increased in size or mass. The layer of ICM cells that faces the blastocyst cavity differentiates into the primitive endoderm, which together with the trophectoderm will make up the parietal yolk sac and, with extraembryonic mesoderm, the visceral yolk sac. The remainder of the ICM, the epiblast or primitive ectoderm, will become the embryo proper.

By this time the embryo has traveled into the uterus, and at 4.5 days post coitum (pc), it implants into the uterine wall. The trophectoderm neighboring the ICM differentiates and proliferates to form the ectoplacental cone and the extraembryonic ectoderm. Consequently, the epiblast is pushed down into the blastocoel to form the egg cylinder, a cuplike structure of embryonic ectoderm covered with the layer of visceral endoderm (Fig. 1C).

By 7 days pc gastrulation begins (Fig. 1D). At this time cells migrate through the primitive streak, an invagination that forms in the embryonic ectoderm. The primitive streak demarcates the posterior end of the anteroposterior axis of the embryo. After migrating through the primitive streak, cells migrate forward and laterally to form the endoderm and mesoderm tissues, thus generating the three

9. Transcription Factors and Mammalian Development 353

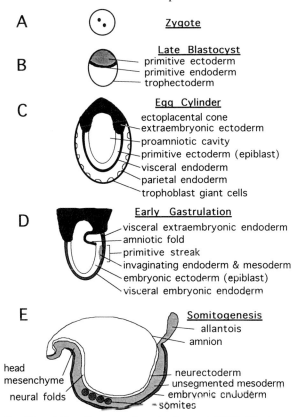

Fig. 1 Schematic representation of mouse embryo development. (A) Fertilized egg or zygote [day 1 postconception (pc)]. (B) Blastocyst stage (day 4 pc). At the late blastocyst stage, the inner cell mass differentiates into primitive endoderm and primitive ectoderm lineages. (C) Egg cylinder stage (day 6 pc). The embryo is now implanted in the uterine wall via the ectoplacental cone and trophoblast giant cells. (D) Early gastrulation (day 7 pc). Cells from the embryonic ectoderm migrate through the primitive streak to form the endoderm and mesoderm. This occurs at the caudal or posterior part of the embryo. Trophectoderm and parietal endoderm are not shown. (E) Somitogenesis (day 8 pc). The somites are formed by condensation of mesodermal tissue and the neural folds form, beginning at the rostral end and progressing caudally. This parasagittal view shows only the embryo and not the extraembryonic tissue except the allantois and amnion.

germ layers of the embryo. Some of the cells that pass through the anterior part of the streak (analogous to Hensen's node in the chick) will form the notochord, a rodlike structure extending forward along the midline. The notochord subsequently induces the neural plate in the ectoderm above it, which will fold and close to form the neural tube, destined to become the spinal cord and brain. Cells that lie at the juncture between the closing neural tube and the surface ectoderm, the neural crest cells, migrate to make up the peripheral nervous system, sympathetic and parasympathetic systems, pigment cells, and certain areas of head cartilage.

The mesoderm on either side of the notochord is the paraxial mesoderm (segmental plate in chick). This tissue condenses into somitomeres, the first obvious example of segmentation in embryos of higher vertebrates (Tam and Meier, 1982). These differentiate into somites, epithelial spheres that lie lateral to the neural tube. Both the neural tube and somites are established progressively in a rostrocaudal order. Later, cells in the ventromedial part of the somite lose their epithelial structure to become mesenchymal sclerotome cells. These cells migrate toward the notochord to form the perichordal zone, initially a uniformly distributed set of cells. As development progresses, periodic alterations of high and low density emerge. The high-density zone contributes to the intervertebral disks, whereas low-density areas develop into prevertebral centers (Keynes and Stern, 1988; Snow and Gregg, 1986). The rest of the somite consists of the dermamyotome, which differentiates into dermatome (which will form the dermis) and myotome (which will form the body wall and limb muscles). The mesoderm that flanks the somites, lateral mesoderm, will make up the segmented nephrotomes of the pronephros, mesonephros, and metanephric kidney and parts of organs such as the lung, liver, and gut. The embryonic endoderm will make up parts of the gut, liver, and lungs.

III. Regulatory Factors in Early Embryogenesis

Not much is known about gene regulation in early embryogenesis of mammals. Most information concerns two members of the POU family of transcriptional factors, Oct 3/4 and Oct 6, and the role of peptide growth factors as inducers of gene expression. Volume 24 of this series covers growth factors in development in depth (Nilsen-Hamilton, 1990).

Two genes that encode novel transcription factors and are predicted to play a role in early patterning of the embryo, *goosecoid* and *forkhead*, have recently been isolated from *Xenopus* cDNA libraries (Blumberg *et al.*, 1991; Altaba *et al.*, in press). *Goosecoid* encodes a homeobox-containing protein that is predicted to have DNA-binding specificity similar to that of *Drosophila bicoid*. In *Drosophila* embryogenesis, *bicoid* acts as an anterior morphogen. Remarkably, *Xenopus goosecoid* is expressed in the dorsal blastopore lip of the embryo, comparable to the archenteron of the mouse embryo, located just rostral to the primitive streak. Furthermore, when micro-injected into the ventral side of the embryo, *Xenopus goosecoid* leads to the formation of a secondary body axis (Cho *et al.*, 1991). This suggests that *goosecoid* plays an important role in early organization of embryo patterning and can act as a Spemann's organizer. *Xenopus forkhead* encodes a protein that contains a DNA-binding domain found

9. Transcription Factors and Mammalian Development 355

in the *Drosophila forkhead* gene, as well as several mammalian transcription factors (Weigel and Jackle, 1990). The *Xenopus forkhead* gene has an expression pattern and an ability to induce a secondary axis similar to *goosecoid*. Murine homologues of these genes are now being sought and an analysis of their expression patterns and functions should provide some information regarding primitive streak formation, organization of the three germ layers and axis determination in the mouse embryo.

A. Octamer-Binding Proteins

Several molecules that may be involved in the early regulation of other developmental control genes have been discovered by virtue of their ability to bind the octamer motif (ATGCAAAT) (Scholer *et al.*, 1989a,b). The transcription factors previously described that bind this motif and thereby regulate gene expression include a ubiquitously expressed factor, Oct-1, a B cell-specific factor, Oct-2, and the pituitary protein, Pit-1 or GHF-1. These proteins, together with another octamer-binding protein, unc-86 from *Caenorhabditis elegans*, constitute the POU family; they share a conserved 150- to 160-amino acid region containing a homeobox-related subdomain, which binds the octamer, and a POU-specific subdomain (Herr *et al.*, 1988). The POU homeodomain, like other homeodomains, contains three well defined alpha helices. In other classes of homeodomains, helix 3 determines DNA-binding specificity, however for POU homeodomains specificity and affinity are also determined by residues outside helix 3. The contribution of the POU-specific domain to DNA binding varies with the DNA binding site and the POU protein. In some instances, such as for Oct-1, the POU-specific domain can make contributions to the DNA-binding specificity. It is also required for protein–protein interactions (Ingraham *et al.*, 1988). In at least one instance this feature has been exploited to provide a negative regulatory function: the *Drosophila* I-POU protein lacks two of the five basic domains that are present in the amino terminal portion of other POU homeodomains. Consequently it cannot bind DNA, but remains as a monomer in solution. When co-expressed with another *Drosophila* POU-protein, Cf1/a, heterodimers are formed and DNA-binding by Cf1/a is inhibited, thus suppressing transcription of the target gene. This dimerization requires the POU domain of Cf1/a (Treacy *et al.*, 1991).

A family of octamer-binding proteins that are differentially expressed during early stages of development was identified by using the immunoglobulin heavy chain octamer as a DNA target (Scholer *et al.*, 1989a,b; Okamoto *et al.*, 1990; Rosner *et al.*, 1990; Meijer *et al.*, 1990; Monuki *et al.*, 1989, 1990). One of these, Oct-6, is expressed in the blastocyst and in embryonic stem cells, which are totipotent cells derived from the ICM. Another, Oct-3/4, is present in maturing and ovulating oocytes and then in the zygote, morula, and the early and expanding blastocyst where it is functioning mainly in the inner cell mass. After embryo implantation, expression of Oct-3/4 continues in the primitive ectoderm and later in the neuroectoderm, but only until day 8 pc. Beyond this time, expression is confined to the primordial germ cells and germ cell derivatives. Therefore, Oct-3/4 expression correlates with cells of a highly undifferentiated phenotype. Oct-3/4, like Oct-6, is also present in undifferentiated embryonic stem and embryonal carcinoma cells.

To see if transcription might be activated early in mouse development by these octamer-binding proteins, the octamer motif was inserted into a thymidine kinase (*tk*) promoter/*LacZ* gene construct, and this construct was used for injection of fertilized mouse oocytes (Scholer et al., 1989a). Expression was detected by β-galactosidase staining and showed that this construct was expressed in the blastocyst and that, more interestingly, expression was restricted to the ICM. This implies that the octamer-binding proteins are present and able to stimulate transcription (via recognition of the octamer motif) in cells of the ICM. It is possible, therefore, that Oct-3/4 and Oct-6 represent transcription factors involved in cell determination in the early embryo and/or initiation of the cascade of gene expression that directs embryogenesis.

Using the approach of identifying genes by their homology to conserved sequences, three new mammalian POU-domain genes were identified (He et al., 1989). Characterization by *in situ* analysis showed that these genes, as well as *Oct-1*, *Oct-2*, *Oct-6*, and *Pit-1*, are all expressed in the developing embryonic nervous system. These seven members of the POU family are therefore expressed in various patterns in the neural tube of the embryo and then with different tissue-specific patterns in the adult.

B. Peptide Growth Factors

Another class of molecules that may be involved in gene regulation during early embryogenesis are the peptide growth factors, which include the families of transforming growth factors (TGFs) and fibroblast growth factors (FGFs). These have been receiving great attention recently, particularly for studies done in the frog: Nieuwkoop (1969) showed that when an animal cap (destined to become epidermis and neural tissue, equivalent to the ectoderm of the mouse embryo) is removed from the blastula of an amphibian embryo and transplanted onto a vegetal mass, it forms mesodermal structures such as muscle and kidney. Subsequently, peptide growth factors have been found to impart the same mesoderm-inducing activity. Specifically, a TGF-like activity purified as XTC-MIF or activin (Smith et al., 1990) is able to induce anterodorsal structures, whereas basic FGF induces posteroventral structures (Ruiz i Altaba and Melton, 1989). Therefore, these growth factors are implicated both in mesoderm induction and in setting up an early polarity along the anteroposterior and dorsoventral axes. This action would be expected to be mediated by plasma membrane receptors that, on ligand binding, cause a cascade of intracellular events, including activation of transcription factors. For example, the recognition of TGF-like factors correlates with appropriate activation of the *XlHbox 1* gene, an anteriorly expressed *Antp*-like homeobox gene; by contrast, basic FGF leads to activation of *XlHbox6* and *Xhox3*, which are posteriorly expressed *AbdB*- and *Eve*-like homeobox genes, respectively (Ruiz i Altaba and Melton, 1989; Cho and DeRobertis, 1990). Therefore, peptide growth factors seem to provide signals for embryo patterning, particularly in mesoderm induction, resulting in spatial-specific activation of transcription factors.

Equivalent murine and chicken TGF-like factors have also been identified, from a mouse macrophage cell line and from chicken cDNA libraries (Sokol *et al.*, 1990; Mitrani *et al.*, 1990). Whether the mouse and chicken peptide growth factors are expressed in a pattern that is consistent with a role for imparting spatial signals during mesoderm induction, as in the frog, is still a matter of some debate. Certainly, characterization of their expression patterns and what and where their receptors are should shed some light on how early events in induction occur.

IV. Transcription Factors through Midembryogenesis

A. Positional Information along the Anteroposterior Axis

With the formation of the primitive streak, the embryo presents us with an obvious physical marker of its anteroposterior axis. How do cells receive their positional information at this time? In *Drosophila* a regulatory cascade of genes is well characterized: Initially, maternal effect genes set up the anteroposterior (A-P) axis, which is then roughly divided by the gap genes and further divided by the pair-rule genes into seven stripes. These 7 stripes are then divided into anterior and posterior halves by the segment polarity genes, and the consequent 14 stripes are given an "address" in relation to where each exists along the A-P axis via homeotic gene expression (Akam, 1987).

Although no equivalents for the maternal effect and gap genes have been identified in the mouse, an initial assignment of anterior versus posterior may be provided by the growth factors described above. Molecules that may subsequently provide positional information along that axis are retinoic acid (RA) and a family of receptors for RA that have been cloned (see also Chapter 8, this volume). In addition, genes that have sequence homology to the later-expressed pair-rule, segment polarity, and homeotic genes have been identified in the vertebrates by using the *Drosophila* genes as probes in low-stringency screening of mouse, frog, chicken, and human cDNA and genomic libraries. Characterization of these genes indicates that a function in imparting positional information has been maintained.

B. Retinoic Acid and Its Receptors

1. Induction by Retinoic Acid

Retinoic acid, a vitamin A derivative, has been of particular interest to embryologists because it was implicated as a possible morphogen in vertebrate development. This was initially discovered in the patterning of the developing chick limb bud in which there is a region of tissue at the posterior margin known as the "zone of polarizing activity" (ZPA). If the ZPA is transplanted from a donor to the anterior margin of the limb of a host embryo, the resultant limb develops with a mirror image duplication of digits along the A-P axis (Saunders and Gasseling, 1968; Tickle et al., 1975). This implies that the ZPA carries a signal of positional information, which it imparts to its neighboring tissue. The simplest model is that the ZPA releases a morphogen that forms a gradient with the highest concentration at the posterior margin. Other tissues that can substitute or mimic the ZPA are Hensen's node, the notochord, and the floor plate (Hornbruch and Wolpert, 1986; Wagner et al., 1990). These tissues could therefore all be determinants along the A-P axis, either in the limb or in the main body axis. In addition, it was found that a bead soaked in RA had the same effect (Tickle et al., 1982; Summerbell, 1983). Also of relevance is the fact that RA is a chemical used to differentiate embryonal carcinoma cells *in vitro* (Strickland and Mahdavi, 1978) and in so doing activates genes expressed in development, most notably the *Hox* genes (Deschamps et al., 1987). Furthermore, endogenous RA was shown to exist in the limb and neural plate, and at least in the case of the limb is present in a shallow gradient (Thaller and Eichele, 1987; Wagner et al., 1990). With all of this evidence it was tempting to postulate that RA acts as a morphogen used to set up positional information along the A-P axis.

2. Does Retinoic Acid Really Qualify as a Morphogen?

Throughout the studies of polarization induced by RA, there were critical voices that caution should be taken in designating RA as a morphogen (Brockes, 1990). Strictly speaking, a morphogen is a signaling substance emitted from an organizing center that at different concentrations causes cells to follow alternative pathways of differentiation, resulting in the characteristic anatomy of the organ concerned. Rather than acting as a morphogen, it is possible that RA actually acts to induce a new organizing center. For instance, when an RA-soaked bead is implanted in a chick limb bud, it could actually act to induce a new ZPA, which then emits some other, as yet unidentified, substance that is the true morphogen. Recently, this has been demonstrated to be the more probable situation. Wanek *et al.* (1991) showed that after transplanting an RA-soaked bead in the anterior margin of the limb, adjacent tissue could be explanted and grafted to another host, where it was capable of acting as a ZPA. The possibility that this effect is due simply to carryover of RA is ruled out by the fact that an incubation period of 16 hr with RA is required before the newly induced ZPA is competent, whereas the applied RA would be at the highest levels early in the application. It could still be possible that the new ZPA induced by RA in turn produces RA, salvaging the role of RA as a morphogen. However, Noji *et al.* (1991) showed that whereas an RA-soaked bead will induce endogenous RA receptor-β (*RAR*-β) transcription in the host limb, the RA-induced ZPA does not induce *RAR*-β transcription in the new host. Therefore if the ZPA actually emits RA as a morphogen, it at least is not enough to stimulate the RA-responsive *RAR*-β gene. Also of relevance is an experiment carried out with mouse transgenic embryos, in this case a construct in which the RA response element of the *RAR*-β gene and basal promoter of the heat shock gene were linked to the *LacZ* gene. This construct was used to detect endogenous RA and responding RARs that might exist in the developing embryo at high enough concentrations to induce expression of the *LacZ* gene. In the transgenic embryos, staining for β-galactosidase revealed expression of the transgene in all three germ layers with sharp rostral and caudal boundaries, but this expression was not induced in a gradient fashion across the rostrocaudal axis, and no expression was detected at all in the limb (Rossant *et al.*, 1991; see also Chapter 8, this volume). Therefore, as measured by an RA-response element, RA does not specify a gradient of activation along the rostrocaudal axis or across the limb buds of developing embryos.

As it now stands, RA is capable of at least inducing ZPA activity, whereas the morphogen that acts in a gradient to specify anteroposterior information is not yet known. Some role for RA in gene regulation probably exists because endogenous RA is present, an entire family of RA receptor genes has been cloned, and target genes that are transcriptionally regulated by these receptors have been identifed. However, RA now appears to act as an on/off switch rather than with a graded response. This on/off switch is nevertheless a complex one, because more than one receptor exists and there is variation among the RA response elements and a diversity of cofactors that modulate the effects of RARs.

3. Cloning and Expression Pattern of Retinoic Acid Receptors

A family of receptors for RA has now been cloned. As members of the steroid receptor family, these proteins function by binding to DNA target elements via a zinc finger DNA-binding domain and thereby regulate transcription. There are at least three families of RA receptors, RAR-α, RAR-β, and RAR-γ, as well as RAR-δ in the newt (Giguere *et al.*, 1987; de The *et al.*, 1987; Brand *et al.*, 1988;

Zelent *et al.*, 1989; Krust *et al.*, 1989; Ragsdale *et al.*, 1989). Because of alternative splicing, isoforms of each receptor also exist (Kastner *et al.*, 1990; Giguere *et al.*, 1990). The RARs exhibit the same modular structure as other members of the steroid receptor superfamily (Fig. 2A). Comparison of the amino acid sequences of the human receptors with the mouse receptors revealed that the interspecies conservation of an RAR subfamily is much higher than the conservation of all three receptors within a given species, suggesting that RAR-α, RAR-β, and RAR-γ each has its own specific function (Fig. 2A) (Zelent *et al.*, 1989; see Chapter 8, this volume, for additional details).

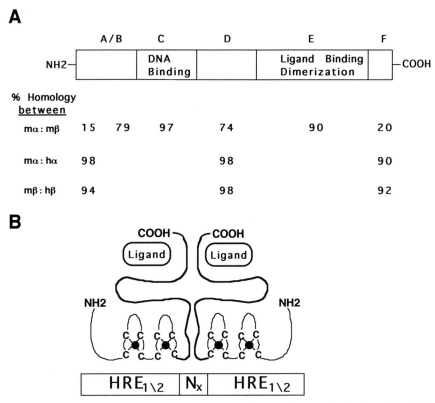

Fig. 2 Retinoic acid receptor (RAR) structure and binding. (A) The protein domains of the RAR are indicated, with the DNA-binding activity lying in domain C and the ligand binding and sequences necessary for dimerization of the receptors lying in domain E. The percentage homology between domains of the murine α (mα) and murine β (mβ), the murine α and human α (hα), and the murine β and human β (hβ) RARs is shown underneath. (B) Schematic illustration of receptor binding to a hormone response element (HRE). Each receptor binds one-half of the palindromic HRE via the zinc finger domain, with its zinc molecules each complexed to four cysteines. Within this domain, the specificity for the spacing between the halves of the palindrome (Nx) is also determined. The ligand binding and dimerization domain lies in the carboxy-terminal portion of the protein.

9. Transcription Factors and Mammalian Development

In situ hybridization analysis was used to study expression of the three receptors throughout embryogenesis (Dolle *et al.*, 1990; Ruberte *et al.*, 1991). At day 7.5 to 8 pc a probe directed against the RAR-α transcript gave a weak, diffuse signal. A stronger signal was seen at day 8.5 to 9 pc, when expression remained ubiquitous, although it was strongest in the migrating neural crest cells. The RAR-β transcript was detected in the lateral mesoderm through days 7.5 to 8.5 pc, as well as the splanchnopleuric mesoderm at day 8.5 pc and the most proximal mesenchyme of the limb at day 9 pc In addition, expression was detected in the neural epithelium up to the caudal hindbrain starting at day 8 pc and continuing to day 9 pc, when its caudal boundary coincided with the boundary between rhombomere 6 and 7. No expression was detected caudal to the neural tube closure. A signal was also detected in the foregut endoderm from day 8 through day 9 pc. An RAR-γ probe yielded a signal starting at day 8 pc through all three germ layers in the region of the primitive streak. On day 8.5 and 9 pc expression was in the open neural tube, therefore caudal to the expression of RAR-β in the neural tube, in presomitic mesoderm and at day 9 pc in lateral and limb mesoderm. Later, during organogenesis, RAR-α transcripts appeared to be ubiquitously distributed, whereas RAR-β and -γ expression in general continued to be mutually exclusive (Dolle *et al.*, 1990). In the same studies it was suggested that two cytosolic retinol and RA-binding proteins, CRBP and CRABP, may interact to establish a gradient of RA, for instance across the limb and the neural tube. In this scenario, CRABP would bind free RA, competing with the RARs for it, and thereby provide the low end of the gradient. These expression patterns do not clarify how a gradient of RA may act, but they do indicate how a diversity of responses to RA may be mediated by the presence of a different combination of receptors in different tissues.

4. Transcriptional Regulation by the Retinoic Acid Receptors

As mentioned above, the RARs belong to the family of steroid receptors (reviewed in Forman and Samuals, 1990; see also Chapter 8, this volume). These are nuclear receptors that bind their ligand through a ligand-binding domain, which increases their affinity to dimerize to a second receptor through the same domain (Fig. 2B) (Kumar and Chambon, 1988; Fawell *et al.*, 1990). They also possess a zinc finger DNA-binding domain (Hollenberg and Evans, 1988; Freedman *et al.*, 1988), part of which determines the sequence specificity for half of a hormone response element (HRE; each half being bound by one receptor of the dimer) and another part of which is used to distinguish the spacing between the two halves of the HRE (half-site spacing) (Mader *et al.*, 1989; Umesono and Evans, 1989; Danielson *et al.*, 1989). The receptors can bind DNA with or without binding their ligand, but typically binding of the receptor without its ligand represses transcription of a gene whereas binding in the presence of ligand enhances transcription (Damm *et al.*, 1989; de The *et al.*, 1990). This is further complicated in two ways: One is that the effect on transcription can vary depending on whether a homodimer or a heterodimer of a steroid receptor binds the HRE (Glass et al., 1989). The second is that different cell types possess different cofactors that bind to the steriod receptors, and these cofactors also determine whether HRE binding will result in enhancement or repression of transcription (Glass et al., 1990). In sum, RA, RARs, the HREs they recognize, and the cofactors that also bind present a multitude of possibilities for the effect they can have on gene regulation. Such complexity would be expected and required in the intricate process of setting up spatial and tissue identities during embryogenesis.

During the course of the cloning of the first RAR to be reported, it was noted that RAR-α could recognize the thyroid response element (TRE), the target of the thyroid receptor, which is a steroid receptor quite homologous to the RARs (Umesono et al., 1988). Indeed, a common ancestor motif for steroid receptors of invertebrates and vertebrates has been determined. The only difference between the optimal invertebrate ecdysone receptor response element and the vertebrate estrogen, thyroid, or RA receptor response elements was found to lie in the spacing between the half-palindromes (Martinez et al., 1991). However, natural gene targets have subsequently been reported that are apparently regulated by the RARs, and the target response elements often only weakly resemble TREs but are perhaps the bona fide RAR elements. For instance, the gene encoding RAR-β is itself induced by RA (de The et al., 1989) and a 27-base pair (bp) fragment that is able to confer RA responsiveness was identified upstream of the gene (de The et al., 1990). This sequence contains a direct repeat of the motif, GTTCAC, reminiscent of the 5' half-palindrome of the TRE (GGTCA). Another gene is the laminin B_1 gene, which is induced 24 to 48 hr after RA treatment. In this case the RAR element was found to lie in a 46-bp region at -477 to -432 (Vasios et al., 1989). This sequence contains four variants of the TRE-like sequence. However, transcriptional activation through this element required expression of exogenous transfected RARs in F9 cells, where endogenous RARs are already expressed. It may be that this sequence is divergent enough that it has become a poor RAR element and therefore the laminin gene takes a relatively long time, 24 hr, to be induced. Other genes that are reported to be transcriptionally activated by RA are genes for AP-2, tissue plasminogen activator (TPA, whose induction appears to be an indirect effect through Sp1-like binding sites) and growth hormone (Luscher et al., 1989; Rickles et al., 1989; Bedo et al., 1989). The family of *Hox* genes is also induced by RA as discussed below.

C. *Hox* Genes

Many developmental control genes that have been identified in vertebrates were cloned on the basis of homology to *Drosophila* genes. The developmental genes

9. Transcription Factors and Mammalian Development   363

of the fly that were the best characterized initially were the homeotic genes, contained in two gene complexes: the antennepedia complex (ANT-C) and the bithorax complex (BX-C) (Lewis, 1978; Mahaffey and Kaufman, 1987). Each member of the homeotic gene family contains a conserved DNA sequence of 183 bp designated the homeobox (McGinnis *et al.*, 1984a; Scott and Weiner, 1984). Using the prototype homeobox from the *Antp* gene as a probe, homeobox genes were identified in *Xenopus*, mouse, and human genomic libraries (Carrasco *et al.*, 1984; Muller *et al.*, 1984; McGinnis *et al.*, 1984b; Colberg-Poley *et al.*, 1985; Joyner *et al.*, 1985; Levine *et al.*, 1984). In the mouse the genes (*Hox* genes) exist as clusters on chromosomes 2, 6, 11, and 15 (Fig. 3B). The existence of two *Drosophila* gene clusters and multiple representations of that array in mouse and higher vertebrates probably came about by local gene duplication initially, followed by diversification and duplication of entire homeobox gene clusters (Kappen *et al.*, 1989).

In the mouse genome, each *Hox* cluster consists of 7 to 11 genes organized in a homologous linear arrangement whereby genes occupying the same position in each cluster (paralogous genes) show the most homology. On this basis, 13 paralogous gene groups have been determined (Fig. 3B) (Kappen *et al.*, 1989) Paralog groups 1 to 5 all contain a homeobox that is most similar to the *AbdB* homeobox (Duboule *et al.*, 1990). These genes also do not encode a conserved hexapaptide, which is present in all remaining genes of the paralogous groups 6 to 13 (Kessel *et al.*, 1987).

1. Patterns of Expression

The possibility that these genes not only represent conserved sequences but might also perform conserved functions was first indicated by RNA blots and from *in situ* hybridization data, which showed that they are expressed with spatial and temporal specificity during embryogenesis. Typical expression of the mammalian *Hox* genes begins during gastrulation (7.5 to 8.5 days pc) and continues through midgestation (Holland and Hogan, 1988; Kessel and Gruss, 1990). The genes are expressed in the presomitic mesoderm and overlying ectoderm; however, the anterior and posterior boundaries of expression vary between the genes. Further on in development, at day 12 pc when the level of most of the *Hox* gene transcripts has peaked, a similar pattern is seen. That is, the ectoderm-derived tissues (brain and spinal cord) continue to express the *Hox* genes with a variation in the anterior boundary ranging from the myelencephalon through to the hindbrain and cervical region of the spinal cord. Likewise, transcripts are detected in the mesoderm compartment, in the sclerotome (prevertebrae), myotome, and in gut tissue derived from mesoderm flanking the somites, but for each gene this expression has an anterior limit that corresponds, but is slightly offset posteriorly, to the limit of expression in the ectoderm-derived tissue.

2. Sequential Activation of Genes within a *Hox* Cluster

An interesting feature of this variation in the anterior limit of expression is that there is a direct correlation between the anterior boundary of expression of a given gene and its position in the gene cluster (Fig. 3C). For example, considering expression of genes of the *Hox-1* cluster in the ectoderm-derived tissue, transcripts of *Hox-1.6* extend to the presumptive myelencephalon of the neural tube, whereas, proceeding upstream in the cluster, for *Hox-1.5*, *-1.4*, *-1.3*, and *-1.2* the limits of expression lie more posteriorly in the hindbrain (Gaunt et al., 1988). Likewise in the mesoderm, *Hox-1.6* expression begins anterior to the first somitic mesoderm, *Hox-1.5* expression in prevertebra 1, *Hox-1.4* expression in prevertebra 2, *Hox-1.3* in prevertebrae 3 to 4, and *Hox-1.2* in prevertebra 8 (Dressler and Gruss, 1989). A similar situation has been reported for the *Hox-2* cluster (Graham et al., 1989). The genes are also turned on over time from the 3'-most to the 5'-most in the cluster, thus correlating with the sequence of embryonic development that proceeds from anterior to posterior.

A correlation between gene position in the cluster and expression pattern was also found for the *Hox-4* cluster in the developing limb (Dolle et al., 1989). In this case, the more 5' the position of a gene is in the cluster, the later and more distal is the expression. This suggests that the gradient of *Hox* gene expression is manifested in the morphology of the limb.

The possibility that RA establishes the gradient of *Hox* gene expression has been suggested in human teratocarcinoma cells, which can be differentiated with RA. The genes of the human *Hox-2* cluster were found to respond sequentially to RA, with the genes at the 3' end of the cluster being expressed more quickly and requiring lower concentrations of RA than more 5'-located genes. Genes of the first five paralogous groups, however, were either activated very weakly, did not respond at all, or were even repressed by RA during differentiation of NT2 cells (Simeone et al., 1990). Likewise, genes of the murine *Hox-1* cluster are sequentially induced in F9 and P19 embryonal carcinoma cells when differentiated by RA treatment (C. G. Lobe and P. Gruss, unpublished observations, 1991). In the mouse cells, as in the human cells, genes of the 5' end of the cluster were more sensitive to RA than genes of the 3' end of the cluster, and paralog groups I and II (*Hox-1.10* and *-1.9*) were not induced by RA. Thus, the sequential activation of *Hox* genes in the embryo can be mimicked by RA treatment of teratocarcinoma cells, although genes of the extreme 3' end of the cluster (the *AbdD*-like genes) are either much less sensitive or require additional signals to RA.

A similar relationship exists between the order of genes within the cluster and the order of anterior boundaries of expression of each gene for the *Drosophila* homeotic genes (see Fig. 3A). However, the A-P axis of the *Drosophila* embryo is represented once in the two complexes of ANT-C and BX-C, whereas in vertebrates multiple copies apparently exist, one per *Hox* cluster. This may reflect the evolution of a more complex body pattern of the vertebrates and the

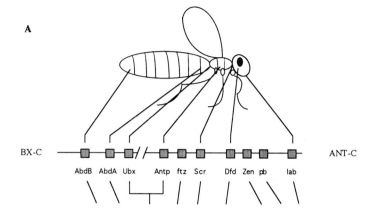

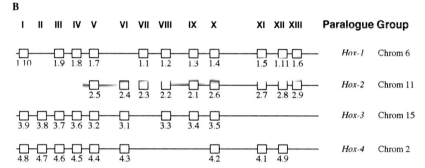

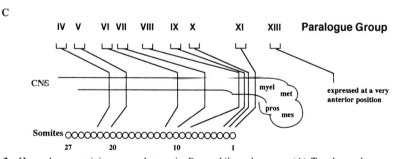

Fig. 3 Homeobox-containing gene clusters in *Drosophila* and mouse. (A) Two homeobox gene clusters exist in *Drosophila*: BX-C and ANT-C. Each gene (indicated with a filled box) has an anterior border of expression in the *Drosophila* embryo as indicated, with *labial* (*lab*) having the most anterior expression and *AbdB* the most posterior. (B) The four *Hox* clusters of the mouse, *Hox-1*, *-2*, *-3*, and *-4*. Each gene is represented by an open box. The chromosome location of each cluster is also indicated at the right. The most closely related genes are aligned vertically and belong to a single paralogue group, indicated at the top (I to XIII). (C) The anterior boundary of expression for each paralogue group in the mouse embryo is shown, both for the developing central nervous system and for the somites. myel, Myelencephalon; met, metencephalon; mes, mesencephalon; pros, prosencephalon.

attendent need to encode and decipher more detailed positional information. The single representation of the A-P axis of the embryo in *Drosophila* and the multiple representation of this information in higher vertebrates is also in accord with the idea of duplication of a primordial gene cluster and the observation that greater homology is shown between genes of different clusters than between genes within the same cluster (Kappen *et al.*, 1989).

3. *Hox* Gene Expression and Vertebral Specification

The *Drosophila* homologs of *Hox* genes, the homeotic genes, specify positional information by the combination of their expression domains. Their inappropriate expression results in altered segmental identities (homeosis). Is it conceivable that a combination of expressed *Hox* genes also specifies segmental identity in vertebrates?

The fate of a somite is specified early, therefore if *Hox* genes act as determinants, they should show autonomy of expression coincident with a stable tissue fate. This implies that if a somite is relocated to a different level in the embryo, it should affect neither the expression of *Hox* genes nor the fate of the somite. The unaltered fate of transplanted somites has been shown to be true in experiments using chicken embryos (Kieny *et al.*, 1972). The stability of *Hox* gene expression has been tested recently by grafting somites of transgenic animals, expressing β-galactosidase under the regulation of the *Hox-1.1* promoter, to heterotopic positions (R. Beddington, A. Puschel, and P. Gruss, unpublished observations, 1991). The data show that independent of the new location of the transplanted somite, *Hox-1.1* promoter activity remains stable. This result is in agreement with the notion that *Hox* genes specify the identity of individual body segments.

A more direct functional analysis to test the concept that a combination of *Hox* genes determines vertebral identity was performed by using a dominant gain-of-function approach in transgenic mice. This experimental strategy was previously used successfully in *Drosophila* by ectopic expression of *Antennepedia* and the *deformed* gene (Schneuwly *et al.*, 1987; Kuziora and McGinnis, 1988). In the transgenic mice, dominant gain of function was accomplished by placing the *Hox-1.1* gene under the control of the ubiquitously active β-*actin* promoter (Balling *et al.*, 1989). Because this promoter expressed the *Hox-1.1* gene ectopically in tissues located rostral to the normal *Hox-1.1* expression domain, the overall combination of expressed *Hox* genes was altered (Fig. 4). This alteration resulted in striking morphological changes of the most rostral vertebrae. In particular malformations of the atlas, the axis and the basioccipital bone were observed, as well as the existence of an additional vertebra, a proatlas (Kessel and Gruss, 1990). In these transgenic animals the atlas, which normally does not contain a vertebral body, did possess a vertebral body, whereas the axis lost its second ossification center, the dens axis. Thus, the ectopic expression of a *Hox* gene could deregulate the finely tuned program so that the base of the skull, the

9. Transcription Factors and Mammalian Development

Normal Expression of Hox 1.1

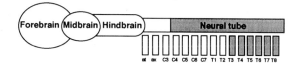

Ectopic Expression of Hox 1.1

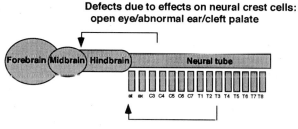

Fig. 4 Ectopic expression of *Hox-1.1*. In transgenic mice that express *Hox-1.1* in more rostral regions of the embryo than normal, abnormalities occur both in mesoderm-derived structures (the vertebrae) and ectoderm-derived structures (the neural crest derivatives). *Hox-1.1* expression is indicated with shading. at, Atlas; ax, axis; c, cervical; t, thoracic.

atlas, and the axis exhibit a morphology that renders them similar to more posterior cervical vertebrae. These changes in the vertebrae can be interpreted as homeotic transformations. The model developed for homeotic genes of *Drosophila* by Lewis (1978) therefore appears to apply to mammalian *Hox* genes. This experiment supports the concept that a defined combination of *Hox* genes is required for specification along the A-P axis, at least for vertebrae.

Loss-of-function experiments are complementary to the dominant gain-of-function approach. The possibility of insertionally inactivating genes in embryonal stem (ES) cells, in conjunction with the generation of chimeric mice by using these ES cells, allows mouse mutants be produced (Thomas and Capecchi, 1987). Using this experimental strategy, two different mouse *Hox* mutants have been generated, one carrying an inactivated *Hox-1.5* gene (Chisaka and Capecchi, 1991) and the other carrying an inactivated *Hox-1.6* gene (Lufkin *et al.*, 1991). These mutants show severe abnormalities in regions that correspond to their rostral domains of expression: the *Hox-1.5* mutant has an absence of the thymus and parathyroids, reduced thyroid and submaxillary tissue, and a wide range of throat, heart, blood vessel, and craniofacial abnormalities. The structures that are affected arise from the mesenchyme of the branchial arches and pharyngeal pouches. The *Hox-1.5* mutant displays normal hindbrain, rhombomere, cranial nerve, spinal cord, lung, gut, stomach, spleen, and kidneys, although all of these tissues express *Hox-1.5* during embryo development. The *Hox-1.6* mutant has defects at the level of rhombomeres 4 to 7, in structures that arise from paraxial mesoderm (bones of the skull), neurectoderm (motor nuclei of cranial nerves), neural crest (cranial ganglia), placodal ectoderm (membranous part of the inner ear, cranial ganglia), and head mesoderm (inner ear). Like the *Hox-1.5* mutant, the *Hox-1.6* mutant has no defects more caudally in the embryo. There is an overlap in the region of neural crest cells that are affected by the *Hox-1.5* and *-1.6* inactivations; however, for *Hox-1.5*, derivatives of the mesenchymal neural crest cells are defective, whereas for *Hox-1.6*, derivatives of the neurogenic neural crest cells are defective (Lufkin *et al.*, 1991). These experiments indicate that the *Hox* genes are important regulators of vertebrate embryogenesis. A "posterior prevalence rule" has also been suggested (Duboule, 1991; Lufkin *et al.* 1991), that is, the identity of a cell in a given region is determined by the Hox protein that corresponds to the most 5' (posterior) *Hox* gene expressed in that cell. This would explain the restriction of the defects of the *Hox-1.5* and *-1.6* mutants to the rostral domains of expression of the genes, because in more caudal regions, more 5' (posterior) *Hox* genes are expressed. It is also worth noting that paralogs of the *Hox-1.6* and *Hox-1.5* genes did not rescue the *Hox-1.6* and *Hox-1.5* inactivations, therefore paralogous *Hox* genes are not completely redundant in function. Furthermore, the lack of a phenotype in rostral regions of the embryo indicates that products of the genes in the 3' region of the *Hox* cluster are not required for subsequent expression of genes in the 5' region of the *Hox* cluster.

Loss of function has also been accomplished in the frog by injecting an anti-Hox protein antibody into *Xenopus* oocytes. In this way the *Xenopus XlHbox 1* (*Hox-3.3*) gene product was inactivated, resulting in an extended hindbrain, indicative of an anterior transformation (Wright *et al.*, 1989). In this case, alterations of individual vertebrae were not examined in detail.

4. *Hox* Gene Products as Transcription Factors

The conserved homeobox sequence that was used to clone the *Hox* genes is a DNA-binding domain, as initially implied by its homology to prokaryotic and yeast gene regulatory proteins and subsequently confirmed by X-ray crystallographic data (Shephard *et al.*, 1984; Laughon and Scott, 1984; Kissinger *et al.*, 1990). A consensus binding site for several *Drosophila* homeobox proteins was identified based on *in vitro* binding to 5'-flanking sequences of homeobox genes themselves (Hoey and Levine, 1988; Desplan *et al.*, 1988). Other homeobox proteins that were shown to bind sequences in their own promoters are the murine Hox-1.5 and Hox-1.3 proteins, which were produced in bacterial and insect expression systems, respectively (Fainsod et al., 1986; Odenwald et al., 1989). A demonstration that the *Drosophila* homeobox proteins can regulate transcription via their DNA-binding activity has been provided by cotransfection experiments (Jaynes and O'Farrell, 1988; Han et al., 1989). Multiple copies of the *Drosophila* homeobox-binding consensus sequence were linked to heterologous promoters and reporter genes and these constructs were cotransfected with expression vectors that produced various homeobox proteins. Cotransfection of some homeobox genes (*ftz, prd,* and *zen*) led to stimulation of transcription of the reporter gene construct, whereas others (*en* and *eve*) had no effect. The proteins that stimulated transcription through the homeobox-binding sequence could also act synergistically, and the proteins that did not activate transcription inhibited the activity of those that did.

The experiments described above suggest that the homeobox proteins recognize common binding sites but are able to impart different effects on promoter activity. It may be that various homeobox proteins compete for the same sites and that a range of protein affinities is provided through deviations from the consensus binding site. Because more than one consensus binding element is usually present in eukaryotic gene promoters, binding of more than one homeobox protein species would allow for synergistic or antagonistic interactions. Whether genes containing the promoter response element(s) are transcriptionally active or silent in that cell would be determined through competition and interactions between the set of homeobox proteins present in a cell.

D. More *Drosophila*-like Genes: *Engrailed, Pax,* and *Evx*

1. *En 1* and *Pax 1* in the Developing Vertebral Column

engrailed is a gene of the segment polarity class, is expressed in a band of cells just posterior to the A-P boundary of each segment in *Drosophila*, and is crucial for the development of the posterior half of each segment. The gene contains a homeobox, distantly related to the *Antp* homeobox, which was used to isolate two mouse genes, *En-1* and *En-2* (Joyner and Martin, 1987). An *in situ* analysis of expression detected transcripts of both genes during the early somite stage in a band restricted to the anterior part of the neual folds. At day 12 pc, both genes are still expressed in a ring at the midbrain hindbrain junction of the neural tube, but *En-1* transcripts are also observed extending down the spinal cord. Additionally, *En-1* is expressed in the mesoderm compartment, in sclerotome-derived prevertebrae and in dermatome-derived cells under the skin throughout the length of the embryo, and in the limb and tail buds (Davis and Joyner, 1988; Davidson et al., 1988). The prevertebrae expression displays a periodic pattern, present in the prevertebrae but not the intervertebral disks. Later, at 17.5 days pc, *En-1* and *En-2* are expressed in different but overlapping areas of the mid- and hindbrain. This expression pattern is consistent with a role, together with the *Antp*-type homeobox genes, in compartmentalization of the early central nervous system (CNS) and later in neural differentiation. A targeted disruption of the *En-2* gene does affect the development of the mid- and hindbrain, but surprisingly does not result in a severe mutant phenotype (Joyner *et al.*, 1989). The periodic expression of *En-1* in the prevertebrae is suggestive of the *engrailed* pattern of expression in *Drosophila*, but apparently occurs too late for establishment of segmentation and may play a role later, in vertebral column formation (see below).

In two other classes of *Drosophila* developmental genes, another conserved sequence is present. This is the paired-box domain, a 128-amino acid sequence shared by the pair-rule gene, *paired* (*prd*), and the segment-polarity genes, *gooseberry-proximal* (*gsb-p*) and *gooseberry-distal* (*gsb-d*) (Bopp *et al.*, 1986). A family of murine genes, designated "*Pax*" for paired box, was identified by using the *Drosophila* paired-box sequence (Deutsch *et al.*, 1988) (Fig. 5). One member, *Pax-1*, was detected by *in situ* analysis in sclerotome cells starting at 9 days pc and therefore, like murine *En-1*, appears to be induced after the primary segmentation of the mesoderm. At 12 days pc this pattern shifted so that expression was seen only in periodic bands of the sclerotome that correspond to the intervertebral disk anlagen. *Pax-1* together with *En-1* might therefore play a role in directing the formation of the vertebrae.

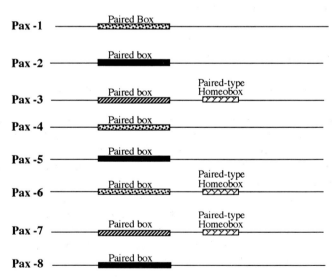

Fig. 5 The murine *Pax* gene family. Eight members of the paired-box-containing gene family have been isolated. Paired boxes that are most similar are indicated by similar shading. *Pax-3, -6,* and *-7* contain a paired-type homeobox in addition to the paired domain.

In accord with such a role for *Pax-1* was the finding that a mouse mutant, *undulated*, has a mutated *Pax-1* sequence (Balling *et al.*, 1988). Mice homozygous for *undulated* exhibit a reduction of the posterior part of the vertebrae, abnormally large intervertebral disks, and small vertebra centers. Analysis of the *Pax-1* gene from *undulated* mice revealed that the sequence contained a point mutation causing a Gly-Ser replacement in a highly conserved part of the paired box. DNA-binding studies have shown that this change causes the Pax-1 protein to lose its ability to bind its DNA target element (Chalepakis *et al.*, 1991). Later it was found by genetic analysis that two other related mutants that exhibit a related phenotype to *undulated*, namely *undulated short tail* and *undulated extreme*, have a deletion of the the *Pax-1* gene and a reduced level of *Pax-1* mRNA, respectively (R. Balling, personal communication). This marks the first correspondence of one of the mouse genes identified through *Drosophila* homology to a mouse developmental mutant. Further proof that the *undulated* phenotype is caused by a *Pax-1* mutation awaits a rescue experiment, in which the wild-type gene introduced into a mutant mouse restores a normal phenotype.

2. Other *Pax* Genes in Mesoderm-derived Tissue

Several other members of the *Pax* gene family are expressed in mesoderm-derived tissue in addition to *Pax-1* in the sclerotome. In other compartments of the somitic mesoderm, *Pax-3* and *Pax-7* are expressed (Goulding *et al.*, 1991; Jostes *et al.*, 1991). This expression begins at day 9 pc in the dermamyotome of the somites. At this stage, the dermatome and myotome are not distinguishable, but as development progresses *Pax-3* and *-7* are restricted to the dermatome and myotome, respectively, and later to tissues derived from those cell compartments. Expression continues until day 14 pc.

Two other *Pax* genes, *Pax-2* and *Pax-8* (Dressler *et al.*, 1990; Plachov *et al.*, 1990), are expressed in the intermediate mesoderm, which flanks the somitic mesoderm. These cells differentiate into the segmented nephrotomes of the pronephros, mesonephros, and metanephric kidney, which later develop into the kidney and excretory system.

3. Neural Tube Expression of *Pax*, *En-1*, and *Evx-1* Genes

Another domain of expression of the *Pax* genes is the developing neural tube, together with *En-1* and *Evx-1*. The neural tube forms when the lips of the neural plate fuse along the dorsal midline and this structure separates from the surface ectoderm. Initially the neural tube consists of only a single layer of epithelial cells, but on closure of the neural tube these cells begin to proliferate. The mitotically active epithelial cells lining the inner lumen form the ventricular zone. After several divisions, a stem cell loses its ability to divide, migrates away from the ventricular zone, and differentiates in a precise ventral-to-dorsal progression. Consequently, motor neurons develop first and are confined to the ventral region of the neural tube, excluding the floor plate. Later, relay neurons form in the midlateral region and neurons for the sensory pathway form in the midlateral and dorsal regions.

Unlike the *Hox* genes, which apparently specify position along the anteroposterior axis, the *Pax* genes are expressed throughout the length of the neural tube but in specific subsets of cells transversally. This suggests that they play a role in neuronal identity or dorsoventral regional specification.

Three of the *Pax* genes, *Pax-3*, *-6*, and *-7*, are expressed in the mitotically active cells of the ventricular zone throughout the neural tube caudal to the rhombencephalon. *Pax-3* expression is limited to cells dorsal of the sulcus limitans, which will give rise to the alar plate and neural crest cells (Goulding *et al.*, 1991). Likewise, *Pax-7* is restricted to the dorsal part of the neural tube, but unlike *Pax-3* is not expressed in the roof plate or neural crest cells (Jostes *et al.*, 1991). *Pax-6*, on the other hand, is expressed in cells of the ventricular zone that are ventral to the sulcus limitans and that will differentiate in the basal plate, but is not expressed in the floor plate of the neural tube (Walther and Gruss, 1991).

Pax-2 and *-8* are expressed starting at day 9 pc at the boundary of the ventricular zone and the intermediate zone, and therefore probably in cells that have already finished their last round of mitosis and are migrating radially outward (Dressler *et al.*, 1990; Plachov *et al.*, 1990). Later on, at day 11 pc, their expression is confined to two populations of cells, one dorsal and one ventral to the sulcus limitans. These may represent early commisural neurons.

Another gene, *Evx-1*, which was cloned on the basis of homology to the *Drosophila Eve* gene and contains an *Eve*-type homeobox distantly related to the *Antp* and *En* homeoboxes, is also expressed in the developing neural tube from day 10 to day 12 pc (Bastian and Gruss, 1990). It is initially expressed in cells at the outer edge of the ventricular zone, ventral to the sulcus limitans. Subsequently, it is expressed lateral to the ventricular zone in the basal plate and later also in the alar plate in cells that may be early interneurons.

Finally, *En-1*, a gene described earlier for expression in the developing vertebral column, is also expressed in the neural tube (Joyner *et al.*, 1985). Initially transcripts are detected in a ventral region in a subset of cells that are the differentiating motor neurons. Expression later follows these cells as they migrate out into the ventral horns.

A clear assignment of each of these genes to a particular cell fate cannot be made because there is not yet a fate map available for the neural tube. Nevertheless, their expression in a complex pattern of domains in the neural tube where commitment and differentiation into different types of neuronal cells is ongoing strongly suggests that they might play a role in some aspect of neuronal differentiation.

Further support of this role has been provided by grafting experiments in chicken embryos. These experiments are based on the principle that it is the

9. Transcription Factors and Mammalian Development 373

notochord that induces and organizes the arrangement of cells in the neural tube (van Straaten *et al.*, 1989; Placzek *et al.*, 1990; reviewed in Lumsden, 1991) (Fig. 6). This initially involves the induction in the neural tube of a floor plate and at a further distance away, in the ventrolateral regions, the differentiation of motor neurons. If a supernumerary notochord is transplanted from a donor to a

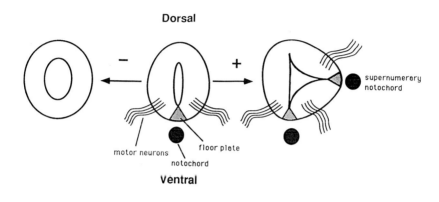

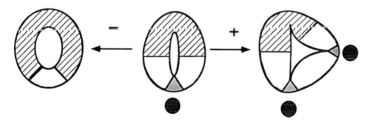

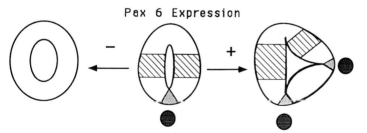

Fig. 6 Schematic representation of notochord transplant experiments. A normal neural tube is shown in the center, with motor neurons extending from the ventrolateral regions. To the left and right is shown the effect of removing the notochord or transplanting a supernumerary notochord, respectively. *Pax* gene expression is depicted by hatching.

host embryo at a position lateral or dorsal to the neural tube, a "floor plate reaction" occurs in which the neural tube develops two ventral poles. Conversely, if the notochord is removed shortly after forming, no floor plate develops. Using an antibody that recognizes only motor neurons, SC1, Yamada et al. (1991) showed that notochord transplants also influenced the production of motor neurons. Namely, a grafted notochord induces motor neurons close to the induced floor plate whereas extirpating the notochord results in a lack of motor neurons.

The consequent *Pax-3* and *Pax-6* expression on notochord grafts has also been investigated (Goulding et al., 1992). In keeping with the expression patterns described above, a supernumerary notochord led to *Pax-6* expression in the regions that neighbor both the normal and the induced, ectopic floor plate and *Pax-3* expression was lost in the region where *Pax-6* was abnormally induced. If the notochord was removed, *Pax-3* expression extended throughout the entire transverse region of the neural tube, whereas almost no signal was observed for *Pax-6*.

The *Pax* gene expression patterns in the neural tube imply that at least two signals exist in the developing neural tube, one originating from the floor plate and one from the roof plate. The former could be a secreted substance that is initially produced by the notochord and subsequently by the induced floor plate. Motor neurons and *Pax-6*-expressing cells are found at a fixed distance away from the notochord or floor plate, which would correspond to a certain concentration of the secreted signal. Because neither *Pax-6* nor *Pax-3* expression could be shifted to the dorsalmost cells of the neural tube, there is probably also a separate, independent signal emanating from the roof plate. This signal could be responsible for inducing *Pax-3* expression. These changes in expression patterns in response to a supernumerary notochord are in keeping with a role for *Pax-3* and *Pax-6* in regional specification of the neural tube.

Further evidence of a role for *Pax-3* in neurogenesis comes from chromosomal localization of the gene. *Pax-3* was mapped to a location near or at the *splotch* (Sp^{2H}) locus on chromosome 1 (Olson et al., 1990). Analysis of genomic DNA and cDNA clones representing transcripts from Sp^{2H}/Sp^{2H} embryos revealed a deletion of 32 nucleotides in the *Pax-3* mRNA transcript and gene (Epstein et al., 1991). The *splotch* mutant displays abnormal neural development, including exencephaly, overgrowth of the neural tissue near the posterior neuropore resulting in spina bifida, and a reduction or absence of neural crest cell-derived spinal ganglia and their derivatives. Thus, a mutation in the *Pax-3* gene correlates with abnormalities of the neural tube and neural crest cells, implicating a role for the *Pax-3* gene product in neural development. It will be interesting to learn what effect a nonfunctional *Pax-3* gene has on the organization of cell types in the neural tube and expression patterns of the other *Pax* genes, *Int-1* and *En-1*.

Chromosomal localization of *Pax-6* also provided a correlation to a pre-existing mouse mutant, *small-eye* (*Sey*) (Hill et al., 1991). In the Sey^H allele, *Pax-6* sequences are missing, while in *Sey*, a single point mutation leads to a truncated

Pax-6 protein product. Homozygous mutants exhibit an absence of eye and nose structures, reflecting *Pax-6* expression during induction of those tissues (Walther and Gruss, 1991). However, no gross defects occur in the neural tube, suggesting that there is some redundancy of function in the proteins which determine neural tube patterning.

The *Pax-3* and *Pax-6* mutants have each been correlated with human syndromes. A mutation in HuP2, the human homologue of the *Pax-3* gene leads to Waadenburg syndrome I, an autosomal dominant combination of pigmentary disturbances and deafness (Tassabehji *et al.*, 1992; Baldwin *et al.*, 1992). The human homolog of the *Pax-6* gene was discovered when the aniridia (AN) locus was cloned using reverse genetics (Ton *et al.*, 1991). Thus, it appears that deletions in the human homolog of *Pax-6* cause the aniridia disorder, characterized by complete or partial absence of the iris.

4. DNA-Binding by the *Pax* Gene Products

As mentioned above, the *Pax* genes were isolated on the basis of their homology to the paired box domain of the *Drosophila paired* gene. In addition to the paired box motif, *Pax-3*, *-6*, and *-7* also contain a paired-type homeobox (Fig. 5). The homeobox recognizes the upstream half of the e5 site, a DNA element located upstream of the *Eve* gene, which contains the ATTA motif, whereas the paired domain binds to the downstream half (Chalapakis *et al.*, 1991). These two domains may cooperate in their binding, so that *Pax-3*, *-6*, and *-7* are able to activate some promoters that *Pax-1*, *-2*, and *-8* cannot. Alternatively, the binding by the two domains may lead to different consequences in the transcriptional regulation of a target gene.

V. Conclusions

A picture of how the complex process of embryo development occurs at the molecular level is now emerging. At the level of transcriptional regulatory factors, a few DNA-binding protein domains are reiterated in a number of genes. These domains include the zinc finger, homeobox, and paired box. Although their definitive roles have yet to be established, their expression patterns and the phenotypes observed when they are inappropriately expressed points to a role for retinoic acid and its receptors and the *Hox* genes in specification along the anteroposterior axis, for *Pax-1* in development of the vertebral column, and other members of the *Pax* gene family in regional specification of the neural tube. Further analysis *in vivo* by gain- or loss-of-function mutants using transgenic mice or targeted disruption by homologous recombination will provide definitive tests of their role in embryogenesis. In addition, experiments utilizing tissue culture cells and yeast and bacterial expression systems will be useful in identifying DNA-binding sites and determining how the protein products regulate their target genes. Together these studies should provide us with some understanding of how cells are regulated at the level of gene expression to divide, migrate, and differentiate in a concerted fashion, thereby generating a healthy newborn animal.

Acknowledgments

I thank P. Gruss for time given generously in helpful discussions and in proofreading the manuscript.

References

Akam, M. (1987). The molecular basis for metameric pattern in the *Drosophila* embryo. *Development (Cambridge, UK)* **101,** 1–22.
Baldwin, C. T., Hoth, C. F., Amos, J. A., da-Silva, E. O. and Milunsky, A. (1992). An exonic mutation in the HuP2 paired domain gene causes Waardenburg's syndrome. *Nature* **355,** 637–683.
Balling, R., Deutsch, U., and Gruss, P. (1988). *undulated*, a mutation affecting the development of the mouse skeleton, has a point mutation in the paired box of *Pax-1*. *Cell (Cambridge, Mass.)* **55,** 531–535.
Balling, R., Mutter, G., Gruss, P., and Kessel, M. (1989). Craniofacial abnormalities induced by ectopic expression of the homeobox gene *Hox-1.1* in transgenic mice. *Cell (Cambridge, Mass.)* **58,** 337–347.
Bastian, H., and Gruss, P. (1990). A murine even-skipped homologue, *Evx-1*, is expressed during early embryogenesis and neurogenesis in a biphasic manner. *EMBO J.* **9,** 1839–1852.
Bedo, G., Santisteban, P., and Aranda, A. (1989). Retinoic acid regulates growth hormone gene expression. *Nature (London)* **339,** 231–234.
Blumberg, B., Wright, V. E., De Robertis, E. M. and Cho, K. W. Y. (1991). Organize-specific homeobox genes in *Xenopus laevis* embryos. *Science* **253,** 194–196.
Bopp, D., Burri, M., Baumgartner, S., Frigerio, G., and Noll, M. (1986). Conservation of a large protein domain in the segmentation gene *paired* and in functionally related genes of *Drosophila*. *Cell (Cambridge, Mass.)* **47,** 1033–1040.
Brand, N., Petkovich, M., Krust, A., Chambon, P., de The, H., Marchio, A., Tiollais, P., and Dejean, A. (1988). Identification of a second human retinoic acid receptor. *Nature (London)* **332,** 850–853.
Brockes, J. (1990). Reading the retinoid signals. *Nature (London)* **345,** 766–768.
Carrasco, A. E., McGinnis, W., Gehring, W. J., and De Robertis, E. M. (1984). Cloning of an *X. laevis* gene expressed during early embryogenesis coding for a peptide region homologous to *Drosophila* homeotic genes. *Cell (Cambridge, Mass.)* **37,** 409–414.
Chalepakis, G., Fritsch, R., Fickenscher, H., Deutsch, U., Goulding, M., and Gruss, P. (1991). The molecular basis of the *undulated/Pax-1* mutation. *Cell (Cambridge, Mass.)* **66,** 873–884.
Chisaka, O., and Capecchi, M. R. (1991). Regionally restricted developmental defects resulting from targeted disruption of the mouse homeobox gene *Hox-1.5*. *Nature (London)* **350,** 473–479.
Cho, K. W. Y., and DeRobertis, E. M. (1990). Differential activation of *Xenopus* homeobox genes by mesoderm-inducing growth factors and retinoic acid. *Genes Dev.* **4,** 1910–1916.

9. Transcription Factors and Mammalian Development 377

Cho, K. W. Y., Blumberg, B., Steinbeisser, H. and De Robertis, E. M. (1991). Molecular nature of Spemann's organizer: the role of the *Xenopus* homeobox gene *goosecoid*. *Cell* **67**, 1111–1120.

Colberg-Poley, A. M., Voss, S. D., Chowdhury, K., and Gruss, P. (1985). Structural analysis of murine genes containing homeobox sequences and their expression in embryonal carcinoma cells. *Nature (London)* **314**, 713–718.

Damm, K., Thompson, C. C., and Evans, R. M. (1989). Protein encoded by v-*erb*A functions as a thyroid-hormone receptor antagonist. *Nature (London)* **339**, 593–596.

Danielson, M., Hinck, L., and Ringold, G. M. (1989). Two amino acids within the knuckle of the first zinc finger specify DNA response element activation by the glucocorticoid receptor. *Cell (Cambridge, Mass.)* **57**, 1131–1138.

Davidson, D., Graham, E., Sime, C., and Hill, R. (1988). A gene with sequence similarity to *Drosophila engrailed* is expressed during the development of the neural tube and vertebrate in the mouse. *Development (Cambridge, UK)* **104**, 315–316.

Davis, C. A., and Joyner, A. L. (1988). Expression patterns of the homeobox-containing genes *En-1* and *En-2* and the proto-oncogene *int-1* diverge during mouse development. *Genes Dev.* **2**, 1736–1744.

Deschamps, J., DeLaaf, R., Joosen, L., Meijlink, F., and Destree, O. (1987). Abundant expression of homeobox genes in mouse embryonal carcinoma cells correlates with chemically induced differentiation. *Proc. Natl. Acad. Sci. U.S.A.* **84**, 1304–1308.

Desplan, C., Theis, J., and O'Farrell, P. H. (1988). The sequence specificity of homeodomain-DNA interaction. *Cell (Cambridge, Mass.)* **54**, 1081–1090.

de The, H., Marchio, A., Riollair, P., and Dejean, A. (1987). A novel steroid thyroid hormone receptor related gene inappropriately expressed in human hepatocellular carcinoma. *Nature (London)* **330**, 667–670.

de The, H., Marchio, A., Tiollais, P., and Dejean, A. (1989). Differential expression and ligand regulation of the retinoic acid receptor α and β genes. *EMBO J.* **8**, 429–433.

de The, H., Vivanco-Ruiz, M. M., Tiollais, P., Stunnenberg, H., and Dejean, A. (1990). Identification of a retinoic acid responsive element in the retinoic acid receptor β gene. *Nature (London)* **343**, 177–180.

Deutsch, U., Dressler, G. R., and Gruss, P. (1988). *Pax-1*, a member of a paired box homologous murine gene family, is expressed in segmented structures during development. *Cell (Cambridge, Mass.)* **53**, 617–625.

Dolle, P., Iypisua-Belmonte, J.-C., Falkenstein, H., Renucci, A., and Duboule, D. (1989). Coordinate expression of the murine *Hox-5* complex homoeobox-containing genes during limb pattern formation. *Nature (London)* **342**, 767–772.

Dolle, P., Ruberte, E., Leroy, P., Morriss-Kay, G., and Chambon, P. (1990). Retinoic acid receptors and cellular retinoid binding proteins. I. A systematic study of their differential pattern of transcription during mouse organogenesis. *Development (Cambridge, UK)* **110**, 1133–1151.

Dressler, G. R., and Gruss, P. (1989). Anterior boundaries of *Hox* gene expression in mesoderm-derived structures correlate with the linear gene order along the chromosome. *Differentiation (Berlin)* **41**, 193–201.

Dressler, G. R., Deutsch, U., Chowdhury, K., Nornes, H. O., and Gruss, P. (1990). *Pax-2*, a new murine paired-box-containing gene and its expression in the developing excretory system. *Development (Cambridge, UK)* **109**, 787–795.

Duboule, D. (1991). Patterning in the vertebrate limb. *Curr. Opin. Genet. Dev.* **1**, 211–216.

Duboule, D., Boncinelli, E., De Robertis, E., Featherstone, M., Lonai, P., Oliver, G., and Ruddle, G. H. (1990). An update of mouse and human *Hox* gene nomenclature. *Genomics* **7**, 458–459.

Epstein, D. J., Vekemans, M., and Gros, P. (1991). *splotch* (Sp^{2H}), a mutation affecting development of the mouse neural tube, shows a deletion within the paired homeodomain of *Pax-3*. *Cell (Cambridge, Mass.)* **67,** 767–774.

Fainsod, A., Bogarad, L. D., Ruusala, T., Lubin, M., Crothers, D. M., and Ruddle, F. H. (1986). The homeo domain of a murine protein binds 5' to its own homeo box. *Proc. Natl. Acad. Sci. U.S.A.* **83,** 9532–9536.

Fawell, S. E., Lees, J. A., White, R., and Parker, M. G. (1990). Characterization and colocalization of steroid binding and dimerization activities in the mouse estrogen receptor. *Cell (Cambridge, Mass.)* **60,** 953–962.

Forman, B. M., and Samuals, H. H. (1990). Dimerization among nuclear hormone receptors. *New Biol.* **2,** 587–594.

Freedman, L. P., Luisi, B. F., Korszun, Z. R., Basavappa, R., Sigler, P. B., and Yamamoto, K. R. (1988). The function and structure of the metal coordination sites within the glucocorticoid receptor DNA binding domain. *Nature (London)* **334,** 543–546.

Gaunt, S. J., Sharpe, P. T., and Duboule, D. (1988). Spatially restricted domains of homeo-gene transcripts in mouse embryos: Relation to a segmented body plan. *Development (Cambridge, UK)* **104** (Suppl.), 169–179.

Giguere, V., Ong, E. S., Segui, P., and Evans, R. M. (1987). Identification of a receptor for the morphogen retinoic acid. *Nature (London)* **330,** 624–629.

Giguere, V., Shago, M., Zirngibl, R., Tate, P., Rossant, J., and Varmuza, S. (1990). Identification of a novel isoform of the retinoic acid receptor gamma expressed in the mouse embryo. *Mol. Cell. Biol.* **10,** 2335–2340.

Glass, C. K., Lipkin, S. M., Devary, O. V., and Rosenfeld, M. G. (1989). Positive and negative regulation of gene transcription by a retinoic acid-thyroid hormone receptor heterodimer. *Cell (Cambridge, Mass.)* **59,** 697–708.

Glass, C. K., Devary, O. V., and Rosenfeld, M. G. (1990). Multiple cell type-specific proteins differentially regulate target sequence recognition by the alpha retinoic acid receptor. *Cell (Cambridge, Mass.)* **63,** 729–738.

Goulding, M. D., Chalepkis, G., Deutsch, U., Erselius, J. R., and Gruss, P. (1991). *Pax-3*, a novel murine DNA binding protein expressed during early neurogenesis. *EMBO J.* **10,** 1135–1147.

Goulding, M. D., Lumsdem, A. G. S., and Gruss, P. (1992). Notochord induced changes to paired box gene expression in the developing spinal cord. *Development,* in press.

Graham, A., Papalopulu, N., and Krumlauf, R. (1989). The murine and *Drosophila* homeobox gene complexes have common features of organization and expression. *Cell (Cambridge, Mass.)* **57,** 367–378.

Han, K., Levine, M. S., and Manley, J. L. (1989). Synergistic activation and repression of transcription by *Drosophila* homeobox proteins. *Cell (Cambridge, Mass.)* **56,** 573–583.

He, X., Treacy, N., Simmons, D. M., Ingraham, H. A., Swanson, L. W., and Rosenfeld, M. G. (1989). Expression of a large family of POU-domain regulatory genes in mammalian brain development. *Nature (London)* **340,** 35–42.

Herr, W., Sturm, R. A., Clerc, R. G., Corcoran, L. M., Baltimore, D., Sharp, P. A., Ingraham, H. A., Rosenfeld, M. G., Finney, M., Ruvkun, G., and Horvitz, H. R. (1988). The POU domain: A large conserved region in the mammalian *Pit-1*, *Oct-1*, *Oct-2* and *Caenorhabditis elegans unc-86* gene products. *Genes Dev.* **2,** 1513–1516.

Hill, R. E., Favor, J., Hogan, B. L. M., Ton, C. C. T., Saunders, G. F., Hanson, I. M., Prosser, J., Jordan, R., Hastie, N. D. and Van Heyningen, V. (1991). Mouse *small eye* results from mutations in a paired-like homeobox-containing gene. *Nature* **354,** 522–525.

Hoey, T., and Levine, M. (1988). Divergent homeobox proteins recognize similar DNA sequences in *Drosophila*. *Nature (London)* **332,** 858–861.

Hogan, B. L. M., Constantini, F., and Lacy, E. (1986). "Manipulating the Mouse Embryo: A Laboratory Manual." Cold Spring Harbor Laboratory, Cold Spring Harbor, New York.

Holland, P. W. H., and Hogan, B. L. M. (1988). Expression of homeo box genes during mouse development: A review. *Genes Dev.* **2**, 773–782.

Hollenberg, S. M., and Evans, R. M. (1988). Multiple and cooperative transactivation domains of the glucocorticoid receptor. *Cell (Cambridge, Mass.)* **55**, 899–906.

Hornbruch, A., and Wolpert, L. (1986). Positional signalling by Hensen's node when grafted to the chick limb bud. *J. Embryol. Exp. Morphol.* **94**, 257–265.

Ingraham, H. A., Chen, R., Mangcelam, H. J., Elsholtz, H. P., Flynn, S. E., Lin, C. R., Simmons, D. M., Swanson, L., and Rosenfeld, M. G. (1988). A tissue-specific transcription factor containing a homeodomain specifies a pituitary phenotype. *Cel,* **55**, 519–529.

Jaynes, J. B., and O'Farrell, P. H. (1988). Activation and repression of transcription by homoeodomain-containing proteins that bind a common site. *Nature (London)* **336**, 744–749.

Johnson, M. H., and Ziomek, C. A. (1981). The formation of two distinct cell lineages within the mouse morula. *Cell (Cambridge, Mass.)* **24**, 71–80.

Jostes, B., Walther, C., and Gruss, P. (1991). The murine paired box gene, *Pax-7*, is expressed specifically during the development of the nervous and muscular system. *Mech. Dev.* **33**, 27–38.

Joyner, A. L., and Martin, G. R. (1987). *En-1* and *En-2*, two mouse genes with sequence homology to the *Drosophila engrailed* gene—expression during embryogenesis. *Genes Dev.* **1**, 29–38.

Joyner, A. L., Kornberg, T., Coleman, K. G., Cox, D. R., and Martin, G. R. (1985). Expression during embryogenesis of a mouse gene with sequence homology to the *Drosophila engrailed* gene. *Cell (Cambridge, Mass.)* **43**, 29–37.

Joyner, A. L., Skarnes, W. C., and Rossant, J. (1989). Production of a mutation in mouse *En-2* gene by homologous recombination in embryonic stem cells. *Nature (London)* **338**, 153–156.

Kappen, C., Schughart, K., and Ruddle, F. H. (1989). Two steps in the evolution of Antennapedia-class vertebrate homeobox genes. *Proc. Natl. Acad. Sci. U.S.A.* **86**, 5459–5463.

Kastner, P., Krust, A., Mendelsohn, C., Garnier, J. M., Zelent, A., Leroy, P., Staub, P., and Chambon, P. (1990). Murine isoforms of retinoic acid receptor gamma with specific patterns of expression. *Proc. Natl. Acad. Sci. U.S.A.* **87**, 2700–2704.

Kessel, M., and Gruss, P. (1990). Murine developmental control genes. *Science* **249**, 374–379.

Kessel, M., Schulze, F., Fibi, M., and Gruss, P. (1987). Primary structure and nuclear localization of a murine homeodomain protein. *Proc. Natl. Acad. Sci. U.S.A.* **84**, 5306–5310.

Keynes, R. J., and Stern, C. D. (1988). Mechanisms of vertebrate segmentation. *Development (Cambridge, UK)* **103**, 413–429.

Kieny, M., Mauger, A., and Sengel, P. (1972). Early regionalization of the somitic mesoderm as studied by the development of the axial skeleton of the chick embryo. *Dev. Biol.* **28**, 142–161.

Kissinger, C. R., Liu, B., Martin-Bianco, E., Kornberg, T. B., and Pabo, C. O. (1990). Crystal structure of an *engrailed* homeodomain-DNA complex at 2.8 Å resolution: A framework for understanding homeodomain-DNA interactions. *Cell (Cambridge, Mass.)* **63**, 579–590.

Krust, A., Kastner, P., Petkovich, M., Zelent, A., and Chambon, P. (1989). A third human retinoic acid receptor, hRAR-γ. *Proc. Natl. Acad. Sci. U.S.A.* **86**, 5310–5314.

Kumar, V., and Chambon, P. (1988). The estrogen receptor binds tightly to its responsive element as a ligand induced homodimer. *Cell (Cambridge, Mass.)* **55**, 145–156.

Kuziora, M. A., and McGinnis, W. (1988). Autoregulation of a *Drosophila* homeotic selector gene. *Cell (Cambridge, Mass.)* **55**, 477–485.

Laughon, A., and Scott, M. P. (1984). Sequence of a *Drosophila* segmentation gene: Protein structure homology with DNA-binding proteins. *Nature (London)* **310**, 25–31.

Levine, M., Rubin, G. M., and Tjian, R. (1984). Human DNA sequences homologous to a protein coding region conserved between homeotic genes of *Drosophila*. *Cell (Cambridge, Mass.)* **38**, 667–673.

Lewis, E. B. (1978). A gene complex controlling segmentation in *Drosophila*. *Nature (London)* **276**, 565–570.
Lufkin, T., Dierich, A., LeMeur, M., Mark, M., and Chambon, P. (1991). Disruption of the *Hox-1.6* homeobox gene results in defects in a region corresponding to its rostral domain of expression. *Cell (Cambridge, Mass.)* **66**, 1105–1119.
Lumsden, A. (1991). Motorizing the spinal cord. *Cell (Cambridge, Mass.)* **64**, 471–473.
Luscher, B., Mitchell, P. J., Williams, T., and Tjian, R. (1989). Regulation of transcription factor AP-2 by the morphogen retinoic acid and by second messengers. *Genes Dev.* **3**, 1507–1517.
McGinnis, W., Levine, M. S., Hafen, E., Kuroiwa, A., and Gehring, W. J. (1984a). A conserved DNA sequence in homoeotic genes of the *Drosophila antennapedia* and *bithorax* complexes. *Nature (London)* **308**, 428–433.
McGinnis, W., Hart, C. P., Gehring, W. J., and Ruddle, F. H. (1984b). Molecular cloning and chromosome mapping of a mouse DNA sequence homologous to homeotic genes of *Drosophila*. *Cell (Cambridge, Mass.)* **38**, 675–680.
Mader, S., Kumar, V., de Verneuil, H., and Chambon, P. (1989). Three amino acids of the oestrogen receptor are essential to its ability to distinguish an oestrogen from a glucocorticoid-responsive element. *Nature (London)* **338**, 271–274.
Mahaffey, J. W., and Kaufman, T. C. (1987). The homeotic genes of the *antennepedia* complex and the *bithorax* complex of *Drosophila*. *In* "Developmental Genetics of Higher Organisms" (G. Malacinski, ed.), pp. 329–359. Macmillan, New York.
Martinez, E., Givel, F., and Wahli, W. (1991). A common ancestor DNA motif for invertebrate and vertebrate hormone response elements. *EMBO J.* **10**, 263–268.
Meijer, D., Graus, A., Kraay, R., Langeveld, A., Mulder, P., and Grosveld, G. (1990). The octamer binding factor Oct6: cDNA cloning and expression in early embryonic cells. *Nucleic Acids Res.* **18**, 7357–7365.
Mermod, N., O'Neill, E. A., Kelly, T. J., and Tjian, R. (1989). The proline-rich transcriptional activator of CTF/NF-1 is distinct from the replication and DNA-binding domain. *Cell (Cambridge, Mass.)* **58**, 741–753.
Mitrani, E., Ziv, T., Thomsen, G., Shimoni, Y., Melton, D. A., and Bril, A. (1990). Activin can induce the formation of axial structures and is expressed in the hypoblast of the chick. *Cell (Cambridge, Mass.)* **63**, 495–501.
Monuki, E. S., Weinmaster, G., Kuhn, R., and Lemke, G. (1989). SCIP: A glial POU domain gene regulated by cyclic AMP. *Neuron* **2**, 783–793.
Monuki, E. S., Kuhn, R., Weinmaster, G., Trapp, B. D., and Lemke, G. (1990). Expression and activity of the POU transcription factor SCIP. *Science* **249**, 1300–1303.
Muller, M. M., Carrasco, A. E., and De Robertis, E. M. (1984). A homeo-box-containing gene expressed during oogenesis in *Xenopus*. *Cell (Cambridge, Mass.)* **39**, 157–162.
Nieuwkoop, P. D. (1969). The formation of the mesoderm in urodelean amphibians I. Induction by the endoderm. *Wilhelm Roux Arch. Entwicklungsmech. Org.* **162**, 341–373.
Nilsen-Hamilton, M. (ed.). (1990). *Curr. Top. Dev. Biol.* **24**.
Noji, S., Nohno, T., Koyama, E., Muto, K., Ohyama, K., Aoki, Y., Tamura, K., Ohsugi, K., Ide, H., Taniguchi, S., and Saito, T. (1991). Retinoic acid induces polarizing activity but is unlikely to be a morphogen in the chick limb bud. *Nature (London)* **350**, 83–86.
Odenwald, W. F., Garbern, J., Arnheiter, H., Tournier-Lasserve, E., and Lazzarini, R. A. (1989). The Hox 1.3 homeo box protein is a sequence-specific DNA-binding phosphoprotein. *Genes Dev.* **3**, 158–172.
Okamoto, K., Okazawa, H., Okuda, A., Sakai, M., Muramatzu, M., and Hamada, H. (1990). A novel octamer binding transcription factor is differentially expressed in mouse embryonic cells. *Cell (Cambridge, Mass.)* **60**, 461–462.

9. Transcription Factors and Mammalian Development

Olson, E., Edmondson, D., Wright, W. E., Lin, V. K., Guenet, J.-L., Simon-Chazottes, D., Thompson, L. H., Stallings, R. L., Schroeder, W. T., Duvic, M., Brock, D., Helin, D., and Siciliano, M. J. (1990). Myogenin is in an evolutionarily conserved linkage group on human chromosome 1q31-q41 and unlinked to other mapped muscle regulatory factor genes. *Genomics* **8**, 427-434.

O'Neill, E. A., Fletcher, C., Burrow, C. R., Heintz, N., Roeder, R. G., and Kelly, T. J. (1988). Transcription factor OTF-1 is functionally identical to the DNA replication factor NF-III. *Science* **241**, 1210-1213.

Plachov, D., Chowdhury, K., Walther, C., Simon, D., Guenet, J.-L., and Gruss, P. (1990). *Pax-8*, a murine paired box gene expressed in the developing excretory system and thyroid gland. *Development (Cambridge, UK)* **110**, 643-651.

Placzek, M., Tessier-Lavigne, M., Yamada, T., Jessell, T., and Dodd, D. (1990). Mesodermal control of neural cell identity: Floor plate induction by the notochord. *Science* **250**, 985-988.

Ragsdale, C. W., Petkovich, M., Gates, P. B., Chambon, P., and Brockes, J. P. (1989). Identification of a novel retinoic acid receptor in regenerative tissues of the newt. *Nature (London)* **341**, 654-657.

Rickles, R., Darrow, A. L., and Strickland, S. (1989). Differentiation-responsive elements in the 5' region of the mouse tissue plasminogen activator gene confer two-stage regulation by retinoic acid and cyclic AMP in teratocarcinoma cells. *Mol. Cell. Biol.* **9**, 1691-1704.

Rosner, M. H., Vigano, M. A., Ozato, K., Timmons, P. M., Poirier, F., Rigby, P. W. J., and Staudt, L. M. (1990). A POU-domain transcription factor in early stem cells and germ cells of the mammalian embryo. *Nature (London)* **345**, 686-692.

Rossant, J., Zirngibl, R., Cado, D., Shago, M., and Giguere, V. (1991). Expression of a retinoic acid response element-*hsplacZ* transgene defines specific domains of transcriptional activity during mouse embryogenesis. *Genes. Dev.* **5**, 1333-1344.

Ruberte, E., Dolle, P., Chambon, P., and Morriss-Kay, G. (1991). Retinoic acid receptors and cellular retinoid binding proteins II. Their differential pattern of transcription during early morphogenesis in mouse embryos. *Development (Cambridge, UK)* **111**, 45-60.

Ruiz i Altaba, A., and Melton, D. A. (1989). Interaction between peptide growth factors and homeobox genes in the establishment of anteroposterior polarity in frog embryos. *Nature (London)* **341**, 33-38.

Saunders, J. W., and Gasseling, M. T. (1968). Ectodermal-mesenchymal interactions in the origin of limb symmetry. *In* "Epithelial-Mesenchyme Interactions" (R. Fleischmayer and R. E. Dillingham, eds.), pp. 78-97. Williams & Wilkins, Baltimore, Maryland.

Schneuwly, S., Klemenz, R., and Gehring, W. J. (1987). Redesigning the body plan of *Drosophila* by ectopic expression of the homeotic gene *Antennapedia*. *Nature (London)* **330**, 816-818.

Scholer, H. R., Balling, R., Hatzopoulos, A. K., Suzuki, N., and Gruss, P. (1989a). Octamer-binding proteins confer transcriptional activity in early mouse embryogenesis. *EMBO J.* **8**, 2551-2558.

Scholer, H. R., Hatzopoulos, A. K., Balling, R., Suzuki, N., and Gruss, P. (1989b). A family of octamer-specific proteins present during mouse embryogenesis: Evidence for germline specific expression of an Oct-factor. *EMBO J.* **8**, 2543-2550.

Scott, M. P., and Weiner, A. J. (1984). Structural relationships among genes that control development: Sequence homology between the *antennapedia*, *ultrabithorax* and *fushi tarazu* loci of *Drosophila*. *Proc. Natl. Acad. Sci. U.S.A.* **81**, 4115-4119.

Shephard, J. C. W., McGinnis, W., Carrasco, A. E., De Robertis, E. M., and Gehring, W. J. (1984). Fly and frog homoeodomains show homologies with yeast mating type regulatory proteins. *Nature (London)* **310**, 70-71.

Simeone, A., Acampora, D., Arcioni, L., Andrews, P., Boncinelli, E., and Mavilio, F. (1990). Sequential activation of *Hox-2* homeobox genes by retinoic acid in human embryonal carcinoma cells. *Nature (London)* **346**, 736–766.

Smith, J. C., Yaqoob, M., and Symes, K. (1990). Purification, partial characterization and biological effects of the XTC mesoderm-inducing factor. *Development* **103**, 591–600.

Snow, M. H. L., and Gregg, B. C. (1986). The programming of vertebrate development. *NATO ASI Ser.* **118**, 301–311.

Sokol, S., Wong, G. G., and Melton, D. A. (1990). A mouse macrophage factor induces head structures and organizes a body axis in *Xenopus*. *Science* **249**, 561–564.

Strickland, S., and Mahdavi, V. (1978). The induction of differentiation in teratocarcinoma stem cells by retinoic acid. *Cell (Cambridge, Mass.)* **15**, 393–403.

Summerbell, D. (1983). The effect of local application of retinoic acid to the anterior margin of the developing chick limb. *J. Embryol. Exp. Morphol.* **78**, 269–289.

Tam, P. P. L., and Meier, S. (1982). The establishment of a somitomeric pattern in the mesoderm of the gastrulation mouse embryo. *Am. J. Anat.* **164**, 209–225.

Tassabehji, M., Read, A. P., Newton, V. E., Haris, R., Balling, R., Gruss, P., and Strachan, T. (1992). Waardenburg's syndrome patients have mutations in the human homologue of the *Pax-3* paired box gene. *Nature* **355**, 635–636.

Thaller, C., and Eichele, G. (1987). Identification and spatial distribution of retinoids in the developing chick limb bud. *Nature (London)* **327**, 625–628.

Thomas, K. R., and Capecchi, M. R. (1987). Site-directed mutagenesis by gene targetting in mouse embryo-derived stem cells. *Cell (Cambridge, Mass.)* **51**, 503–512.

Tickle, C. A., Summerbell, D., and Wolpert, L. (1975). Positional signalling and specification of digits in chick limb morphogenesis. *Nature (London)* **254**, 199–202.

Tickle, C., Alberts, B., Wolpert, L., and Lee, J. (1982). Local application of retinoic acid to the limb bud mimics the action of the polarizing region. *Nature (London)* **296**, 564–565.

Ton, C. C. T., Hirvonen, H., Miwa, H., Weil, M. M., Monaghan, P., Jordan, T., van Heyningen, V., Hastie, N. D., Meijers-Heijboer, H., Drechsler, M. Royer-Pokora, B., Collins, F., Swaroop, A., Strong, L. C. and Saunders, G. F. (1991). Positional cloning and characterization of a paired box- and homeobox-containing gene from the Aniridia region. *Cell* **67**, 1059–1074.

Treacy, M. N., He, X., and Rosenfeld, M. G. (1991). I-POU: a POU-domain protein that inhibits neuron-specific gene activation. *Nature (London)*, **350**, 577–584.

Umesono, K., and Evans, R. M. (1989). Determinants of target gene specificity for steroid/thyroid hormone receptors. *Cell (Cambridge, Mass.)* **57**, 1139–1146.

Umesono, K., Giguere, V., Glass, C. K., Rosenfeld, M. G., and Evans, R. M. (1988). Retinoic acid and thyroid hormone induce gene expression through a common responsive element. *Nature (London)* **336**, 262–265.

van Straaten, H. W. M., Hekking, J. W. M, Beursgens, J. P. W. M., Terwindt-Rouwenhorst, E., and Drukker, J. (1989). Effect of the notochord on proliferation and differentiation in the neural tube of the chick embryo. *Development (Cambridge, UK)* **107**, 793–803.

Vasios, G. W., Gold, J. D., Petkovich, M., Chambon, P., and Gudas, L. J. (1989). A retinoic acid-responsive element is present in the 5′ flanking region of the laminin B1 gene. *Proc. Natl. Acad. Sci. U.S.A.* **86**, 9099–9103.

Verrijzer, C. P., Kal, A. J., and van der Vliet, P. C. (1990). The DNA binding domain (POU domain) of transcription factor Oct-1 suffices for stimulation of DNA replication. *EMBO J.* **9**, 1883–1888.

Wagner, M., Thaller, C., Jessel, T., and Eichele, G. (1990). Polarizing activity and retinoid synthesis in the floor plate of the neural tube. *Nature (London)* **345**, 819–822.

Walther, C., and Gruss, P. (1991). *Pax-6*, a murine paired box gene, is expressed in the developing CNS. *Development (Cambridge, UK)*, **113**, 1435–1449.

9. Transcription Factors and Mammalian Development

Wanek, N., Gardiner, D. M., Muneoka, K., and Bryant, S. V. (1991). Conversion by retinoic acid of anterior cells into ZPA cells in the chick wing bud. *Nature (London)* **350,** 81–83.

Weigel, D. and Jackle, H. (1990). The fork head domain: A novel DNA binding motif of eukaryotic transcription factors? *Cell* **63,** 455–456.

Wright, C. V. E., Cho, K. W. Y., Hardwicke, J., Collins, R. H., and De Robertis, E. M. (1989). Interference with function of a homeobox gene in *Xenopus* embryos produces malformations of the anterior spinal cord. *Cell (Cambridge, Mass.)* **59,** 81–93.

Yamada, T., Placzek, M., Tanaka, H., Dodd, J., and Jessell, T. M. (1991). Control of cell pattern in the developing nervous system: Polarizing activity of the floor plate and notochord. *Cell (Cambridge, Mass.)* **64,** 635–647.

Zelent, A., Krust, A., Petkovich, M., Kastner, P., and Chambon, P. (1989). Cloning of murine alpha and beta retinoic acid receptors and a novel receptor gamma predominantly expressed in skin. *Nature (London)* **339,** 714–717.

Index

A

Activin, role in early embryogenesis, 356–357
Acute promyelocytic leukemia, retinoic acid treatment, 329
Adhesion, cell
  blastomere–blastomere, in marsupial cleavage, 201–202
  blastomere–zona, in marsupial cleavage, 198–199
  fibronectin-coated substrata, 112–113
  in gastrulation, 94–95
Aggregate formation, in migrating mesodermal cells, 63–64
α1T mutation (*Drosophila*), 285
α4T mutation (*Drosophila*), 285
Anaphases, abnormal (*Drosophila*), 293–294
Aneuploidy
  in *Drosophila*, 293–294
  pathogenesis, chimeric studies, 261
Anteroposterior axis, embryonic, and growth factors, 357–358
Anti-cancer agents, retinoids as, 330
Anti-fibronectin IgG, Fab' fragments, 113
Anti-integrin IgG, Fab' fragments, 113–115
Antisense oligodeoxynucleotide probes, for blockage of gastrulation movements, 119–120
Anurans
  bottle cell ingressions, 47–48
  convergence and extension in, 67–68
  fate maps, 92–93
  fibrillar ECM in, 101
  integrin expression, 104
AP-1 transcription factor, retinoic acid regulation of, 333
AP-2 transcription factor, retinoic acid regulation of, 362
Archenteron, elongation, 151–154
Arg-Gly-Asp peptides, 115–116
asp mutation
  in *Drosophila*, 281, 294, 298
  maternal effect of (*Drosophila*), 286

Audiogenic seizures, studies using aggregation chimeras, 263
aurora mutation (*Drosophila*), 285, 294
Autonomous processes, in gastrulation, 94
Axillary buds
  N. silvestris, developmental behavior, 12–16
  N. tabacum, 8–10
Axis formation, in marsupial blastocysts, 226

B

bicoid gene, *Drosophila*, 354
Blastocoel roof
  extracellular matrix (*Xenopus*), 54–57
  fibrillar ECM on basal surface of, 98–101
  fibrils on (*Xenopus*), 55
  fibronectin (*Xenopus*), 55–56
  gastrulation in *Xenopus* lacking, 75
  inversion of, 111
  and mesodermal involution, 77
Blastocysts
  marsupial
    bilaminar
      complete, 217
      embryonic area vs. medullary plate, 216–217
      formation, 211–214
      polarity renewal and, 215
      primary endoderm cells, 215–216
    trilaminar
      cell lineages in, 219–220
      formation, 217–218
      mesoderm formation, 218–219
    unilaminar
      characterization, 205–206
      expansion and growth, 209–211
      formation, 207
      structure, 207–209
    octamer-binding protein expression, 355–356
Blastomere–blastomere adhesion, in marsupial cleavage, 201–202

385

Blastomere–zone adhesion, in marsupial cleavage, 198–199
Blebbing, *see* Protrusive activity
Bone marrow chimeras, immunological tolerance, 259–260
Bottle cells
  formation
    dorsoventral progression, 50
    and evolution, 51
    function during gastrulation, experimental tests of, 50–51
    types, 49
  gastrulation in *Xenopus* lacking, 75
  ingressions
    in anurans, 47–48
    function of, 48–49
    mechanism of, 48
    somitic and notochordal mesoderm in urodeles, 47
  invagination
    apical constriction, 41–43
    behavior, 41
    context dependency of, 43–44
    invasiveness of, 44–45
    pattern of formation, 45
    respreading, 45–46
    tissue interactions, 46–47
bra mutation (*Drosophila*), 294
Bud-rooting assays, for floral determination (*N. tabacum*), 10–11

C

cand mutation (*Drosophila*), 283
Carbonate dehydratase, *in situ* histochemical localization, 237
Carcinomas, in mosaic individuals, 257
Caroxyesterase cell marker, 237
cdc2 gene product, *Drosophila* homologs, 288–289
cdc13 gene product, *Drosophila* homologs, 288–290
Cell behavior
  abnormal, and neural tube defects, 150–151
  molecular analysis of, 120–121
Cell competence, in floral determination, 31–32
Cell cultures, models for retinoic acid-induced differentiation, 325–327

Cell division
  in *Drosophila*, genetic analysis
    genes borrowed from yeast, 288–291
    imaginal development, 291–293
    imaginal tissue, mitotic mutation effects, 293–297
    kinesin gene family, 300–302
    male meiosis, 297–300
    maternal effect, 286–287
    methodology, 278–279
    phenotypic classes, 277–278
    postblastoderm embryonic development, 287–288
    preblastoderm embryonic development, 280–286
    spermatogenesis, 297–300
    tubulin gene family, 300–302
  during marsupial cleavage, 204–205
  and neural plate shaping/bending, 149–150
  order, and marsupial cell fate, 224–225
  during *Xenopus* gastrulation, 157
Cell fate, marsupial embryos, 224–227
Cell interactions
  and cytomechanics, 79
  in mesoderm cell migration, 63–64
  and tissue affinities, 93
Cell intercalation
  boundary polarization of protrusive activity, 72–73
  mediolateral
    convergence and extension by, 68–70
    convergence and extension of NIMZ, 70
    protrusive activity during, 70–71
    simultaneous mesodermal cell migration, 73–74
  within neural plate, 148
  radial
    convergence and extension by, 68–70
    convergence and extension of NIMZ, 70
  during *Xenopus* gastrulation, 156
Cell lineages, in marsupial trilaminar blastocysts, 219–220
Cell markers
  biochemical, 237
  exogenous, 238
  in situ, 237–238
  for mosaic patterns, 237–238
Cell rearrangements
  and convergent extension, 95–96
  and epiboly, 95–96

Index

during gastrulation, 151–154
  protrusive activity in, 161–164
  roles of, 154, 157–158, 160–161
  in sea urchins, 151–154
  in teleost fish, 158–160
  in *Xenopus*, 154–158
  and neural plate shaping/bending, 139–141
  during neurulation, 133–149
Cell shape
  changes during *Xenopus* gastrulation, 157
  changes in neural plate shaping/bending, 149–150
Cell surface, glycoconjugates, 105–106
Cellular adhesiveness, *see* Adhesion, cell
Cellular motility, *see* Motility, cell
Ceratophrys ornata, convergence and extension movements, 75–76
Chemical perturbation, of neurulation, 145–146
Chimeras
  aggregation, 235
    genetic diseases, 260–263
    immunological tolerance, 259–260
    mosaic pattern analysis
      biochemical, 239–240
      patches, 240–245
    organogenesis, 245–246
      computer simulations of, 253–255
      dermis, 247–248
      epidermis, 247–248
      hematopoietic stem cell regulation, 251–252
      kidney, 247
      muscle, 246
      nervous system, 248–250
      retina, 246–247
      sexual differentiation, 250–251
      thymus, 252–253
    procedures for production of, 238–239
    cell markers for, 237–238
    primary, 235
    secondary, 235–236
Chromatid bridges, in abnormal anaphases (*Drosophila*), 294
Chromosome breaks (*Drosophila*), 294
Chromosome condensation, abnormal (*Drosophila*), 294
Chromosome disjunction (*Drosophila*), 283–284
Chromosome nondisjunction, and male meiosis (*Drosophila*), 297

Circus movements, 95
Cleavage, marsupials, 191–205
  blastomere–blastomere adhesion, 201–202
  blastomere regulation, 224
  blastomere–zona adhesion, 198–199
  cell divisions during, 204–205
  cell populations during, 202–203, 222–224
  cytoplasmic emissions, 191
  extracellular matrix emission, 192–193
  patterns of, 194–198
  site of, 193–194
  yolk elimination, 192
Clones, in gynandromorphic *Drosophila*, 242
Cloning, retinoic acid receptors, 359–362
c-mos gene, in fruit flies, 280–281
Collagen, in extracellular matrix (amphibian), 97–98
Common variable immunodeficiency, retinoic acid effects, 330
Conditional mutations, 279
Contact interactions
  mesodermal cells, 72
  migrating mesodermal cells, 63–64
Convergence movements
  in anuran gastrulation, 67–68
  in *C. ornata*, 75–76
  in *H. regilla*, 75–76
  in urodele gastrulation, 67–68
  in urodeles, 75–76
  in *Xenopus* gastrulation, 67
    deep mesodermal cells in, 68
    epithelial cells and, 68
    noninvoluting marginal zone, 70
    in sandwich explants, 67, 74
Convergent extension
  defined, 70
  role of cellular rearrangements, 95–96
  surface epithelial cells, 148
  Xenopus, 154–158
Correlative processes, in gastrulation, 94
Cotransfection, with retinoic acid receptor genes, 336
Cyclin, genes for (*Drosophila*), 288–290
Cyclin A, mutation in gene coding for (*Drosophila*), 288
Cytoplasmic emissions, during marsupial cleavage, 191
Cytoplasmic polarity
  activation effects (marsupials), 188–189
  marsupial oocytes, 180–181

D

(1)d deg-3 mutation (*Drosophila*), 294
(1)d deg-10 mutation (*Drosophila*), 294
(1)d deg-11 mutation (*Drosophila*), 295
Deep cells, of blastoderm during epiboly, 159–160
Dermis, mosaic pattern analysis, 247–248
Determination events
  floral determination, 4, 22–23
  in flowering, 4
Deutoplasmolysis, in marsupials, 192
Dextran sulfate, effects on gastrulation, 117
3,4-Didehydroretinoic acid, in chick wing bud, 313
Differentiation, retinoic-acid induced, cell culture models, 325–327
Dimerization
  in vitro heterodimerization of RXRs with RARs, 336
  retinoic acid receptors, 316–318
D-kinesin, *Drosophila*, 301
DNA binding
  by *Pax* gene products, 375
  sequence specificity of retinoic acid receptors, 318–322
DNA-binding proteins, retinoic acid receptors and, 332–333
DNA synthesis, *asp* and *gnu* effects (*Drosophila*), 281
Dominant-negative repression, of RXR and RAR function, 337–338
dot mutation (*Drosophila*), 286
Drosophila melanogaster
  cell division, genetic analysis
    genes borrowed from yeast, 288–291
    imaginal development, 291–293
    imaginal tissue, mitotic mutation effects, 293–297
    kinesin gene family, 300–302
    male meiosis, 297–300
    maternal effect, 286–287
    methodology, 278–279
    phenotypic classes, 277–278
    postblastoderm embryonic development, 287–288
    preblastoderm embryonic development, 280–286
    spermatogenesis, 297–300
    tubulin gene family, 300–302
  oscillators in, 283

E

Ectodermal cells, adhesion to fibronectin-coated substrata, 112–113
Egg envelopes (marsupial)
  mucoid, 189–190
  shell, 190–191
  zona pellucida, 189
Embryoblasts, development during marsupial cleavage, 202–204
Embryogenesis
  early, regulatory factors in, 354–357
  hox genes, 362–363
  mouse, stages of development, 352–354
  octamer-binding proteins, 355–356
  positional information along anteroposterior axis, 357–358
  retinoic acid receptors and, 358–362
Embryonal carcinoma cell lines, for retinoic acid-induced differentiation, 325–327
Embryos
  arrested hybrid, ECM disruption, 117–118
  gastrulation, *see* Gastrulation
Endoderm, primary, formation (marsupial)
  direct proliferation of, 213–214
  proliferation from another embryonic area, 214
  via endoderm mother cells, 211–213
Endoreduplication (*Drosophila*), 293
En-1 gene
  in developing vertebral column, 369–371
  neural tube expression of, 371–375
Enveloping layer, of blastoderm during epiboly, 159–160
Epiboly
  autonomy of, 76–77
  cell rearrangements during
    evidence for, 156–157, 160
    role of, 95–96, 157–158, 160–161
  in urodeles, 76
  Xenopus, 76, 154–158
Epidermis, mosaic pattern analysis, 247–248
Epithelial cells
  endodermal, cell behavior, 68–69
  marginal zone, role in convergence and extension (*Xenopus*), 68
Evolution, and bottle cell formation, 51
Evx-1 gene, neural tube expression of, 371–375

Extension movements
 in anuran gastrulation, 67–68
 in *C. ornata*, 75–76
 in *H. regilla*, 75–76
 in urodele gastrulation, 67–68
 in urodeles, 75–76
 in *Xenopus* gastrulation, 67
  deep mesodermal cells and, 68
  epithelial cells and, 68
  noninvoluting marginal zone, 70
  in sandwich explants, 67, 74
Extracellular matrix
 assembly, 98
 blastocoel roof (*Xenopus*), 54–57
 composition in amphibian gastrulae, 97–98
 cues for mesodermal cell migration (amphibian), 59–61
 disruption
  in arrested hybrid embryos, 117–118
  by mutation, 118–119
 emission during marsupial cleavage, 192–193
 fibrillar
  composition, 99
  contact guidance by, 109–111
  discovery, 98–99
  distribution, 98–99
  manipulation of oriented fibrils, 107–109
  orientation, 99
  spatial pattern of, 106–107
  temporal pattern of, 106
 gene expression
  fibronectin, 101–102
  integrins, 103–104
 glycoconjugates in, 105
 and neurulation, 146
 receptors for components of, 103–104
 synthesis, 98

F

Fab' fragments
 of anti-fibronectin IgG, 113
 of anti-integrin IgG, 113–115
Fate maps, urodele vs anuran, 92–93
Fertilization, marsupials
 activation effects, on nuclear polarity, 187–188
 in vivo, 186
 polar and radial patterns during, 187
 sperm–egg interactions, 186–187
 timing of events, 183, 185–186
 tubal transport, 185–186
Fibrils, ECM
 artificially aligned, contact guidance by, 109–111
 manipulation of, 107–109
Fibroblast growth factor
 and positional information along anteroposterior axis, 357–358
 role in early embryogenesis, 356–357
Fibronectin
 in blastocoel roof (*Xenopus*), 55–56
 in extracellular matrix (amphibian), 97
 gene expression, 101–102
 and mesodermal cell migration, 57–59
 substrata, cell adhesion to, 112–113
Fibronectin receptors, 103–104
Fish, teleost, gastrulation, 158–164
Floral branches, floral determination in (*N. silvestris*), 17–23
Floral determination
 and cell competence, 31–32
 conceptual framework for, 4, 31, 33
 determination events, 4, 22–23
 in *Henliathus annus*, 29–30
 in *Lolium temulentum*, 27–29
 in *Nicotiana tabacum*
  early state of, 23–24
  in explants from floral branches and pedicels, 17–20
  and genotype, 7–8
  grafting assays, 10
  isolation assays, 10
  late state of, 23
  in organized buds and meristems, 11–17
  in organized vs. stem-regenerated meristems, 23–24
  position-dependency of, 8–10, 21–22
  in regenerated shoots, 20
  root–shoot interplay, 6–8
  terminal buds
   nodes of, 5–8
   rooting assay, 11–12
   types of shoot apical meristems, 16–17
 in *Pharbitis nil*, 24–26
 in *Pisum sativum*, 30
 species differences in, 33
Floral stimulus, defined, 3
Florigen hypothesis, 2–3
Flowering, regulation of, 2–4
forkhead gene, *Xenopus*, 354

Fractal objects
  iterating functions, 254–255
  liver patches in chimeric rats, 242–245
Friend virus-induced leukemia, studies using aggregation chimeras, 263
fs(3)820 mutation (*Drosophila*), 294

G

Gastrulation
  autonomous processes in, 94
  basic patterns of, 78
  bottle cell function in, 41–51
  bottle cell removal during, 50–51
  and cell motility, 95
  cell rearrangements during, 151–164
    protrusive activity in, 161–164
    in sea urchins, 151–154
    in teleost fish, 158–160
    in *Xenopus*, 154–158
  cell surface glycoconjugates, 105–106
  cellular adhesiveness in, 94–95
  correlative processes in, 94
  diversity of cell behavior in, 78–79
  extracellular matrix
    composition of (amphibian), 97–98
    glycoconjugates in, 105
  fate maps in urodeles and anurans, 92–93
  heparin effects, 117
  involution in, 77–78
  in sea urchins, 151–154
  targeted molecular probes for, 119–120
  tenascin effects, 116–117
  tissue affinities in, 93
  in urodeles and anurans, convergence and extension, 67–68
  Xenopus
    convergence and extension, 67, 74–75
    convergent extension, 154–158
    epiboly, 154–158
    function of mesodermal cell migration in, 51–66
    without blastocoel roofs, 75
    without bottle cells, 75
Gene dosage, studies using aggregation chimeras, 261–262
Genetic diseases, studies using aggregation chimeras, 260–263
Genotype, inductive capacity of leaves, 7
Glucosephosphate isomerase marker, 237

β-Glucuronidase, *in situ* histochemical localization, 237
Glycoconjugates
  cell surface, 105–106
  in extracellular matrix, 105
gnu mutation (*Drosophila*), 281
goosecoid gene (*Xenopus*), 354
Grafting experiments
  for floral determination (*N. tabacum*), 10
  leaf–apex communication, 3
Growth hormone, transcriptional activation by retinoic acid, 362
Growth patterns, position-dependent, in *N. tabacum*, 8–10
gsb-p and *gsb-d* genes (*Drosophila*), 370

H

hay mutation (*Drosophila*), 300
Hematopoietic stem cells, reuglation in chimeric mice, 251–252
Henliathus annus, floral determination, 29–30
Hensen's node, 147
Heparin, effects on gastrulation, 117
Hepatomas, in mosaic individuals, 256
Heterodimerization, of RXRs with RARs, 336
HL-60 cell line, retinoic acid-induced differentiation, 327
hox gene family
  clusters, sequential activation of genes within, 364–366
  gene products as transcription factors, 368–369
  identification in various species, 363
  patterns of expression, 363–364
  retinoic acid regulation of, 331
  role in embryogenesis, 362–363
  and vertebral specification, 366–368
Hybrid embryos, arrested, ECM disruption in, 117–118
Hyla regilla
  autonomous and correlative movements, 94
  function of convergence and extension movements, 75–76

I

Imaginal disks, *Drosophila*
  development, 291–293
  effect of mitotic mutations, 293–297

Index 391

Immunological tolerance, in aggregation chimeric mice, 259–260
Ingression, bottle cells
  in anurans, 47–48
  function of, 48–49
  mechanism of, 48
  somitic and notochordal mesoderm in urodeles, 47
In situ hybridization, retinoic acid receptors during embryogenesis, 361
In situ studies, of retinoic acid receptor mRNA, 315
Integrin receptors, 103–104
Integrins, gene expression, 103–105
Intercalation, cell, *see* Cell intercalation
Interkinetic nuclear migration
  and cell shape, 149–150
  model of, 163
Invagination
  bottle cells
    apical constriction, 41–43
    behavior, 41
    context dependency of, 43–44
    invasiveness of, 44–45
    pattern of formation, 45
    respreading, 45–46
    tissue interactions, 46–47
  primary and secondary, 152–153
Involuting marginal zone
  autonomy of convergence and extension movements, 75
  bottle cell anchorage for, 44
  cell behavior during mediolateral intercalation, 70–74
  and involution, 77
Involution, 77–78
Irradiation chimeras, immunological tolerance, 259–260
Isocitrate dehydrogenase cell marker, 237
Isoforms, of retinoic acid receptors, 311
Isolation assays, for floral determination, 10
Isotretinoin, teratogenicity, 328
Iterating functions, and cell positioning during development, 254

K

Kidney, mosaic pattern analysis, 247

L

Laminin
  in extracellular matrix (amphibian), 97
  synthesis in *Pleurodeles* embryos, 56
Lateral neurepithelial cells, rearrangements during neurulation, 147
Leaves
  communication with apex, 3
  root–leaf interplay (*N. tabacum*), 6–7
Lectins, in extracellular matrix (amphibian), 97–98
legless mutation, and retinoic acid regulation, 339
l(3)7m-62 mutation (*Drosophila*), 295
l(3)13m-281 mutation, 284, 293–294
lodestar mutation (*Drosophila*), 285, 294
Lolium temulentum, floral determination, 27–29
l(1)TW-6cs mutation (*Drosophila*), 302

M

Malic enzyme, *in situ* histochemical localization, 237
Marsupials
  blastocysts
    axis formation, 226
    bilaminar
      complete, 217
      embryonic area vs. medullary plate, 216–217
      formation, 211–214
      polarity renewal and, 215
      primary endoderm cells, 215–216
    trilaminar
      cell lineages in, 219–220
      formation, 217–218
      mesoderm formation, 218–219
    unilaminar
      characterization, 205–206
      expansion and growth, 209–211
      formation, 207
      structure, 207–209
    cell fate specification, 224–225
    cleavage, 191–205
      blastomere–blastomere adhesion, 201–202
      blastomere regulation, 224

blastomere–zona adhesion, 198–199
cell divisions during, 204–205
cell populations during, 202–203, 222–224
cytoplasmic emissions, 191
extracellular matrix emission, 192–193
  patterns of, 194–198
  site of, 193–194
  yolk elimination, 192
egg envelopes
  mucoid, 189–190
  shell, 190–191
  zona pellucida, 189
fertilization
  activation effects, on nuclear polarity, 187–188
  in vivo, 186
  polar and radial patterns during, 187
  sperm–egg interactions, 186–187
  timing of events, 183, 185–186
  tubal transport, 185–186
oocytes
  apolar state of, 178–179
  cytoplasmic polarity, 180–181
  nuclear polarity, 179–180
  ovulation, 182–184
  polarity, and investments, 221–222
  vitellogenesis, 181–182
  yolk, 181–182
Maternal determinants, in marsupial cell fate, 223–224, 226–227
Maturation promoting factor, homolog in *Drosophila*, 280
mdg mutation, studies using aggregation chimeras, 263
Median hinged point
  cell rearrangements during neurulation, 147
  foramtion during neural plate bending, 137–139
  origination cranial to Hensen's node, 147
Mediolateral intercalation
  cell interactions during, 71–72
  convergence and extension by, 68–70
  in noninvoluting marginal zone, 70
  protrusive activity during, 70–71
  simultaneous mesodermal cell migration, 73–74
Medullary plate, and marsupial embryonic area, 216–217
mei-9 mutation (*Drosophila*), 286
mei-41 mutation (*Drosophila*), 286

Meiosis, male (*Drosophila*), 297–300
Meristems, *N. tabacum*
  floral determination in, 11–17
  nodes prior to flowering, 5–8
Mesenchymal cells, in sea urchins, 152–153
Mesoderm
  marsupial, formation, 218–219
  tissue, *Pax* gene expression, 371
Mesodermal cells
  adhesion to fibronectin-coated substrata, 112–113
  deep
    cell behavior, 69–7068
    role in convergence and extension (*Xenopus*), 68
  fibrillar matrix interaction, disruption with Arg-Gly-Asp peptides, 115–116
  Fab' fragments of anti-fibronectin IgG, 113
  Fab' fragments of anti-integrin IgG, 113–115
  heparin, 117
  tenascin, 116–117
  involution, 77–78
  migration
    cell interactions, 63–64
    cell motility during, 61–62
    changes at onset of, 62–63
    as coherent stream, 53–54
    extracellular matrix cues for (amphibian), 59–61
    fibronectin role, 57–59
    function in amphibian gastrulation, 65–66
    initiation of, 96–97
    in mesoderm movement, 51–53
    probes for disruption of, 113–119
    simultaneous mediolateral intercalation, 73–74
    substrate of, 57–59
    tenascin effects, 116–117
    in urodeles, 116
    in *Xenopus*, 115
  protrusive activities during medilateral intercalation, 70–71
Messenger RNA
  cyclin, in *Drosophila*, 290
  retinoic acid receptors, 315–316
  retinoid X receptors, 315–316
mgr mutation (*Drosophila*), 294–295, 298–299
mh mutation (*Drosophila*), 282

Index

Microfilaments, in *Xenopus* contracted apices, 42
Microsurgical studies, of neurulation, 145-146
Migration, mesodermal cells
   cell interactions, 63-64
   cell motility during, 61-62
   changes at onset of, 62-63
   as coherent stream, 53-54
   extracellular matrix cues for (amphibian), 59-61
   fibronectin role, 57-59
   function in amphibian gastrulation, 65-66
   initiation of, 96-97
   in mesoderm movement, 51-53
   probes for disruption of, 113-119
   simultaneous mediolateral intercalation, 73-74
   substrate of, 57-59
   tenascin effects, 116-117
   in urodeles, 116
   in *Xenopus*, 115
mit mutation (*Drosophila*), 283
Mitotic index (*Drosophila*), 293
Mitotic mutations, effects on *Drosophila*
   imaginal tissues, 293-297
   kinesin mutations, 300-302
   male meiosis and spermatogenesis, 297-300
   maternal effect, 286-287
   postblastoderm embryonic development, 287-288
   preblastoderm embryonic development, 280-286
   tubulin mutations, 300-302
Monosomy, studies using aggregation chimeras, 262
Morphogens, retinoic acid, 358-359
Mosaics
   cell markers for, 237-238
   computer simulations of patterns, 253-255
   monozygotic, 236
   multizygotic, 235
   neoplasia, 255-257
   pattern analysis in aggregation chimeras
      biochemical, 239-240
      patches, 240-245
Motility, cell
   in gastrulation, 95
   migratory mesodermal cells, 61-62
ms(3)K81 mutation (*Drosophila*), 282
Mucoid envelope, 189-190
Muscle, mosaic pattern analysis, 246
Muscular dysgenesis, studies using aggregation chimeras, 263
mus-101 mutation (*Drosophila*), 285, 294
mus-105 mutation (*Drosophila*), 294
Mus satellite DNA, 237
Mutation
   disruption of ECM by, 118-119
   ECM disruption by, 118-119
Mutations
   conditional, for cell division mutants, 279
   mitotic, effects on *Drosophila*
      imaginal development, 291-293
      imaginal tissues, 293-297
      kinesin mutations, 300-302
      male meiosis and spermatogenesis, 297-300
      maternal effect, 286-287
      postblastoderm embryonic development, 287-288
      preblastoderm embryonic development, 280-286
      tubulin mutations, 300-302

N

ncd mutation (*Drosophila*), 301
Neoplasia, in mosaic individuals, 256-257
Nervous system, development in chimeras, 248-250
Neural plate
   bending
      cell division and, 149-150
      cell rearrangements and, 139-141
      cell shape changes and, 149-150
      characterization, 137-139
   formation
      characterization, 135
      intrinsic/extrinsic neurulation forces, 141-146
   shaping
      cell division and, 149-150
      cell rearrangements and, 139-141
      cell shape changes and, 149-150
      characterization, 135-137
Neural tube
   defects, 150-151
   expression of *Pax, En-1,* and *Evx-1* genes, 371-375
Neuroepithelial cells, shape of, 149-150
Neurulation
   cell rearrangements during, 133-149

extracellular matrix role in, 146
intrinsic and extrinsic forces of, 141–146
neural groove closure, 139
neural plate
  bending, 137–141
  formation, 135
  shaping, 135–137, 139–141
neural tube defects, 150–151
stages of, 135
Nicotiana silvestris terminal buds, developmental behavior, 11–16
Nicotiana tabacum, floral determination
early state of, 23–24
in explants from floral branches and pedicels, 17–20
and genotype, 7–8
grafting assays, 10
isolation assays, 10
late state of, 23
in organized buds and meristems, 11–17
in organized vs. stem-regenerated meristems, 23–24
position dependency of, 8–10, 21–22
in regenerated shoots, 20
root–shoot interplay, 6–8
terminal buds
  nodes of, 5–8
  rooting assay, 11–12
types of shoot apical meristems, 16–17
NIMZ, *see* Noninvoluting marginal zone
Nodes, *N. tabacum*
  axillary buds, 8–10
  terminal buds, 5–8
nod mutation (*Drosophila*), 283, 301
Noninvoluting marginal zone
  convergence and extension, 70
  in sandwich explants (*Xenopus*), 67
  and involution, 77
Northern gel analysis, retinoic acid receptor gene expression, 314–315
Notochord–somite boundary, inhibition of intercalating cells, 72–73
Nuclear polarity, of marsupial oocytes, 179–180
activation effects, 187–188

O

Octamer-binding proteins, 355–356
OCT gene family, retinoic acid regulation of, 332

Onion stage, of spermatids, 299
Oocytes
  marsupial
    apolar state of, 178–179
    cytoplasmic polarity, 180–181
    and investments, 221–222
    nuclear polarity, 179–180
    ovulation, 182–184
    vitellogenesis, 181–182
    yolk, 181–182
  octamer-binding protein expression, 355–356
Oscillators (*Drosophila*), 283
Ovulation, in marsupials, 182–184

P

Papillomas, in chimeric mice, 257
Parathyroid hyperplasia, in chimeras, 257
Patches, mosaic pattern analysis, 240–245
Pattern formation, bottle cells, 45
Pax-1 gene, in developing vertebral column, 369–371
Pax-3 gene, and teratogenic effects of retinoic acid, 328
Pax gene family
  gene products, DNA-binding by, 375
  in mesoderm-derived tissue, 371
  neural tube expression of, 371–375
pcd gene action, studies using aggregation chimeras, 262
Pedicels, floral determination in (*N. silvestris*), 17–23
Pharbitis nil, floral determination, 24–26
Phenotypic classes, of mitotic mutants (*Drosophila*), 277–278
Phosphoglycerate kinase cell marker, 237
Photoperiod, and flowering, history of, 2–4
Pisum sativum, floral determination, 30
Pit-1 protein expression, 355
Pleurodeles, *see also* Urodeles
  extracellular matrix
    components and structure, 54–56
    role in mesodermal cell migration, 59–61
  function of convergence and extension movements, 75–76
  substrate for mesodermal cell migration, 57–59
plu mutation (*Drosophila*), 281
PML–RAR chimeras, 329–330

Index 395

png mutation (*Drosophila*), 281
Polarization
  marsupial oocytes
    activation effects, 187–189
    cytoplasmic, 180–181
    and investments, 221–222
    nuclear, 179–180
    and position signals for cell fate, 224–225
polo mutation (*Drosophila*), 285, 294
Polyploidy, mutation-induced (*Drosophila*), 295
Positional signals, for specification of cell fate in marsupials, 224–225
PP1 87B mutant neuroblasts (*Drosophila*), 296
prd gene (*Drosophila*), 370
Primary bodies, development, 113–115
Primary mesenchymal cells, 152
Probes, for blockage of gastrulation movements, 119–120
Protein kinase p34^{cdc2}, 288
Protrusive activity
  boundary polarization of, 72–73
  cell interactions and, 71–72
  in cell rearrangements during gastrulation, 161–164
  during mediolateral intercalation, 70–71
Protrusiveness, of bottle cells, 44–45
Pulsatile activity, *see* Protrusive activity
Pulse-chase autoradiography, glycoconjugates in ECM, 105
Purkinje cells, chimerism, 249

R

Radial intercalation
  convergence and extension by, 68–70
  in noninvoluting marginal zone, 70
RARs, *see* Retinoic-acid receptors
rdy gene action, studies using aggregation chimeras, 262–263
Respreading, bottle cells, 45–46
Retina, mosaic pattern analysis, 246–247
Retinoic acid
  differentiation, cell culture models for, 325–327
  and embryonic positional information along anteroposterior axis, 357–358
  enzymes regulating, 336
  genes transcriptionally activated by, 362
  and *hox* gene expression, 364–366
  legless mutation, 339
  medical applications, 329–330
  as morphogen, 358–359
  presence in embryo, 335–336
  role in developing chick wing bud, 324–325
  teratogenicity, 328–329
9-*cis*-Retinoic acid, ligand for retinoid X receptor-α, 323
13-*cis*-Retinoic acid, teratogenicity, 328
Retinoic acid receptors
  amino acid comparisons, 312
  and AP-1 transcription factor elements, 333
  classes of, 310–311
  cloning and expression pattern of, 359–361
  dimerization signals, 316–318
  DNA sequence binding specificity, 318–322
  dominant-negative repression of receptor function, 337–338
  function and specificity of, 336–339
  genomic organization, 312
  homologies between receptor classes, 311
  and *hox* gene family regulation, 331
  in situ hybridization, 361
  isoforms, 311
  modular structure, 310–314, 360
  and OCT protein expression, 332
  research concerns, 334–339
  specific expression of, 314–316
  specific functions, 360
  specificity differences for specific retinoids, 322–323
  targeted genes in ES stem cells, 339
  transcriptional regulation by, 361–362
  transgenic studies, 335–336
Retinoids
  availability, 323–324
  specific binding of retinoic acid receptors, 322–323
Retinoid X receptors
  dominant-negative repression of receptor function, 337–338
  function and specificity of, 334–339
  modular structure of, 310–314
  research concerns, 334–339
  specific expression of, 314–316
  specificity differences for specific retinoids, 322–323
  targeted genes in ES stem cells, 339

RNase protection techniques, retinoic acid receptor gene expression, 314–315
rod mutation (*Drosophila*), 285, 294, 298
Roots, *N. tabacum*, leaf–root interplay, 6–7
RXRs, *see* Retinoid X receptors

S

Sandwich explants
  and cell rearrangements during convergent extension, 157
  Xenopus, convergence and extension, 67, 74
Scanning electron microscopy, ECM fibrils, 107–108
Schizosaccharomyces pombe, genes borrowed by *Drosophila*, 288–291
Sea urchins, archenteron formation and elongation, 151–154
Secondary bodies, development, 113–115
Secondary mesenchymal cells, 153
Sexual differentiation, in aggregation chimeras, 250–251
Shell, of marsupial egg, 190–191
Skin tumors, in chimeric mice, 257
Sodium dodecyl sulfate-polyacrylamide gel electrophoresis, glycoconjugates in ECM, 105
Species differences, in floral determination, 33
Spermatogenesis, in mitotic mutants (*Drosophila*), 297–300
Sperm–egg interactions, in marsupials, 186–187
splotch mutation, retinoic acid effects, 328
stg mutation (*Drosophila*), 288, 290

T

Targeted molecular probes, for blockage of gastrulation movements, 119–120
T cells, inductive differentiation, in chimeric rats, 260
T6 chromosomal marker, 237
Teleost fish, gastrulation in, 158–164
Tenascin, effects on gastrulation, 116–117
Terminal buds
  N. silvestris, 11–16
  N. tabacum, 5–8
3-4-5 rule, 321
Thymus, development in aggregation chimeras, 252–253

Tissue affinities, cell morphogenetic rearrangement and, 93
TPA gene, transcriptional activation by retinoic acid, 362
Transcriptional regulation, by retinoic acid receptors, 361–362
Transcription factors
  HOX gene products as, 368–369
  and retinoic acid receptors, 333–334
Transforming growth factors
  and positional information along anteroposterior axis, 357–358
  role in early embryogenesis, 356–357
Trophoblasts, development during marsupial cleavage, 202–204
Ts15 cells, proliferative efficiency in aggregation chimeric mice, 261
α-Tubulin gene mutations (*Drosophila*), 300–302
β-Tubulin gene mutations (*Drosophila*), 300–302
Tumors
  embryo-derived, in chimeras, 257–259
  germ cell, in chimeras, 257–259
  pathogenesis in mosaic individuals, 255–257

U

unc-86 octamer-binding protein, 355
Urodeles
  bottle cell ingressions, 47
  convergence and extension in, 67–68
  convergence and extension movements, 75–76
  epiboly in, 76
  fate maps, 92–93
  fibrillar ECM in, 99–101
  integrin expression, 104–105
  mesodermal cell migration, 116

V

Vertebral column, *En-1* and *Pax-1* in, 369–371
Vertebral identity, *hox* gene role, 366
Vitellogenesis, marsupial, 181–182

W

Wing bud, retinoic acid role in development (chick), 324–325
wrl mutation (*Drosophila*), 300

X

Xenopus, *see also* Anurans
  bottle cells, *see* Bottle cells
  convergent extension, 154–158
  epiboly, 76, 154–158
XTC-MIF (activin), 356

Y

Yolk
  elimination during marsupial cleavage, 192
  marsupial, 181–182
  patterns of distribution, 188–189
Yolk bodies, marsupial, 182
Yolk mass, marsupial, 182
Yolky storage products, marsupial, 180–181

Z

Zona pellucida, of marsupials, 189
Zone of polarizing activity, 324–325, 358